The influence of climate change and climatic variability on the hydrologic regime and water resources

TITLES RECENTLY PUBLISHED BY IAHS

In the last three years the International Association of Hydrological Sciences has produced the following publications:

Hydrological Applications of Remote Sensing and Remote Data Transmission. Proceedings of the Hamburg Symposium, August 1983
Publ.no.145 (1985), price $48

Relation of Groundwater Quantity and Quality. Proceedings of the Hamburg Symposium, August 1983
Publ.no.146 (1985), price $30

Scientific Procedures Applied to the Planning, Design and Management of Water Resources Systems. Proceedings of the Hamburg Symposium, August 1983
Publ.no.147 (1985), price $48

New Approaches in Water Balance Computations. Proceedings of the Hamburg Workshop, August 1983
Publ.no.148 (1985), price $20

Techniques for Prediction of Runoff from Glacierized Areas
Publ.no.149 (1985), price $18

Hydrochemical Balances of Fresh Water Systems. Proceedings of the Uppsala Symposium
Publ.no.150 (1984), *out of print*

Land Subsidence. Proceedings of the Venice Symposium, March 1984
Publ.no.151, price $45

Experiences in the Development and Application of Mathematical Models in Hydrology and Water Resources in Latin America (mostly in Spanish). Proceedings of the HYDROMATH Tegucigalpa (Honduras) Symposium, September 1983
Publ.no.152 (1985), price $30

Scientific Basis for Water Resources Management. Proceedings of the Jerusalem Symposium, September 1985
Publ.no.153 (1985), price $42

Hydrogeology in the Service of Man, volumes 1-4. Proceedings of the IAH/IAHS Cambridge Symposium, September 1985
Publ.no.154 (1985), price $40 the set

Proceedings of the symposia held during the Second IAHS Assembly, Budapest, July 1986:

    Modelling Snowmelt-Induced Processes
    Publ.no.155 (1986), price $40

    Conjunctive Water Use
    Publ.no.156 (1986), price $48

    Monitoring to Detect Changes in Water Quality Series
    Publ.no.157 (1986), price $40

    Integrated Design of Hydrological Networks
    Publ.no.158 (1986), price $40

Drainage Basin Sediment Delivery. Proceedings of the Albuquerque Symposium, August 1986
Publ.no.159 (1986), price $45

Hydrologic Applications of Space Technology. Proceedings of the Cocoa Beach Workshop, August 1985
Publ.no.160 (1986), price $45

Karst Water Resources. Proceedings of the Ankara Symposium, July 1985
Publ.no.161 (1986), price $45

Developments in the Analysis of Groundwater Flow Systems
Publ.no.163 (1986), price $35

Water for the Future: Hydrology in Perspective. Proceedings of the Rome Symposium, April 1987
Publ.no.164 (1987), price $50

Erosion and Sedimentation in the Pacific Rim. Proceedings of the Corvallis Symposium, August 1987
Publ.no.165 (1987), price $55

PLEASE SEND ORDERS AND/OR ENQUIRIES TO:

Office of the Treasurer IAHS	Bureau des Publications	IAHS Press
(Attn: Meredith Compton)	de l'UGGI	Institute of Hydrology
2000 Florida Avenue NW	140 Rue de Grenelle	Wallingford, Oxfordshire
Washington, DC 20009, USA	75700 Paris, France	OX10 8BB, UK

A copy of a catalogue of IAHS publications and information on individual membership of IAHS may by obtained free of charge from any of these addresses.

THE INFLUENCE OF CLIMATE CHANGE AND CLIMATIC VARIABILITY ON THE HYDROLOGIC REGIME AND WATER RESOURCES

Edited by

S.I. SOLOMON
Department of Civil Engineering,
University of Waterloo, Waterloo, Ontario,
Canada N2L 3G1

M. BERAN
Institute of Hydrology, Wallingford,
Oxfordshire OX10 8BB, UK

W. HOGG
Atmospheric Environment Service,
4905 Dufferin Street, Downsview, Ontario,
Canada M3H 5T4

Proceedings of an international symposium held
during the XIXth General Assembly of the
International Union of Geodesy and Geophysics
at Vancouver, British Columbia, Canada,
9-22 August 1987. The symposium was organized
by the IAHS International Commissions on
Surface Water, on Groundwater, and on Water
Resources Systems; and sponsored by the World
Meteorological Organization and the
Atmospheric Environment Service, Canada,
and UNESCO

IAHS Publication No. 168

Published by the International Association of
Hydrological Sciences 1987.
*IAHS Press, Institute of Hydrology, Wallingford,
Oxfordshire OX10 8BB, UK.*
IAHS Publication No. 168.
ISBN 0-947571-26-4.

*The Convenors/Editors are greatly indebted to
Mr Ted Cooper for his assistance in preparing
the camera-ready copy for this volume. Thanks
are also due to Mr Chaitawat Saowapon and
Mr Tao Tao.*

The camera-ready copy for this publication was
produced at the Department of Civil Engineering,
University of Waterloo, Waterloo, Ontario,
Canada N2L 3G1.

Preface

It is eight years since IAHS last turned its attention to the issue of climatic change. At that time the International Commission on Snow and Ice convened a symposium on Sea Level, Ice, and Climatic Change as part of the 17th General Assembly in Canberra. In the intervening period much has happened. In particular the steady upward march of the carbon dioxide concentration in the atmosphere has continued and evidence of a global warming has accumulated in large scale land and ocean surface temperature data sets. Thus the general sign of the change which remained in doubt as recently as the late 1970s has largely been resolved within the last five years.

On the official front the Villach conference in 1985, sponsored by UNEP, WMO and ICSU, in a sense set the scientific seal of approval on studies on impacts of climatic change even in advance of totally unambiguous evidence. The possible consequences of such changes were regarded as so profound as to merit immediate action by governments and regional organizations who were urged to support strongly, monitoring and modelling studies aimed at quantifying and predicting the various impacts.

Of course the hydrologic and scientific water resource communities are very well accustomed to coping with variability and making decisions in the face of considerable uncertainty. Thankfully human tragedies such as the ones caused by the 1970 flooding in Bangladesh which arose from the joint action of an extreme cyclone with a high tide, and the extraordinary sequence of deficit years that has beset the subsaharan zone are rare indeed. However such events and their more everyday counterpart which are the bread and butter of the water scientist leave him with a strong feeling for variability and its consequences.

It is entirely timely that IAHS once again gives attention to this important topic, possibly the biggest problem for mankind in the 21st century. It is also appropriate that on the occasion it is the Surface Water Commission that acts as the lead agency as it is those more immediately reacting water bodies that are most sensitive to the changes, and of course these bodies – streams, watercourses, lakes etc – are major sources worldwide for drinking water, energy, communication, fisheries, industry and agriculture.

Those approaching the subject are confronted by a number of important initial questions that have not thus far received a satisfactory answer. Among them:

– How does one define climatic variability and is it distinct from climatic change?

– Which are the main causes of long and short term climatic change?

– To what extent are people inadvertently causing additional climatic variability or perhaps triggering climatic change?

– What are the best ways of treating our data to detect and quantify variability and change?

– What are the quantitative relationships between climatic and hydrologic variability and change?

— How best could mankind, which is so dependent upon water resources, deal with changes in their characteristics?

Up to now people have not been able to affect, except inadvertently, climate and its time variation. However, water resources have been, in many places, under some local or regional human control. Nevertheless this control is under certain limitations, imposed by extraneous inputs, chief among them climatic variability. Painfully, through the effects of air pollution and acid rain, men and women have learned that there are also qualitative aspects to climatic variability and change, and that the chemical composition of the atmosphere and of water resources are closely linked. Chernobyl has impressed upon all of us the image of the global and rapid repercussions of any toxic release into the atmosphere.

While men and women continue to be subject to the whims of the weather, one should note on the positive side that they have learned to observe it quite well, particularly from space. This has enabled them to obtain some insight into the processes involved in climatic variability and acquire some capability of predicting it, at least for periods of a few days. A recent repetition of the Bangladesh disastrous flood resulted in a much smaller human toll, most of this reduction being achieved due to the timely prediction of the event. Similar, if less dramatic examples, abound all around the world. These successes and some similar ones in the area of water resources give us some ground to hope that progress can be achieved on some of the questions listed above.

The prodigious variety of papers submitted to this symposium indicates clearly the wide front on which variability must be faced. Differences in geography and in technical standpoint make for obvious differences. However we see here an attack on the problem through the entire water cycle from atmospheric processes and feedbacks, through water in its solid form in glaciers, to quantity and quality examinations in rivers and lakes. It is also exciting to see papers which take a global view of hydrology, surely pointing the way to a future growth area for climate oriented studies. On the time axis too we see papers which deal with the remote past, with regional overview of current variability, and with attempts to foreshadow resources in a "warm-world" future.

While it would be extremely optimistic to expect that this symposium will bring satisfactory answers to the questions raised above, it is hoped that it will at least indicate where we stand in the quest for these answers, and perhaps point to some possible avenues which appear more promising for this quest.

Co-convenors:
S.I.SOLOMON
Department of Civil Engineering
University of Waterloo, Waterloo
Ontario, Canada N2L 3G1

M.BERAN
Institute of Hydrology
Wallingford, Oxfordshire
OX10 8BB, UK

Préface

Il s'est écoulé huit ans depuis que l'AISH s'est intéressée aux
changements climatiques. A cette époque la Commission Internationale
des Neiges et Glaces avait réuni un colloque sur le niveau de la mer,
la glace et les changements climatiques, ce colloque constituant une
partie de la 17e Assemblée Générale à Canberra. Dans la période qui
a suivi il s'est passé beaucoup de choses. En particulier la
constante progression de la concentration du bioxyde de carbone dans
l'atmosphère a suivi son cours et les preuves d'un réchauffement
global se sont accumulées dans une série de données de températures
de la terre et de la surface de l'océan. Ainsi les premiers signes
de ce changement qui restait encore douteux jusqu'à la fin des
années 1970 se sont confirmés largement au cours des cinq dernières
années.

Sur le plan officiel la conférence de Villach en 1985 parrainée
par le PNUE, l'OMM et le CIUS apposa, dans un sens, le sceau
scientifique d'approbation sur les études des impacts des changements
climatiques, anticipant même sur les preuves absolument indiscutables.
Les conséquences possibles de tels changements ont été considérées
commes très importantes au point de justifier une action immédiate de
la part des gouvernements et des organisations régionales qui ont été
priés instamment d'aider vigoureusement les études de contrôle et de
modèles ayant pour but de quantifier et de prévoir les divers impacts.

Naturellement, les communautés d'hydrologues et de chercheurs
s'occupant des ressources en eau sont bien habitués à faire face aux
problèmes de variabilité et à prendre des décisions même en présence
d'importantes incertitudes. Des tragédies humaines telles que celles
qui ont été causées par les inondations de 1970 au Bangladesh qui a
résulté des actions concomittantes d'un cyclone extrêmement actif et
d'une forte marée et par la série extraordinaire d'années déficitaires
qui a frappé la zone limitrophe du Sahara, sont par bonheur vraiment
rares. Cependant de tels évènements et leurs contreparties beaucoup
plus courantes qui sont le pain quotidien des spécialistes des
sciences de l'eau leur a donné un sens très net de la variabilité et
de ses conséquences.

Il est tout a fait opportun que l'AISH, encore une fois, dirige son
attention sur ce sujet important, peut être le grand problème pour
l'humanité au 21ème siècle. Il convient également qu'à cette
occasion ce soit la Commission des Eaux de Surface qui prenne
l'initiative dans cette action puisque c'est dans les masses d'eau qui
réagissent le plus immédiatement (fleuves, cours d'eau, lacs etc.)
que les changements climatiques sont les plus sensibles, et dans le
monde entier, ce sont surtout ces masses d'eau qui sont mises à
contribution pour l'alimentation en eau potable, la production
d'énergie, les transports, la pêche, l'industrie et l'agriculture.

Ceux qui abordent ce sujet sont confrontés avec un bon nombre de
questions préliminaires importantes qui n'ont pas encore reçu de
réponse satisfaisante. Parmi celles-ci:
- Comment doit-on définir la variabilité inhérente au climat et en
quoi est elle distincte des changements de climat?

- Dans quelle mesure l'homme peut-il provoquer par inadvertance un renforcement de la variabilité climatique et peut-il même déclencher un changement de climat?

- Quels sont les meilleurs moyens de traiter nos données pour détecter et quantifier la variabilité et le changement de climat?

- Quelles sont les relations quantitatives entre la variabilité climatique et la variabilite hydrologique et entre les changements de climats et de régime hydrologique?

- Comment l'homme, qui dépend dans une si large mesure des ressources en eau, peut-il s'accommoder au mieux des modifications des caractéristiques de ces ressources?

Jusqu'ici l'homme n'a pas pu, sauf par inadvertance, affecter le climat et ses variations temporelles. Cependant les ressources en eau ont été mises en certains endroits sous son contrôle local ou régional. Cependant ce contrôle s'exerce à l'intérieur de certaines limites imposées par des influences extérieures, la principale parmi celles-ci étant la variabilité du climat. Douloureusement l'homme a appris par les effets de la pollution de l'air et les pluies acides qu'il y avait aussi des aspects liées à la qualité dans la variabilité climatique et les changements de climat et que la composition chimique de l'atmosphère et celle des ressources en eau étaient intimement liées. Chernobyl a imposé à tous l'image des répercussion globales et rapides d'un rejet dans l'atmosphère.

Alors que l'homme continue à subir les caprices du temps, on devrait noter du côté positif le fait qu'il a appris à les observer très bien, en particulier depuis l'espace. Ceci lui a permis d'avoir un aperçu sur les processus mis en jeu dans la variabilité climatique et de se rendre capable de la prévoir, au moins pour des périodes de quelques jours. La récente répétition de la crue désastreuse du Bangladesh a donnée lieu à un tribut beaucoup moins lourd en vies humaines, la majeure partie de ce résultat ayant été obtenue par une prévision faite à temps de cet évènement. Des exemples analogues quoique moins dramatiques, sont abondants dans le monde entier. Ces succès et quelques autre semblables dans le domaine des ressources en eau nous donnent quelques raisons d'espérer que l'on pourra faire des progrès pour les réponses à certaines des questions énumérées plus haut.

La prodigieuse variété des communications présentées à ce colloque indique clairement le large front sur lequel on doit faire face à la variabilité. Les différences de situations géographiques, de domaines techniques conduisent à des conclusions différentes. Cependant nous voyons ici qu'on aborde le problème à travers l'ensemble du cycle hydrologique depuis les processus atmosphériques et leurs réactions mutuelles, à travers les différents aspects des problèmes de l'eau, dans sa forme solide dans les glaciers, jusqu'aux recherches de caractère quantitatif et celles concernant la qualité dans les rivières et les lacs. Il est également passionnant de voir des communications qui prennent une vue globale de l'hydrologie, montrer avec sécurité la voie pour une zone de future croîssance dans les études orientées vers le climat. Sur l'axe des temps nous voyons aussi des communications qui traitent du passé reculé, avec un survol régional de la variabilité courante et avec les tentatives pour prévoir les ressources en eau dans un "monde futur plus chaud".

Alors que ce serait se montrer extrêmement optimiste que d'attendre

de ce colloque qu'il apporte des réponses satisfaisantes aux
questions soulevées plus haut, on espère qu'il montrera au moins où
nous en sommes dans la recherche orientée vers ces réponses et peut
être qu'il indiquera des approches qui apparaîtraient plus
prometteuses pour cette recherche.

Co-convocateurs:
S.I.SOLOMON
Department of Civil Engineering
University of Waterloo, Waterloo
Ontario, Canada N2L 3G1

M.BERAN
Institute of Hydrology
Wallingford, Oxfordshire
OX10 8BB, UK

Contents

5   CLIMATIC CHANGE AND WATER RESOURCES SYSTEMS

6 MAN'S INFLUENCE ON THE HYDROLOGIC REGIME

1

**Definition of climate change
and climate variability
in hydrology**

The Influence of Climate Change and Climatic Variability on the Hydrologic Regime and Water Resources (Proceedings of the Vancouver Symposium, August 1987). IAHS Publ. no. 168, 1987.

Runoff variability: a global perspective

T.A.McMahon (1), B.L.Finlayson (2),
  A.Haines (1), & R.Srikanthan (3)
(1) Department of Civil and Agricultural
    Engineering, University of Melbourne, Australia
(2) Department of Geography, University of
    Melbourne, Australia
(3) CSIRO Division of Land and Water Resources,
    Canberra, Australia

ABSTRACT   A data base consisting of monthly and annual flows, peak annual instantaneous flows and monthly precipitation has been assembled from around the world to allow analysis of streamflow characteristics at the continental scale.   At the annual level, the major intercontinental differences are in terms of variability. Australia and Southern Africa are distinguished from the rest of the world by their high variability and the observed differences persist even when the comparisons are made between areas of like climatic characteristics and the same latitudes.   For any given precipitation variability, the runoff variability in Australia and Southern Africa is much higher than for the rest of the world.   Analysis using a single linear storage model suggests that it is the higher variability of effective precipitation in Australia and Southern Africa, which results from high evaporative demand there, which is one of the major causes of the observed differences in runoff variability.

RESUME   Nous avons assemble les donnees de base suivantes relevees dans diverses parties du mond: debits mensuels et annuels, debits annuels maximum, et quantite mensuelle de pluie.   Ces donnees nous ont permis de mesurer les caracteristiques debit/ecoulement a l'echelle continen-tale. Au niveau annuel, les differences intercontinentales les plus marquees se situent au niveau de la variabilite. L'Australie et la partie sud du continent africain se distinguent du reste du monde par leur niveau eleve de variabilite. De plus, les differences que nous avons observees persistent, meme lorsqu'on les comparaisons sont etablies avec des regions de climat semblable et de meme latitude; et de climats similaires. Pour n'importe quelle variabilite de precipitation, la variabilite de l'ecoule-ment en Australie et dans le sud de l'Afrique est beaucoup plus haute que dans le reste du monde.   Une analyse pour laquelle nous avons utilise un modele unique de reservoire lineaire, semble suggerer que la plus grande variabilite de precipitation effective - resultant d'une evaporation intense dans ces regions - est une des causes majeures des différences que nous avons observées dans la variabilité de l'écoulement des eaux.

Introduction

The work of McMahon (1975,1978,1979,1982a,1982b) has shown that significant differences exist between the flow characteristics of Australian streams and those of the rest of the world and that the global relationships postulated by Kalinin (1971) do not fit the Australian data. Though the data sets used by both Kalinin and McMahon were inadequate with respect to the southern hemisphere continents, McMahon's early results indicated the possibility of there being substantial differences between the hemispheres. A data base has been established which enables the questions of interhemispheric and intercontinental differences to be addressed.

The present data set contains the world data as used by McMahon (1982a) extensively supplemented with records from all continents. The streamflow records consist of monthly flows, annual flows, and annual peak instantaneous discharges from 87 countries. Monthly and annual flows are available from 938 gauging stations with an average record length of 33 years (a total of 30,800 station years) and the peak instantaneous flows are available for 921 stations with an average record length of 31 years (28,500 station years). These data have been acquired from a variety of published and unpublished sources and less frequently on magnetic tape from national water authorities. Rainfall records for 424 stations worldwide have been extracted from magnetic tapes from the National Center for Atmospheric Research (Boulder, Colorado). The origins of the data and the structure of the data base are described in more detail in Finlayson et al. (1986) together with a map showing the locations of the stream gauging stations and raingauges. The most serious deficiency in the data set is the relatively small number of rainfall stations and it is planned to increase this substantially in the current phase of the study.

For this paper the data have been analysed in eight continental groups. Europe (EUR), Asia (AS), North America (NAM), South America (SAM) and Australia (AUS) are as normally defined; Southern Africa (SAF) and Northern Africa (NAF) are separated by the Equator; and South Pacific Islands (SP) data are mainly from New Zealand. Where all the data have been analysed together the results are referred to as "World" (WOR), where "Rest of World" (ROW) is specified, this refers to the world data minus Australia and South Africa (jointly ASAF).

Analysis of the data base is planned in several parts. The first, part of which is reported in this paper, consists of broad intercontinental comparisons using the annual flows, annual rainfalls, and the annual peak discharge data. The second main phase of analysis will look at the monthly data and particularly at regional rainfall runoff relationships. Other studies will include time series analysis and the definition of seasonal river regime types at the global scale. Ultimately it is hoped that this work will lead to a better understanding of global hydrology and to the definition of a set of world regions for model transferability.

Intercontinental comparisons

Annual runoff

Figure 1 shows the relationship between annual flow volume and
catchment area by continents and for the world data. In all cases
the correlations are significant at the 1% level and area explains a
high proportion of the variance, the lowest being AUS at 60%. The
regression lines shown in this and all other figures in this paper
are least squared fits and in each case the number of data points,
the value of R , and the level of significance are given. Most of
the relationships cluster around the world one with minor exceptions
for SP, AUS and SAF. Lower flow volumes for AUS and SAF, especially
for the larger catchments, are as would be expected for continents in
their latitudinal positions. NAF lies close to the WOR line even
though it is a predominantly arid area because the data for NAF comes
mainly from the humid area bordering the Gulf of Guinea.

Figures 2 and 3 show the relationships between coefficient of
variation of annual flows (Cvr) and mean annual runoff (MAR) and area
respectively. Cvr is calculated as the standard deviation divided by
the mean. Since at the annual level runoff represents the difference
between precipitation and evaporation, MAR is a climatic indicator
representing the level of aridity or humidity of the climate. In
Figure 2 AUS and SAF, while following the world trend of decreasing
Cvr with increasing MAR, are notable in having higher Cvr than the
other continents. MAR explains considerably more of the variance in
Cvr for AUS and SAF and there is also a substantial difference
between these two. This situation appears anomalous and certainly
needs further explanation.

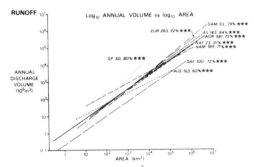

Figure 1 $\log_{10}$ annual volume vs.
$\log_{10}$ area.

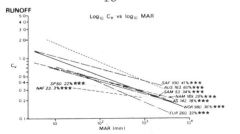

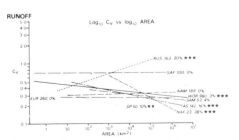

Figure 2 $\log_{10}$ Cv vs. $\log_{10}$ MAR.   Figure 3 $\log_{10}$ Cv vs. $\log_{10}$ area.

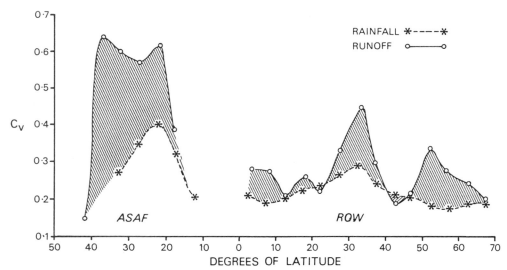

Figure 4   The distribution of variability of rainfall and runoff
            by latitude.

Cvr is poorly correlated with area (Figure 3) though there is a
strong   indication in the data that Cvr tends to decrease as   area
increases   and this is what would be expected on statistical grounds.
It should be noted however that no correlation was found between Cvr
and area for SAF, NAM, EUR and SAM.   AUS is a notable exception to
this trend and is the only continent where Cvr increases with area.
Given the relationship established in Figure 2 where Cvr   increases
as the climate becomes more arid,   the anomalous relationship between
Cvr   and area for AUS can be explained in terms of the   distribution
of climates on the Australian continent.   Humid climatic zones
parallel the coast in a relatively thin strip around the northern,
eastern, south-eastern and south-western coasts. Any large catchments
in Australia must extend into parts of the drier interior causing an
increase in Cvr.

Figure 4 shows the distribution of Cvr and Cvp (the coefficient
of variation of annual precipitation) with latitude.   On the left
hand side of the graph Australia and Southern Africa have been
plotted together and all other continents are shown jointly on the
right hand side.   As would be expected given the latitudinal
distribution of climates Cvr and Cvp tend to peak at around   30°,
the location of the subtropical high pressure cells.   Here again the
anomalous condition of Australia and Southern Africa is evident. The
Cvr's are higher and these high values extend over a wider range of
latitude than is the case for the other continents.

Annual floods

Like the annual flow volume (Figure 1), the mean annual flood ($\bar{q}$),
expressed as a discharge, is strongly correlated with area (Figure 5)
though AUS has the lowest value of R^2 (47%). The relationship between
$\bar{q}$ and area is remarkably similar for all continents. However, when
measures of the variability of flood behaviour are used (Figures 6
and 7) AUS and SAF are distinctly different to the other continents.
In Figure 6, Iv, the coefficient of variation of the annual   peak

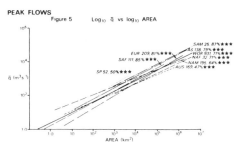

Figure 5 $\text{Log}_{10}\ \bar{q}$ vs. log_{10} area.

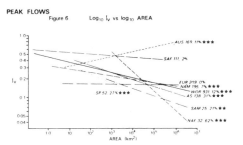

Figure 6 Log_{10} Iv vs. log_{10} area.

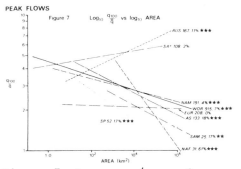

Figure 7 $\text{Log}_{10}\ q_{100}/q$ vs. log_{10} area.

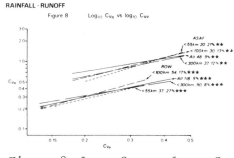

Figure 8 Log_{10} Cvr vs. log_{10} Cvp.

discharges in the log domain, is plotted against area. AUS and SAF have generally higher Iv values than the other continents, and as in the case of Cvr, AUS shows a reversal of the general trend.

The variability of flood behaviour can also be represented using the ratio $q_{100}/\bar{q}$ (Figure 7). q_{100} is calculated assuming the peak instantaneous annual discharges follow a power normal distribution (Chander et al., 1978). Here also AUS and SAF are distinguished by having generally higher ratios than the other continents and AUS again has a reversed trend. This reinforces the fact that AUS and SAF streams are highly variable when compared with those of other continents. While there is an inverse relationship between Cvr and MAR (Figure 2) no similar relationship was found to exist between Iv and $\bar{q}_S$, the specific mean flood, or between $q_{100}/\bar{q}$ and $\bar{q}_S$. Under flood conditions catchment area alone appears more important than the other parameters examined.

Australia and Southern Africa are more variable than the other continents in terms of both annual flows and annual floods. Australian streams also consistently show a typical behaviour when measures of variability are related to catchment area. It should be noted, however, that small catchments (<100 km^2) in Australia have similar Cvr to the rest of the world (Figure 3). These catchments have been discussed by McMahon (1986). Given that MAR is a climate related variable, for any given climate Australian and Southern African streams are more variable (Figure 2) and this can also be shown by comparing streams in similar climatic zones. Table 1 shows, for catchments in the range $1000 - 10,000$ km^2, mean values of both Cvr and Iv for ASAF and ROW and their ratios for Koppen climatic zones. In all cases where sample sizes are large enough for reliable comparisons to be made, ASAF streams are more than twice as variable

Table 1   Annual flow variability and peak discharge variability stratified by climatic type for catchments in the size range 1000 – 10,000 km^2

Climatic region	ASAF No.	C_V	ROW No.	C_V	Ratio	ASAF No.	I_V	ROW No.	I_V	Ratio
Am	–	–	–	–	–	1	.36	3	.12	(3.0)
Aw	1	.76	8	.21	(3.6)	–	–	–	–	–
BSk	9	.86	9	.43	2.0	8	.70	4	.38	1.8
Cfa	24	.95	14	.56	1.7	24	.54	18	.25	2.2
Cfb	15	.72	32	.29	2.5	12	.45	29	.19	2.4
Csa	–	–	–	–	–	1	.45	6	.45	(1.0)
Csb	2	.61	1	.50	(1.2)	5	.55	15	.26	2.1
Cwa	7	.92	5	.29	3.2	6	.57	6	.19	3.0
Cwb	15	.72	2	.39	(1.8)	11	.39	2	.18	(2.2)

(–) Parentheses indicate sample size very small

as ROW streams in the same climate zone. This is true also for catchments <1000 km^2 in area (see Finlayson et al., 1986).

Sources of variability in runoff

While the results presented so far do not support the hypothesis that a significant difference in mean values exists between the hemispheres, they do show that AUS and SAF are different to the other continents when measures of variability are involved. In exploring the sources of variability the data have been split into two groups, Australia and Southern Africa have been combined and the other continents grouped together. Since it is virtually impossible to determine catchment rainfalls for the streams in the data set, catchments larger than 10,000 km^2 have been eliminated and each raingauge was paired with the nearest stream gauge in order to investigate the relationship between rainfall variability and stream-flow variability. Table 2 sets out the characteristics of the data used in this analysis. Raingauge/streamgauge pairs have been grouped into classes based on the distance separating them.

Table 2   Characteristics of raingauge/streamgauge pairs

Group	Australia and Southern Africa No.	Mean distance (km)	R^2 (%)	Rest of the world No.	Mean distance (km)	R^2 (%)
All pairs	48	319	9	118	298	9
<300 km	37	75	12	90	98	8
<100 km	30	50	13	54	44	17
<55 km	20	32	21	37	27	27

The influence of variability of total annual precipitation

Cvr has been plotted against Cvp on Figure 8 for each of the separation classes. Here ASAF and ROW are clearly different with ASAF having much higher Cvr for any given Cvp than ROW. For example, at Cvp of 0.25, the difference in Cvr between ASAF and ROW is 0.3. The form of this relationship is consistent irrespective of the average station separation though as station separation increases the strength of the relationship declines. In statistical terms, for all the regression lines shown on Figure 8 there is no significant difference between the regression coefficients but the intercepts are significantly different between the ASAF and ROW groups. Although the transfer of variability from precipitation to runoff is greater for ASAF than ROW, there is no difference between the two groups in terms of the amount of runoff variability explained by precipitation variability (R^2 in Figure 8 and Table 2).

Other influences

In order to investigate further the source of the differences between ASAF and ROW a storage model analysis has been carried out on those raingauge/streamgauge pairs with station separation of less than 55 km.
One thousand years of rainfall data for each station with parameters based on those of the observed data were generated synthetically using a Markov process as modified by the Wilson-Hilferty transformation (Wilson and Hilferty, 1931). Annual total rainfalls were converted to annual effective rainfalls using a constant runoff coefficient based on the observed data. For each catchment the storage effects were mimicked by routing the 1000 years of annual effective rainfall through a conceptual single linear storage model of the form:

$$S = KR \tag{1}$$

where S is catchment storage and R is annual runoff volume. The storage delay time parameter (K) in the model was optimized for each catchment so that the time series of annual flows reproduced the observed variability. The details of this methodology will be published elsewhere.
Median values of parameters from the observed data and the model results are shown in Table 3 where Cvpe , annual effective precipita-

Table 3   Median values of parameters from the observed data and the model results.   (RC is the mean runoff coefficient from the observed data.)

	MAP (mm)	Cvp	Cvpe	RC	K (yr)	Cvr
ASAF	770	0.23	1.10	0.15	0.50	0.67
ROW	800	0.18	0.40	0.51	0.25	0.30

tion, and K are derived from the model.   The ASAF catchments have higher storage than ROW catchments. While the values shown in Table 3 are only relative they are consistent with known storage times (T.G.Chapman, pers. comm.).   In general it might be expected that higher storage would lead to less variable runoff but this does not occur here because Cvpe for ASAF is so much larger than that for ROW. The high storage in the ASAF catchments leads to a dramatic reduction in variability between Cvpe and Cvr (compared to that for the ROW data) but the Cvr is still substantially above that for ROW.

This result arises because of the lower runoff coefficients for ASAF and the values we observe are confirmed by Korzun et al. (1974). ASAF experiences relatively high evaporative demand on the world scale, partly because of the excess energy advected from their dry interiors to the humid coastal areas, and partly because of the fact that at these latitudes the southern hemisphere has a higher potential evaporation than the northern hemisphere (Baumgartner & Reichel, 1975).   It is this high evaporation which leads to the lower runoff coefficients. Note from Table 3 the difference between Cvp and Cvpe for ASAF compared to ROW.   While intuitively it might have been expected that ASAF would have lower storage values than ROW, because of the absence of features such as substantial snowfields, it is obvious as a result of this model analysis that other factors such as low relief and soils more than compensate for this.   Paton (1978) has made a case for recognizing Australian and African soils as being significantly different to those of the other major continental areas because of the long period of continental stability and the absence of continental glaciation during the Pleistocene.

Conclusion

This paper has briefly described a new data base which has been assembled to investigate streamflow characteristics at the continental scale.   While the data base contains streamflow records at the monthly level, to date only the annual data have been analyzed, both annual totals and annual peak discharges. The data base also includes precipitation records from all continents and it is intended to add to these.

Analyses of annual runoff presented here show that the most important intercontinental differences are in terms of variability. In particular, Australia and Southern Africa show levels of annual variability nearly twice that of the other continents.   The same result appears in the analysis of annual peak instantaneous discharges.   The observed differences persist even when the data are stratified by similar climatic types.   With one exception, climatic zones in Australia and Southern Africa have variabilities of annual and peak flows approximately twice those of the rest of the world.

These differences cannot be ascribed solely to the variability of annual total rainfall though it is noteworthy that for any given precipitation variability, runoff variability in Australia and Southern Africa is significantly higher than in the rest of the world.   A single linear storage model analysis carried out on catchments less than 10,000 sq.km in area for which a precipitation record was available within a 55 km radius indicates that the important factor in determining the high runoff variability in Australia and Southern Africa is the variability of effective precipitation. High variability of effective precipitation is a function of high evaporative demand in the atmosphere. An unexpected outcome of the model analysis was the high storage values for catchments in Australia and Southern Africa. While it might be expected that high storage would be associated with low runoff variability it appears that the high values of variability of effective precipitation more than compensate for the storage effects.

ACKNOWLEDGEMENTS   The work discussed in this paper was supported by a grant from the Australian Research Grants Scheme.

References

Baumgartner, A. & Feichel, E. (1975) The World Water Balance, Amsterdam, Elsevier.

Chander, S., Spoila, S.K. & Kumar, A. (1978) Flood frequency analysis by power transformation, J. Hydr. Div. ASCE, 104(HY11), 1495–1504.

Finlayson, B.L., McMahon, T.A., Srikanthan, R. & Haines, A.   (1986) World hydrology: a new data base for comparative analysis. Hydrology and Water Resources Symposium 1986,   Institution of Engineers Australia, Nat. Conf. Publn. No.   86/13, 288–296.

Kalinin, G.P. (1971) Global Hydrology (Israel Program for Scientific Translations).

Korzun, V.I., Sokolov, A.A., Budyko, M.I., Voskresensky, K.P., Kalinin, G.P., Konoplyantsev, A.A., Korotkevich, E.S. & Lvovich, M.I. (Eds) (1974) Atlas of World Water Balance, (annex to monograph World Water Balance and Water Resources of the Earth), USSR Committee for I.H.D., Moscow, Hydrometeorological Publishing House.

McMahon, T.A. (1975) Variability, persistence and yield of Australian streams. Hydrology Symposium 1975, Institution of Engineers Australia, Nat. Conf. Publn. No. 75/3, 107–111.

McMahon, T.A. (1978) Hydrological characteristics of arid zones. The Hydrology of Areas of Low Precipitation, (Proc. Canberra Symposium), IAHS-IASH Publn. No. 128, 105–123.

McMahon, T.A. (1979) Hydrological characteristics of Australian streams, Monash University Civil Engineering Research Report 3/1979.

McMahon, T.A. (1982a) World hydrology: Does Australia fit? Hydrology and Water Resources Symposium 1982, Institution of   Engineers Australia, Nat. Conf. Pubn. No. 82/3, 1–7.

McMahon, T.A. (1982b) Hydrological characteristics of selected rivers of the world. Technical Documents in Hydrology, (UNESCO Paris).

McMahon, T.A. (1986) An overview of small rural catchment hydrology in Australia. J. Agric. Engng. Soc. (Aust.), (in press).

Paton, T.R. (1978) The Formation of Soil Material. London,   George Allen and Unwin.

Wilson, E.B. & Hilferty, M.M. (1931) Distribution of Chi-square, Proc. Nat. Acad. Sci., Wash., 17, 684–688.

The Influence of Climate Change and Climatic Variability on the Hydrologic Regime and Water Resources (Proceedings of the Vancouver Symposium, August 1987). IAHS Publ. no. 168, 1987.

Variation des débits des cours d'eau et des niveaux des lacs en Afrique de l'ouest depuis le début du 20 ème siècle

Jacques H.A. Sircoulon
Secrétaire-délégué du Comité National Français
des Sciences Hydrologiques
213, rue La Fayette
75480 Paris Cedex 10, France

RESUME   Les ressources en eaux de surface des grands bassins hydrographiques d'Afrique de l'Ouest présentent des fluctuations interannuelles très marquées qui reflètent bien les variations climatiques que connaît cette vaste région. L'évaluation correcte de la ressource se heurte à de nombreuses difficultés et seules de rares chroniques d'observations existant depuis le début du 20ème siècle permettent d'en suivre les fluctuations de façon très globale.
   Au cours des vingt dernières années, la tendance à la baisse que semble présenter les cours d'eaux tropicaux depuis la fin du siècle dernier s'aggrave; les apports des grands fleuves à la zone sahélienne se reduisent de 40% en moyenne et les lacs de cette zone présentent de très bas niveaux, voire même un assèchement total pour certains d'entre eux.

Variation of river discharges and lake levels in West Africa since the beginning of the twentieth century

ABSTRACT   Water surface resources of large river basins in West Africa experience great interannual fluctuations due to the climatic variations occurring in this area.
   A reliable appraisal of the resource encounters numerous difficulties and only scarce time series recorded since the beginning of the twentieth century. However, it appears that the runoff contributing to the tropical rivers has been decreasing since the end of the last century, and particularly during the last twenty years. The water yields to the Sahelian zone have fallen 40 percent resulting in very low lake levels, and in some cases, the disappearance of the lake entirely.

Introduction

Les variations des débits des grands fleuves tropicaux et des niveaux des lacs en Afrique de l'Ouest et Centrale intègrent les Variations spatiotemporelles des régimes pluviométriques et leur abondance annuelle. Alors que l'on observe un besoin croissant de connaissance

des ressources en eau de surface de cette vaste région et qu'il
existe la nécessite impérieuse de mieux gérer une ressource qui va en
s'amenuisant en liaison avec la dégradation climatique très sensible
de ces vingt dernières années, il faut reconnaître qu'ici comme
ailleurs les observations hydrométriques comme pluviométriques ont
connu de grandes vicissitudes liées à la qualité des mesures et à la
détérioration des réseaux, les frais de mesures, de contrôle et
d'entretien, représentant une très lour-de charge pour les pays.

Comme cela est général en la matière les observations
pluviométriques commencent bien avant les mesures hydrométriques avec
un réseau de base fonctionnant dès le début des années 1920 et un
certain nombre de stations suivies depuis le tout début du siècle,
voire avant. Les données recueillies indiquent clairement que
l'Afrique tropicale a connu au cours du 20ème siècle trois périodes
sèches 1910-1916, 1940-1949 et de 1968 à maintenant et deux phases
humides vers 1925-1935 et 1950-1965. Lorsqu'on cherche à connaître
les variations, concomitantes de l'écoulement l'on s'aperçoit très
vite qu'il n'existe que deux stations de référence l'une sur le
fleuve Sénégal, l'autre sur le fleuve Niger et couvrant ces quatre
vingts dernieres années qui soient capable de quantifier globalement
l'influence du climat.

La situation actuelle très déficitaire de l'écoulement demanderait
à ce que l'on puisse étendre ces chroniques au 19ème siècle afin de
mieux apprécier les fluctuations interannuelles de celui-ci et le
phénomène de persistance ac tuel. Les résultats obtenus (Olivry,
Chastanet - 1986 par ex.) restent limités, une grande prudence
s'impose dans la reconstitution de séries anciennes, les sources
historiques, les récits d'exploration étant souvent contradictoires:
de plus le phénomène de zonalité des fortes sécheresses n'est pas
toujours aussi généralisé qu'on pourrait le penser, le bassin du
Logone Chari pouvant avoir, à titre d'exemple, une hydraulicité
certaines années qui est très différente de celles observées sur les
bassins du Sénégal et du Niger.

Historique des mesures sur les grands fleuves

Il faut attendre le début des années 1950 pour voir la mise en place
en Afrique noire francophone de véritables réseaux organisés; par
ailleurs aucune mesure de débit n'a été faite au cours du 19ème
siècle même si des échelles limnimétriques ont pu exister dès la fin
du siècle dernier pour tenter d'évaluer la navigabilité de ces voies
d'eau.

Pour les fleuves tropicaux parvenant à la zone sahélienne, on peut
esquisser une historique rapide:

(a) Sur le fleuve Sénégal, navigable en hautes eaux sur plusieurs
centaines de km, on recense dès le début des années 1890 une
quarantaine d'échelles suivies d'août à décembre et installées sur
les seuils et aux escales des bateaux circulant sur le fleuve. La
station de Bakel qui contrôle l'ensemble des apports provenant à la
basse vallée (218 000 km^2) est installée en 1901; ses relevés sont
utilisables depuis 1903 mais ne sont complets en basses eaux que
depuis 1951.

(b) Sur le fleuve Niger, Mage installe une échelle à Segou qui sera suivie en hautes eaux en 1864 et 1865.   Dès 1899 il existe plusieurs stations suivant la crue du fleuve en amont et en aval de la cuvette lacustre (Lenfant).   La station de Koulikoro qui contrôle l'écoulement du haut Niger (120 000 km^2) est ouverte en 1907 par la Compagnie Générale des Colonies et sera suivie sans aucune lacune jusqu'à présent.   Cette station sert de référence au même titre que celle de Bakel.

(c) Sur le bassin du Logone Chari une première échelle est installée à Ndjamena en 1903, la mission Tilho installe en 1906 une station qui permettra de connaître seulement les crues de 1906 et 1908; les observations, capitales pour la connaissance de l'alimentation du lac Tchad ne reprendront qu'en 1932.

En dehors de ces trois grands ensembles, les plus longues chroniques concernent:

- La station de l'Oubangui à Bangui qui contrôle un bassin de 500 000 km^2 à régime de transition équatoriale.   Des observations sont faites de 1890 à 1987 (Bruel) pour l'étude des possibilités de navigation.   Ces mesures ne reprendront vraiment qu'en 1911 mais avec beaucoup de lacunes de 1920 à 1935.

- Les stations de Kinshasa/Brazzaville qui permettent de constituer une chronique ininterrompue depuis 1902 sur un bassin de 3 500 000 km^2 à prédominance équatoriale.

Données sur l'écoulement

Le tableau I regroupe pour les 5 grands fleuves cités précédemment quelques caractéristiques de leurs modules (jusqu'en 1985 ou 1986 inclus).   Les valeurs sont en m^3/s.

Tableau I

Station	Nbre d'années d'observation	Module interannuel	Coef. de variation	Module max.	An	Module mini.	An
Sénégal à Bakel	84	702	0,38	1247	24	215	84
Niger à Koulikoro	80	1437	0,18	2300	25	636	84
Chari à Ndjamena	54	1115	0,31	1720	55	213	84
Oubangui à Bangui	56	4130	0,19	5560	61	2180	84
Congo à Kinshassa	84	40910	0,10	56000	61	33500	13

Le tableau II fournit pour ces mêmes stations les modules moyens sur 5 ans consécutifs pour les périodes sèches et humides les plus marquées (valeurs en m^3/s).

Ces tableaux regroupent des bassins fluviaux soumis à des régimes climatiques variés et présentant une abondance de l'écoulement et une irrégularité interannuelle très diverse; néanmoins ce rassemblement est très instructif:

- au niveau des années humides, on constate un comportement différent entre les bassins d'Afrique de l'ouest et d'Afrique centrale (avec réserve due au fait que les chroniques d'observations ne sont pas homogènes), pour les fleuves Sénégal et Niger ce sont les années 1924 et 25 qui présentent des valeurs records alors qu'en

Tableau II

Station	Humide		Sec	
	Module	Période	Module	Période
Sénégal à Bakel	1027	54 - 58	285	82 - 86
	941	32 - 36		
	917	24 - 28		
Niger à Koulikoro	2024	24 - 28	804	82 - 86
	1942	51 - 55		
Chari à Ndjamena	1500	60 - 64	533	81 - 85
	1450	52 - 56		
Oubangui à Bangui	5200	60 - 64	2880	81 - 85
Congo à Kinshasa	49800	61 - 65	36900	81 - 85 et 11 - 15

Afrique centrale c'est l'année 1961 (année célèbre pour l'augmentation brutale de niveau des lacs d'Afrique orientale et les fortes hydraulicités des cours d'eau issus de la crête Congo-Nil); même le Chari présente cette année là un module très voisin de celui de son maximum de 1955. Sur cinq années consécutives, on retrouve la même dualité géographique, les plus forts écoulements étant observés au début des années 20, 30 ou 50 à l'ouest et au début des années 60 à l'Est.

- au niveau des années sèches on constate par contre un meilleur synchronisme dans le comportement de ces grands fleuves. L'année 1984 est ainsi la plus faible partout, sauf pour l'ensemble du bassin du Congo ou cette année là vient au 4ème rang avec 34300 m^3/s.   Mais sur cinq années consécutives tous ces fleuves sans exception présentent le plus faible module moyen en 1981-85 ou 1982-86 (le fleuve Congo retrouvant les basses valeurs de la période 1911-15).

Ceci montre que la période actuelle est la plus déficitaire du 20ème siècle en zone tropicale, et qu'en cas de paroxysme de sécheresse l'extension du phénomène se fait sentir en zones équatoriales comme ce fut déjà le cas lors des années 13.

Apports au Sahel et au lac Tchad

Le tableau III récapitule les apports (en 10^9 m^3) des grands fleuves tropicaux à la zone sahélienne. Les apports totaux interannuels sont de l'ordre de 120 milliards de m^3. ls étaient de 130 à 135 * 10^9 m^3 pour la période s'arrêtant à 1967 inclus et tombent à 84 * 10^9 m^3 pour la période de sécheresse actuelle soit un déficit considérable (Fig. 1).

Les apports totaux annuels sont tous déficitaires depuis 1968 (Fig.2.1) sauf l'année 1969 pour laquelle les bassins du Sénégal et du Niger présentent une hydraulicité légèrement positive. Dans ces conditions, les débordements dans les basses vallées du Sénégal, le remplissage de la cuvette intérieure du Niger et l'inondation des plaines au sud du lac Tchad sont de moins en moins assurés.

Tableau III

	Sénégal à Bakel	Niger à Koulikoro	Bani à Douna	Chari à Ndjamena	Total
Début des observ. à 1985	22,3	46,2	17,3	35,1	120,9
Début des observ. à 1967	24,7	48,7	22,1	40,4	135,9
Période 1968 - 1985	13,7	37,7	8,3	24,6	84,3
Année 1984	6,9	20,1	2,2	6,3	35,5

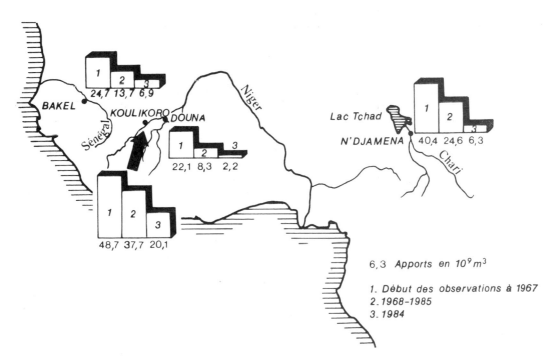

Figure 1   Variation des apports annuels des grands fleuves à la zone sahélienne.

En ce qui concerne le remplissage du lac Tchad lui-même, les déficits de l'ensemble Logone-Chari se font sentir dès 1965 (confer fig.2.2.) expliquent la diminution du niveau du lac depuis 63-64 et son quasi assèchement actuel (confer plus loin). Ces déficits sont de l'ordre de 40% pour la période 1968 à 1985, ils atteignent 83% pour la seule année 1984.

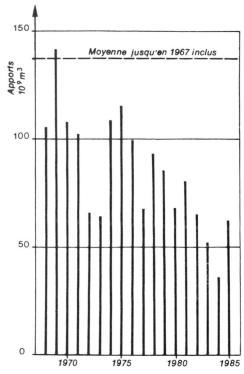

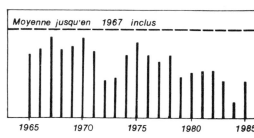

Figure 2.1 Apports annuels des
grands fleuves au
sahel.

Figure 2.2 Apports annuels du
Chari au lac Tchad.

Evolution des modules

Tableau IV

	Début à 1967	1968 à 1985	Déficit (%)
Sénégal à Bakel	782	435	44
Niger à Koulikoro	1521	1192	22
Chari à Ndjamena	1282	781	39
Oubangui à Bangui	4128	3607	13
Congo à Kinshassa	40870	41070	

Les considérations qui précèdent montrent que pour les fleuves
tropicaux ou à transition équatoriale on assiste à une baisse très
sensible alors que le module interannuel du fleuve Congo reste
stable; le plus fort déclin est celui du fleuve Sénégal.
   Les essais d'extension de la période d'observations
hydropluviométriques à partir de données historiques, par Olivry et
Chastanet, 1986, semblent montrer globalement des conditions beaucoup
plus humides entre 1857 et 1902 sur ce bassin. Le module interannuel

aurait pu être voisin de 900 m^3/s au cours de cette période. La période la plus humide se situe de 1860 à 1880 (une concordance assez nette apparaît avec les événements signalés pour cette période en Afrique soudano-sahélienne (Maley, Nicholson) et aux Iles du Cap-Vert (Olivry).

Il semblerait ainsi que l'on assiste à une diminution croissante de la resource en eau depuis le début du siècle pour les fleuves tropicaux.

Evolution de la cuvette lacustre du Niger

Historique

La cuvette lacustre alimentée par le haut Niger et le Bani s'étend aux hautes eaux sur 60 000 km^2 environ. C'est une vaste région d'épandage, fond d'un immense delta qui à son apogée à l'holocène (vers 8000 ans B.P.) couvrait une étendue bien plus considérable (N. Petit-Maire). Dans cette cuvette à très faible pente, bras principaux, émissaires, chapelets de marés et grands lacs cohabitent (fig.3). L'alimentation en eau des lacs est tributaire de l'inondation annuelle du fleuve bien que certains d'entre eux puissent être remplis simplement par des précipitations locales. Sur l'ensemble de la cuvette les observations chiffrées sont très fragmentaires: quelques stations ont été installées vers 1955, un réseau plus dense es t mis en place en 1975 et des tournées regulières ont été effectuées en 83-84 pour améliorer la connaissance de la limnimétrie et de la bathymétrie des lacs et mieux connaître les bilans d'écoulements.

Le lac Faguibine, en rive gauche (Fig. 4) est le plus grand lac d'Afrique de l'ouest après le lac Tchad et peut couvrir 600 km^2. Il fluctue d'environ 1.5 m par an et ses variations pluriannuelles reflètent assez bien le comportement du fleuve mais avec un décalage de plusieurs années, son remplissage demandant plusieurs années à hydraulicité abondante. Découvert par les militaires français en 94-95 il ne beneficie d'observations connues que de 1937 à 1941, de 1958 à 67, et au cours de ces dernières années.

Niveaux dans la cuvette et le lac Faguibine

Le degré de remplissage des lacs de rive droite est connu de façon qualitative depuis 1908 (rapport Bourgues de 1933). Les lacs les plus extrêmes comme le lac Haribongo après avoir été rempli par les fortes crues de 1893-95, semblent être restés à sec jusqu'en 1925. L'abondance exceptionnelle des années 1924 et 1925 suffit à remplir les lacs qui se maintiennent à un niveau assez élevé jusqu'en 1933. En 1955-56 les lacs seront à nouveau pratiquement pleins jusqu'à la fin de la période abondante en 1967. Depuis 1972 les lacs les plus extrêmes sont à nouveau a sec et les autres lacs (comme le lac Korarou) ne sont alimentés que certaines années. En 1983 et 84 la crue du fleuve Niger est si faible qu'aucun seuil d'alimentation en eau des lacs n'a été atteint.

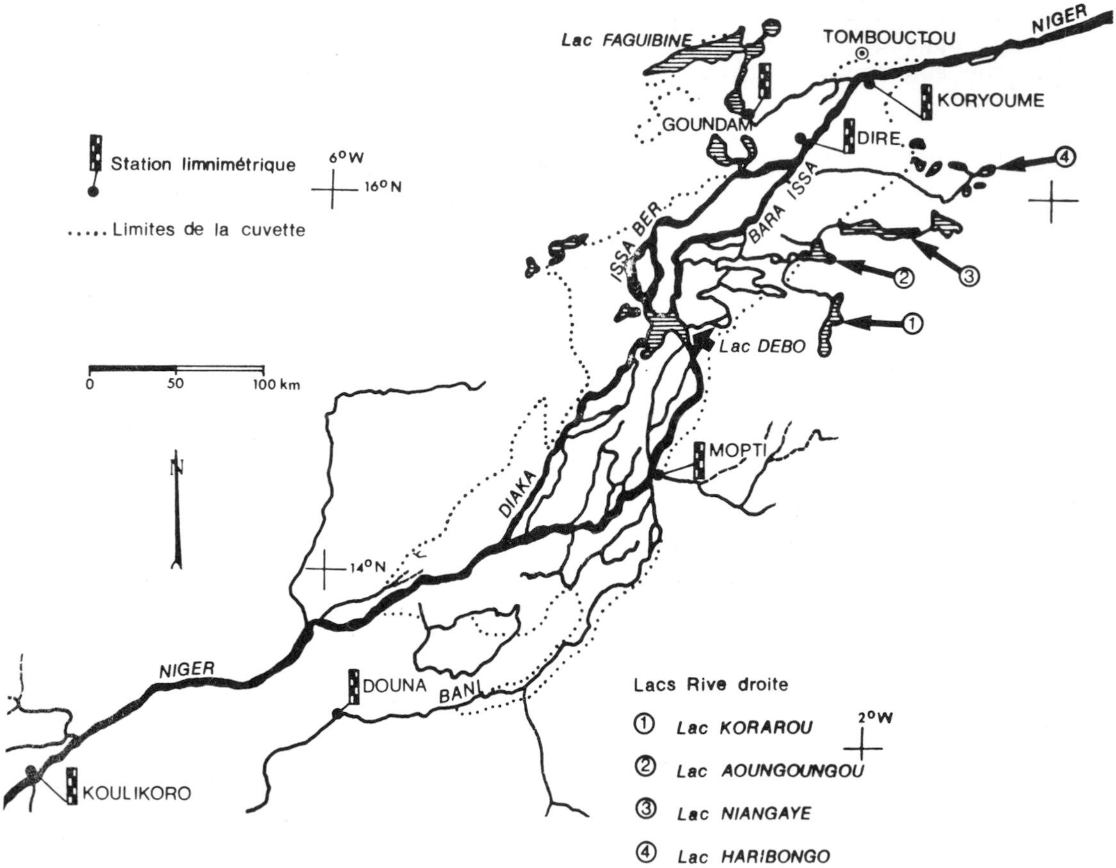

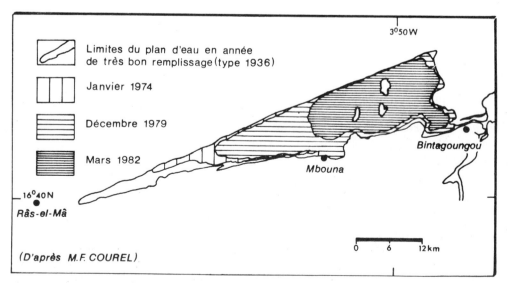

Figure 3   Cuvette lacustre du fleuve Niger.

Figure 4   Le lac Faguibine.

En ce qui concerne le lac Faguibine, plusieurs reconstitutions du niveau du lac ont été faites, par Viguier pour la période 1894-1941, par Vauchel et Guiguen pour la période 1958 à 1984 et par M.F. Courel (reconstitution partielle par imagerie satellitaire).

Il semblerait que le lac ait été complètement asséché en 1910, 1924, 1941 et rempli en 1894, 1917 (?), 1930, 1955 et assez haut en 1969. Après la pointe de sécheresse de 1972-73, le lac régresse régulièrement (fig. 4), il n'atteint plus la station de M'Bouna en 1976 et on peut le considérer à nouveau asséché en 1983 (Vauchel).

Evolution du Lac Tchad

Historique rapide

Le lac Tchad (fig.5.1) plus grand lac d'Afrique de l'ouest et vestige de la grande mer intérieure qui couvrait à l'holocène plusieurs centaines de milliers de km^2 enregistre de façon sensible les variations climatiques que subit cette région. Formé d'une nappe peu profonde ce qui le rend très vulnérable à la sécheresse, l'évaporation dépassant deux mètres par an, il est constitué à l'époque contemporaine par deux cuvettes séparées par une zone de hauts fonds (la Grande Barrière). Ses niveaux suivent une oscillation annuelle provoquée par les variations saisonnières de la crue de l'ensemble du Logone et du Chari, les apports de ces bassins correspondant à environ 82% des apports totaux au lac.

Découvert en 1823 par l'explorateur Denham, un certains nombre de voyageurs célèbres en ont donné une description partielle au 19 ème siecle. La mission Tilho en 1904-1906 en a fait une étude complète et plus scientifique. Les observations limnimétriques du lac sont malheureusement très courtes, la station la plus connue du lac (Bol-Dune) ne fonctionne que depuis 1955 et après la sécheresse de 1972 les hauteurs lues à la station ne sont plus en général représentatives du niveau général du lac.

Evolution

De nombreux essais de reconstitution des niveaux du lac ont été faits (Tilho, Bouchardeau, Maley, Nicholson...). Mais il est très difficile de reconstituer même de façon qualitative les fluctuations du lac. Les témoignages des explorateurs, des militaires, les récits recueillis auprès des habitants ou les études historiques apportent un éclairage souvent contradictoire.

Si l'on s'en tient aux deux siècles écoulés, il semblerait que le lac présente
- ses plus bas niveaux au cours de la seconde partie du 18ème siècle, vers 1830-1840, au début du 20ème siècle (Fig. 5.2.) et actuellement.
- ses plus hauts niveaux au début du 19ème siècle, vers 1880 et au début des années 1960.

Après les apports très abondants du fleuve Chari de 1960 à 1962 (50 milliards de m^3 par an en moyenne), le lac Tchad atteint en janvier 1963 la cote de 283 m a Bol, la surface du lac est alors de 23500 km^2 pour un volume stocké de 105 milliards de m^3. Depuis cette

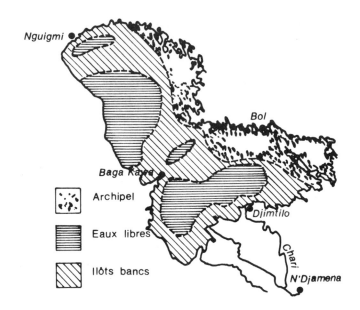

Figure 5.1   Le lac Tchad à la cote 281,50 m.

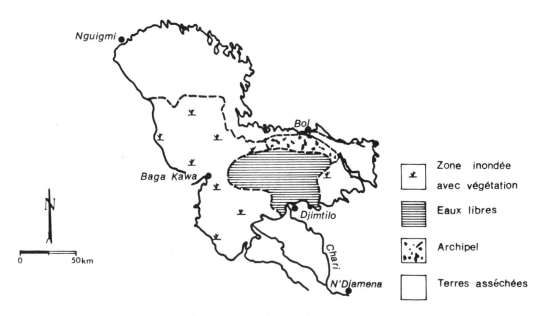

Figure 5.2   Situation début 1908 (Tilho).

date les niveaux vont baisser chaque année régulièrement, les apports
annuels du Chari étant tous déficitaires depuis 20 ans comme nous
l'avons déjà vu.
- En avril-mai 1973 la faiblesse de remplissage du lac aboutit à la
coupure de celui-ci en deux cuvettes avec exondation de la Grande
Barrière (Fig. 5.3.). La médiocrité de réalimentation saisonnière de

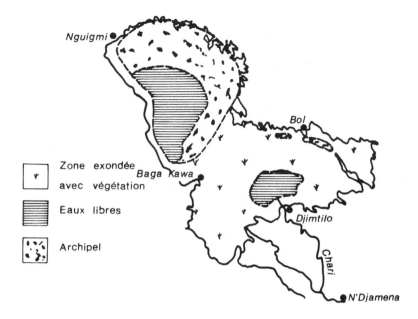

Figure 5.3   Situation début juillet 1973.

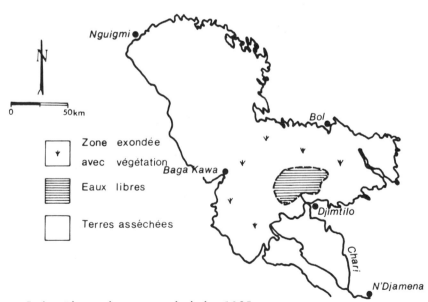

Figure 5.4   Situation en mai-juin 1985.

la cuvette nord provoque en novembre 1975 l'assèchement complet de celle-ci.   Depuis lors on assiste chaque année au même phénomène. En 1983 et en 1984 (apports de $6,3 * 10^9$ m^3) la crue du Chari ne permettra même pas une réalimentation temporaire de cette cuvette nord; en mai-juin 85 la surface d'eaux libres de la cuvette sud se réduit ainsi à 2000 km^2 environ (Fig. 5.4.).   La situation actuelle semble être pire que celle observée au début du siècle, elle a des conséquences catastrophiques pour toutes les populations riveraines et pose de façon aigüe la nécessite d'une concertation entre les pays de la région pour gérer en commun la pénurie.

Conclusion

Les rares chroniques de longue durée disponibles pour les fleuves tropicaux montrent que ceux-ci présentent des variations importantes de l'écoulement annuel.   Il semble que l'on assiste à une diminution des ressources en eau de surface depuis la fin du siècle dernier dans cette zone climatique qui connaît depuis une vingtaine d'années une phase de sécheresse particulièrement sévère et étendue et dont la persistance soulève de très graves questions quant à l'utilisation à terme de cette ressource.   La situation prévalant en zone sahélienne exige une stricte gestion de la pénurie, une étroite concertation entre pays riverains et une intensification des recherches sur les phénomènes qui régissent le climat.   Les zones équatoriales quant à elles sont sensibles aux phases aigües de sécheresse se produisant en zone tropicale mais les valeurs interannuelles de l'écoulement ne semblent pas etre affectees globalement.

Références

Anonyme - 1958 - Note de synthese du projet d'amenagement des lacs Tele et Faguibine. Mission d'etudes et d'amenagement du Niger (Mean).

Brunet-Moret, Y. et al. (1986) Monographie hydrologique du fleuve Niger - Tome II - in collection Monographies hydrologiques Orstom, N 8 Paris.

Chouret, A., Lemoalle J., (1974) Evolution hydrologique du lac Tchad durant la secheresse 1972-74 - Centre Orstom de N'Djamena, 12p. + Graph.

Chouret, A., (1977) La persistance des effets de la secheresse sur le lac Tchad, centre Orstom de n'Djamena, 12p. 10 fig.

Chouret, A., Berthault, C., Pepin, Y. (1986) Persistance de la secheresse au Sahel - Etudes de stations pluviometriques et hydro-logiques de longue duree au Mali. Observations de l'annee 1985. Direction Nationale de l'Hydraulique et de l'Energie et Orstom - Bamako

Courel, M.F., (1984) Etude de l'evolution recente des milieux saheliens a partir des mesures fournies par les satellites. Universite Paris-Sorbonne, These de doctorat d'etat, publication du centre scientifique IBM-France.

Guigen, N., (1984) Tournee hydrologique dans la cuvette lacustre sur les lacs de la rive droite du Niger du 17 au 23/12/84 (2000 km parcourus). Rapport interne Orstom.

Lenfant, Capitaine (1903) Le Niger - Librairie Hachette - Paris.

Maley, J., (1981) Etudes palynologiques dans le bassin du Tchad et paleoclimatologie de l'Afrique nord-tropicale de 30 000 ans a l'epoque actuelle, in travaux et documents de l'Orstom N 129 - Paris.

Nicholson, S.E., (1981) The Historical climatology of Africa; in Climate and History (T.A.L. Wigley, M.J. Ingram and G. Farmer, eds) Cambridge University Press, Cambridge, 249-270.

Olivry, J.C.   (1983)   Le point en 1982 sur la secheresse en Senegambie et aux Iles du Cap vert - Examen de quelques series de longue duree (debits et precipitations).   In Cah. Orstom, serie Hydrol., Vol.XX,N 1, pp. 47-69.

Olivry, J.C., Chastanet, M.   (1986)   Evolution du climat dans le bassin du fleuve Senegal (Bakel) depuis le milieu du 19eme siecle. In colloque sur les changements globaux en Afrique au cours du quaternaire, Inqua, Asequa, Dakar (avril 1986).

Petit-Maire, N.   (1986) Paleoclimatologie du Sahara occidental et central pendant les deux derniers optima climatiques,aux latitudes paratropicales in "Changements globaux en Afrique durant le quaternaire" Symposium Inqua-Asequa, Dakar, 21-28 avril 1986.

Rochette, C.   (1974) Le Bassin du fleuve Senegal. In monographies Hydrologiques Orstom, N 1, Paris.

Sircoulon, J.   (1976) Les donnees hydropluviometriques de la secheresse recente en Afrique intertropicale.   Comparaison avec les secheresses "1913 et 1940".   In cah. Orstom, ser. Hydrol., vol. XIII, N 2, 1976, pp. 75-174.

Sircoulon, J.   (1986) La secheresse en Afrique de l'Ouest. Comparaison des annees 1982-84 avec les annees 1972-73.   In cah. Orstom, ser. Hydrol., vol. XXI, N 4, 1984/85, PP. 75-86.

Tilho, J. (1910) Documents scientifiques de la mission Tilho 1906-1909. Imprimerie Nationale Paris, Tome 1, 412 p., Tome II 598 p.

Vauchel, P., Guiguen, N. (1984) Etude hydrologique complementaire de la cuvette lacustre du Niger. Direction Nationale de l'Hydraulique et de l'Energie, Bamako.

The Influence of Climate Change and Climatic Variability on the Hydrologic Regime and Water Resources (Proceedings of the Vancouver Symposium, August 1987). IAHS Publ. no. 168, 1987.

An analysis of the variation trend of the annual runoff on the northern slope of Qilian Shan

Lai Zuming
Lanzhou Institute of Glaciology and Geocryology
Academia Sinica, Lanzhou, China

ABSTRACT   The relationship between climatic change and runoff variation of the streams in this region is analysed in conjunction with information on the atmospheric circulation and the year-to-year change of temperature and precipitation both inside and outside this mountainous area.   The analysis shows that the occurrence of a year with low water in the eastern section of the mountains is usually related to a strengthening of the zonal circulation in high-mid latitudes in Eurasia; the influence of convective precipitation in the mountain area on the variation of annual runoff cannot be ignored.   The variation of the annual runoff is simulated using time series analysis.   A forecast for the next five to ten years is made based on the models and taking the changes of the circulation and climate into account.

Analyse de la tendance sur la variation de l'eau de ruissellement annuel dans le flanc du nord de Quilianshan

RESUME   La relation entre le changement climatique et la variation de l'eau de ruissellement des rivières dans cette region est analysée avec la circulation atmosphéri-que ét avec le changement d'année en année de la tempéra-ture et de la précipitation à l'intérieur et à l'extérieur de ces montagnes.   Cet analyse montre que les années où le niveau d'eau est faible dans la section est des montagnes correspondent souvent à l'intensité de la circulation zonal dans les régions de hautes et moyennes latitudes en Eurasie.   La précipitation de convection dans les montagnes exerce une influence importante sur la variation de l'eau de ruissellement annuelle.   Nous avons simulé cette variation par la méthode analyse de la série des temps.   Sur la base du modèle obtenu, nous avons fait une prévision sur la tendance de la variation de l'eau de ruissellement pour les 5 a 10 années à venir, en tenant compte du changement de la circulation et du climat.

Introduction

Quilian Shan is a large mountain system, rising along the northeas-tern periphery of the Qing-Zang (Tibet) Plateau and the boundary between Qinghai and Gansu Provinces (94°-104° E, 36°-40° N). It runs

from Wyshaoling in the east to Altun Shan in the west, in a direction
of ESE-WNW, adjoining the Hexi Corridor at its northern slope.   The
rivers originating from the northern slopes of Qilian Shan can be
divided into three river systems; Shiyanghe, Heihe and Shulehe
streams (Figure 1).   Glaciers, which increase from the east to the
west, occupy the alpine zone over 4000 m a.s.l. and are one of the
important sources for these rivers.

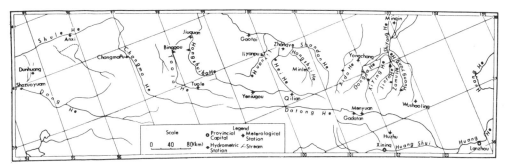

Figure 1   The distribution of stream systems on the
            north slope of Qilian Shan.

There is a broad alpine zone in Qilian Shan.   According to Wang
(1980), about 30% of the total mountain area is over 4000 m a.s.l.
The mean elevation of the basin gradually rises from the east to the
west.   The area above 3000 m a.s.l. increases from about 60% of the
total basin area down to the outlet of the Shiyanghe in the eastern
section up to over 90% of the Shulehe and Danghe rivers in the west.
The change in the ratio of drainage area to elevation gives rise to a
change in the components of nourishment for streams, as Lai (1985)
has found.
Qilian Shan is located in the interior of the Eurasia continent
where the climate is arid.   Its eastern section is affected by
monsoon.   In summer when the thermal low on Qing-Zang Plateau
intensifies, the southwest airflow can carry moisture from the Indian
Ocean and Bay of Bengal into the Hexi Corridor.   In addition, in the
height of summer, the southeast airflow can also carry a little
moisture northwestward, while the western section is usually affected
by westerlies.   This difference in the synoptic regime results in
little synchronization in the multiyear variation in runoff from the
eastern and the western portions of the mountain range.

Analysis of secular variation in annual runoff

Characteristics of annual runoff

The basic statistics of the records, shown in Table 1, indicates that
the annual runoff is stable in this region with a coefficient of
variation C_V below 0.3.   The maximum among all the mean annual
discharges does not exceed four times the minimum among all the mean
annual discharges.   The minimum annual runoff in the eastern section
occurred in 1962, and the maximum in the 1950's.   In the western

Table 1   The characteristics of the runoff for main streams
on the northern slope of Qilian Shan   (1955-1984)

Stream	Station	Drainage area (km²)	C_v	Mean multi-yearly discharge	Maximum among mean annual discharges		Minmum among mean annual discharges		(6)/(8)
				$Q(m^3s^{-1})$	$Q(m^3s^{-1})$	year	$Q(m^3s^{-1})$	year	
(1)	(2)	(3)	(4)	(5)	(6)	(7)	(8)	(9)	(10)
Zamuhe	Zamusi	851	0.27	7.7	15.7	1958	4.4	1965	3.6
Xiyinghe	Jiutiaoling	1077	0.16	10.5	13.6	1967	7.8	1962	1.7
Dongdahe	Shagousi	1614	0.16	9.8	13.2	1955	7.4	1962	1.8
Xidahe	Chajianmen	811	0.23	5.0	7.8	1956	3.1	1962	2.5
Heihe	Yingluoxia	10009	0.16	48.1	67.0	1983	32.4	1973	2.1
Liyuanhe	Liyuanbao	2240	0.25	7.0	10.4	1983	4.0	1968	2.6
Taolaihe	Bingou	6883	0.15	20.4	28.9	1972	16.4	1956	1.8
Changmahe	Changmabao	10961	0.24	26.9	44.2	1972	13.1	1956	3.4
Danghe	Shazaoyuan	16970	0.08	9.3	11.6	1982	7.9	1976	1.5

section, however, the minimum annual runoff occurred in 1956 and the
maximum in 1972 (Table 1).   The unobtrusive change in annual runoff
is due to stable precipitation in the mountain area and regulation of
the glacier meltwater.   The latter contributes more to the streams
lying to the west of Taolaihe River than to those lying to the east.

A year with a frequency of exceedence of annual runoff below 30%
is regarded as a wet year, and a year above 70% as a dry year.   A
year between 30% and 70% is considered a normal year.   The distributions of wet and dry years for six streams are shown in Figure 2.
One can see from Figure 2 that a four or five-year continuous wet
period occurred twice in the Danghe stream in the far west, also a
four-year continuous dry period occurred twice in the Heihe Stream in
the middle section.   Usually, a wet year or a dry year lasts two or
three years in other streams.

Secular variation in annual runoff and its climatic background

In general, the annual runoff varied as follows: During the 1950's
the runoff of the streams lying to the east of the Taolaihe was
abundant while those lying to the west was low.   During the 1960's,
the runoff of all streams in this region, especially the Shiyanghe,
was rather low.   In contrast with the 1950's, during the 1970's, the
runoff of the rivers lying to the east of Taolaihe Stream remained
low, especially in the Heihe, while in the Changmahe Stream and the
Danghe Stream runoff was abundant.   Since 1980, the runoff of most
streams has become abundant.   Even those lying to the west of the
Taolaihe experienced the maximum runoff among the last 30 years.
During 1980-1984, the mean discharge of rivers in the Shiyanghe Basin
was higher than that in the 1960's or 1970's, with the exception of
the Zamuhe.

A statistical analysis (Table 2) was carried out according to the
classification of circulation patterns for high-mid latitudes   in

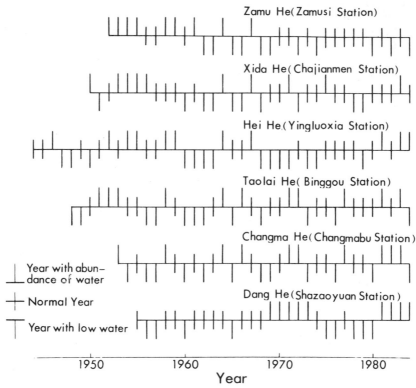

Figure 2 Temporal distribution of water levels.

Table 2 Occurrence percentage of the three
circulation types, W, E, C

Interval	For 5 entire years			For 5 summers (May to Sept.)		
	W	E	C	W	E	C
1950–1954	30.9	41.5	27.6	32.2	35.0	32.8
1955–1959	30.6	36.0	33.4	31.2	36.6	32.2
1960–1964	36.8	30.7	32.5	42.5	32.9	24.6
1965–1969	49.9	25.8	24.3	51.3	32.0	16.7
1970–1974	44.5	29.8	25.7	43.0	37.8	19.2
1975–1979	38.2	32.9	28.9	40.9	32.8	26.3
1980–1984	42.6	29.8	27.6	41.7	32.7	25.6

Eurasia suggested by Wangengeim (Girs, A.A. 1960).
    The analysis indicates:
    (a) During the 1950's the meridional circulation was enhanced,
among the three patterns, Pattern E occurred most frequently, and
Pattern C was more active than at any time since. The enhancement of
meridional circulation was beneficial to the meridional exchange of
air.   The thermal low on the Qing-Zang Plateau was strengthened with
the weakening of the westerlies intensity, allowing more moisture
from the Indian Ocean and Bay of Bengal to reach the eastern section

of Qilian Shan.   Therefore, the precipitation increased and the
runoff rose accordingly in the eastern section.

(b)   During the 1960's and the first half of the 1970's the
meridional circulation weakened, especially in the second half of
the 1960's; Pattern C fell into its weakest period since the 1950's
(occurring only 16.7% of the days between May and September),
meanwhile Pattern W became more prominent (occurring 51.3% of the
days between May and September) and the westerlies increased in
strength.   Runoff in the western section, affected by westerlies,
increased beyond that in the 1950's, and the Danghe in particular
experienced a continuous wet period lasting for 5 years, from 1969 to
1973.   The development of the thermal low on Qing-Zang Plateau was
limited by the strength of the westerlies.   Thus the moisture
transported to Qilian Shan was greatly reduced, so that the Shiyanghe
Basin, fed mainly by precipitation, suffered the lowest streamflow.

(c)   From the second half of 1970 up to the present, Pattern W
still dominates, but the number of days of occurrence has been
reduced obviously by comparison with the 1960's and the first half of
the 1970's.   Therefore, runoff in the eastern section, except Zamuhe
Stream, is increasing more or less. Particularly in the Heihe basin,
in the middle section of the mountains, runoff has been very abundant
since 1980.

Statistical analysis of the climatic data shows:

(a)   Temperature From the 1950's to the mid 1960's the temperature within the mountain range was lower, but higher outside.   From
the late 1960's to the early 1970's, the temperature inside the
mountain range rose, while outside the mountains it fell.   In the
late 1970's, the temperature both outside and inside the mountain
range generally reduced.   Since 1980 the temperature in the western
section has again risen but the temperature has not changed obviously
in the eastern section (Figure 3).

(b)   Precipitation   It is quite clear from Figure 3 that in the
western section the precipitation was rather deficient during the
1950's, followed by a marked increase from the 1960's to the present.
In the eastern section, however, the precipitation within the
mountain range was quite abundant in the 1950's. Particularly in the
alpine and intermediate zones above 2500 m a.s.l. this was the most
abundant period, although outside the mountains it was dry.   From the
late 1960's to the early 1970's it was driest inside the mountains,
meanwhile outside the mountains it became rainy. In the late 1970's,
the precipitation above intermediate zones was similar to that during
the 1950's, while it was quite abundant below the intermediate zone.
Since 1980 the precipitation above the intermediate zone has been
close to that in the late 1970's and the precipitation reduced
markedly in and below the intermediate zone. (Table 3).

It should be noted that 1967 was a turning point for climatic
change.   During the ten years from 1967 to 1976 both the annual
average temperature and the average temperature from May to September
dropped markedly outside the mountains, while the temperature varied
slightly inside the mountain range, and it even rose a little at the
stations of Qilian and Yeniugou (Table 4).   As a result, the
convectional circulation was weakened due to the decrease in the
temperature differences between the inside and outside of the
mountain range, and then the coefficient of inner cycle of moisture

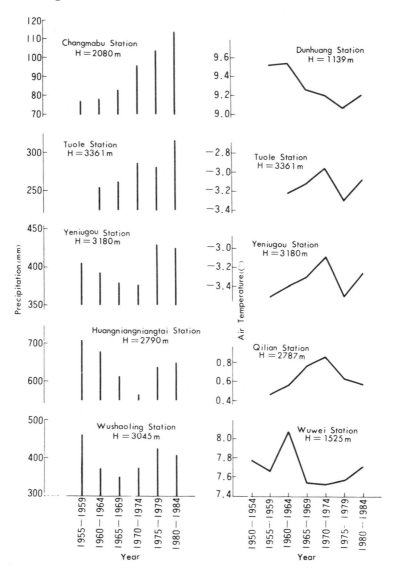

Figure 3   Comparison of 5 year averages of precipitation and
temperature with altitude.

reduced (Tang M. and Xu M., 1984). Such a situation is quite
unfavourable to convective precipitation over the mountain range.
Meanwhile, the runoff of the streams lying to the east of the
Liyuanhe reduced markedly from 1968 to 1974. The mean seven year
discharges, 1954-1960, in comparison to that over 1968-1974 for the
main streams are shown in Table 5. Table 5 reveals that the runoff
over the later seven years was 16-51% less than that over the earlier
seven years. In the meanwhile it is heard that the forest damage and
the open wasteland took place gravely somewhere in the mountains.

The question arises: What is the main reason for the decrease in
runoff: human activity or the climate? Analysis shows that:

(a) The long term water deficit during the 1960's and 1970's, and
the drought from 1968 to 1974 in the eastern section, were regionally

Table 3 Mean precipitation in the eastern section of Qilian Shan relative to the period 1955-1959

Station	Altitude (m)	Precipitation differences* (mm)				
		ΔP_1	ΔP_2	ΔP_3	ΔP_4	ΔP_5
Wushaoling	3054	-89.1	-109.9	- 88.4	-36.4	-53.2
Huangniang-niangtai	2790	-74.6	- 93.5	-140.1	-68.3	-59.4
Kongjia-zhuang	2520	-26.1	-49.3	- 68.5	- 4.5	-31.1
Zamusi	2010	-11.5	-34.9	- 51.6	12.2	-50.4
Wuwei	1525	16.1	52.0	31.5	27.2	6.4
Minqin	1354	- 7.1	12.0	19.9	24.7	-10.5

*ΔP_1, ΔP_2,....ΔP_5 are the precipitation averaged over 1960-1964, 1965-1969, 1970-1974, 1975-1979 and 1980-1984 minus that averaged over 1955-1959, respectively.

Table 4 Temperature variation inside (A) and outside (B) the mountains

	Station	Altitude (m)	$T_2 - T_1$ (°C)		$T_3 - T_2$ (°C)	
			Yearly	Summerly	Yearly	Summerly
A	Minqin	1367	-0.62	-0.42	0.53	0.13
	Zhangye	1483	-0.38	-0.42	0.28	0.05
	Jiuquan	1477	-0.59	-0.39	0.52	0.14
	Dunhuang	1139	-0.60	-0.54	0.22	-0.08
B	Wushaoling	3045	-0.39	-0.17	0.07	-0.02
	Qilian	2787	0.06	0.12	-0.08	-0.24
	Yeniugou	3180	0.03	-0.04	-0.06	-0.04
	Yuole	3361	-0.06	-0.09	-0.01	-0.02

* T_1, T_2 and T_3, are temperature averaged over 1957-1966, 1967-1976 and 1977-1984, respectively.

Table 5 Comparison of discharge of various streams

Stream	Station	Q_1	Q_2	Q_2-Q_1	$(Q_2-Q_1)/\bar{Q}$	Interval
Huangyanghe	Shajintai	5.14	3.11	-2.03	-51.0	1954-1984
Zamuhe	Zamusi	10.10	7.22	-2.88	-36.0	1954-1984
Dongdahe	Shagousi	10.90	9.33	-1.57	-16.1	1955-1984
Xidahe	Cajianmen	6.24	4.68	-1.56	-30.5	1954-1984
Xiyinghe	Jiutiaoling	11.50	9.77	-1.73	-16.5	1955-1984
Heihe	Yingluoxia	51.70	41.30	-10.40	-21.5	1954-1984
Liyuanhe	Liyuanbu	7.33	5.60	-1.73	-24.8	1954-1984
Taolaihe	Binggou	19.70	21.20	1.50	7.4	1954-1984
Changmahe	Changmabu	23.50	28.00	4.50	16.8	1954-1984
Danghe	Shazaoyuan	8.99	9.76	0.77	8.3	1955-1984

* $\bar{Q}$ — Discharge averaged over the whole statistical interval (m^3s^{-1}).

Q_1— Discharge averaged over 1954-1960 (m^3s^{-1}).

Q_2— Discharge averaged over 1968-1974 (m^3s^{-1}).

extensive.   It   occurred not only in the Shiyanghe and Heihe Basins
but also in the Datonghe Basin on the southern flanks of Qilian Shan,
as   well   as   Huangshui Basin and the upper reaches   of   the   Huanghe
(Yellow River) to the south of the Datonghe. For example, the runoff
in the 1970's was 12.4% less than that in the 1950's at the Gadatan
gauge on the Datonghe Stream, above which there was almost no human
economic activity.

(b) Similar low water occurred continually at both Xunhua gauges
in the Huangshui Basin and also at Minhe on the upstream part of the
Huanghe during 1969–1974. By contrast, an abundance of water occurred
continually in the streams in the western section of Qilian Shan and
in the streams south of Qaidam Pendi in the same period.

(c) Precipitation reduced in the mountain area, especially the
area above the intermediate zone, during the 1960's and 1970's. It is
shown in Table 3 that the precipitation differences in the mountain
area above the intermediate zone had a large negative value in the
1960's and 1970's, while there was a positive value outside the
mountains. It is evident that the decrease in runoff during the
1960's and 1970's was caused mainly by a decrease in precipitation
above the intermediate zone, in relation to convective precipitation
to some degree.

In the western section, which is affected by the westerlies to a
certain degree, the precipitation inside the mountains during the
1960's and 1970's was more than that in the 1950's due to the
increase in westerlies. On the other hand, the temperature in the
mountains increased obviously from the late 1960's to the early
1970's and has accelerated the melting of snow and ice, so the runoff
of the Danghe and Changmahe Rivers increased when compared to the
1950's.

The trend prediction of variation in annual runoff

Streamflow is a natural process and varies continuously in time, thus
it can be treated as a changing stochastic variable (stochastic
process). The time series analysis builds on the theory of stochastic
processes.

In general, the time series, $Y(t)$, is composed of a trend term,
$H(t)$, periodic term $P(t)$, and stochastic term, $\varepsilon(t)$:

$$Y(t) = H(t) + P(t) + \varepsilon(t) \tag{1}$$

Using a short series to obtain the trend period, the following
equation is employed according to Huang (1983):

$$f(t) = C_o + C_1 \sin[(2\pi t/T_1) + \phi_{1,o}] + C_2 \sin[(2\pi t/T_2) + \phi_{2,o}]\ldots$$
$$+ C_m \sin[(2\pi t/T_m) + \phi_{m,o}] \tag{2}$$

Where $N < T_1 < T_2 < \ldots < T_m$, N is number of samples, T_1, T_2, $\ldots$, T_m
are arbitrary positive integers, representing various long periods,
C_o, C_1, $\ldots$, C_m are amplitudes of harmonics, $\phi_{1,o}$, $\phi_{2,o}$, $\ldots \phi_{m,o}$ are
initial phase angles.

The periodic term is determined by the following Fourier analysis.

$$E(K) = (1/m) \left[R(0) + 2\sum_{\tau=1}^{m-1} R(\tau)\cos(K\pi/m) + R(m)\cos(K\pi) \right] \qquad (3)$$

Where the autocorrelation function

$$R(\tau) = [1/(N-\tau)] \sum_{j=1}^{N-\tau} Q'_j \; Q'_{j+\tau}, \qquad (4)$$

m is maximum time lag (year), $\tau = 0,1,\ldots,m$. Q'_j is a sample series of departures. The stochastic term is obtained by means of autoregression analysis.

The time series analysis for annual runoff of a few streams has been carried out by methods such as regression analysis, variance analysis and the wave spectrum analysis. Taking the Zamuhe and Danghe Rivers as examples, we have the general trend equation representing fluctuation of the time series as follows:

$$Q(J)_{zu} = 10.1 + 3.1 \sin\{[2\pi(J-1)/91] + 181°\}$$
$$+ 0.77 \sin\{[2\pi(J-1)/21] + 39°\} + 1.42 \sin\{[2\pi(J-1)/3] + 99°\} \qquad (5)$$

$$Q(J)_{Dong} = 9.15 + 0.39 \sin\{[2\pi(J-1)/13.5] + 37°\}$$
$$+ 0.48 \sin\{[2\pi(J-1)/9 + 144°\} + 0.32 \sin\{[2\pi(J-1)/3.9] + 162°\} \qquad (6)$$

The correlation coefficients between calculated and observed values are 0.73 for Eq.(5) and 0.79 for Eq.(6) with significance levels of 0.05.

From the trends of climatic changes and circulation development, it is inferred that the zonal circulation which has predominated for twenty years will weaken, and that meridional circulation will be enhanced, as a result, it will benefit the movement of moisture to reach the eastern section of Qilian Shan. In addition, the climate has recently become warmer and the warming rate outside is greater than that within the mountain range, thus it is favourable to convectional precipitation in the mountain area. The warming climate in the western section will contribute to the melting of snow and ice, increasing the snow melt component. On the other hand, the weakness of the zonal circulation, following the weakened westerlies, will be unfavourable to the precipitation of the western section. It seems that future possible climate and circulation changes are advantageous to the streams in the eastern section of Qilian Shan.

From the calculated result and future possible climatic changes and circulation development, it is inferred that the runoff will tend to increase in the east section of Qilian Shan but to decrease in the west section in the next five to ten years.

References

Girs, A.A. (1960) <u>Osnovy dolgoscrochnykh prognozov pogody</u> (Principles of long-germ weather forecast). Gidrometeoizdat, Leningrad.

Huang Zhongshu, 1983, <u>Method of wave spectrum analysis and its application to hydrology and meteorology.</u> Meteorological press, Beijing. (in Chinese)

Lai Zuming, 1985, Variations of runoff supply and discharge of rivers with elevation in Qilian Mountains. Memoirs of Lanzhou Institute of Glaciology and Cryopedology, Chinese Academy of Sciences, No. 5. (in Chinese)

Tang Maocang and Xu Manchun, 1984, Climatic variations over Qilian Mountain area. <u>Plateau meteorology,</u> Vol.3, No.4 (in Chinese with English abstract)

Wang Zongtai et al. 1980, Glacier inventory of China I. Qilian Mountains. (in Chinese)

The Influence of Climate Change and Climatic Variability on the Hydrologic Regime and Water Resources (Proceedings of the Vancouver Symposium, August 1987). IAHS Publ. no. 168 1987.

Rainfall trends in West Africa, 1901-1985

Oyediran Ojo
Department of Geography, University of Lagos
Lagos, Nigeria

ABSTRACT This paper attempts to examine the characteristics of rainfall variations between 1901-1985 in West Africa using instrumental records. It discusses the characteristics of rainfall, particularly, periodicities and variabilities, emphasizing decadal, quinquennial and annual characteristics. The paper also discusses some hydroclimatologic consequences of climatic change and climatic variations with respect to streamflow characteristics. From the various analyses, it was observed

(a) that the present century began with a relatively long period of drought persistence (1900-1926). This was followed by a wet period which lasted between 1927-1960;

(b) that the present drought conditions have been characteristically persistent since about 1961 and particularly since about 1965;

(c) that in spite of the above generalizations, no regular patterns can be observed in trends, periodicities and persistence of hydrologic consequences of rainfall variations to allow for predictability of these consequences in relation to rainfall variations; and

(d) that because of the large spatial and temporal variations in the characteristics of rainfall, and the resulting hydroclimatologic consequences, there is urgent need to improve the availability and reliability of the data.
Note: A French abstract of this paper can be found at the end of the text.

Notation

P	precipitation
$\bar{p}_i$	mean precipitation at station i
I	Index of variability
σ	standard deviation
i	station
j	month/year
n_i	number of occurrences at station i
N	number of stations
N_i	total number of occurrences at station i
J	number of months/year
Y	normalized departures of rainfall
F	probability of occurrence

Introduction

The success of any applications, impact studies and research on
climatic change and climatic variability and their relationships with
hydrology and water resource systems depend on the development of an
adequate reliable data base.   Consequently, a great deal of research
effort has been devoted to examining climatic records in many parts
of the world with the aim of determining climatic characteristics and
their relationships with human society.   For example, in Africa in
general and West Africa in particular a lot of studies have been
carried out on the characteristics of climatic change and climatic
variability, (Grove, 1972, 1973; P. Lamb, 1980; Kerr, 1985; Ojo,
1986a, 1986b; Nicholson, 1981, 1982 and 1983).   However, the results
of these studies have shown divergences of opinion about the nature
and characteristics of climatic variability and climatic change on
the continent.   Thus, many different controversial statements on the
status of the characteristics and the possible future trends of
climate in the region have been made by several scientists.   For
example, some scientists have concluded that the Sahelian droughts of
West Africa have persisted since 1969 and would even persist into the
next century, thus indicating climatic change to drought conditions
(Lamb, 1973; Winstanley, 1973).   According to Lamb (1973), the
Sahelian droughts "have already a long history...it is not likely to
disappear in the near future." Winstanley (1973) also noted that the
downward trend in rainfall in West Africa will continue for 50 years.
In contrast to these conclusions that the Sahelian droughts will
persist for some time, some scientists regard the drought condition
as being somewhat unusual in terms of the recent past but not
necessarily deviating from the longer term probabilities.   In fact,
to these scientists, the recent Sahelian droughts are part of the
normal climate rather than an indication of climatic change.   For
example, Landsberg (1975) concluded that the Sahelian drought of the
1970's has to be accepted as part of the normal climate of the
region.   The available data do not indicate a trend, and there is not
an indication of a climatic change.
   Most of the conclusions made by the scientists were based on
rainfall characteristics of the early 1970's, when the consequences
of the climatic events were so disastrous that greater and more
widespread concern for climatic variations were demonstrated than for
any other climatic event on the continent.   Less has been done to
examine the nature and characteristics of climatic variabilities for
periods earlier than 1969 and these, in addition to the variabilities
since 1969, are of significance for understanding the dynamics of
hydroclimatologic components and predicting the sensitivity of water
resource systems to climatic change and climatic variations.
   In the present paper, therefore, an attempt is made to examine the
characteristics of rainfall variations between 1901-1985 in West
Africa in general and the Sahelian region in particular, using
instrumental records.   The paper discusses some aspects of rainfall
characteristics, particularly trends, periodicities and variabilities,
emphasizing decadal, quinquennial, annual and seasonal characteristics.   The paper also discusses some hydroclimatologic consequences
of climatic change and climatic variations, for example, with respect
to streamflow characteristics within some river basins in West

Africa, persistence, severity and widespread nature of droughts and floods and aspects of spatial and temporal variations of some water balance components in the region.

Data and methodology

The rainfall data used in this study consisted of 60 stations in West Africa, twenty-four of which were in the Sahel region while 36 stations were outside the Sahel savanna. Data from the Nigerian stations were obtained from the Nigerian Meteorological Department in Lagos while data for the other countries were obtained through: Professor Gregory of the Department of Geography, University of Sheffield; Dr. Lamb of Climatology Section, Illinois State Water Survey; and through Mr. Semenya of the Department of Geography, University of Lagos. Data obtained through Professor Gregory and Mr. Semenya were those published in "Resume Mensuel D'Observations Meteorologiques" by ASECNA, Dakar. Because of the problems associated with data collection in the region, the number of years for which data were available varies from station to station. For example only 5 of the 24 stations in the Sahel region have the complete 85 year (1901-85) record of rainfall while another seven stations have records for about 80 years. All the stations used in the present study, however, have rainfall records for periods of at least sixty years.

The mean annual rainfall for n years, at each station (i) was computed by using the equation which can be expressed in the form:

$$\bar{P} = \frac{1}{J} \sum_{j=1}^{n_i} P_{ij} \tag{1}$$

where j refers to the year.

Although there are many approaches to defining droughts, the present paper emphasizes droughts as related to rainfall variations because climatic variations in West Africa can virtually be equated to rainfall variations (see for example, Oguntoyinbo and Odingo, 1979; Ojo, 1986a, 1986b). The index used in the study thus emphasizes rainfall variability and is the time series of the normalized annual departures of rainfall in the region. Using the various data, the climatic index (equation 2) was computed for the five climatic zones in West Africa, namely, the Tropical Rainforest, Guinea Savanna, Sudan Savanna, Sahel Savanna and the Southern Sahara (Ojo, 1986a; Church 1980). Quantitatively, the climatic index used can be expressed in the form:

$$I_{ij} = \frac{1}{N} \sum_{j=1} \frac{P_{ij} - P_i}{\sigma_i} \tag{2}$$

if calculated for one year j, at station i; or

$$I_{ij} = \frac{1}{N} \left[\frac{P_{ij} - P_i}{\sigma_i} \right] \tag{3}$$

Rainfall variabilities

Figure 1 shows the rainfall variability indices in West Africa for
1901–85 averaged for all the 60 stations used in this study using the
results obtained from equations 2 and 3. In general, three major
periods can be observed since 1901.   The century began with a
relatively long period of drought which lasted until about 1926 with
breaks of relatively normal or wet periods which lasted one or two
years each.   Over the period 1901–1926 only three years may actually
be regarded as wet, with climatic indices greater than + $\sigma/2$. In
contrast there were about 10 years with climatic indices equal to or
less than −$\sigma/2$. On the whole about 52% of the period may be regarded
as normal with climatic indices between ±$\sigma/2$.

CLIMATIC INDICES FOR WEST AFRICA : 1900–1985 (Average for 60 Stations)

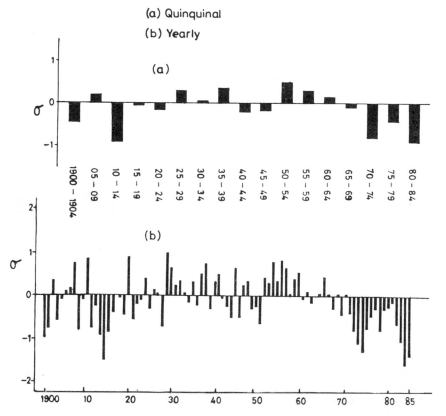

Figure 1   Rainfall variability indices in West Africa.

The second period is a relatively wet period which lasted from
1928 to 1960.   For most of this second period, two or three
relatively wet years were generally followed by another two or three
years characterized by near normal, but slightly drier conditions.
About 11 years of the 33-year period have climatic indices which are
greater than +$\sigma/2$ while only three years have indices which are −$\sigma/2$
or less. The remaining nineteen years have standard deviations which
are between ±$\sigma/2$.   During the six years from 1961 to 1966 conditions

were near normal with indices between $+\sigma/2$.   From 1967 droughts have been relatively persistent and the indices for most of the years are equal to or less than $-\sigma/2$.

Temporal variation also occurred on the decadal and quinquennial scales.   For example the decades 1900-09, and 1930-39 were near normal.   Similarly, the 1960-69 decade was near normal with slight negative index.   In contrasts, 1950-59 was a relatively wet decade while 1940-49 and 1970-79 were relatively dry decades.   The worst conditions occurred during the 1970-79 decade, with an index of -0.62.

On the quinquennial (five-year) scale, seven of the seventeen quinquennial periods have positive indices. Of the periods with negative indices, the worst were 1910-14, 1970-74 and 1980-84 with climatic indices less than $-\sigma/2$.   No regular periodicities may be discerned visually for either the decadal or the quinquennial periods.

Rainfall variations were also examined on a regional basis, using the five climatic regions already noted.   The analysis shows that a lot of variation also occurs between the different climatic zones and even within the same climatic zone.   For example, drought conditions were more persistent and more widespread in the Sudan Savanna, the Sahel Savanna and the Southern Sahara than in either the tropical rainforest region or the Guinea Savanna.

Hydroclimatic consequences

Rainfall variations in West Africa have many hydroclimatic consequences, probably the most obvious of which are the occurrences of droughts and floods discussed above, and in addition, variations in the discharges of the region's rivers. Unfortunately in West Africa runoff data exist for only a very sparse network, for very short periods, and in many cases the data are not continuous.   Thus, it is very difficult to analyze and discuss adequately these aspects of the hydroclimatologic consequences of rainfall variations.

From the few data available, it is evident that a lot of variation in both space and time occurred in these hydroclimatologic consequences.   For example, between 1969-77, runoff data for the river Niger at both Koulikoro and Niamey, show that only 1969 experienced a positive variability index in the annual runoff. Negative variabilities prevailed between 1970-77.

More detailed analyses of the variations of the occurrence, severity and widespread nature of the hydroclimatic characteristics show that it is difficult to generalize for relatively large areas using information from a relatively small area.   For example, in contrast to conditions at Save Bridge on River Queme and at Bakel on River Senegal where the driest years of the 1971-80 decade were 1973, 1977 and 1978, the driest year at Quessaba on River Sassandra was 1974.   Variations sometimes occur in the characteristics of the discharges along the same river.   For example, although most of the years between 1969 and 1977 were characterized by drought conditions at Maradi and Dobel on river Goulbi, the year-to-year variabilities in the percentage of deficiency of annual discharges at both locations show that droughts were more persistent at Dobel, occurring approximately between 1970-74 and 1976 with slight recovery in 1975. At Maradi on the other hand, droughts occurred in 1972-73 and 1976-

77.   Also the severity of droughts were greater at Maradi than at
Dobel.   It may also be noted that conditions at Maradi show that the
years 1974-75 were characterized by very wet conditions in contrast
to Dobel where the years showed either negative variabilities or
positive conditions near normal.

Conclusions

West Africa, like many other parts of the continent, has experienced
pronounced climatic variations, with their accompanying climatic
events and the hydroclimatologic consequences of these events. The
patterns of the rainfall variabilities since 1901 in the region shows
that the 20th century began with a relatively long period of dry
climatic conditions which were fairly persistent until about 1926
with occasional breaks of normal or wet conditions.   This was
followed by a relatively wet period which lasted until about 1960.
From about 1961 droughts have been relatively persistent in West
Africa and particularly since 1969, droughts have been so persistent
that the climatic indices for most of the years have been equal to or
less than $-\sigma/2$.
     More detailed analysis however shows a lot of spatial and temporal
variations in the characteristics of rainfall since 1901, so that it
becomes difficult to generalize for relatively large areas, using
information from a relatively small area.   The study also reveals
that the validity of the empirically determined data and information
on hydrology and hydroclimatology is considerably limited particularly for the purpose of the application of these data for research and
development.   Any empirical data and information obtained for any
area may be valid within reasonable bounds only in such an area for
which the data and information were derived. Any generalizations for
larger regions need careful error analysis.
     The above conclusion raises the important issue of obtaining
adequate and reliable data.   At present, no country in West Africa
has accurate and continuous information on the spatial and temporal
characteristics of rainfall or any other hydroclimatic components.
Indeed, Africa in general and West Africa in particular, are in
serious need of a vast variety of information on water resources, to
improve the capacity to plan for the future of the various countries.
     There are, however, data sets of varying degrees of completeness
on rainfall, archived by many national agencies and individual
research centres in the region.   A number of institutions are also
engaged in research projects which are only indirectly related to
hydrology and water resources, but whose results and data sets can be
useful in advancing the cause of hydrology. The co-ordination of the
efforts of these various groups will considerably improve the
validity of studies in hydrology and water resources, and rainfall in
particular.

References

Church, H.R.J. (1957)   West Africa: A Study of the Environment and
     Man's Use of It. Longmans, Green and Co., London.

Grove, A.T. (1972) "Climatic Change in Africa in the last 20,000 Years," Les Problemes de Development dudu Sarhara. Septentrional, Vol. 2, Algier.

Grove, A.T. (1973) "A Note on the Remarkably Low Rainfall of the Sudan in 1913, Savanna 2, 133-138.

Lamb, H.H. (1973) "Some Comments on Atmospheric Pressure Variations in the Northern Hemisphere. In Drought in Africa (Dalby, D. and Harrison Church R.J. Editors) School of Oriental and African Studies, London.

Kerr (1985) "Fifteen Years of African Drought". Science Vol. 227, 1453-1454.

Lamb, P.J. (1980) "Sahelian Drought," N.Z.J. Georgr. 68, 12-16.

Landsberg, H.E. (1975) "Sahel Drought: Change of Climate or Part of Climate?" Arch., Met., Geoph. Biocl., Ser. B. 193-200.

Nicholson, S.E. (1981) "Rainfall and Atmospheric Circulation During Drought Periods and Wetter Years in West Africa," Mon. Weather Rev., 109 2191-2208.

Nicholson, S.E., and Chervin, R.M. (1982) "Recent Fluctuations in Africa -- Interhemispheric Telecommunications" in Variations in the Global Water Budget, Street -- Perrott et al (eds.) Reidel/Dordrecht, 495 pp.

Nicholson, S.E. (1983) "Sub-Saharan Rainfall in the Years 1976-80: Evidence of Continued Drought." Mon. Weather Rev., 111 1614-1654.

Oguntoyinbo, J.S. and Odingo, R.S. (1979) "Climatic Variability and Land Use: An African Perspective," Summaries of Papers Presented at the WMO World Climate Conference, Geneva 262-272.

Ojo, O. (1969) "Potential Evapotranspiration and the Water Balance in West Africa: An Alternative Method of Penman," Arch. fur Met. Biok., Ser. B. XVII, 239-260.

Ojo, S.C. (1983) "Recent Trends in Aspects of Hydroclimatic Characteristics in West Africa." Hydrology of Humid Tropical Regions with particular Reference to the Hydrological Effects of Agriculture and Forestry Practice, IAHS Publ. No. 140 97-104.

Ojo, O. (1986a): "The Sahel Droughts: 1941-1984" Proc. ISLSCP Conference, Rome, Italy, ESA SP-248 411-415.

Ojo, O. (1986b): "Drought Persistence in Tropical Africa since 1969," Programme on Long Range Forecasting Research WMO Report Series, No. 6, Vol. 1, WMO/TD 87, 73-85.

Winstanley, D. (1973) "Rainfall Patterns and the General Atmospheric Circulation." Nature 245, 190-194.

The Influence of Climate Change and Climatic Variability on the Hydrologic Regime and Water Resources (Proceedings of the Vancouver Symposium, August 1987). IAHS Publ. no. 168, 1987.

Some effects of climate variability on hydrology in western North America

D.H. Peterson
U.S. Geological Survey, Menlo Park, CA 95025, USA
D.R. Cayan
Scripps Institution of Oceanography, La Jolla,
CA 92093, USA
J. DiLeo-Stevens
U.S. Geological Survey, Menlo Park, CA 95025, USA
T.G. Ross
U.S. Geological Survey, Reston, VA 22092, USA

ABSTRACT   The strong north-south gradient in precipitation along the West Coast makes this region an interesting laboratory for studying the influence of climate on runoff variability in general and riverine chemistry in particular. Interannual fluctuations in large-scale atmospheric circulation and associated precipitation and runoff can produce major disruptions in the "average" climatologic picture. Such fluctuations can be inferred and simplified from the time-averaged atmospheric pressure field and large-scale patterns of stream flow anomalies (eg., high or low stream flow). Further, the effect of the climate gradient along western United States on the total dissolved solids concentrations in rivers is summarized as a highly idealized force-response model of total dissolved solids concentrations as a function of river flow. The response in wet years is more like the wetter climate response curves observed to the north and the response in dry years is more like the drier response curves observed to the south.

Introduction

Space-time characteristics of climate variability offers an interesting and revealing subject for hydrologic research. Two perspectives of this broad subject are presented herein. The first briefly introduces the interests of a multidisciplinary group of scientists characterizing climate variability of the eastern north Pacific and western North America (PACLIM). The second gives a preliminary example linking climate and riverine chemistry.

A series of ongoing meetings, called the PACLIM Workshop (Mooers et al., 1986) has brought together oceanic, atmospheric and terrestrial scientists, including hydrologists, who are discussing the interrelations that link the physical, chemical and biological variability in the PACLIM region. These discussions and their subsequent collaboration efforts will provide the basis for an interdisciplinary monograph series describing observed and expected responses of these regimes to anomalous (e.g., wet and dry) behavior in climate. Where possible the connections between the oceanic, atmospheric and terrestrial realms will be interpreted in terms of variations in the hydrologic cycle. Future PACLIM activities including dates of forthcoming meetings and requirements for participation, have been published (Mooers et al., 1986). Included here are two examples representing extremes in time scales that are interesting to hydrologists.

On an event time scale, a key to understanding the origin and nature of floods is to identify atmospheric circulations leading to major floods. One of the largest recorded floods of the PACLIM region occurred in January 1862 (Roden, 1966). Although data are scanty, it is clear that the 1862 flood was associated with an abnormal shift of the large scale atmospheric circulation. Premonitory to this flood a remarkable outbreak of an Arctic air mass surged south along this western corridor, on the east side of a high air pressure "blocking ridge" in the Gulf of Alaska in western Canada (Figure 1 upper panel). This event was so extreme, the largest river of western United States, the Columbia River, froze, (Roden, 1966). From these observations and from our detailed knowledge of similar flooding events in the modern record, we can hypothesize that the subsequent heavy precipitation episode initiated when the blocking high pressure migrated westward towards the Aleutians. This migration probably induced a sequence of severe storms with westerly and southwesterly winds which penetrated California. Numerous other records of abnormal large scale atmospheric circulation coincide historically with the greatest recorded floods of coastal streams along the western United States (Roden, 1966; Cruff & Rantz, 1965; note also Figure 1 lower panel).

In leaping from short time scales of extreme episode events to long time scales of millenia over the past 10,000 years, a useful way to understand hydrological fluctuations over northwestern North America is to interpret them as shifts from the present seasonal cycle of atmospheric circulation and associated precipitation. Work of Heusser et al. (1985) on pollen samples from peat bogs suggests that the climate of northwestern North America has shifted from a temperature maximum and precipitation minimum about 8,000 years ago, toward the present cool and wet regime that has persisted since about 5000 years before present. Heusser et al. (1985) hypothesized that during this dryer period the subtropical north Pacific high (anticyclone) dominated the west coast climate for a longer period of the year at higher latitudes than presently so that southern Alaska was warmer and dryer (Figure 2). Presumeably, the climate shifted towards the present regime as the subtropical Pacific high pressure center weakened and the Gulf of Alaska-Aleutian low pressure center strengthened. The broad scale nature and possible interdisciplinary implications of this phenomena are suggested from a ca. 5,000 YBP transition which appears in other proxy records including lake sediments (Yang, 1986), the terrestrial biological record (Spaulding, 1985), and marine continental slope sediments of western North America (Gardner & Hemphill-Haley, 1986).

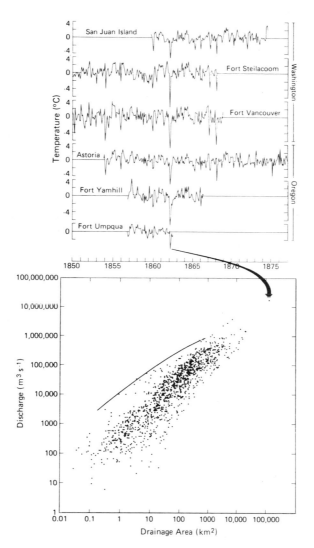

Figure 1   Upper panel: Mid - 19th century air temperature anomalies
in western United States record the extremely cold winter
of 1861-1862 in the Pacific Northwest (adapted from Rodden
1966).
Lower panel: The maximum peak discharges in California in
relation to drainage area (adapted from Waananen & Crippen
1977). Dot in far upper right-hand corner is a very rough
estimate based on anecdotal information for the 1861-1862
California flood (see Peterson et al., 1985). The curved
line is the envelope of maximum floods in conterminous
United States (after Crippen & Bue, 1977; as cited in
Riggs, 1985).

The west coast hydrological climate

Some prominent effects of climate variability on riverine chemistry

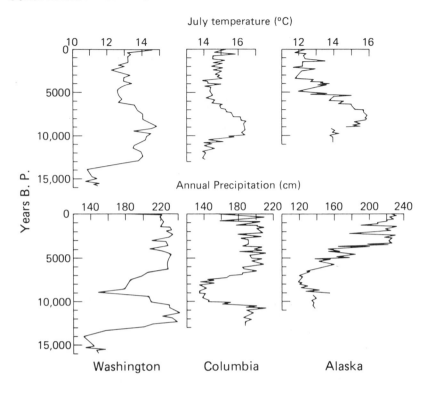

Figure 2   Late-Quaternary mean July Temperature and annual
           precipitation estimated from proxy pollen records
           from peat cores (adapted from Heusser et al., 1985).

are illustrated below for the corridor of western North America
extending from southwestern Canada to the California-Mexico border
(Figure 3). Here the effects of climate are dramatic in both space
and time. There is a strong north-south gradient in precipitation
(cf.Figure 4) which results from the shielding effects of the
subtropical "North Pacific High" pressure cell, which tends to
deflect Pacific storms to the north. In winter the contraction of the
high and the intensification of the Gulf of Alaska-Aleutian low
pressure pattern to the north allows North Pacific storms to invade
the West Coast, resulting in the rainy season (Figure 5). Because the
low pressure cell and associated storm tracks lie to the north, over
an annual cycle precipitation is more prevalent to the north. On an
interannual time scale the anomalous strength and position of the
high and low pressure cells is a key to interpreting the fluctuations
in rainfall.

Climate driven riverine chemistry

The strong north-south gradient in precipitation along the West Coast
makes this region an interesting laboratory for studying the
influences of runoff variations on riverine chemistry. To the north
are coastal rain forests and mountains with very high annual precipitation. Under this extremely wet regime, for example in the northern
Cascade Mountains of Washington, high precipitation and runoff

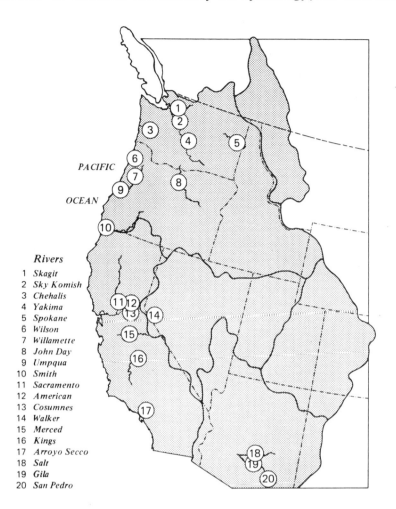

PACIFIC

OCEAN

Rivers
1 Skagit
2 Sky Komish
3 Chehalis
4 Yakima
5 Spokane
6 Wilson
7 Willamette
8 John Day
9 Umpqua
10 Smith
11 Sacramento
12 American
13 Cosumnes
14 Walker
15 Merced
16 Kings
17 Arroyo Secco
18 Salt
19 Gila
20 San Pedro

Figure 3   Locations of rivers with discharge records and, if available, total dissolved solids concentration records of this
study. Data from U.S. Geological Survey, Water Resources
Division.

dilutes, transports and maintains low ionic concentrations. Stream
water ionic concentrations (activities) are calculated to be so low
that they are often in thermodynamic equilibrium with a secondary
clay mineral gibbsite, a silica depleted end product of chemical
weathering (Reynolds & Johnson, 1972); Drever 1982). In sharp
contrast to this northern setting, in the southern sectors riverine
(and soil) ionic activities increase (over an order of magnitude)
due to low precipitation and the greater role of evapotranspiration.
An extreme example is the evaporative concentration of interior
stream waters leading to mineral precipitation of the highly soluable
salts in repository saline desert lakes (Felmy & Weare, 1986).
    Intuitively, then, lie a continuum of river basin soil and stream
salinities between these two extremes in regional precipitation and
associated ion activities. This simple view of the north-south
pattern in soil and stream salinities is, of course, complicated by

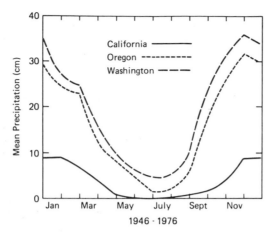

Figure 4   Annual cycle of coastal precipitation (adapted from Cayan
& Roads, 1984). California = 43, Oregon = 200 and Washington = 240 cm per year of precipitation (1931-1970 avg.)

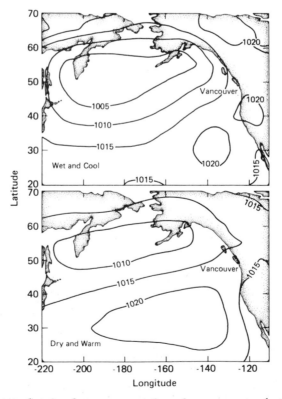

Figure 5   Climatological mean sea level pressure (millibars) based
on 1947 to 1974, monthly mean for winter (Dec.,Jan.,Feb.)
and spring (Mar.,Apr.,May). Adapted from Namias (1975).

other factors including major variations in topography and, therefore, local climate.   For excellent reviews and a more traditional introduction to riverine chemistry of this region (and elsewhere) consult the examples cited by Drever (1982) and recent syntheses (Colman & Dethier, 1986; Trudgill, 1986).

Year to year fluctuations in precipitation can produce major disruptions in this average climatologic picture. In certain extreme episodes, the strong north-south gradient can be overwhelmed so that almost the entire region behaves in the same (relative) mode. These exceptionally wet or dry periods occur when large scale atmospheric circulation in winter over the North Pacific is strongly developed (Cayan et al., 1986). Recent examples of both extremes in precipitation, and runoff, are the winters 1976-1977 and 1982-1983.   These years exhibited tremendous differences in streamflow (Figures 6 and 7) and water chemistry in rivers throughout the west.

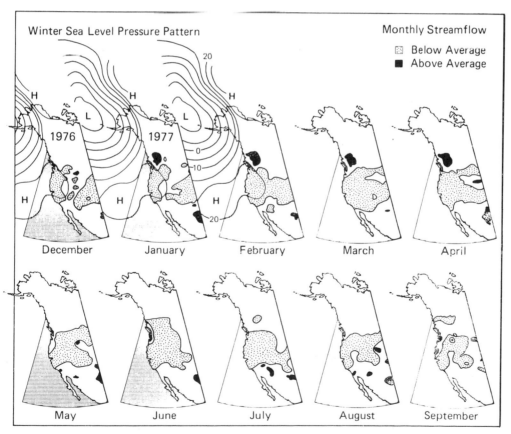

Figure 6   Mean 1976-1977 winter (Dec.,Jan.,Feb.) sea level pressure (millibars - 1000 millibars) from Namias 1975, and stream flow patterns (cf., U.S. Geological Survey, 1986; above normal is the upper quartile flow and below normal is the lower quartile flow.)

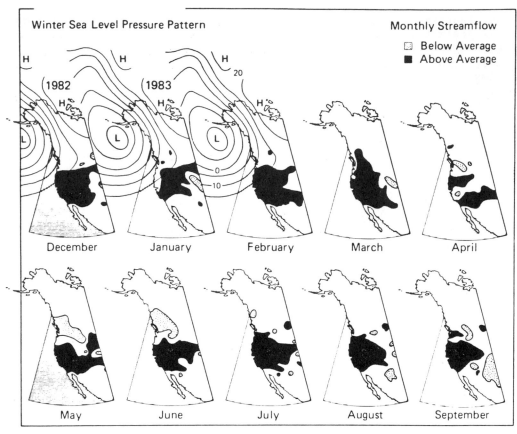

Figure 7   Mean 1982-1983 winter (Dec.,Jan.,Feb.) sea level   pressure
            (millibars - 1000) and streamflow pattern (cf., U.S. Geol.
            Survey, 1986, above normal is the upper quartile flow and
            below normal is the lower quartile flow.)

    Climate, the major control of precipitation and snowmelt, forces
variations in river flow over a broad band of space and time scales.
For this reason an especially long record of specific river flow
provides an interesting and remarkable index of climate. A related
but lesser studied effect of climate is its influence on the
chemistry of the river water which can be striking. Herein we broadly
define river water chemistry to include all dissolved substances in
river water which contribute to its salinity or mineral content,
generally observed as the concentration of total dissolved solids
(TDS). The variation in TDS concentrations with river flow provides
an interesting link between riverine chemistry and climate.

Modeling the effect of streamflow on stream chemistry

To relate TDS concentrations to river flow a simplifying assumption
is that TDS concentrations approach a limiting value, A_o, at very
high river flows (TDS concentrations never go to zero). Also, as an
approximation over a full range of river flows, we assume that

typical decreases in TDS concentrations caused by increases in river
flow are large when TDS concentrations are high (or, conversely,
that the decreases are small when TDS concentrations are low). A
simple model of such TDS concentrations as a function of river flow
is given:

$$C_{TDS} = A_o + A_1 e^{-\beta Q_{sp}} \tag{1}$$

where C_{TDS} is the concentration of TDS, Q_{SP} is specific river flow,
A_o is the TDS concentration at high river flow, and $A_o + A_1$ is the TDS
concentration at zero Q_{sp} (Figure 8). The parameter β is a measure of
how sharply the TDS curve breaks (e.g., how sharply the TDS
concentrations diminish with increasing flow); some less obvious
properties of β are note later. Derived parameters for equation 1
from observed values of TDS vs. Q_{sp} can be computed using an
exponential least-squares method (Tektronix, 1975, see also Jennrick
and Ralston, 1979).

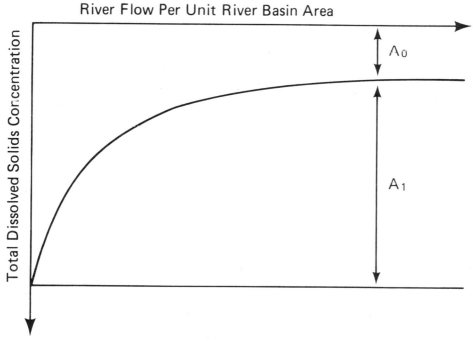

Figure 8   Total dissolved solids as a function of river flow (eq.1)
           Note the vertical axis is downward.

Unlike river flow observations, detailed and complete time series
of riverine water chemistry are rare. One strategy to overcome this
difficulty, although not entirely satisfactory, is to analyze a
composite of data, often representing a sparse collection of observations scattered over several years. Such a composite of monthly or
bimonthly observations made during January, 1978 to December 1985
from the coastal Smith River, California, are used here to exhibit
the variability of equation (1).

Note that when all of the data from the Smith River composite are
analyzed together the results are, perhaps, adequate but not
impressive (Figure 9, upper panel). However, a much better fit is

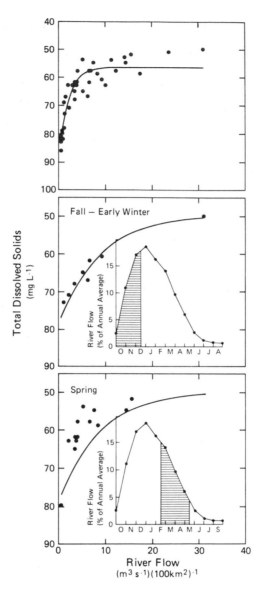

Figure 9   Upper panel: Total dissolved solids concentration as a
           function of river flow (dots), Smith River, Calif.(1978 to
           1985).
           Middle panel: Same data set as per upper panel but for Oct
           Nov., Dec., observations.
           Lower panel: Same data as per upper panel but for Mar.,
           Apr., May observations.
           Curves in upper and a middle panels are exponential least
           squares fit to eq.(1); curve in lower panel is an exponen-
           tial least squares fit of data in middle panel to eq.(1).

obtained if we focus only on values representing the annual rise in
river flow (Figure 9, middle panel, October to December). Similarly,
for the subset of observations during declining river flow that

occurs later in spring, TDS concentrations tend to remain at lower
values for the same flow (Figure 9, lower panel, March to May) in
comparison to the period of increasing flow. An interpretation of
this feature for many riverine systems is that the source or supply
of easily soluble salts is greater at the onset of the rainy season
(the "wash cycle") than afterwards (the "rinse cycle").

Equation (1) appears to be useful for describing the Smith River
basin response to a) rising and b) declining river flow. Further,
this exercise suggests strong influences of seasonal cycles of
climate in driving the stream chemistry. The nonunique behavior of
TDS concentrations with river flow found for the coastal Smith River
is clearly illustrated with more detailed data from an inland high
Sierra Mountain Merced River, California. For the Merced, and for
rivers in general, annual variations in river flow and TDS concentrations are not in phase. Thus, equal values of river flow are observed
to have different TDS concentrations (Figure 10). Similar nonlinear
relations between TDS concentrations and river flow have been
described on time scales varying from storm events (Webb & Walling,
1983) to annual hydrologic cycles (Davis & Keller, 1983). Later we
consider this nonlinear response feature in more detail, for now it
is sufficient to appreciate that TDS--river flow relations often
differ over a broad band of time scales especially during rising
(gradual response) and falling (sharp response) river flow. Further,
as might be expected, the rise or fall response can markedly differ
on an interannual time scale; wet years, for instance, can depress A_o
to significantly lower values than in dry years (Figure 11).

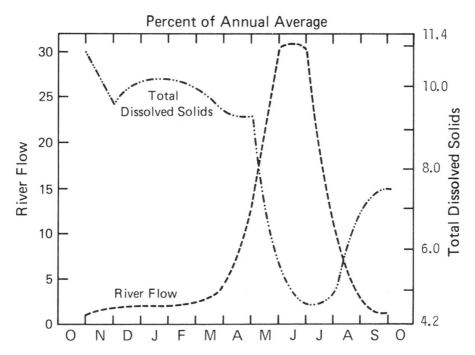

Figure 10   Annual cycle of mean monthly river flow and total
dissolved solids concentrations (as a percent of the
annual avg. value), Merced River, Calif., (1968 to 1981).

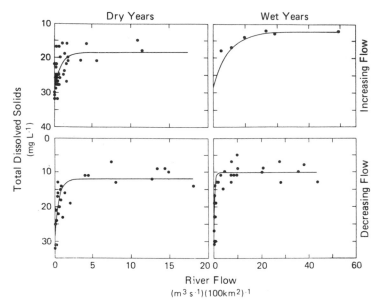

Figure 11 Total dissolved solids concentrations as a function of
river flow (dots), Merced River, Calif. for increasing and
decreasing flows in the annual cycle (Wet winter years =
1969, 1973, 1974, 1978, 1979; Dry winter years = 1971, 1972
1975, 1976, 1977, 1981, see Fig. 10) Curves are exponential
least squares fit to eq. (1).

The discrepancy in TDS concentrations with rise and fall of the
annual river flow and with interannual fluctuations of the flow
illustrate the importance of temporal climatic variability on river
chemistry. The effects of spatial variability in climate on stream
chemistry can similarly be characterized with equation (1) whereby an
A_O value in a higher rainfall climate to the north (e.g., Figure 12,
upper panel Yakima River, Washington) is much lower than in a more
arid climate to the south (e.g., Figure 12, lower panel, Gila River,
Arizona). Further, the response in wetter years is more like the
wetter climate response curves observed to the north and the response
in dry years is more like the drier response curves observed to the
south. This effect of the climate gradient along western United
States on the TDS concentrations in rivers is summarized as a highly
idealized "force-response model" of TDS concentrations as a function
of river flow (Figure 13).
    Finally, we find it instructive to simulate TDS concentration and
river flow data assuming equation (1) is a cumulative exponential
distribution function and using inverse transform methods (Naylor et
al., 1966; Gordon 1978). One interesting property in such an
assumption is that β^{-1} is defined as the mean value of an exponential
probability distribution. In this regard a preliminary unifying
picture of the chemical response of streams to specific flow is
provided by considering some of these rivers together on the same
plot. A plot of observed estimates of β^{-1} versus the long term (e.g.,
approximately 50 years) mean river flow (Figure 14) provides insight
into the possible hydrologic significance of β within the context of
the assumptions above. As a caution it should be emphasized that the

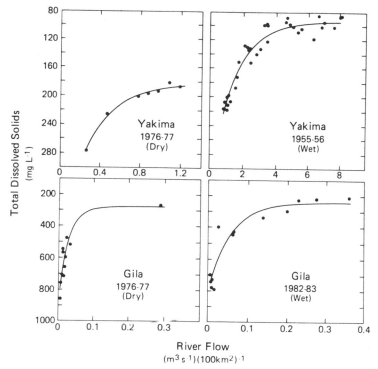

Figure 12   Total dissolved solids concentrations as a function of
river flow (dots), Yakima River, Wash., and Salt River,
Arizona. Curves are an exponential least sqares fit to eq.
(1). Note changes in horizontal scale.

extremely close relation between β^{-1} and mean annual river flow in
Figure 15 is fortuitous. What seems meaningful, however, is that
estimated values of β^{-1} for the Gila River in wet and dry years are
consistent with the overall pattern. The estimated values of β^{-1} are
lower in dry years as well as in regions of dryer climates. Earlier,
we noted that the results using equation (1) were consistent in
defining spatiotemporal variability in the sense that the distribution in wet years were more similar to that of wet climates and vice
versa. This does not, however, explain the cyclical annual (and storm
event) behavior of TDS concentrations with river flow where β^{-1} values
are often high during the wet cycle and lower during the rinse cycle.

Implications

The above is a broad-brush attempt to link the spatiotemporal
variability of riverine chemistry to climatic variability. This
coarse analysis shows that climate stands out as a strong force
controlling riverine (Hem, 1985; Peters, 1984) and, apparently, soil
(Folkoff & Meentemeyer, 1985) chemistry. An intermediate step in such
analyses includes characterizing the link between large scale
atmospheric circulation and regional stream flow (Cayan et al.,
1986). Further, process-oriented studies frequently identify the
importance of wet and dry intervals in influencing chemical fluxes

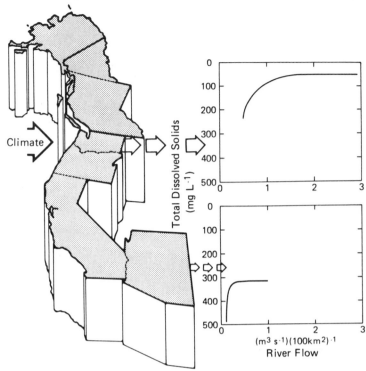

Figure 13   Highly   idealized   response   of   total   dissolved   solids
concentrations as function of specific river flows   along
the climate gradient of western North America.

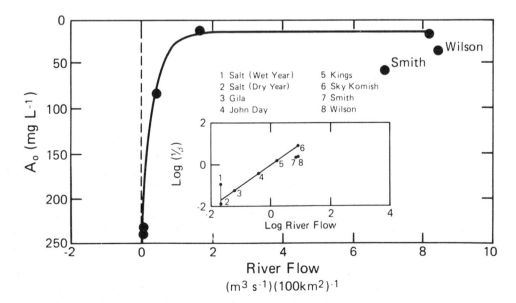

Figure 14 A_o (see eq.(1)) as a function of long-term   mean   annual
river flow for a variety   of rivers   from the study   area.
Inset   panel: β^{-1} (see eq.(1)) as a function of   long-term
mean annual river flow for the same rivers.

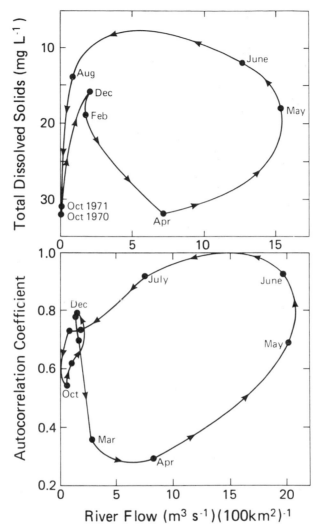

Figure 15   Upper panel: Annual cycle of total dissolved solids
concentrations as a function of river flow Oct., 1971.
Lower panel: Long term mean annual cycle of
autocorrelation coefficients in mean-monthly river flow
anomalies (from their mean-monthly value) at lag 1 month
as a function of the long-term mean monthly river flow.

1981). A related matter for future investigation is the similarity
between the pattern of the annual cycles of TDS concentrations and
river flow autocorrelations (Figure 15; cf., Moss & Bryson, 1974;
Vecchia, 1975).
    A logical next step is to evaluate the general response of
individual ions to climatological variability. Equation (1), for
instance, has been found adequate in a study of river flow vs.
specific ion concentrations (Feller & Kimmins, 1979) although
presented in a somewhat different mathematical formulation. As a
final note, it should be appreciated that distinguishing human from
natural effects in riverine chemistry and even in ground water

chemistry can benefit from a more complete knowledge of the role of climate over a broader band of spatiotemporal scales than is generally considered (cf., Konikow & Patten, 1985).

ACKNOWLEDGEMENTS   We would like to thank Vance Kennedy, Jurate Landwehr, Fred Nichols, James Slack and Frank Triska for reviewing the manuscript and Martha Nichols for technical assistance.

References

Cayan, D.R. & Roads (1984) Local relationships between United States west coast precipitation and monthly mean circulation parameters. Mon. Weath. Rev. 112, 1276–1286.

Cayan, D.R., Peterson, D.H., DiLeo-Stevens, J. & Ross, T.G. (1986) Large-scale North Pacific atmospheric circulation patterns and western North American streamflow anomalies. EOS trans. AGU. 67 (44), 932.

Colman, S.M. & Dethier, D.P. (Eds.) (1986) Rates of Chemical Weathering of Rocks and Minerals. Academic Press, Inc., Orlando, Florida, USA.

Crippen, J.R. & Bue, C.D. (1977) Maximum flood flows in the conterminous United States. USGS Water-Supply Pap. 1887

Cruff, R.W. & Rantz. S.E. (1965) A comparison of methods used in flood-frequency studies for coastal basins in California. USGS Water-Supply Pap. 1580.

Davis, J.S. & Keller, H.M. (1983) Dissolved loads in streams and riverine-discharge and seasonally related variations. In: Dissolved Loads of Riverine and Surface Water Quantity/Quality Relationships, 79–89. IAHS publ. no. 141.

Drever, J.I. (1982) The Geochemistry of Natural Waters. Prentice-Hall Inc., Englewood Cliffs, New Jersey, USA.

Feller, M.C. & Kimmins, J.P. (1979) Chemical characteristics of small streams near Haney in southwestern British Columbia. Wat. Resour. Res. 15(2), 247–258.

Felmy, A.R. & Weare, J.H. (1986) The precipitation of borate mineral equilibria in natural waters: application to Searles Lake, California. Geochimica et Cosmochimica Acta. 50, 2771–2783.

Folkoff, M.E. & Meetemeyer, V. (1985) Climatic control of assemlages of secondary clay minerals in the A-horizon of the United States soils. Earth Surf. Processes. 10, 621–633.

Gardner, J.V. & Hemphill-Haley, E. (1986) Evidence for a stronger oxygen-minimum zone off central California during late Pleistocene to early olocene. Geology 14, 691–694.

Gordon, G. (1978) System Simulation. Pretice-Hall, Inc., Englewood Cliffs, New Jersey, USA.

Hem, J.D. (1985) Study and interpretation of the chemical characteristics of natural water. USGS Water-Supply Pap. 2254.

Heusser, G.J., Heusser, L.E. & Peteet, D.M. (1985) Late-quaternary climate change on the American North Pacific Coast. Nature. 315, 485–487.

Jennrich, R.I. & Ralston, M.L. (1979) Fitting nonlinear models to data. Annual Review in Biophysics and Bioengineering. 8, 195–238.

Klein, M. (1981) Dissolved material transport - the flushing effects in surface and subsurface flow. Earth Surf. Processes. 6, 173-178.

Konikow, L.F. & Patten, E.P., Jr. (1985) Groundwater forecasting In: Hydrological Forecasting (ed. by M. G. Anderson & T.P. Burt), 221-270. John Wiley & Sons, Chichester, Great Britain.

McColl, J.G. (1972) Dynamics of ion transport during moisture flow from a Douglas Fir forest floor. Soil Sci. Soc. Amer. Proc. 36, 668-674.

Mooers, C.N.K., Peterson, D.H. & Cayan, D.R. (186) The Pacific climate workshops. EOS Trans. AGU. 67(52), 1404-1405.

Moss, M.E. & Bryson, M.C. (1974) Autocorrelation structure of monthly streamflows. Wat. Resour. Res. 10(4), 737-744.

Namias, J.(1978) Multiple causes of the North American abnormal winter 1976-1977. Mon. Weath. Rev. 106, 279-295.

Namias, J. (1975) Northern hemisphere seasonal sea level pressure and anormaly charts, 1947-1974. California Cooperative Oceanic Fisheries Investigations, Atlas No. 22. Scripps Institution of Oceanography, La Jolla, California, USA.

Namias, J. & Cayan, J.R., (1984) El Nino The implications for forecasting oceanous. Vol. 27, 41-47.

Naylor, T.H., Balinfy, J.L., Burdick, D.S & Chu, K. (1966) Computer Simulation Techniques. John Wiley & Sons, Inc., New York, USA.

Peters, N.E. (1984) Evaluation of environmental factors affecting yeilds of major dissolved ions of streams in the United States. USGS Water-Supply Pap. 2228.

Peterson, D.H., Smith,R.E., Hager, S.W., Harmon, D.D., Herndon, R.E. & Schemel, L.E. (1985) Interannual variability in dissolved inorganic nutrients in northern San Francisco Bay estuary. Hydrobiologia. 129, 37-58.

Quiroz, R.S. (1983) The climate of the "El Nino" winter of 1982-83 -- A season of extraordinary climatic anomalies. Mon. Weath. Rev. 111, 1685-1706.

Reynolds, R.C. & Johnson N.M. (1972) Chemical weathering in the temperate glacial environment of the northern Cascade Mountains. Geochim. Cosmochim. Act. 36, 537-544.

Riggs, H.C. (1985) Streamflow characteristics. Elsevier Science Pub. B.V., Amsterdam, The Netherlands.

Roden, G.I. (1966) A modern statistical analysis and documentation of historical temperature records in California, Oregon, and Washington, 1821-1962. J. Appl. Meteorol. 5, 3-24.

Spaulding, W.G. (1985) Vegetation and climates of the last 45,000 years in the vicinity of Nevada Test Site, South-Central Nevada. USGS. Prof. Pap. 1329.

Tektronix, Inc. (1977) Program five exponential least-squares. Tektronix R Plot 50 Statistics, 4. Beaverton, Oregon, USA.

Trudgill, S.T. (1986) Solute Processes. John Wiley & Sons Ltd., Chichester, Great Britain.

US Geological Survey (1986) National Water Conditions. (ed by T.G. Ross).

Vecchia, A.V. (1985) Periodic autoregressive-moving average (PARMA) modeling with applications to water resources. Wat. Resour. Bull. 21(5), 721-730.

Waananen, A.O. & Crippen, J.R. (1977) Magnitude and frequency of floods in California. USGS Water Resource Investigation. US Government Printing Office, Washington, D.C., USA.

Webb, W. & Walling, D.E. (1983) Stream solute behavior in the river
    Exe basin, Devon, U.K. In: Dissolved Loads of rivers and Surface
    Water Quantity/Quality Relationships. 153-169. IAHS Publ. no. 141.
Yang, I.C. (1986) Climate changes implied from organic carbon and
    carbon-14 analyses of lake-sediment cores, Walker lake, Nevada.
    EOS Trans. AGU. 67(44), 935-936.

The Influence of Climate Change and Climatic Variability on the Hydrologic Regime and Water Resources (Proceedings of the Vancouver Symposium, August 1987). IAHS Publ. no. 168, 1987.

Evaluation of runoff changes in the Labe River basin by simulating the precipitation-runoff process

Josef Buchtele & Martin Zemlicka
Czech Hydrometeorological Institute
151 29 Prague 5, Holeckova 8, Czechoslovakia

ABSTRACT   The rainfall-runoff simulations based on a daily time interval have been carried out for the Czech part of the Labe River basin (area approx. 52,000 sq. km) for the periods from 1895 to 1933 and from 1939 to 1953. The deterministic Tank model has been used for this purpose. The goal was to evaluate the changes for the longest possible span of time -- but outside the interval in which the reservoirs could have a marked effect. Before deterministic modelling, statistical time series analysis had been made. Several time intervals were used (Δt = 1 day, 1 month, 3 months, 1 year). The results indicate that the most problematic input is evaporation. The analysis also seems to offer some results from which some changes in runoff are apparent: cross-correlation between daily precipitation and runoff were investigated, for each summer half-year separately. A distinctly sharper form of this function in that period seems to be an indication of the basin's faster response. Variations of the simulated runoff due to the changes in some inputs (precipitation, air temperature, snow conditions) were also investigated, and some comparisons are given that illustrate the responses of basins with different areas.

Evaluation de modifications de'écoulement dans le bassin d'Elbe par la simulation des relations "pluies-débits"

RESUME   Les simulations des relations "pluies-débits" ont été effectuée pour la partie tcheque du bassin d'Elbe (la surface environ 52,000 km^2) a l'aide du modele déterministe "Tank." Le but de ce travail est d'évaluer les modifications éventuelles du débit dans une période on ne peut plus longue suaf celle pendant laquelle l'influence des barrages peut se manifester. Dans les analyses statitiques des séries temporaires réalisées d'avance, les divers intervalles ont été utilisés (Δt - 1 jour, 1 mois, 1 trimestre et 1 an). Les résultats donnent a entendre que l'élément le plus problématique soit l'évaporation. Meme l'analyse statistique donne des résultats d'ou l'on peut tirer les conclusions de la modification dans le processus d'écoulment; on a examiné la fonction de la corrélation "pluies-débits" séparément pour chaque période d'été de l'année. Une forme plus

aigue de cette fonction dans la période récente signale
une réponse plus vite du bassin versant.   La variabilité
des débits simulés par les modeles déterministes a été
examiné. Les entrées du modele ont été modifiées (pluies,
température, neige). A la fin, on a comparé les réponses
de bassins versants de diverses surfaces.

Introduction

The most apparent variations of water resources are often connected
with the year's climatic cycle.   Because it is so dominant, it is
necessary to take it into account even when the intention is to
analyze the long-term changes.   The phenomena participating in the
rainfall-runoff process have usually different phases, some
interferences may distort the normal course and oscillations are
possible. Therefore, when a coarse time unit is used the results may
be unreliable. A slightly different but important problem in
statistical analysis may also occur when the time unit is very short:
one can identify short-term tendencies without the ability to see
what could be valid for a longer span of time.   On the other hand, a
drawback of deterministic modelling can be seen in that it usually
assumes attenuation in the governing processes only.

    The aim of this study has been to evaluate the precipitation-
runoff process using parallel statistical time series analysis and
deterministic modelling.   Besides the above reasons, this approach
can be advantageous for checking the quality of the data.

    The Czech part of the Labe River basin coincides approximately
with one administrative unit within Czechoslovakia.   This provides
the opportunity to compare the changes in runoff due to the climate
variability and those which have occurred as a consequence of human
activities.

Data used for analysis and simulation

The character of the data available and the processes studied can be
seen from Figures 1 and 2.   The longest time series to have been
analyzed statistically were yearly series of precipitation, air
temperature and saturation deficit observed from 1804, 1775 and 1945
respectively. Simultaneous monthly areal precipitation and discharge
records have been evaluated for a period of more than 100 years. More
than 600 stations have been used to calculate mean daily areal
precipitation as early as between 1895 and 1905.   It was not for the
whole span of time that the daily values were prepared using so many
points; however, for those periods when only some 40 to 60 stations
were used for the whole basin, the choice of stations was arranged in
such a manner that the monthly averages were essentially identical
with the data offered by the complete network.   Daily data have been
used mostly for deterministic modelling.

    The project has not been completed yet; simulations have been
possible for periods from 1895 to 1933 and from 1939 to 1953. (Since
the reservoirs could have a marked influence in the past few decades,
this interval was omitted from considerations). The 1895-1953 period

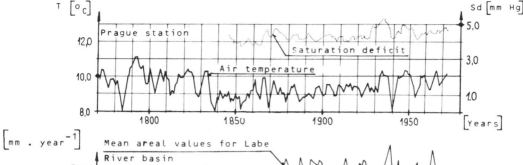

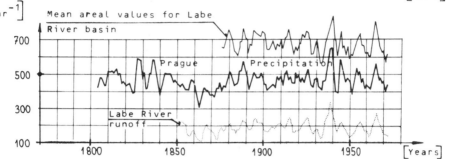

Figure 1   Analysed long time series smoothed   by three year
moving averages.

is considered to be the most suitable time since in this period the
population did not grow but considerable development took place. For
instance, grain production rose from 14 to 22 t / ha, and the total
volume of agricultural production doubled, etc.

As additional data, observations have been included on the two
sub-basins inside the Labe River basin. Both of these sets were
prepared in connection with the studies of reservoir operation.   The
Orlik (Small Eagle) Reservoir has a basin area of approximately
12,000 sq. km and the Husinec (Gooseville) Reservoir basin covers
200 sq. km.   The set of observations on the Orlik basin   provides
monthly data over a period of 80 years, from 1893 to 1973. The
Husenic basin has daily data for several years which has been
prepared for deterministic modelling.

Statistical analysis

Independent series analysis

Preliminary analysis processed data on the Orlik Reservoir. Seasonal
inflow predictions for the reservoir had been the goal in some
preceding investigations,   and some sort of autoregression and
relations similar to ARMA models had been employed -- separately for
individual seasons (i.e. for spring, summer, autumn and winter, each
comprising three months).   As the input, precipitation was used in
two forms; (a) as measured values (P) and (b) as effective
precipitation.   Evaporation (E) observed with the help of a simple
weighing instrument (Wild) was subtracted in the latter case.
(Evaporation observations were carried out for only little more than

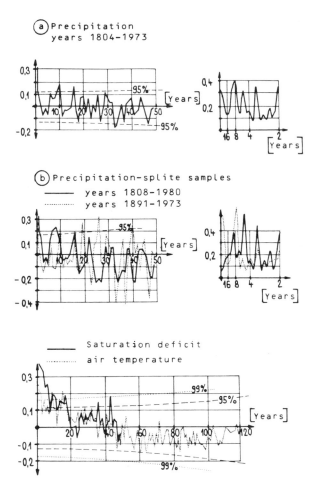

Figure 2   Autocorrelation and spectral functions of
           observations in Prague.

25 years; for the remaining part of those 80 years its values were
derived using the correlation between evaporation and saturation
deficit values.) Autocorrelation functions and spectral density
functions of annual totals of these time series and of runoff (R) are
shown in Figure 3. The "recharge" (W) was similarly analyzed
(W – P – E – R).

Separate autocorrelation functions and spectral density functions
(s.d.f.) of series consisting of quarterly quantities in each year
(ie. series of spring, summer, autumn and winter values) have been
calculated. Only graphs of s.d.f. are presented in this paper
(Figure 4a); one can judge by them that evaporation is a rather
problematic input. The trend indicated by the spring season s.d.f.
is probably due to the changes in the station location. The
functions for annual values of W (Figure 3) also have an interesting
form. Attempts were made to remove the indicated components with the
3.25 year periods and to compare the effect with the one caused by
the removing of the one-year periodicity. It turned out that the
annual cycle is much more significant -- approximately 2% and   15%

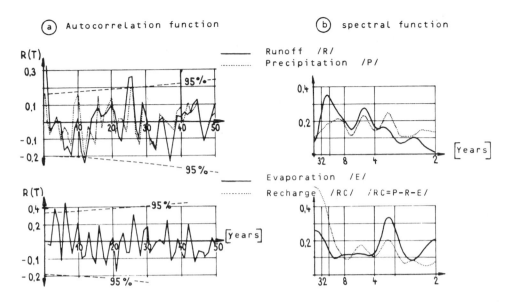

Figure 3   Functions derived for the Orlik reservoir from
           observations from 1892-1973.

respectively of explained variability.   Another result of these
analyses is depicted in Figure 4b.   The dominant role of
autoregression members representing runoff in spring and autumn
periods should not be underestimated.

The problems associated with estimating evaporation led to the
analysis of long time series, some results of which are shown in
Figure 2.   This was the starting point of the analysis for the whole
Czech section of the Labe River basin.   Because the correlation
functions in the foregoing analyses and in the 100-year long series
of precipitation and runoff for the whole River Labe basin showed
some common features, the suspicion arose that these common forms
result from the length of the period employed and the 170-year long
precipitation series (1804-1973) was split into two sets that were
analyzed separately, see Figure 2a, b.   Inspecting the graphs
obtained, the anticipation seems to be valid.

The assumed series for evaporation calculations -- air temperature
and saturation deficit -- were also treated.   Distinct trends are
noticeable in these presentations (Figure 1), as well as in the
correlation and spectral functions.   The course of long-term mean
precipitation and of other series in Figure 1 may also serve as an
illustration of why the assumption about an 11-year cycle is so
popular.   But one can hardly decide with certainty which portion of
it is the result of the so-called small ice period, what has been
caused by instrument changes, and what has been the influence of
urban growth.

Cross-correlations

These functions were investigated independently for each year.   The
series of daily precipitation and runoff for the Labe River in the

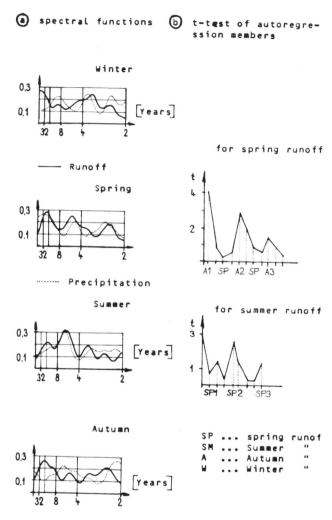

Figure 4   Periodical components in seasonal series at Orlik Reservoir.

1895-1933 and 1939-1953 periods were used. However, due to the presence of noise during winter seasons (snow cover), these functions have been derived separately for summer half-year periods only (21 May to 20 November, i.e. 182 values). The autocorrelation function of precipitation is practically zero for $\tau \neq 0$, and therefore it can be expected that the course of the cross-correlation is, in accordance with the Winer-Hoff equation, a characteristic of the transformation capability of the basin. The non-linearity of the rainfall-runoff process causes the average values of these functions to sometimes yield somewhat distorted notions. Some experiments in which non-linear relations have been applied, Rao A.R. & Rao, R.G.S. (1977), indicate that it is more difficult to interpret such results. Nevertheless, the distinctly sharper form of the averages in more recent periods might be considered to substantiate the fact that acceleration of the runoff has occurred (Figure 5). As for the

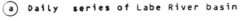

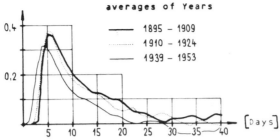

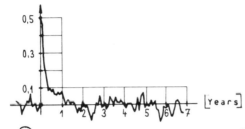

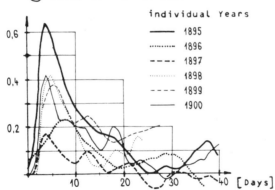

Figure 5   Cross-correlation function of precipitation-runoff.

basin's memory, it should be borne in mind that the results are affected by the time unit used; its role is illustrated by part b in Figure 5. (Monthly series were standardized for this purpose.)

As a by-product of the daily data statistical analysis, the spectral density functions for precipitation might be considered. Examples of these functions and their average shape, derived using individual years' functions, are given in Figure 6. They can serve to determine how meaningful synoptic cycles are in areal evaluation, in view of statistics. A more precise investigation would clearly need the removal of the annual cycle.

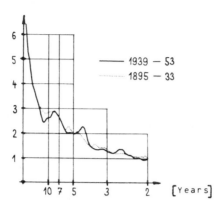

(a) Averages of Years:

1939 — 53
1895 — 33

[Years]

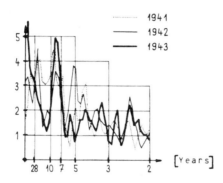

(b) individual Years

1941
1942
1943

[Years]

Figure 6   Spectral function of the mean areal daily precipitation in the Labe River basin.

Deterministic modelling

Snow melt plays an important role in the total annual runoff in the Labe River basin.   For this reason, models which incorporate a snowmelt algorithm can be considered more appropriate for this basin. The Tank model, Sugawara et al. (1984), which incorporates a snowmelt algorithm, has been used in our simulations.   Its algorithms for snow cover and soil moisture are simpler than those included in NWSRFS.   In spite of that, the results seem to offer some insight into the analyzed variations.

Inside the two data sets given (1895–1933 and 1939–1953), even shorter intervals were assessed independently, most of them eight years long.   The summary is listed in Table 1 .   The lack of data for 1934–1938 can be considered a certain drawback in these investigations and, consequently, also an approximate estimation of initial conditions for the latter period.   The experience gained in many trials has shown that it can really have some effect on the

Table 1   The deviations between simulated and observed runoff

Annual values in years		1895-1902	1903-10	1911-19	1920-27	1928-33	1939-46	1947-53
$\overline{O}$ [mm]		211.5	183.9	194.0	211.8	166.6	241.7	147.4
ΔO	[mm]	18.3	16.2	11.0	15.2	9.3	45.1	14.8
max ΔO		-38.2	-31.9	-25.7	-36.4	-20.6	-115.2	+28.7
ΔO	[%]	8.7	8.8	5.7	7.2	5.6	18.7	10.1
max ΔO		16.4	16.6	12.6	17.2	12.5	28.5	25.9

appraisal of outflow from the bottom tank, which explicitly simulates the long-term runoff component. However, the water balance would not be expected to change significantly over some two or three years.

Despite the doubts about the evaporation data quality the initial trials with the Tank model had been made using its values derived from long-term observation of saturation deficit. But, gradually, we have come to the conclusion that better results can be obtained when using long-term monthly averages. The model includes means for decreasing evapotranspiration in dry periods, and rough estimation of water balance for several years has proved that the actual evapotranspiration is far from being constant. Depending on individual years and due to the model's effect, it varies within a range of at least 100 mm per year.

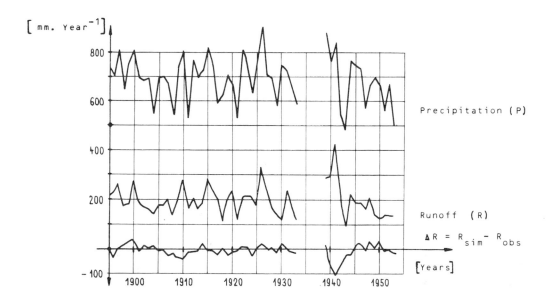

Figure 7   Deviations between observed runoff and runoff simulated
            using the Tank Model.

The simulated and observed yearly runoffs in Figure 7 do not indicate any noticeable tendency in the first period, but quite exceptional discrepancies are apparent between 1939 and 1943.

Successive occurrences of extreme runoff values in those years   make
it difficult to find out what effects were the most   important.   The
following are suggested as plausible explanations:
     (a) retention capacity of the basin and its changes;
     (b) model performance under these rather exceptional circumstances.
(Some doubts can also arise from the fact that the precipitation
station network was affected by World War II events, but deviations
of such magnitude seem to be improbable after checking thoroughly
several times.)
     The sensitivity of simulated runoff to changes in different inputs
was tested for the annual cycle.   The significance of the following
phenomena was investigated: precipitation, evapotranspiration,

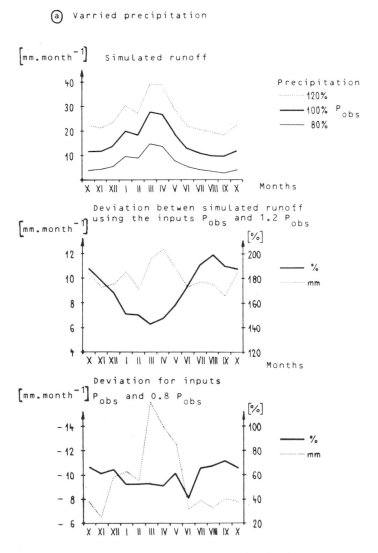

Figure 8a   Monthly long-term averages of simulated runoff
            in the Labe River basin.

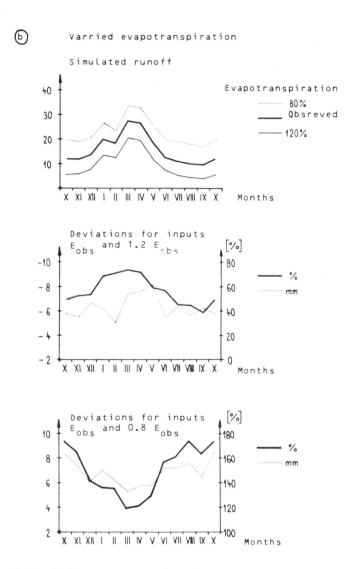

Figure 8b Varied evapotranspiration.

degree day factor for snowmelt and air temperature (for the same
purpose only). Constant 20% changes have been considered throughout
the whole year cycle. The outcome is apparent from Figures 8a and b.
The most significant variations are encountered in precipitation
changes in the late summer and in the autumn, but evapotranspiration
changes also exert significant influence on water balance and
redistribution. The change in the snowmelt parameters results mainly
in a time shift; the response of basins of different sizes to the
same impulses and/or parameter changes is depicted in Figure 9. The
greater persistence of large basins seems to be evident.
    It is believed that the results are an indication that a
deterministic model is able to simulate the changes in similar
circumstances with acceptable reliability. As stated earlier, the
project is not considered to have been completed. The simulated

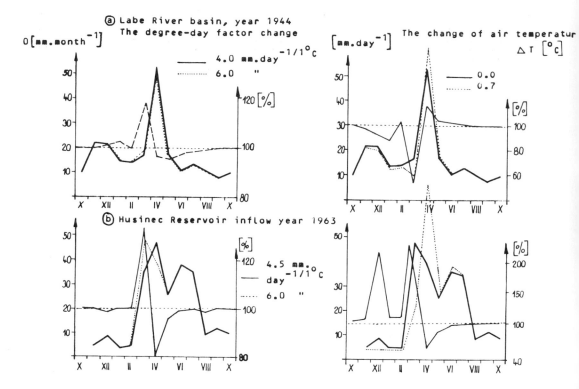

Figure 9   Monthly snowmelt runoff changes due to variation of inputs.

maximum   and   minimum   discharges   and   flood   volumes   are   being
statistically   evaluated   and   the   goal   is   a   comprehensive   simulation
for   the   period   given without interruption.   Another   task   is   to
examine   evapotranspiration   in   greater   detail   and   to   perform
simulations   for other sub-basins of the Labe River,   perhaps   using
another model.

ACKNOWLEDGEMENTS   The   simulations   with the Tank model   have   been
possible thanks to the help of Dr.   Sugawara in the initial stage   of
work   with   his   model.   His generosity and kindness   are   very   much
appreciated.

References

Nemec,   J.   &   Schaake,   J.   (1982)   Sensitivity of   water   resources
     systems to climate variations.   Hydrol. Sci. Bull.   27(3), 327–
     343.
Rao,   A.   Ramachandra & Rao,   R.G.   Srinivasa (1979)   Analysis of the
     effects   of   urbanization   on runoff by the   nonlinear   functional
     series   model   of   rainfall-runoff   process.   Proc. International
     Symposium on Urban Hydrology, Hydraulics and Sediment Control (ed.
     by D.T.   Kao),   Univ. of Kentucky, Lexington, Kentucky, July 1977,
     209–220.

Sugawara, M., Watanabe, I., Ozaki, E. & Katsuyama, Y. (1984) Tank model with snow component. Research Note No. 65, National Research Center for Disaster Prevention, Science and Technology Agency, Japan.

The Influence of Climate Change and Climatic Variability on the Hydrologic Regime and Water Resources (Proceedings of the Vancouver Symposium, August 1987). IAHS Publ. no. 168 1987.

Climatic fluctuations and runoff from glacierised Alpine basins

David N. Collins
Department of Geography
University of Manchester
Manchester, M13 9PL, UK

ABSTRACT   Records of discharge from gauges on rivers draining from glacierised basins in the Swiss Alps for the period 1922-1983 were analyzed together with climatic data in order to describe climatic variation and to determine the impact on total annual flows. Mean May-September air temperature and annual discharge show considerable yet markedly parallel year-to-year variations at all stations. Generally warmer summers in the late 1920s, mid-1930s, 1940s and late 1950s are reflected in higher discharges, with a maximum in 1947. Cooler temperatures and reduced flows characterised the 1960s and 1970s before slight recovery in the 1980s. A fall in ten-year mean temperature of 1°C produced a reduction of about 25% in ten-year mean discharge. Five-year running means of annual totals of precipitation increased from the 1940s to the 1970s. Glacier area reduced between the 1920s and 1950s before stabilising. High levels of explanation of variance of discharge were obtained using multiple regression on mean ablation season air temperature and winter precipitation accumulation, the degree of fit being greater for basins with larger percentage glacierisation.

Objectives

The aims of this paper are to describe year-to-year variations and underlying longer-term trends in flow of rivers draining from extensively glacierised Alpine basins, and to examine contemporary fluctuations of climatic conditions. The manner in which glacierisation influences the climate runoff relation is first described. Relationships between climatic variables and with runoff are then investigated with a view to identifying which linkages might be used to predict the effects of future climatic changes on meltwater yield. The intention is to relate runoff directly to climate, initially without reference to glacier mass balance which can be viewed as the outcome of the same set of climatic variables (Collins, 1985), and measurements of which are available for only a few glaciers. Several catchments in the upper Rhône basin in Switzerland with climatic and hydrologic records were selected for investigation through the period 1922 to 1983.

Variations in runoff in Switzerland expressed as decadal means have been described by Walser (1960) but only for basins of less than 32 per cent glacierisation. Mean runoff was greatest in the decade 1920-1929 and at a minimum in 1940-1949 except for basins with about

30 per cent glacierisation which showed secondary maxima in 1940–1949.

In the period 1910 –1920 glaciers advanced, followed by widespread recession until the mid-1960s. Subsequently termini of many glaciers have readvanced, although increases in planimetric area have been small when compared with previous changes (Kasser 1981, Patzelt 1985).

Influence of glaciers on climate-runoff relationship

Year-to-year climatic fluctuations affect both the amounts of snow and ice stored in, and the quantities of meltwater runoff arising from, glacierised high mountain basins. Since much of the streamflow from highly-glacierised areas is derived from melting of ice, annual meltwater yield is directly related to summer energy input (Collins 1985), but the relationship with precipitation is more complex. On an annual basis, runoff may be reduced when winter snow accumulation has been above average, delaying the rise of the transient snowline and limiting the extent and duration of exposure of underlying ice to melting. Some compensation for the loss of icemelt is provided, however, by the larger contribution of snow melt to runoff (Krimmel & Tangborn, 1974). In an ice-free basin annual total runoff will be directly related to, and always less than, precipitation received. On the other hand, runoff from glacierised basins can be either greater or less than total precipitation according to the change in glacier storage.

Several studies have indicated, without reference to stationarity, that runoff variability is less in glacierised catchments than in ice free basins (Rasmussen & Tangborn, 1976; Collins, 1982; Fountain & Tangborn, 1985). In less glacierized basins this variability reduces with increasing glacierisation (Krimmel & Tangborn, 1974; Tvede, 1982) but appears to remain constant above about 50 percent cover (Collins 1985). Non-stationarity during reference periods as well as any subsequent changes or secular trend will lead to erroneous assessments of yield. Glacier recession, for example, will lead to runoff additional to that which would have been produced if glaciers existed in steady state; conversely, sustained deglaciation results in diminished flows. Kasser (1973) has shown that runoff was significantly reduced in an Alpine catchment of 5220 km^2 in which the glacierised proportion declined from 16.8 to 13.6% in the 52 years to 1968.

Climatic and hydrologic data records

Characteristics of five basins with glacierisation extending from 35.9 to 68.2 per cent and with long discharge records are listed in Table 1, and their locations are shown in Figure 1. Records of the discharge of the meltwaters of the Vispa at Visp were free from the influence of hydropower abstraction until 1956. Flow of the Saaser Vispa tributary, gauged at Zermeiggern until 1963, prior to completion of the Mattmark dam, is derived largely from the Schwarzberggletscher and Allalingletscher.

Discharge of the Massa, arising principally from Grosser Aletschg-

Table 1   Physical characteristics of selected extensively glacierised basins in the upper Rhône catchment area, Switzerland

Characteristic Basin/Gauging station	Basin area km^2	Glacierisation %(year)	Gauge elevation m a.s.l.	Highest point m a.s.l.
Vispa/Visp	778.0	35.9(1950)	659	4634
Lonza/Blatten	77.8	40.6(1983)	1520	3897
Saaser Vispa/Zermeiggern	65.2	44.6(1876) 41.5(1960)	1740	4199
Rhône/Gletsch	38.9	56.4	1754	3634
Massa/Blatten-bei-Naters	194.7	66.02(1973)	1446	4195
Massa/Massaboden	202.0	68.2(1967) 67.6(1934) 64.1(1957)	687	4195

Percentage glacierisation for Massa basin calculated from glacier area given by Kasser (1981)

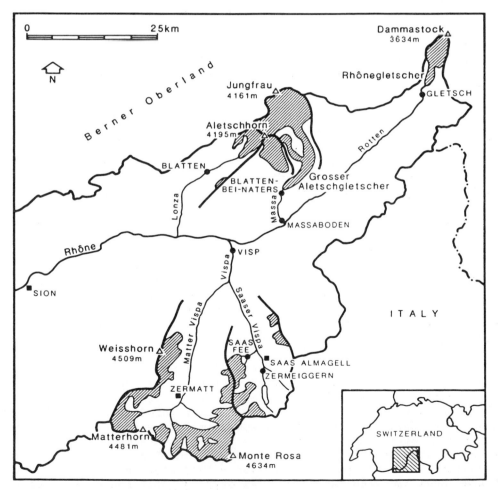

Figure 1   Locations of study basins and meteorologic stations (Saas
Almagell, Sion and Zermatt). Only those glacierised areas
within the gauged basins are indicated.

letscher and Oberaletschgletscher, was measured at Massaboden from 1923 to 1929, and then from 1931 to 1964, before the gauge was relocated upstream at Blatten-bei-Naters, reducing catchment area by 7.3 km^2 (3.6 per cent). Meltwater flows from small glaciers in the Lotschental are gauged in the Lonza at Blatten. Continuous records of flow from Rhônegletscher have been measured at Gletsch since 1956.

Long-term background variations in climatic conditions were inferred from meteorologic stations at Sion (542 m a.s.l.) in the trunk Rhône valley, and within the Vispa catchment area at Zermatt and Saas Almagell (Figure 1). Records of temperature and precipitation were collected routinely from 1922 until 1977 at Sion, which, while at low altitude, is less likely to be affected by topographic microclimate than mountain stations. At Zermatt, measurements were undertaken between 1922 and 1958 (at 1609 m a.s.l.), 1960–1965 (at 1635 m) and then continuing from 1965 at 1632 m, but with precipitation only recorded from 1972 until 1981. The station in Saas Almagell (1669 m) has been operational since 1968. Both Zermatt and Saas Almagell are situated in deep high Alpine valleys. These stations were selected from several available in the upper Rhone basin on the basis of length and continuity of records, availability for the same station of temperature measurements in addition to precipitation data, elevation and proximity to the glacierised basins. Mean summer temperatures for the period 1 May –30 September (T_{5-9} $^\circ$C) are used as indices of ablation season energy input. Annual precipitation is calculated for 12 month periods from 1 November (P_{11-10}mm) whereas discharge totals relate to calendar years (Q_{1-12}x10^{6}m^3) as discussed by Collins (1985).

Variations of climatic conditions 1922–1983

Longer-term secular variations in climatic conditions in the upper Rhone basins are illustrated by the curve of mean ablation season air temperature (T_{5-9}) at Sion, and the plot of annual total precipitation (P_{11-10}) for Zermatt, together with five-year moving averages (Figure 2). Corresponding climatic series for the station at Saas Almagell, presented in Figure 3, overlap for 16 years with precipitation at Zermatt and with temperature and precipitation at Sion for 10 years. Annual fluctuations of mean summer temperature at Saas Almagell and Zermatt are reasonably correlated with those at Sion (Table 2). Comparison of the overlapping section of temperature curve shows close similarity in the temporal pattern of deviation in direction if not in terms of extent. These observations suggest that generally accordant year-to-year variations of summer temperature throughout the area can be represented by the trend recorded at Sion. Precipitation is much more variably distributed spatially in high mountain areas, but there exists some similarity in annual variations of P_{11-10} between the Alpine valley sites Saas Almagell and Zermatt, stronger than that between those stations and Sion (Table 2, Figures 2 and 3). In the period 1967/1968 – 1982/83, the signs of first differences between annual values of P_{11-10} for the series at Saas Almagell were usually the same as those of respective years at Zermatt, for example, precipitation in 1968/1969 was greater than in 1967/68, and also exceeded that in 1969/70. Abnormally high precipi-

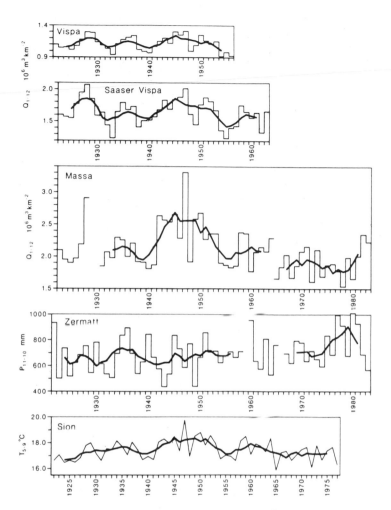

Figure 2   Year to year variations and five-year moving averages of total annual discharge (Q_{1-12}) of the Vispa at Visp, Saaser Visper at Zermeiggern, and Massa at Massaboden to 1964 and at Blatten-bei-Naters from 1965; total annual precipitation (P_{11-10}) at Zermatt; and mean summer air temperature (T_{5-9}) at Sion.

tation at Zermatt in 1979/80 followed by a low total in 1982/83 however, interrupted the pattern of differences.

Characteristics of the climatic series are summarised in Table 3. Mean quinquennial T_{5-9} at Sion increased from the 1920s to the mid 1930s, and after a relatively cool period, reached a maximum of 18.36°C centred on 1949. Such temperatures were maintained until the early 1950s, followed by a sudden decrease in the mid-1950s. After short recovery around 1960, summer temperatures at Sion declined slowly to quinquennial means of about 17.1°C. This trend has been reversed (see Saas Almagell, Figure 3) by the exceptionally warm

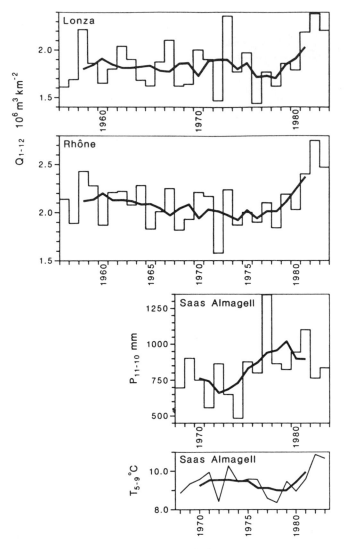

Figure 3   Year to year variations and five-year moving averages of total annual discharge (Q_{1-12}) of the Lonza at Blatten and the Rhône at Gletsch, and total annual precipitation (P_{11-10}) and mean summer air temperature (T_{5-9}) at Saas Almagell.

summers of 1982 and 1983.

Precipitation at Zermatt also varies irregularly from year to year (Figure 2). Quinquennial means of P_{11-10} were below average in the 1920s, slightly above in the mid-1930s, and precipitation was generally reduced during the warm 1940s. Subsequently, five year means fluctuated close to the overall period mean, before increasing to a maximum in 1976/77 – 1980/81. Precipitation at Saas Almagell was also above average between 1975 and 1981 (Figure 3).

Table 2   Matrix of correlation coefficients for the relationships between meteorological variables recorded at Sion (S), Saas Almagell (SA) and Zermatt (Z)

		SA T_{5-9}	S T_{5-9}	Z T_{5-9}	SA P_{11-10}	S P_{11-10}
S	T_{5-9}	0.93 (n=10)	-	-	-	-
Z	T_{5-9}	-	0.83 (49)	-	-	-
SA	P_{11-10}	-0.38 (16)	-	-	-	-
S	P_{11-10}	-	-0.23 (56)	-	0.23 (10)	-
Z	P_{11-10}	-	-	-0.27 (49)	0.78 (10)	0.65 (56)

n = number of years for which records of variables are overlapped.
Coefficients underlined are significant at the p = 0.05 level.

Table 3   Ranges of variability of climatic conditions in the upper Rhône basin, Switzerland

	P_{11-10}				T_{5-9}		
	Saas Almagell	Sion	Zermatt		Saas Almagell	Sion	Zermatt
	1967/68-1982/83	1921/22-1976/77	1921/22-1982/83	1967/68-1982/83	1968-1983	1922-1977	1922-1971
Mean	828	617	700	755	9.47	17.42	10.44
Range	482-1346	349-858	437-1011	568-1011	8.38-10.84	15.89-19.76	9.16-11.98
Coefficient of variation	0.245	0.201	0.196	0.191	-	-	-
Minimum year	1973/74	1924/25	1942/43	1982/83	1978	1965	1965
Minimum as % of mean	-41.8	-43.4	-37.6	-24.8	-11.5	-8.8	-12.3
Maximum year	1976/77	1944/45	1979/80	1979/80	1982	1947	1947
Maximum as % of mean	62.6	39.1	33.9	33.9	14.4	13.5	14.8

Variability of runoff 1922–1983

Annual total runoff (Q_{1-12}) derived from these glacierised basins is shown in Figures 2 and 3. The temporal patterns are approximately synchronous and correlated to a high degree (Table 4). The Saaser Vispa contributed between 10.5 and 13.7 per cent of the Vispa runoff, correlation therefore being spurious.

Large annual totals of runoff were discharged in 1928, 1945, 1947,

Table 4   Matrix of correlation coefficients for the relationships
total annual discharge (Q_{1-12}) recorded at the gauging
stations listed in Table 1

	Vispa	Saaser Vispa	Nassa	Lonza
Saaser Vispa	0.87 (n=34)			
Massa	0.76 (26)	0.76 (33)		
Lonza	-	-	0.78 (19)	
Rhône	-	-	0.89 (19)	0.75 (28)

n = Number of years for which gauging stations are overlapped
All the correlation coefficients are significant at the p = 0.05 level

1950, 1958, 1964, 1982 and 1983, from all the basins with contemporary operational gauges, and additionally, from the Massa in 1942 and
1949. Runoff in 1973 was particularly high in the Lonza.
   Relatively low flows occurred in 1940, 1948, 1954-1957, 1965 and
1972 in all the basins. Flow reached a minimum in 1976 for the Lonza
and in 1978 for the Massa. Characteristics of the variability of
annual discharge from the five basins are summarised in Table 5. The
greatest positive departure occurred at Massaboden during 1947, coin-
ciding with the warmest summer.

Table 5   Ranges of variation of total discharge (Q_{1-12}) from
glacierised basins within the catchment area of the upper
Rhône, Switzerland

	Vispa Visp 1922-1956	Lonza Blatten 1956-1983	Saaser Vispa Zermeiggern 1923-1963	Rhône Gletsch 1956-1983	Massa Blatten-bei-Naters 1965-1983	Massa Massaboden 1931-1964
Mean $10^6 m^3$	860.5	145.0	105.2	81.95	366.80	442.70
Range $10^6 m^3$	702.7-1012.6	111.2-187.3	79.5-132.8	61.48-107.0	297.5-454.2	355.1-665.5
Coefficient of variation	0.102	0.132	0.120	0.117	0.121	0.155
Minimum year	1954	1976	1955	1972	1978	1960
Minimum as % of mean	-18.3	-23.3	-24.4	-25.0	-18.9	-19.7
Maximum year	1947	1982	1928	1982	1982	1947
Maximum as % of mean	17.7	29.2	26.2	30.6	23.8	50.3

Periods during which quinquennial discharges remained above average occurred in the late 1920s (maximum 1928, 26.2% deviation), mid-1930s (deviation up to 11%), and sustained high deviations between 1942 and 1953 (with the exception of 1948 and also 1949 in the Vispa basin). Generally declining total annual flows reached minima in 1958-1959. High runoff since 1981 has apparently reversed the downward trend (Figures 2 and 3).

Use of runoff data between 1931 and 1960 in estimation of water yield for subsequent years, without taking into account climatic fluctuations and glacier area changes, would have predicted levels that subsequent flows have not reached. For example, in the decade 1941-1950, mean Q_{1-12} of the Massa was 2.49 x 10^6 m^3km^{-2} whereas between 1966 and 1975, it was reduced by 24.5 per cent. For the same ten-year periods mean summer temperatures at Sion were 18.08 and 17.13 degrees Celsius respectively. Glacier area in the Massa basin declined from 137.90 to 129.50 km^2 between 1927 and 1957 (6.1%) and to 128.56 km^2 by 1973 (Kasser 1981).

Relationships between climatic variables and runoff from glaciers

The extent to which climatic variation affects runoff from glaciers can be assessed by an examination of the proportions of variance of flow explained by hydrometeorologic variables. The markedly parallel records of annual variations of T_{5-9} and Q_{1-12} suggest that energy input exerts a strong direct influence on discharge. Both form and timing of precipitation will have considerable influence on runoff. In wetter summers, rainfall over glacier ablation areas and ice-free areas will be returned as additional runoff compensating to some extent for reduced icemelt contributions during cooler overcast periods. Summer precipitation on the ablation areas in the form of snow, even in small quantities of water equivalent, will raise albedo, reducing melting of ice and hence runoff (Collins, 1982). Winter snow accumulation contributes a component of summer runoff, yet may also retard the upward rise of the transient snow line over the lower glacier-covered areas, preventing ablation of ice and reducing meltwater production. The physical basis for any relationship between P_{11-10} and Q_{1-12} is less clear. Effects of P_{11-10}, precipitation between November and May (P_{11-5}) and June through September (P_{6-9}) on Q_{1-12} have been analysed.

Correlation coefficients computed between Q_{1-12} and hydrometeorologic variables are given in Table 6. Only values greater than 0.20 or the largest two for each basin are listed. For all basins, relationships between precipitation and discharge are weak and statistically not significant at the p= 0.05 level. Discharge is strongly related to T_{5-9}, except for the Vispa and Lonza basins. In these catchments, glacierisation is relatively low and precipitation totals might be expected to exert greater influence on annual runoff. In the more highly-glacierised basins, between 62.4 and 84.6 per cent of the variance in Q_{1-12} is explained by T_{5-9}, but a maximum of only 10 per cent by precipitation variables. Multiple regression models using one precipitation variable in addition to T_{5-9} usually increased explanation, but the addition of a second precipitation variable in no case provided further improvement due to multicolli-

Table 6   Correlation coefficients showing relationships between climatic variabels at Saas Almagell (SA), Sion (S) and Zermatt (Z) and total annual runoff

River	Period	T_{5-9}	P_{11-10}	P_{11-5}	P_{6-9}
Vispa	1922-1956	S 0.62		S 0.20	
Lonza	1956-1977	S 0.39	S 0.25	S 0.31	
Saaser Vispa	1923-1963	Z 0.55	Z 0.23		
	1956-1977	S 0.79	Z -0.22	S 0.24	
Rhône	1968-1983	SA 0.81			SA 0.18
	1931-1964	S 0.91	Z -0.28	Z -0.20	Z -0.31
Massa	1965-1977	S 0.87		S 0.18	
	1968-1983	SA 0.92		SA -0.22	

Underlined values are significant at the p = 0.05 level

Table 7   Goodness of fit (r^2) and parameters of multiple regression models of the form $Q_{1-12} = a + b_1T_{5-9} + b_2X_2 (+ b_3X_3)$ where X_2 and X_3 are precipitation variables recorded at Saas Almagell (SA), Sion (S) and Zermatt (Z)

Equation	Basin	T_{5-9}	X_2, X_3	Period	a	b_1	b_2	b_3	r^2
1	Vispa	S	S P_{11-5}	1922-1956	-476	72.5	0.196		0.45
2		S	S P_{11-10}		-596	76.3	0.190		0.46
3		S	S P_{11-5}, P_{6-9}		-559	74.9	0.207	0.174	0.47
4	Lonza	S	S P_{11-5}	1956-1977	-57.6	10.4	0.057		0.28
5	Saaser Vispa	Z	Z P_{11-10}	1923-1963	-72.9	14.5	0.037		0.43
6	Rhône	S	S P_{11-5}	1956-1977	-86.7	9.17	0.024		0.71
7		S	S P_{11-10}		-104	9.69	0.027		0.74
8		S	S P_{11-5}, P_{6-9}		-101	9.60	0.028	0.021	0.74
9		SA	SA P_{11-10}	1968-1983	-81.3	15.1	0.024		0.82
10		SA	SA P_{11-5}, P_{6-9}		-81.2	15.0	0.024	0.025	0.82
11	Massa	S	Z P_{11-5}	1931-1964	-988	82.1	-0.059		0.84
12		S	S P_{11-5}	1965-1977	-490	48.8	0.053		0.78
13		SA	SA P_{11-10}	1968-1983	-274	63.8	0.046		0.88
14		SA	SA P_{11-5}, P_{6-9}		-274	63.8	0.046	0.047	0.88

nearity amongst predictor variables. Such models are presented in Table 7, in which the regression coefficients (b_k) are significant at the p = 0.01 level, with the exception of p = 0.05 for equation 4.

Additions of P_{11-5} and P_{11-10} appear to enhance explanation. Considering less glacier-covered basins, r^2 is increased from 0.38 to 0.45 for the Vispa by the addition of P_{11-5}, for the Lonza from 0.15 to 0.28, and for the Saaser Vispa from 0.30 to 0.43 by $P_{11-10} \cdot P_{11-5}$ and P_{6-9} at Sion together raise the explained percentage only to 47% at Visp.

For the Rhône at Gletsch in the period 1956-1977, a substantial improvement in fit from 62% with T_{5-9} to 74% is brought about by addition of P_{11-10} but no further increment accrues when P_{6-9} is included. During the period 1968-1983 equations 9 and 10 reflect 7 and 8 for 1956-1977 but T_{5-9} with P_{11-10} account for 82% of the variance of Q_{1-12} of the Massa. For the catchment most covered with glacier ice Q_{1-12} shows strong dependence on T_{5-9} (75-85% explanation), and precipitation explains an additional 3 per cent of total value. With the exception of equation 11, regression coefficients for precipitation variables are positive, although several simple correlation coefficients have negative signs (Table 6).

Discussion

Goodness of fit of multiple regression equations 6-14 is high even though mean air temperature is probably a poor substitute for radiation measurements, and precipitation records are for stations outside catchment boundaries and at relatively low elevations. The level of explanation is more surprising given reducing glacier areas through the calibration period. It is clear that climatic variations and glacier recession have affected runoff from glacierised areas, particularly in the markedly higher than average flows during the much warmer period from early 1940s to early 1950s, contrasted with low annual discharge totals in the generally cool 1970s. From the late 1950s after earlier contractions, glacier area has not changed significantly in spite of what appear large snout fluctuations (Grosser Aletschgletscher retreated by 567.1 m between 1959 and 1980). Towards the end of that period some glaciers had advanced slightly (e.g. in the Vispa and Saaser Vispa basins). Had a reference period on which to base estimates of future water yield, adjusted using climatic variables alone, included the 1940s, and areal ice-cover continued to reduce, then flows in later years would have been overestimated. The best fit model predicting Q_{1-12} for the Massa with T_{5-9} and a precipitation variable calibrated with 1931-1945 as reference period:

$$Q_{1-12} = 72.7T_{5-9} - 0.626P(Z)_{11-5} - 796$$

($r^2 = 0.87$), overestimates Q_{1-12} in every year from 1952 to 1964, excepting 1957 a cool summer with exceptionally high snowfall in May, by between 0.2 and 15.7 per cent ($\bar{x} = 8.9\%$). The influence of area loss is thus of much less importance than that of temperature and precipitation. With more stable glacier areas, a reference period in the 1960s might be expected to lead to improved prediction for subsequent years. Calibration of equation 10 using the years 1968-1977 for the Rhône at Gletsch yields:

$$Q_{1-12} = 14.3T_{5-9} + 0.021P_{11-5} + 0.007P_{6-9} - 69.8$$

with $r^2 = 0.71$. This equation, however, underestimates Q_{1-12} from 1978 to 1982 (5.2 - 9.9%, $\bar{x} = 7.5\%$).

This result may be due to the likelihood that the rate of expansion of area of exposed ice in glacier ablation zones and the final area bared at the maximum elevation of the transient snowline are also important variables, and considerably more volatile on a year to year basis than changes of glacier area. Effective heat input for melting is regulated by average albedo, which depends on the proportion of glacierised surface comprised of snow, firn and ice. Kasser (1973) utilised changes of glacierised area in a multiple regression analysis of runoff in the Rhone basin above Porte du Scex but the addition of change of glacierised area to the model provides no additional explanation over a dependence on winter precipitation.

Changes in actual ice-covered area, in addition to those of effective ablation area, must also distort the assumption of linearity in the relationship between air temperature and discharge. Relatively poor explanation of variation in Q_{1-12} in the basins with lower percentage glacierisation probably reflects some dependence of runoff on both variables, which have little tendency to vary together consistently. At lower percentage glacierisation, winter precipitation totals will determine runoff totals and in more glacierised basins thermal conditions will dominate with limited influence of the other variable in both cases.

Conclusions

Climatic variation has a marked impact on the flow of rivers draining from Alpine glaciers. Considerable fluctuations of runoff occurred during the period 1922-1983, in which period glaciers experienced considerable loss of mass and area. The range of variability of annual runoff usually extends from 25 per cent less than to 30 per cent more than the mean, with an outstandingly high annual total in the basin with the largest glacierised area, deviating by 50.3 %. Energy inputs dominate year-to-year variations of runoff in basins with more than 35 per cent glacierisation indexed here by mean summer air temperature. Temporal patterns of runoff variations from all the catchments are consequently similar. A change of 1°C in ten year mean air temperature will result in a significant shift of ten-year mean runoff. Some of this variation must however be related to changes in effective planimetric area of ice. Relationships between precipitation variables, such as total annual input and winter precipitation accumulation, with runoff are weak.

Multiple regression equations relating total annual discharge to mean summer air temperature and precipitation variables provide a high degree of explanation in basins with higher proportions of ice cover. The effect of winter precipitation accumulation added to that of mean summer air temperature provides the highest degree of fit. Separation of total annual into winter and summer precipitation sub-totals produces no further increase in explanation. This method of analysis at least provides an approach to predicting runoff yield under given scenarios for future variations in temperature, for basins with more than about 50 per cent glacier cover. Predictive ability will be improved by the use of more conceptually-based models, radiation data and measures representing seasonal and annual variations of effective ablation area.

ACKNOWLEDGEMENTS   The author gratefully acknowledges the following assistance: Landeshydrologie und-geologie, Bundesamt fur Umweltschutz, Bern for supplying discharge records; Schweizerische Meteorologische Zentralanstalt for making available climatic data; Versuchsanstalt fur Wasserbau, Hydrologie und Glaziologie, Eidg.Technische Hochschule, Zurich for the provision of a stimulating working environment during a period of sabbatical leave from University of Manchester, and the Royal Society and Schweizerischer Nationalfonds zur Förderung der wissenschaftlichen Forschung for an award in the European Science Exchange Programme. R. Beecroft assisted with preparation of data.

References

Collins, D.N. (1982) Temporal variations of meltwater runoff from an Alpine glacier. In: Proc. Symp. Hydrolog. Research Basins, Bern. Sonderheft, Landeshydrologie: 3, 781-789

Collins, D.N. (1985) Climatic variations and runoff from Alpine glaciers. Z. Gletscherk. Glazialgeol. 20, 127-145.

Fountain, A.G. & Tangborn, W.V. (1985) The effect of glaciers on streamflow variations. Wat. Resour. Res. 21(4), 579-586.

Kasser, P. (1973) Influence of changes in the glacierised area on summer runoff in the Porte du Scex drainage basin of the Rhone. In: Symposium on the Hydrology of Glaciers (Proc. Cambridge Symp., September 1969), 221-225. IASH Publ No. 95.

Kasser, P. (1981) Rezente Gletscherveränderungen in den Schweizer Alpen. In: Jahrbuch der Schweizerischen Naturforschenden Gesellsc- haft, wissenschaftlicher Teil, 1978 Birkhauser, Basel, 106-138.

Krimmel, R.M. & Tangborn, W.V. (1974) South Cascade Glacier: the moderating effect of glaciers on runoff. Proc. Western Snow Conference 42, 9-13.

Patzelt, G. (1985) The period of glacier advances in the Alps, 1965 to 1980. Z. Gletscherk. Glazialgeol. 21, 403-407.

Rasmussen, L.A. & Tangborn, W.V. (1976) Hydrology of the North Cascades region, Washington. I. Runoff, precipitation and storage characteristics. Wat. Resour. Res.12, 187-202.

Tvede, A.M. (1982) Influence of glaciers on the variability of long runoff series. In: 5th Northern Research Basin Symp., Ullmsvang, Norway, 179-189.

Walser, E. (1960) Die Abflussverhältnisse in der Schweiz wahrend der Jahre 1910 bis 1959. Wasser -und Energiewirtschaft 8/9/10, 197-214.

The Influence of Climate Change and Climatic Variability on the Hydrologic Regime and Water Resources (Proceedings of the Vancouver Symposium, August 1987). IAHS Publ. no. 168, 1987.

Spatial and temporal variability of rainfall and potential evaporation in Tunisia

Ronny Berndtsson
Department of Water Resources Engineering
University of Lund
Box 118, S-221 00 Lund, Sweden

ABSTRACT Different aspects of mainly temporal and spatial variability of rainfall in Tunisia are presented and discussed. Long-term trends seem to follow a general pattern of consecutive dry or wet periods of 5 to 30 years. Spatial patterns of shorter rainfall series (annual, monthly) are governed mainly by topography and coastal influence. Short highintense rainstorms occur with cellular patterns, with a typical cell size of about 6-7 sq.km., cell size being defined as the area within the 0.7 correlation isoline.

RESUME Des aspects différents sur surtout la variabilité spatiale et temporale de la pluie en Tunisie sont présentés et discutes. Des tendances à longtemps paraissent suivre une façon générale des périodes consécutives sèches ou humides de 5 à 30 ans. Des façons spatiales des séries de pluie courtes (annuelles et mensuelles) sont conduis surtout par topographie et d'influence côtière.

Introduction

How to describe spatial and temporal patterns of hydrological processes accurately is still an acute issue in many arid and semi-arid regions with a sparsely spaced or irregular raingauge network. The problem of few or uncertain observations is usually also coupled with an extreme temporal and spatial variability in rainfall. Tunisia is a typical country in this respect. For a country whose main national income is derived from agricultural production, being situated in the center of a meteorological transition zone is something of a problem. In Tunisia apparently small fluctuations may lead to devastating rainstorms and floodings. The extreme rainstorms, which occurred in 1969 and 1973, are typical examples of this (Clarke 1973 or Winstanley 1970). During 24 hours on September 25, 1969, 400 mm rain fell at Gabés in central Tunisia. This volume is about 5 times the mean annual rainfall in the area. After this, intermittent heavy rainfall continued for 38 days. The effects of heavy rainfall are often exacerbated by poorly developed drainage patterns and gypsum encrusted surfaces or sub-surfaces, leading to severe floodings and heavy loss of life and property. These problems make studies of rainfall variability a matter of high priority for hydrologists.

91

This paper sums up findings regarding different aspects of rainfall variability in Tunisia. Some spatial characteristics of monthly potential evaporation are also presented.

General remarks on long-term rainfall variability

Tunisia borders on the Sahelian zone, severely affected by drought, in the south and on the Mediterranean in the north. Consequently, the climate of Tunisia varies from extremely arid in the south to semi-arid and almost humid in the north. Generally speaking, the amount of rainfall decreases with increasing distance from the northern Mediterranean coast. This condition, however, is also greatly influenced by orographic effects, increasing altitude entailing larger amounts of rainfall. The annual average precipitation in Tunisia amounts to about 1,580 mm in the Atlas mountains in the north-west, down to well below 50 mm in the south.

The longest and most reliable rainfall records available refer to Tunis in the north of Tunisia. During the period between 1890 and 1970, the annual rainfall in Tunis varied between 165 mm and 805 mm (Gaussen and Vernet 1940, Riou 1977, Rodier et al 1981). The relation between these yearly extremes is roughly 1 to 5. With regard to Susah, situated about 100 km to the south on the coast, the same figures are 1 to 8 (Dutcher and Thomas 1967). Even on a yearly basis, then, there is a marked difference and the variability will, however, be much more marked when a shorter time step is considered.

A measure of annual variability and long-term trends may be achieved by plotting the cumulative departures from the mean for the period of record. An example of this is given in Figure 1 for annual rainfall at Tunis and Susah. The figures indicate quite dissimilar patterns for the two stations. The diagram for Tunis shows a 30-year downward trend from 1895 to 1925. After 1925, upward and downward trends of a duration of about 10-15 years seem to alternate. The diagram for Susah, on the other hand, suggests a positive trend from about 1900 to 1920, and subsequently, successive downward and upward trends of 5 to 20 years. The general conclusion is that periods with annual rainfall above or below the mean seem to succeed each other in periods of 5 to 30 years. No marked similarities seem to emerge even for stations only about 100 km apart.

The two annual rainfall series, for Tunis (1890-1970) and Susah (1890-1960) respectively, were further used in order to study theoretical probability distributions in respect of annual rainfall. Figure 2 shows the empirical and theoretical distribution functions of these two annual rainfall series. The Weibull formula (Yevjevich 1972) was used when plotting the empirical distributions in Figure 2. Table 1 summarizes the statistical properties of these rainfall series. From the figure and Table 1, it can be seen that the Normal distribution seems to fit the empirical probability distribution for Tunis best, whereas Log-normal and Gamma-2 distributions produce a better agreement with regard to Susah. The annual rainfall at Susah displays a higher positive skewness indicating the predominance of an annual rainfall depth which is less than average, and typical for rainfall in arid conditions.

The variation coefficient (standard deviation divided by mean)

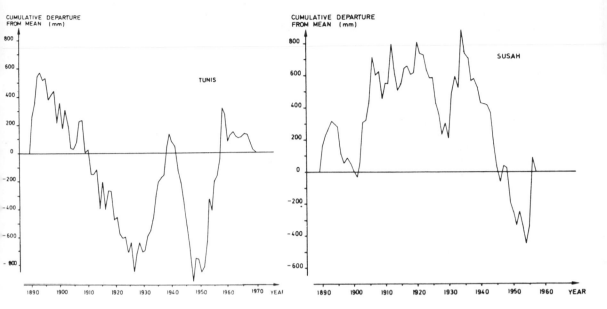

Figure 1 Cumulative departures from the mean of annual
rainfall at Tunis and Susah.

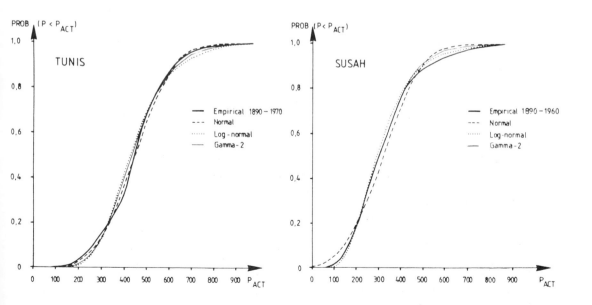

Figure 2   Empirical and theoretical probability distribution
functions for annual rainfall at Tunis and Susah.

shown   in   Table 1 is a non–dimensional measure of variation   between
different   years.   The   station   at Susah shows   a   larger   variation
coefficient   (0.42)   than   the   station   in   Tunis   (0.30).   A   greater
variation   in   respect   of rainfall depth at Susah also indicates   more

Table 1   Statistics pertaining to the annual rainfall and
theoretical probability distribution functions
of Tunis and Susah

Tunis

N	= 82	Chi square:		
Mean	= 446	Normal	= 13.6	**
Median	= 454	Gamma-2	= 14.2	**
St dev	= 133	Log-Normal	= 16.0	*
Max	= 808	Pearson III	= 19.7	*
Min	= 165			
Skew coeff	= 0.23			
Variation coeff	= 0.30			

Susah

N	= 67	Chi square:		
Mean	= 317	Log-Normal	= 4.7	***
Median	= 300	Gamma-2	= 7.8	***
St dev	= 134			
Max	= 750			
Min	= 100			
Skew coeff	= 1.06			
Variation coeff	= 0.42			

where:

*** significance level of 0.001
** significance level of 0.01
* significance level of 0.05

pronounced arid conditions than in Tunis.

Spatial patterns of monthly rainfall and potential evaporation

The spatial character of rainfall or potential evaporation can be
studied by means of spatial correlation patterns. The spatial
correlation patterns may then be interpreted in γ_a meteorological
context including prevailing wind system γ_s, coastal influence and
topographical features (Berndtsson 1987a, Berndtsson 1987b or Sumner
1983). The use of cross correlation coefficients between pairwise
combined stations will tell something about the spatial variability
as influenced by topography and the coast line, for example. This
technique has been applied in studies based on a five-year series of
monthly data from 67 rainfall stations and 49 stations for observation of potential evaporation throughout Tunisia (Berndtsson et al
1986, Berndtsson 1987b). Figure 3 displays a general comparison of
the cross correlation coefficients as a function of the inter-station
distances. It is obvious from the figure that both monthly rainfall
and potential evaporation appear with a large scatter, which is
indicative of a highly anisotropic correlation structure. The values
of the correlations which refer to the potential evaporation are
higher, generally speaking, than those which refer to rainfall. The
coefficients of the potential evaporation, however, seem to appear
with a larger variability. In order to show the influence of the
coast and the effects of topographical divides, correlation linkage

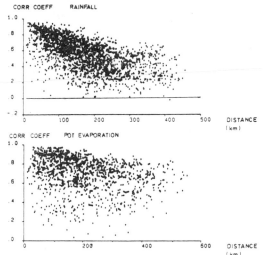

Figure 3   Lag-zero cross correlation coefficients for monthly
           rainfall (67 stations) and potential evaporation
           (49 stations) vs. the interstation distances (1979-83).

analysis as described by Sumner (1983) was used. Figure 4 shows
correlation linkages on different levels of monthly data corresponding to the coefficients in Figure 3.   High correlations as regards
rainfall occur almost only in the northernmost parts of the country.
High correlations seem to be typically oriented parallel to the
coastline. Factors decisive for the spatial pattern are obviously
both topography and coastal influence (Berndtsson 1987a).   The high
Kroumirie Range, in the northernmost parts, fits closely with the
correlation pattern. Apparently, it functions as an effective
topographical barrier for rainfall . South of the range, there are
few high correlations to be found. Further to the north east, the
topography becomes less important (lower altitudes) and the coastal
influence becomes relatively more decisive as regards to the
correlation pattern.

The coastline appears to be less important as regards to potential
evaporation.   High correlations (>0.90) occur in a rather confined
area in the valleys and plains between the Kroumirie Range and the
High Tell and the High Steppes. When lower correlations are also
considered (0.80 - 0.90), this impression is reinforced by correlations typically oriented parallel to the topographical divides.
This is in accordance with findings in previous studies. Branigan and
Jarrett (1969) remark that the High Tell and High Steppes form a
climatic divide between the Mediterranean north of the country and
the Sub-Saharan areas to the south.

A more detailed correlation analysis of daily data for the five-
year period confirms the general rainfall pattern of the monthly
data. Individual months usually evidence typical coastal influence
and effects of the major topographical divides. Based on the correlation linkage analysis, three main rainfall regions may be distinguished (Berndtsson 1987a). The northern region is delimited in the
south mainly by the Kroumirie Range. Due to the coastal influence it
also extends to the southeast along the coastline. The second region
(the inland region) is influenced mainly by topography (the High

96

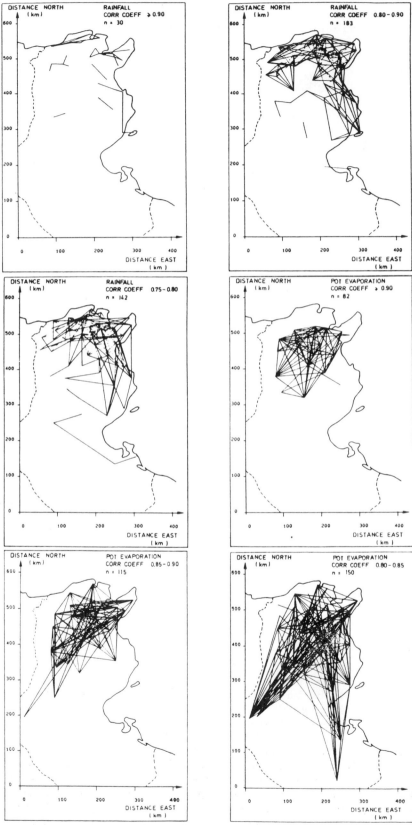

Figure 4 Correlation linkages of monthly rainfall and potential
evaporation (1979–83).

Tell and High Steppes). The third region (the coastal region), is influenced mainly by the distance to the eastern coastline.

High-intensive rainfall

The short-term variability of high-intensive rains in the region has not been studied extensively. Short-term high-intensive rains are, however, interesting from an urban hydrological point of view. The fact that a great many of the rains usually fall in short periods of time also make studies of high-intensive rainfall interesting for other purposes. The spatial and temporal properties of ten high-intensive storms, recorded by a dense gauging network in northern Tunisia, were investigated by Berndtsson and Niemczynowicz (1986). The ten one-hour rainstorms represented, depending on the gauge, between about 10 to 30 percent of the annual rainfall in 1982-83. Again using spatial correlation, it was shown that the typical cell size, as defined as the area within the 0.7 correlation isoline, was about 6 to 7 sq. km for a one-hour duration. Spatial correlation of monthly rainfall, on the other hand, displayed no cellular patterns as shown in Figure 5.

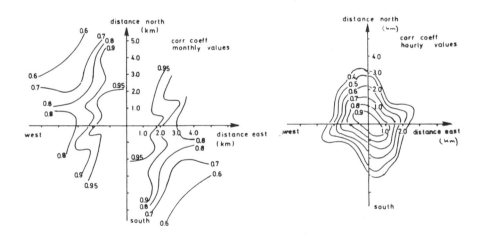

Figure 5 Correlation fields for monthly rainfall and
         ten one-hour rainstorms.

The rainstorms were further analyzed regarding intensity-duration relationships, Figure 6 shows point and areal intensity-duration curves for the ten storms. Areal intensities (10 sq. km and 20 sq. km) were calculated with maximum point intensity. For purposes of comparison the theoretical point intensity-duration curves with a return period of 1,2 and 5 years for Tunis are also shown in Figure 6. As is seen from the figure, the rainfall intensity is quickly reduced when a larger area and a longer duration are considered. Regarding to a 5-minute duration, the mean rainfall intensity for 10 sq. km is only about 60% compared to the point value. For 20 sq. km, the mean areal intensity is about 40% compared to the point value.

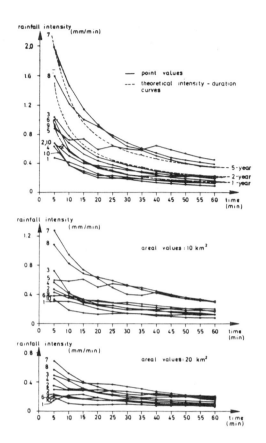

Figure 6   Point   and   areal intensity – duration   curves   for ten
           rainstorms (1,2,..., 10 refer to the rain storm number).

For   longer durations (60 min.) the same values are about 80% and 60%
for 10 and 20 sq. km respectively. The rainfall intensity does   not
invariably decrease with increasing duration (rain No 5). This is due
to   the   occurrence of more than one raincell or   rainfall   intensity
maximum   for the given duration.   These local intensity maxima become
more   important   the   larger   the area or   the   longer   the   duration
considered, causing a local rise in the intensity–duration curves.

Summary and discussion

Some aspects of temporal and spatial rainfall variability for Tunisia
have   been   presented   and   discussed.   The   long–term   variability   of
annual   rainfall seems to follow a general pattern of successive   dry
or   wet periods of 5–30 years.   Stations which are quite close to one
another   do   not display very large similarities as regards   to   this
periodicity. Preliminary rainfall regions may be delimited as regards
to   the   correlation   structures   of   daily   rainfall,   influenced   by
topography and coastline (Berndtsson 1987a).

    These   rainfall regions need to be studied further with regard   to the statistical character of the rainfall within each region.

The correlation pattern of monthly potential evaporation seems to be affected mainly by topographical features. Coastal influences on the potential evaporation seem negligible.

The findings regarding the spatial and temporal variability of high-intensive rainstorms are in many respects similar to those in studies referring to a humid climate. The rain-producing cell size, defined as the area within the 0.7 correlation isoline, has about the same extension, as found both in Sweden and Tunisia (Berndtsson and Niemczynowicz 1986). The areal reduction of point rainfall has also, on average, approximately the same value. The occurrence of multiple rain-producing cells, however, seems to be more frequent in Tunisia, occasionally causing larger areal variability.

ACKNOWLEDGEMENTS   Most of the data used in this study were supplied by the Service de la Meterologie Nationale, Tunis. Data were also used from a Tunisian-Swedish co-operation project on urban hydrology (see Ennabli et al 1984). The access to these data is gratefully acknowledged. The study was partly funded by the Swedish Natural Science Research Council, which is also gratefully acknowledged.

References

Berndtsson, R, Niemczynowicz, J (1986) Spatial and temporal characteristics of high-intensive rainfall in northern Tunisia, Journ Hydrol, 87 (1986), pp 285-298.
Berndtsson, R, Larson, M, Nieczynowicz, J (1986) Spatial characteristics of monthly rainfall in Tunisia, Proc of Nordic Hydrological Conf, August 11-13, 1986, Reykjavik.
Berndtsson, R (1987a), On the use of cross correlation analysis in studies of patterns of rainfall variability, Paper to be published in Journ Hydrol.
Berndtsson, R (1987b) Spatial variability of rainfall and potential evaporation in Tunisia, Paper to be published.
Branigan, J J, Jarrett, H R (1969), The mediterranean lands, MacDonald & Evans Ltd, London.
Clark, F E (1973) The great Tunisian flood, Journ Research U S Geological Survey, Vol 1, No 1, pp 121-124.
Dutcher, L and Thomas, H (1967) Surface water and related climate features of the Sahal Susah Area, Tunisia, Geological Survey Supply Paper 1757-F. U S Dept of the Interior, Washington, D.C.
Ennabli, M, Ennabli, N, Hogland, W, Niemczynowicz, J (1984) Application du Storm Water Management Model a Tunis, Rapport No 3078, Department d'Hydrologie et d'Alimentation en Eau, Institut National Polytechnique/Université de Lund, Lund.
Gaussen, H, Vernet, A (1940) Cartes des Precipitations moyennes annuelles des annees 1900 a 1940, Secretariat d'état à l'Agriculture Tunisie, Tunis.
Riou, C (1977) Evaporation du sol nu et repartition des pluies, relation etavlies en Tunsie a partir des resultats des cases lysimetriques. Annales de l'Institut National de la Recherche

Agromique de Tunisie, Vol 50, Fasc 4, Tunis.

Rodier, J, Colombani, J, Claude, J, Kallel, R (1981) Le bassin de Mejerdah, monographies hydrologiques ORSTOM No 6, Office de la Recherche Scientifique et Technique Outre Mer, Paris.

Service de la Meterologie Nationale 9167) Climatologie de la Tunisie, 1-Normales et statistiques diverses, Service de la Meterologie Nationale, Tunis.

Winstanley, D (1970) The north African Flood Disaster, September 1690, Weather, Vol 25, No 9, pp 390-403.

Yevjevich, V (1972) Probability and statistics in hydrology. Water Resources Publications, Fort Collins, Colorado.

The Influence of Climate Change and Climatic Variability on the Hydrologic Regime and Water Resources (Proceedings of the Vancouver Symposium, August 1987). IAHS Publ. no. 168, 1987.

The climatology of dry and wet periods over western Canada in a general circulation model

Neil E. Sargent
Canadian Climate Centre
Downsview, Ontario, Canada

ABSTRACT   Soil moisture anomalies over the period 1961–
1980 at eight stations in western Canada are estimated by
supplying observed temperature and precipitation to a soil
moisture model.   Temperature and precipitation simulated
by a 20 year run of the Canadian Climate Centre general
circulation model over the same area are treated in the
same way.   These series are normalized and examined from
the point of view of dry and wet run lengths and
severities.   For equal 20 year lengths of record the GCM
has about 11% fewer wet and dry periods then the
observations.   The range and trequency of severities of
dry and wet periods is quite similar between the
observations and the GCM. Calculations of the mean square
tendency of normalized soil moisture anomaly show it is
smaller in the general circulation model data than in
observations.   This results from a smaller than observed
variability in modelled evapotranspirative processes.

Une climatologie des périodes sèches et humides
dans l'ouest canadien dans un modèle de
circulation générale

RESUME   On a éstimé pour la période s'étendant de 1961 à
1980 l'anomalie de l'humidité du sol a huit stations, dans
l'ouest canadien a partir d'un modèle d'humidité du sol
utilisant   les   observations   de température et de
précipitation. On traite pareillement les températures et
précipitations simulées générées par l'execution de 20 ans
du'un modèle de circulation générale (MCG) du Centre
Climatologique Canadien pour le même endroit.   Ces séries
sont normalisées et étudiées par rapport aux durées et
intensités des périodes sèches et humides.   A périodes de
relevées de données égales et de 20 ans de long, on trouve
que les périodes humides et sèches générées par le MCG
sont inférieures d'environ 11% à celles observées. Les
extrêmes et la fréquence des intensités des périodes
sèches et humides du MCG sont similaires a ceux résultant
des observations.   La moyenne quadratique de la tendance
de l'anomalie normalisée de l'humidité du sol calculée
pour le modèle de la circulation général_ est plus faible
que celle de l'observation.   Ceci est le résultat d'une
modélisation du processus d'évapotranspiration générant de
valeurs   plus   faibles   de   la   variabilité   que   celles
observées.

101

Introduction

As a measure of the value of the current version of the Canadian Climate Centre (CCC) General Circulation Model (GCM) (Boer et al. 1984a) in actual applications we consider in this study how realistically it models the frequency, duration and severity of dry and wet spells over the Canadian Prairies.

We define "dry" and "wet" in terms of soil moisture. We have available data from a 20-year run of the CCC GCM at points over the Prairies and would like to compare it with 20 years of observations over the same region. To obtain a 20-year series of estimates of actual (hereafter referred to as "observed") soil moisture values we supply 20-year temperature and precipitation records to a two layer, Palmer type model (AES Drought Study Group, 1986) to estimate soil moistures. We process temperature and precipitation records from the GCM in the same fashion, supplying them to the same soil moisture model to obtain soil moisture values.

To concentrate upon some real measure of relative dryness or wetness we remove the annual cycle from the soil moisture series and normalize the anomaly at each interval by the standard deviation of soil moisture appropriate for that time of year as derived from the original series. This procedure results in a measure Z of soil moisture similar to the Palmer Index (Palmer, 1965) and permits more meaningful intercomparison of relative moistures at different points.

The soil moisture model is not trivial but the treatment of the budget of moisture in the soil is sufficiently coherent that it can be described by a few differential equations. This allows us to write an equation for the normalized soil moisture anomaly Z in terms of contributing factors. That, in turn, permits us to trace back some of the causes of differences of behaviour of Z between the GCM and observed data. We shall see that soil moisture anomalies change less quickly and that deviations from average soil moisture are somewhat less frequent in the 20 years of GCM data we have than in the observations. The severity and duration of anomalies, however, is very similar between the two cases.

Soil moisture model summary

Processes

Documentation of the soil moisture model we use is available elsewhere (AES Drought Study Group, 1986). Here we only summarize the structure and processes of the model.

The core of the model is an upper and a lower soil layer, assumed to hold 1/8 and 7/8 respectively of the total available water capacity of the modelled soil column. Water supplied to the soil surface infiltrates the soil. Some water may run out the bottom of the soil column. Moisture also leaves the soil column by evapotranspiration. Potential evapotranspiration is computed using Method I of Baier and Robertson (1965). This method uses a linear combination of daily maximum temperature, daily temperature range and total daily solar radiation falling on a horizontal surface at the top of the atmosphere to estimate potential evapotranspiration. Actual evapotranspiration is computed as in Baier and Robertson, 1966, Figure 1. (Curves C and E in upper and lower layers respectively.)

Elaborations have been added to this model core as follows. A vegetative crown may intercept, store and evaporate some precipitation. When temperatures fall below freezing the soil freezes and no moisture movement or evaporation occurs and precipitation reaching the surface accumulates as snow. In the spring the accumulated snow melts. Soil type and temperature determines how much of the water supplied to the soil surface actually enters it. The rest is assumed to run off.

Normalized anomaly Z

Once the soil moisture model has generated a series of soil moistures these must be processed into a moisture index. For the rest of this study soil moisture is indexed by a measure Z defined as the deviation of moisture from average divided by the standard deviation of soil moisture for that time of year. To implement this we break each year in the soil moisture series into 37 periods, 36 of 10 days and one of 5 days at the end of the year. For each period we compute the average A and standard deviation D of soil moisture S. A soil moisture index Z for each period in each year can then be computed as

$$Z(year, period) = (S(year, period) - A(period)) / D(period) \qquad (1)$$

Now the equation for total soil column moisture S can be written

$$d S / dt = R - U - E - T \qquad (2)$$

where   S = total soil column moisture,
          R = rate of water supply to surface,
          U = rate that water runs out bottom of column,
          E = rate of evaporation from upper soil layer,
          T = rate of transpiration from lower soil layer.

Using (2) we can write

$$S = \int R \, dt - \int U \, dt - \int E \, dt - \int T \, dt \qquad (3)$$

from which we see the annual cycle A of S may be expressed as a sum

$$A = A_R - A_U - A_E - A_T \qquad (4)$$

So (1) may be rewritten as

$$Z = \frac{(S - A)}{D} = \frac{\int R dt - A_R}{D} - \frac{\int U dt - A_U}{D}$$

(con'd)
$$- \frac{\int E dt - A_E}{D} - \frac{\int T dt - A_T}{D} \tag{5}$$

Equation (5) may now be differentiated to give an equation for the rate of change or tendency of Z:

$$\frac{dZ}{dt} = \frac{d}{dt} \frac{\int R dt - A_R}{D} - \frac{d}{dt} \frac{\int U dt - A_U}{D} \tag{6}$$

$$- \frac{d}{dt} \frac{\int E dt - A_E}{D} - \frac{d}{dt} \frac{\int T dt - A_T}{D}$$

This will be useful in a later section when we pin down contributions to the mean square tendency of Z.

Data

The CCC general circulation model has been documented by Boer et al. (1984a, 1984b). It is basically a forecast model run for very long periods. The aim, of course, is not to make precise long range forecasts but rather to generate a long series of reasonable atmospheric states having a good approximation to observed climates. Boer et al. (1984b) shows that average temperatures and precipitations are reasonably well modelled over the Canadian prairies with the possible exception of higher than observed precipitations over northern Manitoba. In this study of soil moisture anomalies we must go beyond that information to look at how day to day and year to year deviations from average conditions are modelled. To do this we require a long series of model temperatures and precipitations.

The data we have from the CCC GCM are values of simulated ground temperature and accumulated precipitation every 18 hours for 20 years on a grid of more than 2000 points over the globe. Screen level temperature is not available from the GCM. Ground temperature is chosen as the best GCM approximation to that observed variable. In this study we use only data from eight points over western Canada. The location of these points is shown in Figure 1.

Eight real observing stations have been chosen for comparison with the records from the eight GCM grid points. The locations of the real stations are also shown in Figure 1. They were chosen as near the grid points as possible not with any idea of comparing actual point values but rather to obtain an ensemble of eight observational records over prairie Canada as nearly equivalent as possible to that available from the GCM. Observational records of daily maximum and minimum temperature and precipitation were taken from these stations for the period 1961–1980.

Two sets of eight 20-year soil moisture estimates were then generated by applying the soil moisture model to the observed and the GCM simulated temperature and precipitation records.

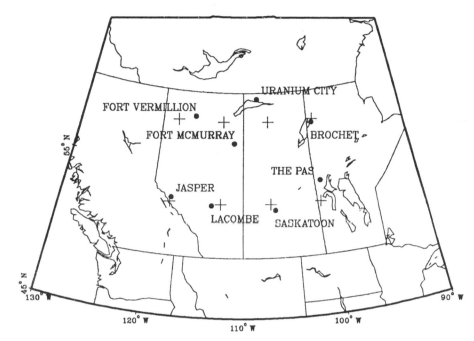

Figure 1   Map showing locations of the GCM grid points (crosses) and observing stations (dots) used in the study.

Statistics

Following Yevjevich (1967) we define the duration of a dry or wet period to be the time between two successive zero crossings of the Z curve (Yevjevich's run-length). The severity of an anomalous period is defined as the integrated area between the curve and the zero line between the two successive zero crossings defining the period (Yevjevich's run-sum).

In Figure 2 we show the cumulative frequency of period lengths: the number of periods less than a given length (in years) is plotted versus the common logarithm of that length. It is clear immediately that there are fewer wet and dry periods in the GCM data than in observations (320 vs 355). Because little moisture change occurs in frozen soil the value of Z is almost constant during winter. This prevents periods of anomalous soil moisture from starting or ending then; anomalies which have not ended before the soil freezes last at least until the soil thaws again. This explains why anomalous periods of length around 6 months are less frequent in both observational and GCM derived data. This appears to affect GCM length frequencies more than observed ones and especially wet period lengths.

DISTRIBUTION OF DRY PERIOD LENGTHS   DISTRIBUTION OF WET PERIOD LENGTHS

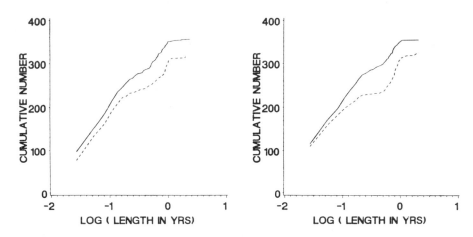

Figure 2   Graph of the number of dry and wet period lengths
           shorter than a given length versus the common logarithm
           of that length for the observations (solid line) and
           the GCM (dashed line).

DISTRIBUTION OF DRY PERIOD SEVERITIES   DISTRIBUTION OF WET PERIOD SEVERITIES

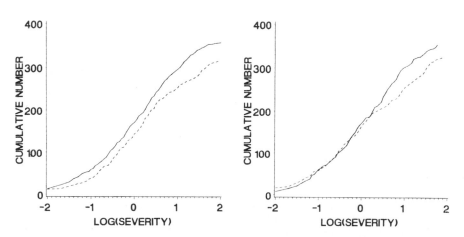

Figure 3   Graph of the number of dry and wet period severities
           less than a given severity versus the common logarithm
           of that severity for the observation (solid line) and
           the GCM (dashed line).

    Graphs of cumulative frequencies of wet and dry period severities
are shown in Figure 3.   Reflecting the smaller number of anomalous
periods in GCM data, there are in general more periods of any given
severity in the observed data.   In relative terms, however, the
percentage of all periods having less than any given severity is

quite comparable between the observations and the GCM.

The duration and severity of dry and wet periods were chosen as measures relevant to climatologists and geographers unfamiliar with the details of general circulation models but wishing some appreciation of their performance in simulating conditions affecting soil moisture. From the statistics just presented the GCM behaves surprisingly well in reproducing the observed distribution of wet and dry period lengths and severities. A larger difference between the GCM and observations appears, however, when we look at the mean square value of Z tendency. Rather than the duration and severity of Z deviations, this measures the rapidity with which Z changes. Before looking at the figures let us return to the equation (6) for Z tendency to derive an equation for $[(dZ/dt)^2]$ (The square brackets around the expression indicate the average over all locations and times.)

Multiplying (6) by dZ/dt and averaging we have

$$\left[(dZ/dt)^2\right] = \left[\frac{dZ}{dt} \quad \frac{d}{dt} \quad \frac{Rdt - A_R}{D} \right] - \left[\frac{dZ}{dt} \quad \frac{d}{dt} \quad \frac{Udt - A_U}{D} \right]$$

$$- \left[\frac{dZ}{dt} \quad \frac{d}{dt} \quad \frac{Edt - A_E}{D} \right] - \left[\frac{dZ}{dt} \quad \frac{d}{dt} \quad \frac{Tdt - A_T}{D} \right] \quad (7)$$

The terms on the right-hand side of equation (7) represent contributions to mean square Z tendency from fluctuations in, respectively, moisture supply to the soil, loss of moisture through the bottom of the soil column, evaporation from the upper layer and transpiration from the lower soil layer. It is possible to compute the value of each of these terms averaged over all locations and times in the observed and GCM generated data sets to get an idea of how $[(dZ/dt)^2]$ compares between the two and which of the terms on the right hand side of (7) contribute to the differences. Table 1 giving the mean values of terms (from left to right) in equation 7 summarizes the results. In the table the terms are referred to as Total, Supply, Loss, Evap and Transp.

Table 1   Contributions to mean square Z tendency

	TOTAL	SUPPLY	LOSS	EVAP	TRANSP
Observations	3.29	0.24	0.02	1.71	1.31
GCM	1.95	1.16	-1.04	0.85	0.97

All values in units of 0.001 X (standard deviations/day)^2

From the first column of Table 1 we see that $[(dZ/dt)^2]$ is larger in the observations than in the GCM.   The contribution to $[(dZ/dt)^2]$ from fluctuations in supply to the soil (Supply) is almost entirely balanced by an increase and a change of sign in the contribution from fluctuations in loss out the bottom of the soil column (Loss).   Taken together these two processes account for 6% of mean square Z tendency in the   GCM compared to 8% in the observations.   In each case   the balance   comes   from   fluctuations in evaporation   and   transpiration (Evap,   Transp).   In   the   GCM   these   latter   processes   contribute respectively   only   50%   and   74%   as   much   to   as   they   do   in   the observations.

In   summary then,   although supply to the soil is almost   5   times more   variable   in the GCM this has little effect upon soil   moisture variability   because   loss out the bottom of the column is   also   more variable   and in phase.   The resultant determinants of soil moisture variability   in the GCM are then evaporation and transpiration   which are less variable than in the observations.

References

AES Drought Study Group   (1986)   An applied climatology of drought in the prairie provinces,   Canadian Climate Centre Report No. 86-4, Atmospheric Environment Service, Downsview, Ontario, Canada.

Baier,   W.   and   Geo.   W.   Robertson   (1965)   Estimation of latent evaporation from simple weather observations, Can. J. Plant Sci. 45, 276-284.

Baier W. and Geo. W. Robertson   (1966)   A new versatile soil moisture budget, Can. J. Plant Sci. 46, 299-315.

Boer,   G.J.,   N.A.   McFarlane,   R.   Laprise,   J.D.   Henderson and J.P. Blanchet   (1984a)   The   Canadian   Climate   Centre   spectral atmospheric general circulation model, Atmos-Ocean, 22(4), 397-429.

Boer,   G.J.,   N.A. McFarlane and R. Laprise   (1984b)   The climatology of   the   Canadian Climate Centre general   circulation   model   as obtained from a five-year simulation, Atmos-Ocean,   22(4), 430-473.

Palmer, Wayne   (1965)   Meteorological Drought, Research Paper No. 45, Weather Bureau, U.S. Dept. of Commerce, Washington, D.C.

Yevjevich,   Vujica   (1967)   An objective approach to definitions and investigations of continental hydrologic droughts,   Colorado State University Hydrology Paper No. 23, Fort Collins, Colorado.

2 Climate change proxy data

The Influence of Climate Change and Climatic Variability on the Hydrologic Regime and Water Resources (Proceedings of the Vancouver Symposium, August 1987). IAHS Publ. no. 168, 1987.

Paleohydrologic studies using proxy data and observations

H.J. Liebscher
Bundesanstalt fur Gewasserkunde
Postfach 309, 5400 Koblenz, FR Germany

Introduction

At the present level of knowledge it is a known fact that climate cannot be regarded as constant. Extraterrestrial effects may contribute to changes in climatic conditions just as much as the concentration of radiatively active gases and aerosols in the atmosphere. It is a further important finding that climatic changes occur within much shorter periods of time than supposed before. Such changes may occur within a few decades' time (Flohn 1980).

Today, mankind is worried first of all about the rapid increase in CO_2 in the atmosphere, caused by burning of fossil fuels and extensive land clearing activities. It has been estimated that a doubling of CO_2 content will occur within the next 50 years; this will be associated with a rise of mean temperature in the earth's Northern Hemisphere by 1.5° to $4.5^{\circ}K$, as indicated by results of global climate models.

The expected rise in temperature due to the increase in CO_2 concentration is not only a problem of temperature but also a problem of water. Warming will result not only in increased evaporation, which means a reduction in available water resources, but will also lead, owing to displacements of the atmospheric circulation, to a regional as well as seasonal redistribution of precipitation and thus to a redistribution of the water resources. There will be favoured and disadvantaged regions, which will inevitably lead to serious consequences in water supply. A change in the structure of precipitation, e.g. as a result of more frequent occurrence of convective or orographic precipitation, may make both the local and regional flood danger more critical.

Using results from global climate models, several institutions attempt today to quantify the impact of these changes on a global and mesoscale basis. As we have become aware of potential future changes in climatic conditions, the issue of assumed weather and flow regimes in comparable past times has assumed increasing importance. For this reason investigations on the behaviour of climate-dependent variables in earlier climate epochs are being conducted in order to be able to assess, the conditions that may pertain after the climatic changes to be expected. This paper reviews some of this work.

In this connection, the fact that the causes leading to previous changes in temperature had been quite different must be taken into consideration. Yet, the investigation of earlier climates has its justification today, because it indicates the natural, anthropogenically uninfluenced climate changes and permits validation of scenarios for the future.

Studies of the streamflow and sediment transport regimes of the distant past constitute part of the subject of palaeohydrology (Berglund 1985, Gregory 1983). In addition, hydrology is concerned with the situation of water-courses, formation of river systems and extent of lakes.

As mentioned above, man has effected through his intervention in nature regime transformations whose impacts have to be estimated. Thus, for example, the question of how the streamflow regime will change under the effects of increasing forest deterioration is of great interest today. An indication of the effects to be expected might be gained from a comparison of the present state with that in the late Middle Ages in Middle Europe when the forests were subjected to intensive clearing. Special attention should therefore be given to the reconstruction of flood conditions in the last thousand years.

General proxy data and observations

A distinction is made between direct and indirect climate evidence. Field data pertain to the group of indirect climate evidence; these are generally designated as proxy data. Examples of these data are moraines, pollen profiles, annual tree rings and isotope conditions. Proxy data can also be found in historical documents such as reports and pictorial representations of glaciers, information on times of grain and grape harvests as well as data on their quality and quantity.

Direct climate evidence comprises observations and instrumental measurements. Observations have been recorded in chronicles, weather diaries and ship logs. Other examples of this category are high-water and low-water marks recording particular extreme events. Long-term records have special significance because the evidential value of other indicators can be derived from them by making a synchronous comparison. On the basis of instrumental measurements, historic data and documents can be calibrated, and evidence of climatic events in past centuries can be preserved.

Figure 1 shows the lengths of records registered by the individual data types. The longer the period registered, the less normally is the time resolution. While this latter is less than one day for instrumental measurements, an accuracy of only 10 to 100 or more years can be attained for field data. The greater the resolution accuracy, the shorter is the period to be interpreted.

The different data types exhibit different problems of interpretation. Field data have a poor time resolution and react with more than one weather element simultaneously (e.g. with temperature and precipitation). Historical documents and data often suffer from observation gaps. It is only the last five to eight centuries that have evidential power, and most observations are subjective. Instrumental measurements are available only for the last two or three centuries. One exception is the river stage record for the Nile. Before the mid-19th century the density of observation stations was very low; many measurement series show inhomogeneities. A synthesis of different data types permits the development of a realistic and detailed picture of climates reaching far back in time.

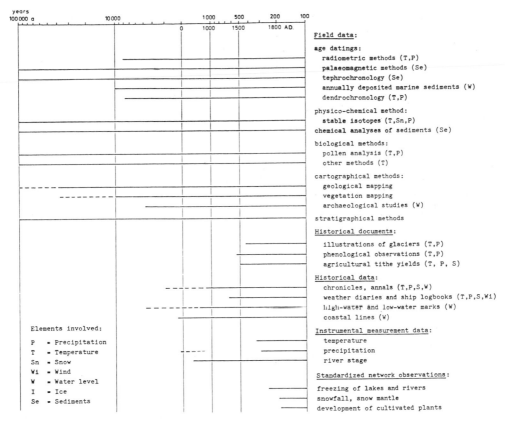

Figure 1   Period included in the reconstruction of climate
and weather history for different data types.

In meteorology climate conditions have been systematically studied
using proxy data for many years (Flohn 1985), in hydrology, however,
only few such investigations have been undertaken. Relevant research
activities in meteorology have concentrated first of all on the
earth's temperature regime. In examining this problem. meteorologists
make use of inter alia, marine deposits, annual tree rings, pollen
analysis, ice drill cores, etc. They have succeeded in retracing the
fluctuations of mean air temperature as far back as 100,000 years
(Mitchell 1980). Other examples of the use of proxy data in climate
reconstruction include: the interrelation between temperature, ice
extent and sea level; evidences of past altered vegetation such as in
the Sahara Desert; and even settlement patterns of human development.
Examples of the latter relate to development of the first very
advanced civilizations along the Nile, Euphrates and Tigris and the
Indus rivers. People who lived in the neighbouring arid regions as
hunters or gatherers, retired to the then fertile river valleys,
where they learned quickly how to use water (Garbrecht 1985). This is
why important proxy data go as far back as those times.

Hydrological proxy data and observations

As for hydrological proxy data and observations, a distinction must likewise be made between direct and indirect data. For field data a stronger differentiation is made; age dating, stratigraphical, physico-chemical and biological methods. Age datings include radiometric techniques, paleomagnetic methods, tephrochronology (volcanic ash), annually laminated lake sediments and dendrochronology (annual tree rings).

These methods offer a valuable aid in classifying the evidence. In stratigraphic methods, deposits in lakes, mires and watercourses and flood-plains are investigated, and comparisons are drawn with similar finds. The physico-chemical methods analyse stable oxygen and carbon isotopes in sediment, and soils. Biological methods include pollen analysis, charred particle analysis (burnt material), and analysis of a wide variety of organisms including spores, algae, seeds and molluscs (Berglund, 1985).

Hydrological proxy data also include a great amount of information taken from chronicles and other sources in which particular events, such as flood, drought, ice cover in surface waters, etc. are reported, or else from historical pictures or maps of glacier tongues in which certain circumstances have been recorded.

Methodical approach to reconstruction of
past hydrological regimes

Reconstruction of past hydrological regimes begins with providing the appropriate sources. Sources in writing, containing climatic and hydrologic history, which can be found in archives and libraries, are classified in types according to genesis and form (Potter, 1978). The sources are integrated into their personal, institutional and spatial scope, and their reliability is tested in compliance with the rules of historical source criticism. Subsequently, the information elements included in the source material are standardized on the basis of a numerical code and classified according to data types.

The purely descriptive elements are subjected to data acquisition and storage in accordance with the methods of data processing; then they should be sorted, together with the proxy data and measurement data, taking the chronological, typological and thematic aspects into consideration. Retransformation into readable language yields a chronology that can then be used for the interpretation of human-ecological data.

Direct and indirect data are subsequently related to one another and interpreted in correlation. Both data types support each other mutually. A statement based on direct data permits in most cases a more refined interpretation of indirect data. Hydrologic conditions for each month can be assessed in the form of a weighted and non-weighted index. This makes it possible to condense the historical hydrological information into numerical values.

The last step consists of forming the ten-year mean of the indices and calibrating these with the recorded measurements. In this procedure the estimated values for the ten-year periods are connected to the new ten-year means obtained by instrumental measurements. This

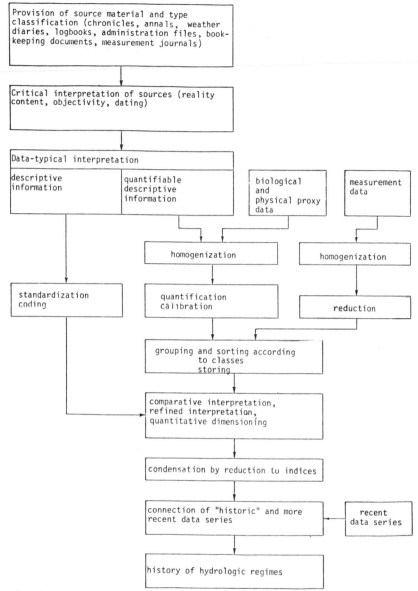

Figure 2   Steps in the reconstruction of past hydrologic regimes
based on different data types.

approach is represented schematically in Figure 2. and was used by
Pfister in several of his studies referred to later.

Present state of studies of past hydrologic regime in past time

The   following sections review selected sources for past information
on hydrologic elements.

(a) General works

Information from chronicles on notable hydrologic and meteorologic events in Central Europe was systematically gathered by Weikinn (1958) until 1750. Unfortunately, the 1750 end date means that there is no connection to the instrumental period for calibration purposes. Hamm (1976), is another example which includes chronicled information on extreme hydrologic events.

(b) Precipitation

The first direct measurements on precipitation depth were carried out in China as early as 400 B.C. In Palestine the first rain gauges were installed at the turn of the era, but regrettably the results are not now available. The first precipitation measurements in modern times date back to the mid-18th-century.

In old chronicles there are frequent indications of droughts or local, mostly torrential rainstorms. Equally common are reports on good or bad harvests from which precipitation conditions can be inferred. Data on the streamflow regime of rivers or on water levels of lakes also give some information about the precipitation conditions. For example, Pfister (1985) has constructed a time series of monthly precipitations for Switzerland from 1525 to 1979. Beran & Rodier (1985) reported the fluctuations of annual rainfalls in Central Africa between 1750 and 1978 from the variations of water levels of Lake Chad.

(c) River-stages

The earliest known flood level marks on the Nile river date from about 1800 B.C. The earlist regular stage registrations on the Nile go back to the 7th century A.D.; these included first only the lowest and highest annual stages. Tousson (1925) has compiled these river stage data for the years 622 to 1921. Many scientists have attempted to interpret these registrations (Hurst et al. 1965, Bell 1970, Riehl & Meitin 1979, Hassan 1981). In Central Europe systematic observations of river stages started about 1750. Complete stage data series do not extend beyond the beginning of the 19th century.

In most cities in Central Europe lying on larger streams exist numerous flood marks that date back in part from the Middle Ages. Further, there are many reports of earlier floods in old chronicles. The same applies to low-water levels. In 1784, G. Potzsch determined the flood levels in Dresden and Meissen on the Elbe from 1501 in his work "Chronological History of Great Floods on the Elbe River for Thousand and More Years". He further described 71 floods during the period from the 6th to the 15th century (Fugner 1982).

In Germany the river stage series going back to the beginning of the 19th century have been set up in the so-called "Stream Works". In these, attempts were also made to compile and assess flood level marks of earlier floods. Compilations of historical flood levels have been included in a large number of publications: e.g. Champignon 1863, Belgrand 1873, Sonklar 1883, Reis 1884, Mitcher 1888, Stuhl and le Blanc 1932, Rodier and Roche 1984. Pfister (1985) has succeeded in compiling the high-water and low-water levels in the stream and lakes of the Swiss Mittelland for the period from 1525 to 1880.

Earlier flood level marks can be also be inferred from sediment samples of flood deposits (Baker et al. 1979).

Many attempts have been made to include historic flood events derived from old chronicles or other sources in flood probability calculations often in combination with more recent series (Benson 1950, Chen et al. 1974, Gerard & Karpuk 1979, Tasker & Thomas 1978, Cohn 1984, Fugner & Schirpke 1984, Sutcliffe 1986, Deisenhofer et al. 1986). Theoretical work concerning the benefit of inclusion of historic flood events in flood probability calculations have come from Condie and Lee (1982), Condie and Pilon (1983), Hosking and Wallis (1986), Stedinger and Cohn (1984).

Just as low-water and high-water levels of the streams, variations in water levels of lakes and oceans are also of interest to the hydrologist.

By analysing sediments the past shore lines of lakes and seas can be traced, and from these the lake-levels and sea-levels can be deduced. e.g. Butzer et al. 1972, Mehringer et al. 1979, Lamb 1977. Archaeological, historical and instrumental sources are all of value in reconstructing water level time series from such water bodies. Examples of long-term sea-level series at coastal gauges are given by Fairbridge (1961), Shepard (1963), Schofield and Thompson (1964), Schell (1970), Rossiter (1972), Morner (1973), and Teh-Lung Ku et al. (1974).

(d) Discharges

The earliest discharge measurements in Central Europe were made at the close of the 18th century. The longest discharge data series therefore extend back no farther than the beginning of the 19th century. For this reason, statements on the streamflow regime of our present rivers in past times can be obtained only on the basis of earlier river stage registrations. Thus, several attempts were made to allocate discharges to earlier high-water level marks (Fugner and Schirpke 1984), Sutcliffe 1986, Schiller 1987).

The literature contains a number of attempts to reconstruct discharges of former times from annual tree rings, e.g. Cook and Jacoby (1983) who constructed a discharge series for the Potomac River (USA) for the period 1730 to 1880.

(e) Evaporation

As determination of evaporation is fraught with certain difficulties even today, there is no direct information available concerning evaporation data in past times. They can be estimated merely from the pattern of the temperatures. Examples of long temperature series can be found in numerous publications (Lamb 1977, Mitcher 1980, et al.).

(f) Soil moisture, groundwater

Systematic measurement of soil moisture dates back only a few decades, and observations on groundwater go back to the mid-19th century (Beran and Rodier 1985). Direct observations from past times are not available. Sometimes there are data to be found in chronicles, reporting on drying up of wells. Information on droughts permit at least indications of past soil moisture contents (Wigley and Atkinson 1977).

(g) Glaciers

The precipitation – evaporation difference can be estimated from analysis of drill cores of glaciers; it cannot be excluded in all cases, however, that there might have been surface runoff from rainwater or glacial water. The residual term of the water balance can be quantified on the basis of the O^{18}/O^{16} ratio. Many glacial drill cores from inland ice or glaciers have been analysed. These analyses have permitted the temperature pattern of the past 100,000 years to be retraced.

Ice extent on land surface can be estimated through glacier formations or moraines. Pictorial illustrations of glaciers go as far back as the 16th century. Glacier tongue formations for the Grindelwald Glacier in Switzerland during the period 1590-1978 have been compiled and described by Pfister (1985).

(h) Ice on the surface of rivers and lakes

Ice formations cannot yield evidence, as is generally known, concerning moisture regime. They may give information, however, about the temperature regime. In chronicles, annals, weather records and logbooks there are often indications of ice covers on the surface of streams, lakes or on seashores. Several authors attempted to document past ice sheets, e.g. Pfister (1985) for the lakes and rivers of Switzerland in the period 1525-1970, and Jansen (1980), for the Lower Rhine in the period 1750-1980.

(i) Snow

Depths of snow and water contents of snow mantles have been observed only in most recent times. Chronicles refer sometimes to particularly heavy snowfalls or particularly prolonged snow covers. For example, Pfister (1985) has composed a series of extreme snow durations for the deeper Mittelland of Switzerland from 1525 to 1970.

Summary

Paleohydrology is yet a relatively young subsection of hydrology.
However, it has gained significance through the expected temperature
changes induced by increasing CO_2 concentrations and through the
effects of forest dieback. This subject has been included as a
project in the WMO's World Climate Programme's Water Component. In
this way scientists all over the world should be encouraged to
disclose the sources available in their respective countries and to
perform studies on the hydrologic regimes in earlier centuries and
millennia. In this way it is hoped that one will be able to obtain a
detailed picture, i.e. a temporally and spatially differentiated one,
of the moisture, and streamflow regimes in past times.
   This paper attempts briefly to describe and assess the current
state of knowledge and information on hydrologic proxy data. However,
the paper does not claim to be complete. There are undoubtedly many
other information sources about hydrologic proxy data which,
however, have not been accessible to the author because of the widely
scattered range of publications. Within the scope of activities of
the International commission for Surface Waters (ICSW) of the IAHS an
attempt is now being made to contribute to the World Climate
Programme by gathering and arranging the source material that yields
information on hydrological proxy data. For this reason, the author
requests all hydrologists to support the Project and to bring to the
author's attention further sources not mentioned here.

References

Baker, V., R.C. Kochel and P.C. Patton (1979) Long-term flood
   frequency analysis using geologic data. Proceedings of the
   Canberra symposium on the Hydrology of Areas of Low Precipitation,
   IAHS-AISH Publ. 12, 3-9.
Bell, B. (1970) The oldest records of the Nile floods. Geographical
   Journal, Vol. 134, 569-573
Benson, M.A. (1950) Use of historical data in flood frequency
   analysis. EOS, Trans. AGU, 31(3), 419-424.
Beran, M. and J. Rodier (1985) Hydrological aspects of drought.
   UNESCO-WMO-Studies and reports in hydrology, No. 39, 139 pp.,
   Paris.
Berflund, B.E. (ed) (1985) Handbook of Holocene Paleoecology and
   Paleohydrology. 896 pp., Wiley (Chichester).
Butzer, K.W., G.L. Isaac, J.L. Richardson and C. Washbourn-Kaman
   (1972) Radiocarbon dating of East African lake levels. Science,
   175 (4027), 1069-1076, New York.
Chen, C.-C., Y. Yeh and W. Tan (1974) The important role of
   historical flood data in the estimation of spillway design flood.
   Report, 15 pp., Eng. Bur., Min. of Water Conservancy and Electr.
   Power, Peking.
Cohn, T.A. (1984) The incorporation of historical information in
   flood frequency analysis. M.S. thesis, 79 pp., Cornell Univ.,
   Ithaca, New York.
Condie, R. & K.Lee (1982) Flood frequency analysis with historical
   information. J. Hydrol. Amsterdam, 58(1/2), 47-61.

Cook, E.R. and G.C. Jacoby (1983) Potomac River streamflow since 1730 as reconstructed by tree rings. J.of Climate and applied meteorology. 22, 10, 16591672.

Fairbridge, R.W. (1961) Eustatic changes in the sea level. In: Physics and Chemistry of the Earth, 4, pp. 99-185, Pergamon Press (New York).

Flohn, H. (1980) Modelle der Klimaentwicklung im 21. Jahrhundert. In: Das Klima, pp.3-17, (ed. by H. Oeschger, G. Messerli, M. Svilar).

Fugner, D. (1982) Uber die "Chronoligische Geschichte der groben Wasserfluten des Elbestromes seit tausend und mehr Jahren" von Christian Gottlieb Potzsch anlablich seines 250. Geburtstages. Wasserwirtschaft-Wassertechnik, 32. Jg., Nr. 6, S. 203-205.

Fugner, D. (1982) und H. Schirpke (1984) Neue Ergebnisse der Hochwasserberechung fur den Elbestorm in Dresden. Wasserwirtschaft-Wassertechnik, 34. Jg., Nr. 8, S. 189-191.

Garbrecht, G. (1985) Wasser-Vorrat, Bedarf und Nutzung in Geschichte und Gegewart. Deutsches Museum, Kulturgeschichte der Naturwissenschaften und der Technik. 279 S., Rowohlt (Reinbech).

Gregory, K.J.(ed.) (1983) Background to paleohydrology. 502 pp., Wiley (Chichester).

Hamm, F. (1976) Naturkundliche Chronik Nordwestdeutschlands. 370 S., Landbuch (Hannover).

Hosking, J. and J. Wallis (1986) Paleoflood hydrology and flood frequency analysis. Water Resources Research, Vol. 22, No. 4, 543-550.

Hurst, H.E., Black R.P. and Simaika, Y.M. (1965) Long-term storage. Constable (London).

Lamb, H.H. (1977) Climate, Present, Past and Future. Vol. 2 Climate History and the Future. Methuen (London).

Mehringer, P., Peterson, K. and F. Hassan (1979) A pollen record from Birket Qarum and the recent history of the Fayum, Egypt, Quaternary Research 11, 238-256.

Mitchell, J.M. (1980) History and mechanisms of climate. In: Das Klima, (ed. by Oeschger, B.Messerli, M.Silver, pp. 31-42

Pfister, C. (1985) Snow Cover, snow-lines and glaciers in central Europe since the 16th century. In: The climatic Scene (ed. by Tooley M.J. and Sheail G.M. 154-174, Allen & Unwin (London).

Potter, H. R. (1978) The use of historic records for the argumentation of hydrological data. Rep. 46, Inst. of Hydrology, Wallingford, Oxon, England.

Reis, P. (1884) Uber die Hochwasser des Rheins. Deutsch. Rev. 9,1, Breslau.

Riehl, H. and J. Meitin (1979) Discharge of the Nile river: a barometer of short-period climatic variation. Science, Vol. 206, No. 7, pp. 1178-1179.

Rodier, J. & M.Roche (1984) World Catalogue of maximum observed floods. IASH-AIHS Publication. No. 143, 354 pp., Wallingford (UK).

Schiller, H. (187) Ermittlung von Hochwasserwahrscheinlichkeit am schiffbaren Main und uberregionaler Verglaich der Ergebnisse. Beitrage zur Hydrologie. Sonderheft 6, 79-101, Kirchzarten.

Stedinger, J.R. and T.A. Cohn (1986) Flood frequency analysis with historical and palaeoflood information. Water Resources Res., 22, 785-793.

Stuhl und le Blanc (1932) In: Die Wasserstandsstatistik, insbesondere am Rhein (Brauler L.). Rheikunde, H.8.

Toussoun, O. (1925) Memoire sur l'historie du Nile. Memoires de'l Institut d'Egypte, Vol. 8, Cairo.

Weikinn, C. (1958) Quellentexte zur Witterungsgeschichte Europas von der Zeitwende bis zum Jahre 1850. Quellensammlung zur Hydrogeographie und Meteorologie. Akademischer Verlag (Berlin)

Wigley, T.M.L. and T.C. Atkinson (1977) Dry years in south-east England since 1698. Nature, No. 265, 431-444 (London).

*The Influence of Climate Change and Climatic Variability on the Hydrologic
Regime and Water Resources* (Proceedings of the Vancouver Symposium,
August 1987). IAHS Publ. no. 168, 1987.

Paleoflood hydrology and hydroclimatic change

Victor R. Baker
Department of Geosciences
University of Arizona
Tucson, Arizona
85721 U.S.A.

ABSTRACT   Important recent advances have been made in the
reconstruction   and   interpretation   of   ancient   floods,
particularly in the use of slackwater deposits   and
paleostage indicators (SWD-PSI).   For certain appropriate
geomorphic settings, relatively accurate estimates of
paleoflood discharges and ages can be made over time
scales of centuries and millennia.   New statistical tools
are   available to extract the maximum information   content
from this unconventional hydrologic data. Preliminary SWD-
PSI study results from the southwestern United States
indicate that certain time intervals in the last   several
thousand   years have been characterized by occurrences   of
extraordinary floods, while other intervals have been
relatively free of such events.   Hydroclimatic change is a
likely cause of this nonstationarity.

Introduction

Paleoflood hydrology concerns the study of past or ancient   flow
events using physical or botanical information,   irrespective of   any
direct   human   observation.   The flow events usually have   occurred
prior to the possibility of direct measurement by modern   hydrologic
procedures,   although paleoflood hydrologic techniques can be applied
to modern floods at ungaged sites (Baker <u>et al</u>., in press).   Recent
advances in geochronology, flow modeling, and statistical analysis of
paleoflood data have greatly increased the ability to extract useful
hydrologic information from one variety of paleoflood   investigation:
slackwater deposit-paleostage indicator (SWD-PSI) studies (Stedinger
& Baker, 1987).   SWD-PSI investigations can provide reconstructions
of discharges and magnitudes for multiple paleofloods with remarkably
high accuracy over time scales of centuries and millennia.   However,
such SWD-PSI studies require special combinations of geological
circumstances that must be carefully evaluated in each application.

An outline of SWD-PSI paleoflood hydrology

The methodology of SWD-PSI paleoflood hydrology is discussed by Baker
<u>et</u> <u>al</u>. (1983) and by Baker (in press).   This section will   briefly
review   important aspects of that methodology, emphasizing recent
research developments.

Figure 1 Photograph of an accumulation of slackwater deposits downstream of a bedrock spur on the Salt River in central Arizona.

(a)   Slackwater deposits consist of sand and silt (sometimes gravel) that accumulate relatively rapidly from suspension during major floods, particularly at localities where flow boundaries result in markedly reduced flow velocities (Figure 1).

(b)   Other important paleostage indicators include silt lines, high level scour marks, and flood-modified vegetation.

(c)   Sites of slackwater sediment accumulation occur at the following locations: (i) tributary mouths, (ii) abrupt channel expansions, (iii) in the lee of bedrock flow obstructions, (iv) in channel-margin caves and alcoves, (v) at meander bends, and (vi) upstream of abrupt channel expansions.

(d)   Regional factors useful in locating river reaches appropriate for SWD-PSI studies include the following: (i) adequate concentrations of sand and silt in transport by floods, (ii) resistant-boundary channels not subject to appreciable aggradation, (iii) depositional sites with high potential for preservation of SWD-PSI features, and (iv) narrow, deep canyons or gorges in resistant geological materials.

(e)   Although initially developed and applied in arid and semiarid regions (Baker et al., 1979; Kochel & Baker, 1982; Kochel et al., 1982), SWD-PSI paleoflood hydrology has been extended to the study of humid-region rivers (Kochel & Baker, in press; Patton, in press).

(f)   Computer flow models for step-backwater analysis are used to calculate water surface profiles for various discharges in appropriate SWD-PSI study reaches. Paleodischarges are determined by comparing elevations of the various paleostage indicators to the water surface profiles.

(g)   Recent research has concentrated on strategies for reducing error in paleodischarge estimation. Important concerns in this regard include: (i) paleoflow cross-sectional stability, (ii) relatively deep paleoflows, and (iii) relatively uniform reaches.

(h)   Long-term channel stability is necessary for accurate hydraulic calculations. This can be assured for reaches developed in bedrock, immobile sediment, or other resistant boundary materials.

(i)   Narrow-deep channel cross sections are most useful, since increasing flood discharge results in relatively large stage increases (Baker, 1984).

(j)   Accuracy of the predicted water-surface profiles can be improved when relatively large flows in a systematic gage record are available to test and calibrate the flow model (Ely & Baker, 1985; Partridge & Baker, 1987).

(k)   At ideal SWD-PSI sites thick sequences of multiple sedimentation units record numerous paleofloods (Figure 2). Individual flood units are distinguished by sedimentologic properties such as the following: (i) silt-clay or organic drapes, (ii) buried paleosols, (iii) organic layers, (iv) intercalated tributary alluvium or slope colluvium, (v) abrupt vertical grain size variations, (vi) mudcracks, (vii) color changes, and (viii) induration properties.

(l)   Recent advances in geochronology, particularly radiocarbon analysis (Baker et al., 1985), provide excellent opportunities to determine paleoflood ages. As little as 1 to 2 mg of elemental carbon can be analyzed by the new technique of tandem accelerator mass spectrometry (Taylor et al., 1984).

(m)   The usual "worst case" end member for SWD-PSI paleoflood information content is a single, vertically-stacked sequence of slackwater deposits (Figure 2). In this case, an informational censoring level (the elevation of each succeeding deposit) increases with time.

(n)   Most commonly, SWD-PSI sequences provide much more paleoflood information than in the worst-case scenario. This is achieved by lateral tracing of individual flood deposits to their highest elevations, by correlation of flood deposits among multiple sites, by documenting evidence of limiting high-water levels, and by studying inset stratigraphic relationships.

(o)   The information content in SWD-PSI sequences can be structured for flood-frequency analysis through the concept of censoring levels. Flood experience for various time intervals is then analyzed in terms of exceedances or nonexceedances of the censoring levels or threshold discharges (Stedinger & Baker, 1987).

(p)   The goal of stratigraphic analysis in SWD-PSI studies is to reconstruct a complete catalog of discharges exceeding censoring levels over specified time periods.

(q)   New statistical tools are now available to make optimum use of the information content in appropriately structured paleoflood data (Stedinger & Cohn, 1986; Stedinger & Baker, 1987).

Figure 2   Photograph of The Alcove slackwater sedimentation site (Webb, 1985) on the Escalante River in south-central Utah.

Flood hydroclimatology

Conventional flood-frequency analysis relies on the following assumption: "... the array of flood information is a reliable and representative time sample of random homogeneous events" (U.S. Water Resources Council, 1981, p. 6). Two possible violations of this assumption may be induced by (1) a mixed underlying parent distribution for the flood events, and (2) variation through time in the mean of the underlying probability distribution for flood recurrence (non-stationarity). Both of these situations may derive from climatologic causes (Hirschboeck, in press). Although short-term systematic records are generally ambiguous with regard to such interpretive problems, SWD-PSI paleoflood hydrology provides excellent opportunities to test assumptions. In southern Arizona, for example, annual flow peaks are dominated by floods induced by regional snowmelt, local summer convective storms, and winter frontal storms (Hirschboeck, 1985). More rarely, incursions by tropical storms lead to extraordinary floods that appear as outliers in the systematic flood records. Here the systematic flow record is biased toward one hydroclimatologically induced distribution: that controlling the relatively common, smaller annual floods. Only with the expanded time base provided by paleoflood hydrology can an adequate sample be achieved for the unusually large and rare floods related to another hydroclimatologically induced distribution.

Of course, paleoflood hydrology generally cannot identify the hydroclimatic cause for a given paleoflow event. Nevertheless, the time base of centuries or millennia is ideal for evaluating long-term trends. Knox (1985) documented a pronounced nonstationarity for upper Mississippi Valley floods over the past 9500 years. Early Holocene alluvial fills indicate very low probabilities for large floods between 6000 and 9500 yr B.P. Increased probabilities for large floods are evidenced by boulder gravel in overbank sediments deposited in the following age intervals: (1) 6000 to 4500 yr B.P., (2) 3000 to 1800 yr B.P., and (3) 1000 to 500 yr B.P. (Knox, 1985). Similarly, Patton & Dibble (1982) presented evidence from the Pecos River of western Texas that floods were relatively infrequent during an arid interval between approximately 9000 and 3000 yr B.P., but the extraordinary floods occurring in this interval were unusually large. Between approximately 3000 and 2000 yr B.P. a humid interval resulted in more frequent flooding, but flood magnitudes were moderated. The last 2000 years has been most similar to the early Holocene arid interval.

On a shorter time scale, detailed SWD-PSI studies also have an immense potential for evaluating nonstationarity. For the Columbia River in central Washington, Chatters & Hoover (1986) showed that during the approximate interval 1000 to 1400 A.D. large floods were three to four times more common than at present. Flood frequency characteristics similar to those at present prevailed from approximately 200 to 1000 A.D. and from approximately 1400 A.D. to present. This use of paleoflood hydrology illustrates the fallacy of overly simplistic characterizations of paleoflood records as illustrated by the computer simulations of Hosking & Wallis (1986). Rather than a vague rationalization with which to criticize paleoflood hydrologic studies (Hosking & Wallis, 1986), nonstationarity can be an object of scientific study utilizing the remarkable capability of SWD-PSI studies to generate accurate and complete paleoflood records.

Applications in the southwestern United States

Since 1981 the new procedure of SWD-PSI paleoflood hydrology has been used in a regional study of ancient floods in the southwestern United

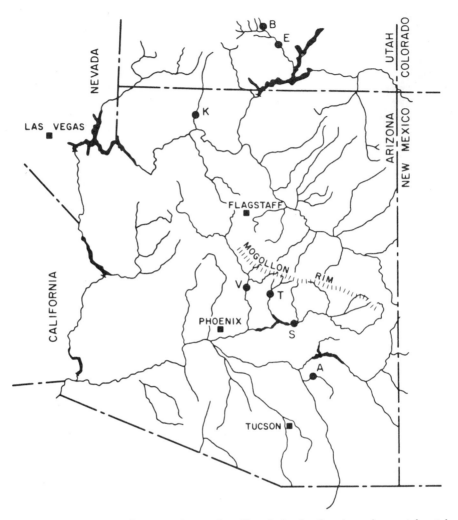

Figure 3   Location of SWD-PSI paleoflood hydrologic investigations
           in the southwestern United States.   Studies were done   in
           south-central Utah on Boulder Creek (B) and the Escalante
           River (E).   In Arizona, studies were done on Kanab Creek
           (K), Tonto Creek (T), Aravaipa Creek (A), and on the Salt
           (S) and Verde (V) Rivers.

States (Figure 3). One goal of this regional analysis has been to
identify temporal and spatial patterns in extraordinary floods.
    The paleoflood record of the Escalante River of south-central Utah
(Figure 3) illustrates the trends seen at other study sites.
Paleofloods documented by Webb (1985), Webb et al. (in press), and
O'Connor et al. (1986) fall into major time groupings.   During the
period 2000 to 1300 yr B.P. floods were relatively small.   Three
major floods occurred between 1200 and 1000 yr B.P., including the
largest flood of the record.   This period was also one of major
arroyo cutting and is well-documented throughout the southwestern
U.S.   No floods were recorded between 900 and 600 yr B.P., but three

floods were recorded between 600 and 400 yr B.P. The next major phase of flooding occurred in the last century, which is the period of most extensive arroyo formation in the region (Webb and Smith, 1986).

The most detailed long-term record of paleofloods in the Southwest occurs just southeast of Phoenix, Arizona (Figure 3). Prehistoric irrigation canals constructed by the Hohokam indians are filled with flood deposits (Masse, 1981). Current research by J.E. Fuller (written communication, 1986) documents that, since 1100 yr B.P., the Hohokam canals recorded a minimum of 25 and a maximum of 30 floods that exceeded 5000 m^3s^{-1}. Of these the largest (>12,000 $m^3 s^{-1}$) occurred about 1100 yr B.P. during a 250-yr period of pronounced flooding. Large floods again appeared in the last 400 years, including three exceedences of 7000 m^3s^{-1}. The last of these was the 1891 flood with a discharge of between 7000 and 8000 m^3s^{-1}.

Essentially the same timing of paleoflood events is observed on upstream reaches of the Salt River (Partridge & Baker, 1987) and the Verde River (Ely & Baker, 1985). Additional work on these streams and Tonto Creek (Figure 3) by J.E. O'Connor and J.E. Fuller (written communication, 1986) confirms the same sequence. The largest flood occurred approximately 1000 yr B.P. on both the Salt and Verde Rivers. Unusually large floods also occurred during the last century.

Discussion

All SWD-PSI paleoflood studies conducted thus far in Arizona and adjacent areas (Figure 3) reveal a remarkably consistent record. Certain time intervals during the past few millenia have been characterized by occurrences of extraordinarily large floods, while other intervals have been relatively free of such events. Major episodes of flooding occurred from approximately 1000 to 1200 yr B.P. and during the past century or two. A somewhat less intense phase of flooding occurred between approximately 400 and 600 yr B.P. Time intervals between these flood phases were characterized by fewer, smaller floods. In addition, there are many indications that channel entrenchment on alluvial streams (arroyo formation) was coincident with flood phases, while aggradation was generally coincident with phases of reduced flooding (Webb, 1985).

The regional coincidence of flood phases in the southwestern United States suggests a hydroclimatologic cause. A possible mechanism is the variable influence of tropical moisture in the region. Work on evaluating this mechanism is currently in progress.

Considerable potential exists for combining SWD-PSI paleoflood studies with other paleoclimatic indicators. For example, tests of nonstationarity in long-term flood series might be achieved by evaluating other paleohydrologic indicators. Long-term tree-ring series and regime-based paleoflow estimates (RBPE) both can be related to various measures of mean streamflow or mean floods. RBPE studies are accomplished in alluvial channels, which are much more common than the resistant-boundary (non-alluvial) channel conditions required for accurate SWD-PSI studies. Accurately dated mean flow estimates plus chronologies of other paleoclimatic indicators, such as pollen records, plant macrofossils, and isotopic records, can be used to evaluate nonstationarity in paleoflood records and interpret the role of climate change in generating such records. Past climatic change may serve as a guide to the potential for future climatic change. Precise data on the magnitudes of past hydroclimatic change may prove useful in testing models intended to predict future change.

ACKNOWLEDGEMENTS   My regional studies of paleoflood hydrology in the southwestern United States were initially supported by the Division of Earth Sciences, Surficial Processes Program, National Science Foundation Grant EAR 81-19981. Subsequent work was supported by the U.S. Department of Interior Water Resources Research Institute Program and by the Salt River Valley Water Users' Association. Contents of this publication do not necessarily reflect the views and policies of the United States Department of the Interior, nor does mention of trade names or commercial products constitute their endorsement by the United States Government.

References

Baker, V.R. (1984) Flood sedimentation in bedrock fluvial systems. In: Sedimentology of Gravel and Conglomerates (ed. by E.H. Koster and R.J. Steel), 87-98. Canadian Soc. of Petroleum Geologists Memoir 10, Calgary, Alberta, Canada.

Baker, V.R. (in press) Paleoflood hydrology and extraordinary flood events. J. Hydrol.

Baker, V.R., Kochel, R.C. & Patton, P.C. (1979) Long-term flood frequency analysis using geological data. In: The Hydrology of Areas of Low Precipitation (Proc. Canberra Symp., December 1979), 3-9. IAHS Public. no. 128.

Baker, V.R., Kochel, R.C., Patton, P.C. & Pickup, G. (1983) Paleohydrologic analysis of Holocene flood slack-water sediments. In: Modern and Ancient Fluvial Systems: Sedimentology and Processes (ed. by J. Collinson & J. Lewin), 229-239. International Assoc. of Sedimentologists Spec. Publ. no. 6.

Baker, V.R., Pickup, G. & Polach, H. (1985) Radio carbon dating of flood events, Katherine Gorge N. Territory, Australia Geology Vol. 13, 344-347.

Baker, V.R., Pickup, G. & Webb, R.H. (in press) Paleoflood hydrologic analysis at ungaged sites, central and northern Australia. In: Flood Frequency and Risk Analysis (ed. by V. Singh). D. Reidel Publ. Co. Dortrech, Holland.

Chatters, J.C. & Hoover, K.A. (1986) Changing late Holocene flooding frequencies in the Columbia River, Washington. Quatern Res. 26, 309-320.

Ely, L.L. & Baker, V.R. (1985) Reconstructing paleoflood hydrology with slackwater deposits: Verde River, Arizona. Phys. Geogr. 6(2), 103-126.

Hirschboeck, K.K. (1985) Hydroclimatology of flow events in the Gila River basin, central and southern Arizona. Ph.D. dissertation, University of Arizona, Tucson, Arizona, USA.

Hirschboeck, K.K. (in press) Flood hydroclimatology. In: Flood Geomorphology (ed. by V.R. Baker, R.C. Kochel & P.C. Patton). John

Wiley and Sons, Inc., N.Y.

Hosking, J.R.M. & Wallis, J.R. (1986) Paleoflood hydrology and flood frequency analysis. Water Resour. Res. 22(4), 543–550.

Knox, J.C. (1985) Response of floods to Holocene climate change in the upper Mississippi Valley. Quatern. Res. 23, 287–300.

Kochel, R.C. & Baker, V.R. (1982) Paleoflood hydrology. Science 215(4531), 353–361.

Kochel, R.C. & Baker, V.R. (in press) Paleoflood analysis using slackwater deposits. In: Flood Geomorphology (ed. by V.R. Baker, R.C. Kochel & P.C. Patton). John Wiley and sons, Inc., N.Y.

Kochel, R.C., Baker, V.R. & Patton, P.C. (1982) Paleohydrology of southwestern Texas. Water Resour. Res. 18(8), 1165–1183.

Masse, W.B. (1981) Prehistoric irrigation systems in the Salt River Valley, Arizona. Science 214, 408–415.

O'Connor, J.E., Webb, R.H. & Baker, V.R. (1986) Paleohydrology of pool and riffle pattern development, Boulder Creek, Utah. Geol. Soc. America Bull. 97, 410–420.

Partridge, J.B. & Baker, V.R. (1987) Paleoflood hydrology of the Salt River, Arizona. Earth Surf. Processes and Landforms. Vol. 12, 109–125.

Patton, P.C. (in press) The geomorphic response of streams to floods in the glaciated terrain of southern New England. In: Flood Geomorphology (ed. by V.R. Baker, R.C. Kochel & P.C. Patton). John Wiley and Sons, Inc., N.Y.

Patton, P.C. & Dibble, D.S. (1982) Archeologic and geomorphic evidence for the paleohydrologic record of the Pecos River in west Texas. American J. Sci. 282, 97–121.

Stedinger, J.R. & Baker, V.R. (1987) Surface water hydrology: historical and paleoflood information. Reviews of Geophysics. Vol. 25, 119–124.

Stedinger, J.R. & Cohn, T.A. (1986) The value of historical and paleoflood information in flood frequency analysis. Water Resour. Res. 22(5), 785–793.

Taylor, R.E., Donahue, D.J., Zabel, T.H., Damon, P.E. & Jull, A.J.T. (1984) Radiocarbon dating by particle accelerators: an archaeological perspective. In: Archaeological Chemistry –– III (ed. by J.B. Lambert), 333–356. American Chemical Society Advances in Chemistry Series, No. 205.

U.S. Water Resources Council (1981) Guidelines for determining flood flow frequency. Bull. No. 17B, U.S. Water Resources Council, Washington, D.C., USA.

Webb, R.H. (1985) Late Holocene flooding on the Escalante River, southcentral Utah. Ph.D. dissertation, University of Arizona, Tucson, Arizona, U.S.A.

Webb, R.H., O'Connor, J.E. & Baker, V.R. (in press) Paleohydrologic reconstruction of flood frequency on the Escalante River, southcentral Utah. In: Flood Geomorphology (ed. by V.R. Baker, R.C. Kochel & P.C. Patton). John Wiley and Sons, Inc., N.Y.

Webb, R.H. & Smith, S.S. (1986) Evolution of arroyos in southern Utah. Geol. Soc. America Abstracts with Programs 18(6), 783.

The Influence of Climate Change and Climatic Variability on the Hydrologic Regime and Water Resources (Proceedings of the Vancouver Symposium, August 1987). IAHS Publ. no. 168, 1987.

Variability in periodicities exhibited by tree ring data

A. Ramachandra Rao
School of Civil Engineering
Purdue University
W. Lafayette, IN 47907, USA
A. Durgunoglu
Illinois State Water Survey Division
Department of Energy and Natural Resources
2204 Griffith Drive
Champaign, Il 61820, USA

ABSTRACT   Tree ring data have been widely used to study climatic variability. An interesting characteristic of the results from these studies is the apparent "periodicity" in these time series. The regional consistency in the periodicities in tree-ring series is investigated in this study. If several series from a region have the same "periodic" behaviour then the argument for systematic climatic variability in that region would be quite strong. Consequently, investigation of "periodicities" in tree-ring data in a region is important.
Four tree-ring series in Salt and Verde river basins in Arizona are analyzed in this study. The Blackman-Tukey and a variant of the maximum entropy spectral analysis proposed by Marple are used. The data exhibit consistent periodic behaviour although the periodicity is weak.

Introduction

Because of the lack of long term climatologic and hydrologic data, proxy data such as tree-ring and varve series are being analyzed to draw inferences about climatic variability. Although there has been a gradual accumulation of evidence during the past two decades that systematic variations in climate exist, these variations have not followed general patterns which may be linked to causal physical mechanisms.
Although there are well documented "periods" in proxy climatic series, the regional consistency in these periodicities is an important characteristic. If several time series from a given region ex- hibit similar behaviour then the support for regional climatic change would be quite strong. Otherwise it would be difficult to use the evidence based on proxy data to support hypotheses of systematic variability in climate.
A related question is of course about the strength of these systematic climatic changes. Although there have been numerous claims about periodicities in these proxy climatic series, very few of the periodicities are strong enough to be used for prediction. The data are noisy and extraction of a weak signal from a noisy background is

a well known formidable problem.

Four tree-ring series from the Salt and Verde river basins in
Arizona are analyzed in this study to address the above two aspects.
If the spectral characteristics of these four series are similar then
they would indicate a regional climatic change in these watersheds.
If the "periodicities" in these series are strong then they may be
used to predict climatic changes in the future.

Spectral analysis, a technique formalized by Blackman and Tukey
and developed further by many other investigators (Jenkins and Watts,
1968) has many well known problems. In order to overcome these, many
parametric spectral analysis methods have been developed recently.
One such method developed by Marple (1980) is a variant of the
maximum entropy spectral analysis developed by Burg (1967). The
Blackman-Tukey and Marple's techniques of spectral analysis are used
in the present study.

The paper is organized as follows. Marple's spectral analysis
method is discussed in the next section followed by a brief
description of the data used. The results of spectral analysis are
presented and discussed in the last section.

Marple's method of spectral analysis

Marple's method is a recursive least squares (LS) AR parameter estimation technique. An Mth order AR model is fitted to the data
sequence $x_1, x_2, \ldots x_N$. The AR model is driven by a white noise ε_n and
has the form:

$$x_n = - \Sigma \, a_{M,m} \, x_{n-m} + \varepsilon_n \qquad n = m+1, \ldots, N \qquad (1)$$

where $a_{M,m}$ is the mth AR parameter of the Mth order AR process.

In his study, Burg(1967) used a constrained LS method to estimate
the AR parameters that minimize the sum of squares of the backward
and forward model errors with the constraint that the entropy in the
data be a maximum. Sometimes Burg's method exhibits poor spectral
resolution (ability to resolve two close spectral peaks) and
frequency estimation bias (shifting of spectral peaks from their true
locations) and spectral line-splitting. Spectral line-splitting is
the occurrence of two or more closely-spaced peaks in an AR spectral
estimate where only one spectral peak should be present. To avoid
spectral line splitting and shifts in spectral peaks which may occur
with Burg's algorithm (especially with large M), Marple developed the
LS method without the constraint in which backward and forward prediction errors are used. In Marple's method the matrix inversions of
Burg's method are avoided by exploiting the special structure of the
correlation matrix. Numerical illconditioning and matrix sin-
gularities are also tested in Marple's algorithm. The model order is
selected by using the relative change in prediction error variance in
successive model orders and by the ratio of the prediction error
variance to the variance of the signal. The algorithm stops either
when the prediction error variance differences in two successive
models are smaller than a pre-specified value or when a pre-specified

model order is reached.

Once the model order is determined and the AR parameters are estimated, the spectral estimate, $\hat{S}_x(\omega)$, is calculated by using Equation 2.

$$\hat{S}_x(\omega) = \frac{\hat{S}_e(\omega)}{\left| 1 + \sum_{k=1}^{m} \hat{a}_k \, e^{-ik\omega} \right|^2} \tag{2}$$

Data used in the study

Four annual tree-ring series from the Salt and Verde river basin are analyzed in this study. The data are taken from Smith(1981) and the location of tree-ring sites are shown in Figure 1. The series analyzed are called Showlow Ponderosa Pine, Nutrioso Pinyon Pine, Red Mountain Canyon Pine and Slate Mountain Ponderosa Pine. None of the tree-ring series indicate any obvious periodicities, as shown in Figure 2.

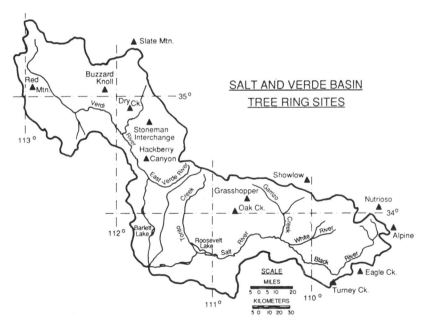

Figure 1   Location of tree-ring sites in the Salt and Verde River basins, Arizona (based on Smith, 1981).

The name, type, period and the source of the data used in the study are listed in Table 1. The elementary statistics of the data are given in Table 2. All the data are normalised so that the mean is zero and variance unity.

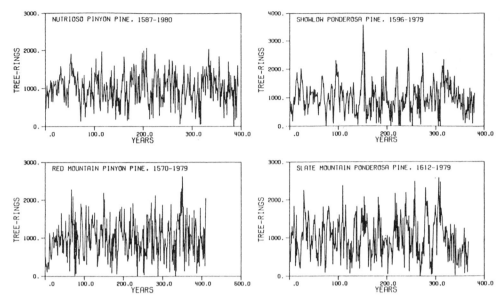

Figure 2   Annual tree-ring thickness series.

Table   1   Data used in the study

Data	Number	Period	Source
Nutrioso Pine	394	1587-1980	Smith(1981)
Showlow Pine	380	1596-1979	Smith(1981)
Red Mtn. Pine	410	1570-1979	Smith(1981)
Slate Mtn. Pine	368	1612-1979	Smith(1981)

Table   2   Statistical characteristics of data used in the study

Data	Mean	Std. Dev.	Coeff. of Variation	Skewness	Kurtosis
Nutrioso	998.5	412.1	0.41	0.03	-0.34
Showlow	1012.0	558.0	0.55	0.69	1.22
Red Mtn.	983.1	490.6	0.50	0.27	-0.25
Slate Mtn.	993.3	528.3	0.53	0.35	-0.37

The correlograms of tree-ring data are shown in Figure 3. All the tree-ring series show significant autocorrelation values at the first few lags. At higher lags the autocorrelation values stay within the two standard error limits. Only the Showlow tree-ring series indicates some periodicities around lag 23.

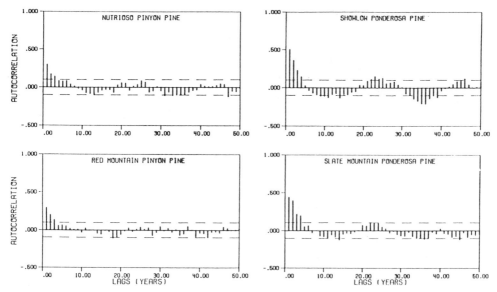

Figure 3   Correlogram of tree-ring thickness data.

Results and discussion

As mentioned earlier, two types of spectra are estimated using the given data. The first is the Blackman-Tukey spectra estimated by using a maximum lag of 0.1N, where N is the number of observations and a Hamming window. The details of Blackman-Tukey spectral estimation are well known (Jenkins and Watts, 1968) and hence are not discussed here further.

Selection of models for the Marple's method

The second type is the maximum entropy spectrum estimated by using Marple's method. In order to use this method and Equation 2, the maximum order M must be estimated first. Three criteria for order selection, namely Akaike Information (Akaike, 1969) the Bayesian (Kashyap, 1977) and the Hannan-Quinn (Hannon and Quinn, 1979) criteria are used in this study.

The statistics used in these criteria are given below where k is the number of parameters in the model and N the number of observations:

$$AIC_k = N * \ln(\rho) + 2k \qquad \text{Akaike Information Criterion}$$

$$B_k = N * \ln(\rho) + 2\ln(N) \qquad \text{Bayesian Criterion}$$

$$H_k = N * \ln(\rho) + 4k*\ln(\ln(N)) \qquad \text{Hannan-Quinn Criterion}$$

The model which gives the minimum criterion value is selected ·as the optimum model for the method tested. The spectral estimate of each series is then obtained by using the optimum model.

The residuals from the selected model are tested by using Portmanteau goodness of fit test. If p is the number of parameters in the model, L is the maximum lag, N is the sample size and r_k is the correlogram of the residuals, then the statistic Q is given by:

$$Q = N \sum_{k=1}^{L} r_k^2 \qquad (3)$$

Q is approximately chi-squared distributed with L-p degrees of freedom. The lack of correlation in residuals is tested by using Q and the chi-squared value $\chi^2(L-p)$.

The models selected in the study are listed in Table 3. The first column in Table 3 identifies the series, the second column shows the candidate models. In the next three columns AIC_k, B_k, and H_k are given. The last column in Tables 3 shows the white noise probability of the residuals.

Table 3   Models Selected from Marple's Method

Data	Model	AIC_k	B_k	H_k	W.N.P.
Nutrioso	AR(6)	4750	4774*	4781*	0.67
Showlow	AR(4)	4740*	2756	4761	0.23
Red Mtn.	AR(7)	5084	5112	5121	0.88
Slate Mtn.	AR(4)	4556*	4572*	2577	0.99

* = minimum criterion value

The models selected for Marple's method are: AR(6) for Nutrioso series, AR(4) for Showlow series, AR(7) for Red Mountain series and AR(4) for Slate Mountain series.

Results of spectral analysis

The estimated spectra for the tree-ring series are shown in Figure 4 through 6.

Nutrioso Pine: The BT spectra of Nutrioso series show a significant peak around 25 year as shown in Figures 4 and 5. The M(6) spectrum of Nutrioso series does not show any significant spectral peaks.

Showlow Series: The BT(40) and BT (60) spectra of the Showlow series indicate spectral peaks at 27 and 24 year periods, respectively, as shown in Figures 4 and 5. The M(4) spectrum does not show any significant peaks.

Red Mountain series: No significant spectral peaks are observed for the Red Mountain series, but some power accumulation is indicated at the low frequencies.

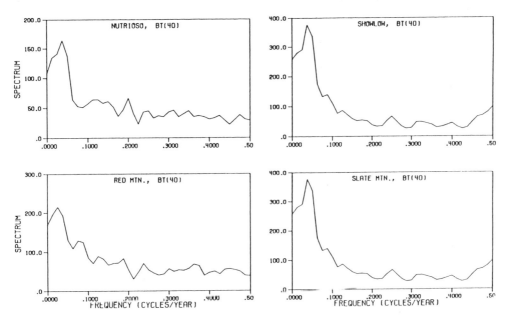

Figure 4   Spectra of tree-ring series estimated by using
B-T method (maximum lags used = 0.1N).

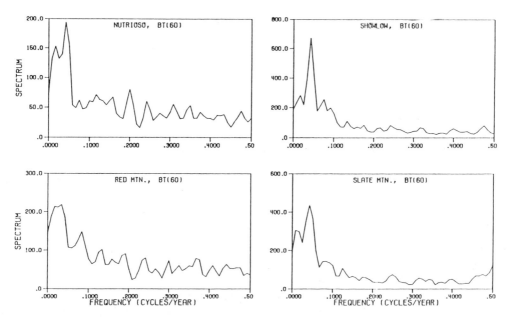

Figure 5   Spectra of tree-ring series estimated by using
B-T method (maximum lags used = 0.15N).

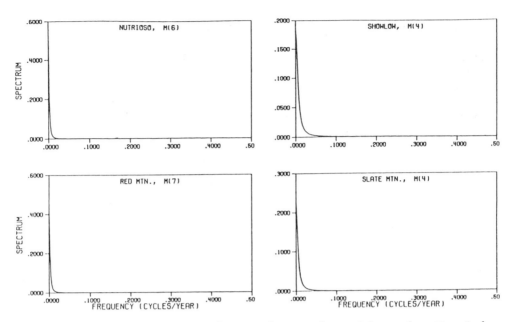

Figure 6   Spectra of tree-ring series estimated by using Marple's
method.

Slate Mountain Series: The BT(40) and BT(60) spectra of the Slate
Mountain series show spectral peaks at 24 and 27 year periods as
shown in Figures 4 and 5. The M(4) spectrum does not indicate any
significant peaks but shows some power accumulation at the low
frequencies.

Conclusions

Except for the Red Mountain series, all the tree-ring series indicate
spectral peaks around a 25 year period. However, the noise level in
Nutrioso Pine and Slowlow Series is rather high. The results from
Marple's method are not very good in delineating the dominant
periodicities. Because of this poor performance, other spectral
methods were used and these confirmed the periodicities in the BT
method. These details are found in Durgunoglu and Rao (1985).

    In conclusion, the variability in climate in the Salt and Verde
River basins as revealed by tree-ring series is consistent over the
basins. This variability has a "periodicity" of about 25 years. The
strength of variation in climate in these basins is rather weak.

REFERENCES

Akaike, H. "Fitting A.R. Models for Prediction", Ann. Inst. Stat.
    Math, 21, pp. 243-247, 1969.

Burg, J.P., "Maximum Entropy Spectral Analysis", Proc. 37th Meeting, Society of Exploration Geophysics, Oklahoma City, Okla. 1967.

Durgunoglu, A. A.R. Rao, "ARMA Spectral Analysis of Hydrologic Time Series", Tech. Rept. CE-HSE-85-13, School of Civil Eng. Purdue University, W. Lafayette, IN 47907, pp 207, 1985.

Hannon, E. J. and B.G. Quinn, "The Determination of the Order of an Autoregression", Jour.Roy.Stat.Soc.,B.,41(2),pp 190-195, 1979.

Jenkins, G.M. and D.G. Watts, "Spectral Analysis and its Applications", HoldenDay, San Francisco,CA, 1968.

Kashyap, R.L., "A Bayesian Comparison of Different Classes of Dynamic Models Using Empirical Data", IEEE Trans. on Aut. Control, AC-22, No.5, pp. 715-727, 1977.

Marple, S.L., Jr., "A New Autoregressive Spectrum Analysis Algorithm", IEEE Trans, A. Speech and Signal Processes, Vol. ASSP-28, 1980, pp. 441-454.

Smith, L.P., "Long Term Streamflow Histories of the Salt and Verde Rivers, Arizona, as Reconstructed From Tree Rings", Lab. of Tree Ring Research, Univ. of Arizona, Tucson Arizona, 85721, 1981.

The Influence of Climate Change and Climatic Variability on the Hydrologic Regime and Water Resources (Proceedings of the Vancouver Symposium, August 1987). IAHS Publ. no. 168, 1987.

Paleorecharge, climatohydrologic variability, and water-resource management

William J. Stone
New Mexico Bureau of Mines and Mineral Resources
Campus Station, Socorro, NM 87801, USA

ABSTRACT   Changes in recharge through time document climatohydrologic variability. Modern and ancient recharge rates can be estimated and dated using a chloride mass-balance approach.   An area studied in South Australia was shown to have been wetter before 13,500-16,000 years ago. In west central New Mexico, conditions were wetter prior to 7,000-17,000 years ago.   Data from northwestern New Mexico show it was wetter there at the same general time and at younger invertvals ending 1,000, 900, and 400 years ago.   Dry periods preceded these wet intervals (occurring 1,000-2,000 and 10,000-14,000 years ago).   Most of these times correspond to paleoclimatohydrologic regimes recognized in previous studies.   These differences in recharge rates are important in water-resource management and waste-disposal planning.   The rates can be used in modeling and to avoid groundwater mining.   Higher previous recharge rates may be taken as worst-case values in designing waste-disposal facilities.
Note: A French abstract of this paper can be found in the back of the text.

Introduction

An area is characterized by its meteorologic and site characteristics. Specific combinations of meteorologic parameters, such as temperature and precipitation, define climates.   Unique combinations of site variables (geology, soils, topography, and vegetation) define landscape settings.

The hydrologic balance of an area is controlled by both its climate and landscape setting.   If one of these controls is fixed, the hydrologic balance will vary as the other is varied.   That is to say, for a given climate, the hydrologic balance varies only with landscape setting.   Likewise, for the same landscape setting, hydrology varies only with climate.

In a setting that has been reasonably stable in the Quaternary, it should be possible to assess climatic variability by comparing modern and paleo values for any of the hydrologic-balance parameters.   But what parameters should be used?   How can modern and paleo values for the parameter be obtained?

Recent studies, using a chloride mass-balance approach, have yielded modern and paleorecharge values in areas of Australia and the American Southwest.   Results document both the timing and trends of climatohydrologic variability.   The purposes of this paper are to 1) review the cloride mass-balance method, 2) describe its application to paleorecharge, 3) present case histories of its use, and 4) discuss implications for water-resource management and waste-disposal planning.

Chloride method

Chloride is a natural tracer that is conservative with respect to water movement. It has been used in various ways by previous workers (for exam.le, Allison & Hughes, 1978; Edmonds & Walton, 1980; Peck et al., 1981; Domenico & Robbins, 1985; and Claassen et al., 1986). Most of these studies employ a mass-balance approach.

a) Modern recharge

Present day recharge can be estimated from the relationship

$$P * Clp = R * Clsw \quad \text{(Allison \& Hughes, 1978),} \tag{1}$$

where P = mean annual precipitation (mm/yr), Clp = mean annual chloride content of precipitation and dry fall (mg/L), R = recharge (mm/yr), and $Clsw$ = mean soil-water chloride content (mg/L). Rewriting for recharge, this becomes

$$R = \frac{Clp}{Clsw} * P \tag{2}$$

This gives a velocity style, local or point recharge value.

P and Clp are either obtained from the literature or determined from samples from the study area. $Clsw$ is determined from samples of the unsaturated zone, taken with a hollow-stem auger rig. More specifically, chloride content of the soil water is determined from the amount of chloride in a salt/water extract, the amount of water added in extraction, and the original moisture content of the soil. $Clsw$ is calculated over an interval representing equilibrium on a chloride vs depth profile for the hole.

The method assumes 1) recharge is from precipitation, 2) recharge is by piston flow, 3) chloride is from precipitation (and dryfall), and 4) precipitation and chloride content of precipitation have been more or less constant through the period represented by the sampled interval. Because these are not always met in practice, recharge values obtained are considered estimates. However, where results have been checked by other methods, they were found to be reasonable (Stone, 1984a, and 1986a).

Chloride vs depth profiles are normally characterized by an increase in chloride content to a peak through the root zone. Below the peak, chloride content either remains constant to water table or drops off to a value that is then maintained to the water table (Figure 1a). Such profiles represent more or less constant recharge conditions during the time represented by the soil water sampled.

(a)

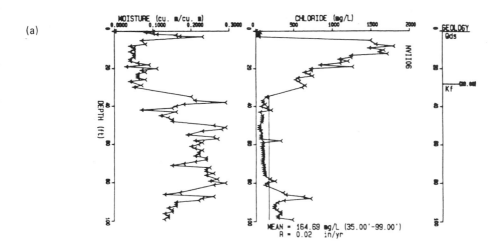

(b)

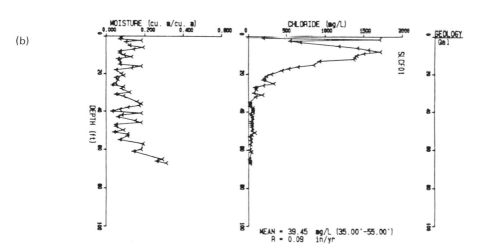

Figure 1   Sample profiles of results: a) with chloride content more
            or less constant with depth (Hole 6, Navajo Mine), b) with
            chloride decreasing with depth (Hole 1, Salt Lake coal
            field).

b) Paleorecharge

By contrast, some profiles show a continuous decrease in chloride
content below the peak (Figure 1). In this case, one of two
conditions can be assumed to have occurred. Either nonpiston flow
has occurred or recharge was greater in the past. In nonpiston flow,
fresh water short circuits the profile via fractures, root tubes, or
burrows, and dilutes the chloride content of the soil water. If
recharge was higher in the past, and decreased slowly, the soil-water
chloride would decrease steadily downhole. If only one profile has
this anomalous shape, nonpiston flow is probably to blame. However,

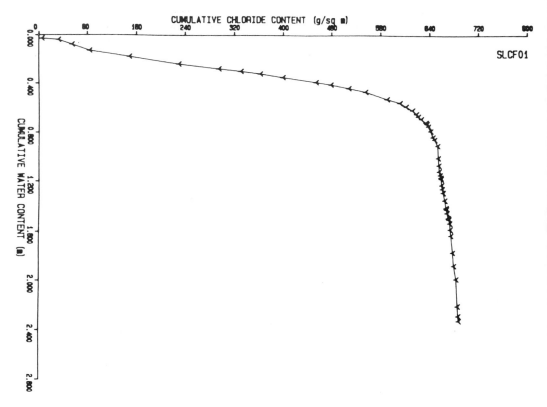

Figure 2   Sample cumulative plot for profile in Figure 1b (Hole 1,
           Salt Lake coal field).

if all profiles in the study area have the anomalous shape, higher
recharge is a more likely explanation.

    To evaluate the possibility that paleorecharge was higher, a
cumulative chloride vs cumulative water plot is made (Figure 2).   If
no change in recharge rate has occurred, this should give a straight
line. However, if recharge has been higher, the curve should be
convex upward, the break in slope marking the time when drying to
modern conditions set in. Similarly, a concave upward slope indicates
recharge was lower.

    These curves may be divided into straight-line segments and the
age of their end points estimated using

$$A = \frac{Clcum}{(Clp * P)} \tag{3}$$

where A = age (yrs BP) and Clcum = cumulative chloride content at
that point.   Paleorecharge for such segments of the curve can be
estimated by the recharge equation, using modern values for Clp and p
or adjusted values as may be possible from available paleoclimate
evidence.

Table 1 Summary of results from case histories

Hole	Setting	Interval (yrs BP)	Recharge (mm/yr)
MURRAY BASIN			
1	undisturbed calcrete	16,000 - Present 30,000 - 16,000	0.10 0.17
2	older sinkhole	13,500 - Present 30,000 - 13,500	0.07 0.09
SALT LAKE COAL FIELD			
1	thick alluvium	9,000 - Present 10,270 - 9,000	0.08 1.52
2	ephemeral lake	7,130 - Present 12,594 - 7,130	0.08 0.43
3	thin alluvium	17,356 - Present 21,190 - 17,356	0.08 0.38
4	bedrock/ grass	4,259 - Present 5,208 - 4,259	0.05 0.51
5	bedrock/ grass	12,937 - Present 15,315 - 12,937	0.05 0.15
NAVAJO MINE			
5	upland flat	404 - Present 1,009 - 404 1,931 - 1,009	1.27 3.05 0.51
6	upland flat	6,507 - Present 9,560 - 6,507 14,210 - 9,560	0.51 0.76 0.25
26	arroyo terrace	925 - Present 1,329 - 925	2.29 2.29
27	upland flat	1,690 - Present 2,426 - 1,690	0.51 2.92

Case histories

These methods have been applied in three study areas: one in South
Australia (Allison et al., 1985) and two in New Mexico (Stone, 1984a,
b, and 1986).   In all cases, paleorecharge values indicate climatohydrologic conditions varied through time (Table 1).

a) Murray basin

The area studied in South Australia lies 130 km northeast of Adelaide, at the western edge of the shallow, early Tertiary structural depression known as the Murray Basin (Figure 3a). Tertiary marine carbonate rocks as well as Quaternary marine, lacustrine and eolian deposits and calcrete crop out in the area. The climate is semiarid with a mean annual precipitation of 300 mm and a potential evaporation rate of 1800 mm. Greatest precipitation occurs in winter (June – August in this southern hemisphere location). Clp = 4.2 mg/L.

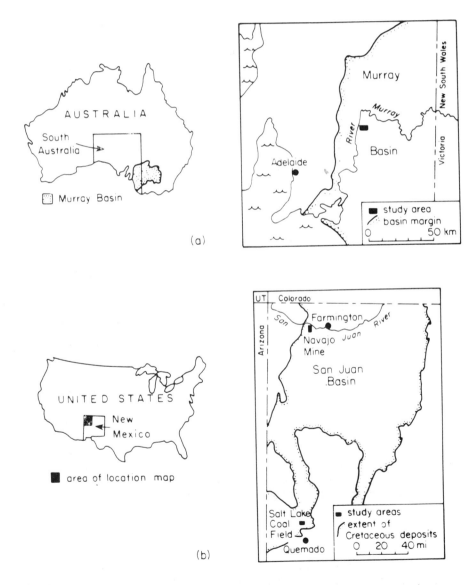

Figure 3 Location of study areas: a) in Murray basin, b) in San Juan basin.

Because of anomalous chloride vs depth profiles, cumulative plots were prepared for two holes on a bare calcrete flat, one where a sinkhole had formed through the calcrete and the other in an undistrubed calcrete setting. The hole in the sinkhole was cored to a depth of 24.3 m. The cumulative plot showed three straight-line segments, with breaks in slope at 30,000 yrs BP and 13,500 BP. Between 30,000 and 13,500 yrs age, recharge was 0.09 mm/yr. From 13,500 yrs ago to present, recharge has been 0.07 mm/yr.

The hole in undisturbed calcrete was cored to a depth of 28.6 m. The cumulative plot showed soil water at this depth to be 30,000 yrs old. A distinct break in slope occurs at 16,000 yrs ago. Recharge between 30,000 and 16,000 yrs ago was higher than at present at 0.2 mm/yr. Between 16,000 yrs ago and present, recharge has been 0.1 mm/yr.

Results from both holes are consistent with published paleoclimate observations. The recharge changes indicated coincide generally with possible climate changes described by Lawrence (1976), Macumber (1976), and Galloway & Kemp (1981).

b) Salt Lake coal field

The first area in New Mexico where paleorecharge rates were determined by the chloride method is located in the west-central part of the state, approximately 25 km north of the village of Quemado (Figure 3b). Geologically, the area lies in a southwestern extension of the San Juan Basin, a Laramide depression at the eastern edge of the Colorado Plateau. Cretaceous coal measures, Tertiary volcanics, and Quaternary alluvium lie at the surface. The climate is semiarid with a mean annual precipitation of 245 mm and a potential evaporation rate of 787 mm/yr. Half of the annual precipitation occurs in the period July through September. Clp used = 0.375 mg/L. Holes were 12.19-20.42 m deep.

As all profiles showed an increase in chloride with depth, cumulative plots were prepared. All plots indicate higher recharge in the past. The timing of the major decrease in recharge ranges from 7,000 to 17,000 yrs ago.

Results compare well with published paleoclimate reconstructions. Markgraf et al. (1983) determined that conditions in a nearby pluvial lake (San Agustin Lake) were more mesic between 10,000 and 8,500 yrs BP. At 8,000 yrs BP drier conditions set in and by 5,000 yrs ago the lake was desiccated. Similar chronologies have been obtained from packrat middens (Van Devender, 1977; Van Devender & Spaulding, 1979) and soil ages of shoreline features of another pluvial lake in southwestern New Mexico (Fleischhauer & Stone, 1982).

c) Navajo Mine

The other New Mexico case history is also from the San Juan Basin. The Navajo Mine is located approximately 30 km southwest of Farmington, in northwestern New Mexico (Figure 3b). Cretaceous coal measures as well as Quaternary alluvial and eolian deposits dominate the landscape. The climate is arid, with a mean annual precipitation

of 145 mm and a potential evaporation rate of 1,422 mm/yr.
Precipitation occurs mainly in the months of August through October.
Clp = 0.60 mg/L. Holes were 5.79-30.18 m deep.

Cumulative plots for four sites indicate variation in recharge
through time. Three of the sites represent upland flats (eolian sand
over coal measures) and the other is on the terrace of an ephemeral
stream (alluvium).

Plots for two of the upland flat sites differ from those of other
sites here and at the Salt Lake Coal Field in that they show an
increase in recharge above the bottom of the profile. In one, this
occurred at 9,500 yrs ago. This was followed by a decrease at 6,500
yrs ago, similar to the drying at the Salt Lake Coal Field. At the
other site, the increase occurred 1,000 yrs ago. Latest drying here
occurred at 400 yrs ago.

The time of drying or recharge decrease at the other sites was
intermediate, hovering about approximately 1,000 yrs BP. At the
upland site this was 1,690 yrs BP, whereas, at the terrace site it
was 925 yrs BP.

Cumulative plots for a fifth site (Badlands) gave a more or less
straight line, indicating little change in recharge there.

Validity of results

One might expect a decrease in recharge at the end of the Pleistocene
(10,000 yrs ago) as that time is generally conceded to have been
wetter. However, ages of recharge decrease at Navajo Mine are
considerably younger than this. At one site (Hole 6) the decline was
in the mid Holocene. Work on ancient lakes in the Animas Valley of
southwestern New Mexico (Fleischhauer & Stone, 1982) indicates that
they formed in latest Pleistocene and persisted, or returned at lower
stands, into the Holocene (3,000-6,000 yrs BP). Timing of the
indicated decrease in recharge at Hole 6 coincides fairly well with
the maximum age for the youngest lake in the Animas Valley. At
another site (Hole 27) recharge decreased in the year 292 or about
the time pit houses were reportedly being built on the floor of Chaco
Canyon (Love, 1977). At Hole 26 the decrease in recharge would have
occurred in the year 1060 or 10 years after major construction of
dwellings in Chaco Canyon began and 67 years before it ceased. One
theory as to the cause for abandonment of Chaco Canyon is that it was
prompted by a climatic change which affected agriculture in the
region. A drying trend would be such a change. The most recent
decrease (Hole 5) would have taken place in the year 1581 or 41 yrs
after Coronado visited New Mexico.

Modern recharge rates at Navajo Mine have been checked by stable
isotopes and tritium (Stone, 1986). Stable-isotope data corroborate
the chloride results, but tritium data do not. This is apparently
due to movement of tritium in the vapor phase in this arid region.
By contrast, chloride moves only in the liquid phase of water.

A study employing the chloride method in southern Nevada provides
an opportunity to check dates obtained by the cumulative chloride
method (Fouty, in progress). In this study, Quaternary ash beds
encountered in sampling were dated for comparison with ages obtained
for soil water at that point on the profile.

Implications

An appreciation for the variation in recharge rates or climate is critical in water-resource management. In a given area the water resources may be the result of higher recharge rates during the Pleistocene or early Holocene. To avoid ground-water mining, withdrawal rates should be regulated in accordance with the modern (reduced) recharge rates. Appropriate recharge estimates are also essential in modeling modern and ancient hydrologic systems.

Recharge is also important in siting waste-disposal facilities. Modern recharge rates and climatic conditions are useful in assessing the volume of water that may come in contact with waste as well as the normal rate of movement in the unsaturated zone. Paleorecharge rates, especially if higher, can serve as worst-case estimates of climatohydrologic variability.

The modern and paleorecharge values obtained by the chloride mass-balance method are only estimates. However, they seem a reasonable indication of climatohydrologic variability where checked by other methods or compared with available paleoclimate data.

References

Allison, G.B., & Hughes, M.W. (1978) The use of environmental chloride and tritium to estimate total recharge to an unconfined aquifer. Aust. J. Soil, v. 16, p. 181-195.

Allison, G.B., Stone, W.J., & Hughes, M.W. (1985) Recharge through karst and dune elements of a semiarid landscape. J. Hydrol., v. 76, p. 1-25.

Claassen, H.C., Reddy, M.M., & Halm, D.R. (1986) Use of the chloride and tritium to estimate total recharge to an unconfined year case study in the San Juan Mountains, Colorado, U.S.A. J. Hydrol., v. 85, p. 49-71.

Domenico, P.A., & Robbins, G.A. (1985) The displacement of connate water from aquifers: Geol. Soc. Am. Bull., v. 96, p. 328-355.

Edmonds, W.M., & Walton, N.R.G. (1980) A geochemical and isotopic approach to recharge evaluation in semi-arid zones: Proc. Advisory Group Meeting on the Application of Isotope Techniques in Arid Zone Hydrology, Vienna, 1978, IAEA, p. 47-68.

Fleischhauer, H.L., Jr., and Stone, W.J. (1982) Quaternary geology of Lake Animas, Hidalgo County, New Mexico: New Mexico Bur. Mines and Min. Res., Circ. 1974, 25 p.

Fouty, Suzanne (in progress) Containment of low-level nuclear waste in the unsaturated zone of alluvial fill in arid environments: M.S. thesis, University of Arizona.

Galloway, R.W., and Kemp, E.M. (1981) Late Cainozoic environments in Australia. Ecological Biogeography of Australia, v. 1, p. 53-80.

Lawrence, C.R. (1976) in Geology of Victoria, J.G. Douglas and J.A. Ferguson (eds.), Geol. Soc. Aust., Spec. Pub. no. 5, p. 276-288.

Love, D.W. (1977) Dynamics of sedimentation and geomorphic history of Chaco Canyon National Monument, New Mexico. New Mexico Geol. Soc. Guidebook, 28th Field Conf., p. 291-300.

Macumber, P.G. (1976) in Geology of Victoria, J.G. Douglas and J.A. Ferguson (eds.), Geol. Soc. Aust., Spec. Pub. no. 5, p. 288-290.

Markgraf, V., Bradbury, J.P., Forester, R.M., McCoy, W., Singh, G., and Sternberg, R. (1983) Paleoenvironmental reassessment of the 1.6-million-year-old record from San Agustin Basin, New Mexico. New Mexico Geol. Soc. Guidebook, 34th Field Conf., p. 291-297.

Peck, A.J., Johnston, C.D., and Williamson, D.R. (1981) Analysis of solute distributions in deeply weathered soils. Agric. Water Management, v. 4, p. 83-102.

Stone, W.J. (1984a) Preliminary estimates of Ogallala aquifer recharge using chloride in the unsaturated zone, Curry County, New Mexico. Proc., Ogallala Aquifer Sympos. II, Lubbock, p. 376-391.

Stone, W.J. (1984b) Preliminary estimates of recharge at the Navajo Mine based on chloride in the unsaturated zone. New Mexico Bur. Mines and Min. Res., Open-file Rept. 213, 60 p.

Stone, W.J. (1984c) Recharge in the Salt Lake coal field based on chloride in the unsaturated zone: New Mexico Bur. Mines and Min. Res. Open-file Rept. 214, 64 p.

Stone, W.J. (1986a) Comparison of ground-water recharge rates based on chloride, stable-isotope, and tritium content of vadose water at the Navajo Mine, Northwest New Mexico (abs.). New Mexico Geol. Soc., Ann. Spring Meeting, p. 39.

Stone W.J. (1986b) Phase-II recharge study at the Navajo Mine based on chloride, stable isotopes, and tritium in the unsaturated zone. New Mexico Bur. Mines and Min. Reg., Open-file Rept. 216, 244 p.

Van Devender, T.R. (1977) Holocene woodlands in the southwestern deserts. Science, v. 198, p. 189-192.

Van Devender, T.R., and Spaulding, W.G. (1979) Development of vegetation and climate in the southwestern United States. Science, v. 204, p. 701-710.

The Influence of Climate Change and Climatic Variability on the Hydrologic Regime and Water Resources (Proceedings of the Vancouver Symposium, August 1987). IAHS Publ. no. 168, 1987.

Lake ice formation and breakup as an indicator of climate change: potential for monitoring using remote sensing techniques

J.A. Maslanik & R.G. Barry
Cooperative Institute for Research in
Environmental Sciences and Department of Geog.
University of Colorado, Boulder, CO 80309, USA

ABSTRACT: Freezeup and breakup dates of lakes, which can serve as indicators of climatic change, are analyzed by remote sensing techniques. Visible-wavelength images from the DMSP satellite are used. A comparison is made of the dates of freezeup and breakup of lakes in Finland and Canada as determined by ground observers and by visual imagery interpretation. For breakup, the dates from image interpretation are several days later than breakup dates from ground observations; the differences decrease inland and vary with lake size and shape. Cloud cover in autumn in both regions prevents the use of satellite imagery to assess lake freezeup. The results emphasize the possibilities for satellite data analysis in the study of regional climate variations.

Indication d'un changement de climat par gel et dégal des lacs: potentiel de contrôle par méthodes de télédétection

RESUME Les dates du gel et dégel des lacs, qui peuvent servir comme indicatrices d'un changement de climat, sont analysées en employant des méthodes de télédétection. Les images dans le visible du satellite DMSP ont été utilisées. Pour des lacs en Finlande et au Canada on a fait une étude comparative entre les dates du gel et du dégel indiquées par des observateurs au sol et par l'interprétation visuelle des images. Pour le degel, les dates prévues par télédétection sont retardées de quelques jours par rapport à celles observées sur le terrain; les differences diminuent vers l'interieur, et varient avec les dimensions et la forme du lac. En automne la nébulosité au dessus des deux régions rend impossible, l'emploi de l'imagerie satellite pour évaluer le gel des lacs. Les résultats soulignent les possibilités de traitement des données satellites pour l'étude des variations du climat régional.

Introduction

Modelling studies which predict global warming due to an increase of carbon dioxide in the atmosphere generally suggest that the warming trend will be greatest in high latitudes (National Research Council,

1982). Because of the relative scarcity of meteorological stations in these latitudes, the early detection of a statistically significant warming trend will be difficult. As a means of increasing the number of monitoring sites, recent studies have considered the utility of using the dates of formation and breakup of lake ice as a proxy measure of trends in air temperature (Tramoni et al., 1985; Palecki & Barry, 1986). Remote sensing techniques offer the potential to extend significantly the data set available for statistical reconstruction of temperature trends (Dean & Ahlnaes 1984).

The objectives of this study are to examine further the spectral properties of lake ice as they relate to remote sensing, and to determine the feasibility of interpreting ice freezeup and breakup dates for large geographic areas over several seasons using remotely-sensed data.

Information requirements

The temporal and spatial requirements of remotely-sensed data for the detection of freezeup and breakup have been defined by statistical studies (Tramoni 1984; Palecki & Barry, 1986). Temporal coverage of the data must be frequent enough to determine the timing of freezeup and breakup to within a few days. The data must also provide adequate spatial resolution to observe lakes of less than 1000 sq. km. Operational satellite imagery which potentially meets the spatial and temporal requirements noted above is limited to polar orbiters that provide daily coverage at moderate resolution. Such sensors include the Defense Meteorological Satellite Program (DMSP) Operational Linescan System (OLS) and the NOAA Advanced Very High Resolution Radiometer (AVHRR) sensors. In order to assess whether the spectral sensitivity of these sensors is adequate for lake ice detection, spectral information on ice and water are compiled from several sources (Figure 1). As expected, snow cover is clearly distinguishable from open water and snow-free ice. However, potential for confusion exists between water and dark ice types. A closer examination of the limited data that are available for the spectral response of lake ice suggests that differences between ice and turbid water occur near 0.4 micrometers, where the ice response stays fairly high while the water reflectivity decreases. It can also be expected that some spectral difference between water and ice might exist in the reflected infrared range, where liquid water is known to have very low reflectivity. The reflectivity of ice in this part of the spectrum has not been well-documented. A study of freezeup and breakup of lakes in Alaska (Dean and Ahlnaes, 1984) also note the potential for confusion between turbid water and ice, and between open water and clear ice.

To provide more data on the reflectivity of lake ice, Thematic Mapper (TM) data in digital form for seven channels were analyzed for a portion of central Wyoming. The TM image contained snow-covered land, lakes with open, clear water, partially frozen lakes, and frozen lakes with snow cover of various types. Spectral responses of these surface types for TM channels 1-4 are shown in Figure 2.

In this TM scene, several frozen lakes without snow cover are not clearly distinguishable from turbid water in the spectral range of

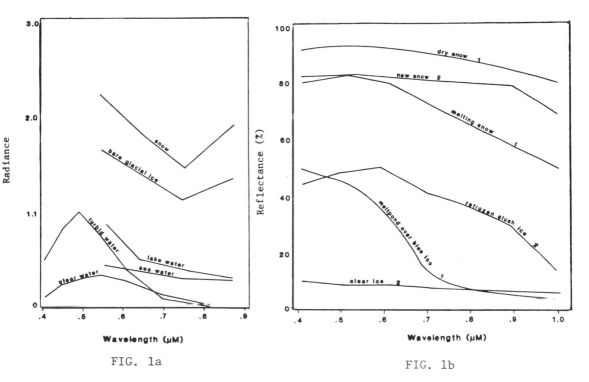

FIG. 1a   FIG. 1b

Figure 1   Observed spectral responses of snow, ice, and water
surfaces. FIG. 1a – data from Moore (1978) and Dowdeswell
(1984).   FIG. 1b – data from Grenfell & Maykut (1977,
Item 1) and Bolsenga (1983, Item 2). Radiance given in
$mW\ cm^{-2}SR\ \mu m^{-1}$.

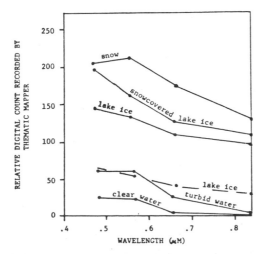

Figure 2   Mean digital counts recorded by Landsat Thematic Mapper
channels 1 – 4 for   snow, ice, and water surfaces in
Wyoming scene.

0.48–0.65 µm. This range corresponds closely to AVHRR Channel 1 and
to the lower third of the DMSP visible channel. A significant
difference between liquid water and blue ice is apparent in the far
visible and near infrared part of the spectrum, as hypothesized
above. This spectral region corresponds closely to AVHRR Channel 2,
and includes the longer-wavelength portion of the DMSP visible
channel. This difference in the near infrared wavelengths suggests a
threshold which, along with an albedo threshold at any visible
channel for snow-covered lakes, could be used to discriminate between
ice and shallow or sediment-loaded water.

The DMSP sensor provides only one channel in the visible
wavelengths, but the spectral response range of the DMSP channel
indicates that a large increase in albedo due to snow cover should be
readily detectable. The lack of a separate near infrared channel,
however, might mean that a cover of black ice or blue ice could not
be distinguished from shallow water or sediment-laden water.

The above analysis demonstrates that both DMSP and AVHRR sensors
should be capable of detecting the presence or absence of lake ice in
mature form. The following discussion describes an extended
interpretative study of lake ice freezeup and breakup in Finland and
Canada using DMSP data.

Method of analysis

Visible-band DMSP positive transparencies are available in 2.7 km
resolution and 0.6 km resolution on a daily basis from the National
Snow and Ice Data Center, University of Colorado, Boulder. These were
manually interpreted to estimate the dates of lake freezeup and
breakup. Images were analyzed for calendar years 1977–1981 and 1983
for Finland, and 1978–1980 for western Canada. Due to the poor
success of interpretation of autumn freezeup, as discussed below,
only results for spring are discussed in detail.

For both Finland and Canada, a sample of lakes visible in cloud-
free, summer DMSP imagery were chosen for study. Lakes were selected
to represent several climatic regimes in each area (Figures 3 and 4).
To the degree possible, lakes were selected for which ground-observed
information on ice condition was also available. Thirty-six lakes
were interpreted for Finland; ground-observed ice condition was
available for 23 of these lakes. In Canada, 72 lakes were
interpreted, with ground-observed information available for 25 lakes.
In total, about 250 DMSP scenes were analyzed for Finland, and about
600 scenes for Canada, with all interpretation performed by the same
interpreter. Physical problems with the data were primarily caused
by the deteriorating condition of the older transparencies, and by
missing data due to satellite sensor malfunctions.

A small sample of 0.6 km resolution (direct-readout) DMSP imagery
was compared to the standard 2.7 km resolution DMSP data to determine
the minimum size of lakes detectable in the two data types. This
comparison showed that a lake of approximately 9 sq. km can be
resolved in enough detail using the 0.6 km resolution DMSP imagery to
detect the presence of lake ice. On the coarser resolution (2.7 km)
DMSP imagery, the smallest lake interpretable is about 30 sq. km in
size.

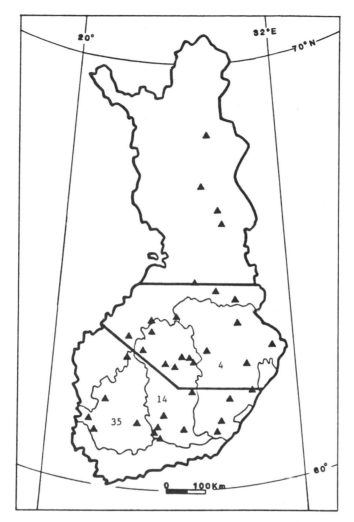

Figure 3 Locations of lakes analyzed in Finland. Regions 4, 14, and
35 are physiographic regions. (▲ lakes photointerpreted)

Results

Table 1 compares photo-interpreted mean breakup dates with ground-
observed mean dates for Finland and Canada.   Too few images were
available for 1979 over Canada, due to satellite coverage problems,
to permit statistical assessment. With the exception of 1978 in
Canada, mean breakup dates from ground observations were earlier than
the dates derived from interpretation of the DMSP imagery.
    A test of difference between photo interpreted and ground observed
breakup dates for paired observations shows significant difference
between the data sets for all years of Canadian and Finnish data
studied.   When considered by physiographic region for Canada,
differences are smallest in Region III, and greatest in Regions II
and IV.   For the Finnish data, physiographic Regions 4 and 14 show
the smallest differences between photo-interpreted and ground-
observed breakup dates.   In Finland, differences are greater nearer

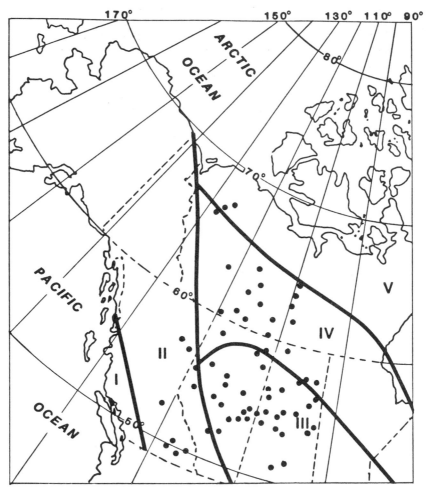

Figure 4   Locations of lakes analyzed in Canada.   Regions I-IV are
            physiographic regions.   (● lakes photointerpreted)

the coasts, while differences of 2 days or less are common in the
central part of the peninsula. Photo-interpretation of lake freezeup
in Finland was not possible for the years studied due to extensive
cloud cover which permitted only infrequent views of individual
lakes.

     Greater differences between photo-interpreted and ground-observed
breakup dates were observed for Canada than for Finland.   Lakes in
Region III, the most continental of the Canadian climatic regimes
studied, generally showed the lowest standard deviation within the
region and the lowest difference between photo-interpreted and
ground-observed dates.   As is the case for Finland, differences
between photo-interpreted and ground-observed dates appear to
decrease as the maritime influence is reduced.   How much of this
effect is due to increased cloud cover hindering interpretation, and
how much is due to the physical effect of maritime weather patterns
on breakup, could not be determined in this study.

     To investigate further the coherence of lake response by region,

Table 1 Comparison of mean lake breakup dates (Julian day) and standard deviation of the mean (in days) from satellite image interpretation and ground observations

FINLAND

YEAR	PHOTO-INTERP.	GROUND-OBSERVED
1977	mean = 292 st.dev. = 3.00	mean = 281 st.dev. = 2.98
1978	mean = 291 st.dev. = 7.48	mean = 284 st.dev. = 9.20
1979	mean = 290 st.dev. = 5.11	mean = 285 st.dev. = 3.96
1980	mean = 282 st.dev. = 7.86	mean = 279 st.dev. = 3.14
1981	mean = 296 st.dev. = 2.05	mean = 292 st.dev. = 2.92
1983	mean = 286 st.dev. = 4.28	mean = 276 st.dev. = 5.29

CANADA

YEAR	PHOTO-INTERP.	GROUND-OBSERVED
1978	mean = 262 st.dev. = 14.9	mean = 263 st.dev. = 19.3
1980	mean = 261 st.dev. = 22.0	mean = 254 st.dev. = 23.5

discriminant analysis was performed using the 1978 and 1980 photo-interpreted and ground-observed breakup dates for Canada. The discriminant analysis shows that lakes tend to be grouped by location, although these groupings do not always correspond with climatic regions. The general south-north trend of breakup described by Allen (1977) is shown, with a tendency for the lakes near the central Alberta and Saskatchewan border to break up earlier than their latitude would suggest.

Discussion

The results of this study indicate that the interpretation of breakup is much simpler than the detection of freezeup for both Finland and western Canada. For Finland, an explanation for this is that cloud cover for the times studied was too persistent and extensive to permit image interpretation. For Canada, cloud cover was less extensive, and the problem of freezeup detection may involve an inability of the DMSP sensor to resolve dark ice types. Also, the variability in tone and contrast of the transparencies themselves (either due to atmospheric effects, different enhancements used in image production, or degradation of the transparencies) could lead to considerable ambiguity in the determination of freezeup.

The interpretation of breakup was considerably more straightforward since its occurrence was accompanied by relatively abrupt changes in albedo from that of a snow or white-ice covered lake, to open water. Determining the point at which the lake is considered no longer ice-covered is complicated for lakes of 1000 sq. km and larger, since these lakes appear to break up in stages, with ice persisting in the center of the lake for several days after other portions of the lake have become ice-free. Although an analysis of the variability of lakes classified according to surface area showed no significant trend, the interpreter's experience indicated that lakes greater than 40 sq. km and smaller than about 800 sq. km showed the most consistent behavior. The discriminant analysis results, although limited to only two breakup periods for Canada, indicate some natural grouping of lakes by breakup date. Averaging within these groups should provide an improved estimate of breakup dates for regression analysis than would be achieved using an average of lakes within more strictly-defined climatic regions.

Conclusions

Based on these analyses, the interpretation of lake ice breakup for large geographic areas using remote sensing techniques appears feasible with reasonable effort and cost. The estimates of breakup dates for inland lakes are typically a few days later than ground-observed dates, with regional differences apparent in the timing and variability of breakup.

More investigation of the spectral properties of lake ice are needed to determine whether difficulties in the interpretation of freezeup were due to limitations of the DMSP sensor or to the interpretation procedures used. A regional analysis comparing freezeup detection using DMSP and digital AVHRR data would perhaps resolve this question.

Acknowledgments

This study was supported by U.S. Department of Energy, Carbon Dioxide Program Contract DE-AC02-83ER60106. Thanks are due to M. Grim for additional data analysis, to T. Wiselogel for drafting, and to M. Strauch for typing help.

References

Allen, W.T.R., 1977. Freeze-up, break-up, and ice thicknesses in Canada. Fisheries and Environment Canada, CL1-1-77.

Bolsenga, S.J., 1983. Spectral reflectances of snow and freshwater

ice from 340 through 1100 nm.   Journal of Glaciology,   29, 102,
   pp. 296-305.
Dean, K.G. and K. Ahlnaes, 1984.   A satellite derived climatic
   indicator.   Geophysical Institute   report, University of Alaska,
   Fairbanks, 13 pp.
Dowdeswell, J.A., 1984.   Remote sensing studies of Svalbard glaciers.
   Ph.D. thesis, University of Cambridge.
Grenfell, T.C. and G.A. Maykut, 1977.   The Optical properties of ice
   and   snow   in   the   Arctic   Basin.   Dept.   of Atmos.   Sciences,
   University of Washington, Contrib. No. 406.
Moore, G.K.   1978   Satellite surveillance of physical   water-quality
   characteristics.   Proceedings   of the 12th Symposium on the Remote
   Sensing of the Environment VI 20-26 April 1978, 445-462.
National Research Council, 1982.   Carbon dioxide and climate a second
   assessment.   Climate   Board,   National   academy   of   Sciences,
   Washington, DC.
Palecki, M.A. and R.G. Barry, 1986.   Freeze-up and break-up of lakes
   as   an   index   of   temperature   changes   during   the   transition
   seasons: a   case   study for Finland.   Journal of Climatology and
   Applied Met., Vol. 25, No. 7, pp. 893-902.
Tramoni, F., 1984.   Lake ice occurrence as a climatic indicator in
   studies of carbon dioxide warming: a Canadian case study.
   Masters thesis, Dept. of Geography, University of Colorado.
Tramoni, F., Barry, R.G., and J. Key, 1985.   Lake ice cover as a
   temperature   index   for   monitoring   climate   perturbations.
   Zeitschrift fur Gletscherkunde und Glazialgeologie,   Vol.   21, pp.
   43-49.

The Influence of Climate Change and Climatic Variability on the Hydrologic Regime and Water Resources (Proceedings of the Vancouver Symposium, August 1987). IAHS Publ. no. 168, 1987.

Mass balance of North Cascade Glaciers and climatic implications

Mauri S. Pelto
Institute for Quaternary Studies
University of Maine, Orono Maine 04469
Foundation for Glacier & Environmental Research
Seattle, Washington 98109, USA

ABSTRACT   A mass balance inventory of North Cascade, Washington glaciers was conducted to determine their response to recent climatic change, and the consequent changes in glacier runoff. The changing reservoir capacity and mean annual mass balance was determined on 47 North Cascade Glaciers. On ten of these glaciers the annual mass balance was measured for 1984, 1985 and 1986. The annual mass balance is a function of three climatic parameters: accumulation season cyclonic activity, ablation season temperature, and summer anticyclonic activity. Since 1977 an increasing frequency of anticyclonic conditions has caused a 1.1°C rise in ablation season temperature and a 15% decrease in winter precipitation. The result has been a mean annual balance of −0.30 m to −0.55 m water equivalent, for the 1977–1986 period. The decrease in annual winter balance and glacier accumulation area has also caused a 15% to 24% decrease in glacier reservoir capacity, the amount of meltwater a glacier can store.
Note: A French abstract of this paper can be found in the back of the text.

Introduction

The North Cascades, Washington extend from Snoqualmie Pass to the Canadian Border, covering 20,000 sq.km. (Figure 1). There are 756 glaciers covering 267 sq.km in the North Cascades (Post et al., 1971). Glaciers for the purpose of this study are bodies of perennial snow and ice with an area of at least 0.1 sq.km. The region is heavily forested up to 1500 m, and the treeline at 1500–2100 m is often above the accumulation zone of the glacier. This is indicative of the regions strongly maritime climate. Annual precipitation is from 2.0 m to 4.0 m on the west side of the range and from 1.0m to 2.0m on the east or dry side of the range. Annual temperatures are moderated by the strong maritime air flow, the mean annual temperature at the glaciation threshold ranging from 0.5°C to 2°C.

Washington's glaciers store as much water as all of its lakes, rivers and reservoirs, three-quarters of which is stored in the North Cascades (Meier, 1969). Annual glacier runoff from the North Cascade glaciers is 800 million m³. Proper water resources management requires an understanding of the changing contribution of runoff by

163

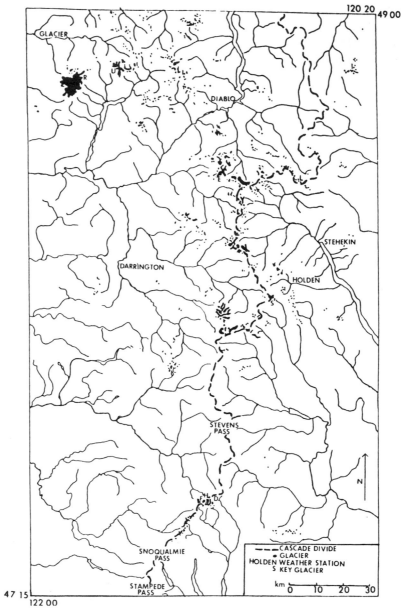

Figure 1 The North Cascades of Washington, indicating known
glaciers weather stations used in this study and key
glaciers, where the annual mass balance was measured in
1984, 1985 and 1986. C=Columbia, D=Daniels, E=Eldorado,
F=Foss, H=Lynch, R=Rainbow, S=Spider, U=Lower Curtis,
Y=Yawning.

glaciers.

The United States Geological Survey (USGS) has monitored the mass
balance of the South Cascade Glacier since 1955. The extensive
hydrologic and mass balance measurements and their correlation with

climatic and hydrologic records have greatly increased the understanding of North Cascade glacier hydrology, movement and the climatic
parameters that control mass balance(Tangborn et al., 1975; Tangborn,
1980). North Cascade glaciers fall into six categories based on
varying sensitivity to specific climatic parameters(Pelto, 1987). The
South Cascade Glacier represents only one small group; thus, no
quantitative climatic or hydrologic conclusions can be drawn for the
vast majority of North Cascade glaciers. For this reason in 1984 the
Foundation for Glacier and Environmental Research(FGER) established a
system for annually monitoring the mass balance of North Cascade
glaciers.

Mass balance

The mass balance of North Cascade glaciers has been calculated from
detailed field observations, and identification of the activity index
and accumalation area ratio (AAR). The activity index is the budget
gradient in the vicinity of the equilibrium line (Meier and Post,
1962).The AAR is the percentage of a glacier's area in the accumulation zone (Meier and Post, 1962). The mass balance of North Cascade
glaciers is a surface mass balance only, with the exception of a few
cases where periodic large scale avalanches occur. The annual balance
(bn) is the difference of winter balance (bw) and summer balance
(bs).

$$bn = bw - bs \qquad\qquad\qquad (1)$$

The glacier mass balance is assessed at the end of the ablation
season only. Thus, only the annual mass balance is measured. Winter
and summer balances are estimated from hydrologic, meteorologic and
mass balance data.

It has been found on the South Cascade Glacier (Meier and Tangborn,
1965) and Columbia Glacier, that annual mass balance is a function of
absolute altitude, as the activity index is not annually variable. In
other words the difference between mass balance at any two points on
the glacier does not change only the actual mass balance at each
point. This would allow calculation of the mass balance from measurement at a single representative point, unfortunately there is no such
point. A glacier's activity index and area elevation distribution
determine the mass balance associated with any AAR value for that
glacier. If the activity index and area elevation distribution are
known then only annual observation of the AAR is neccessary for
reasonable estimates of glacier mass balance(Meier and Tangborn,
1985). The AAR has been observed on 47 glaciers during the 1984-1986
period. The activity index was measured on each of these glaciers by
observing the rate of rise with time of the snowline in comparison
with measured snow depth above the snowline. The mass balance for
each of the forty-seven glaciers was then calculated for the 1984-
1986 period; this data will be examined later.

Detailed measurement of annual mass balance was completed on ten
of the above glaciers in 1984, 1985, and 1986 to more accurately
ascertain the annual mass balance and to determine the accuracy of
the above mass balance calculation methods. At least one glacier from

each of the six glacier climatic sensitivity types was selected.
Field measurements were carried out at the end of the ablation season.

In the accumulation zone, annual accumulation layer thickness was
determined using crevasse stratigraphy, snowpits and probing, with a
density of approximately 200 points/sq.km. Techniques used were based
on those of Ostrem and Stanley (1969), who stated that an acceptable
density of measurements in the accumalation zone was 100
points/sq.km. The density is measured in two vertical profiles in
snowpits.The accumulation layer thickness and density were checked at
several locations the following year to determine changes in mass
balance in the last week or two of the ablation season.The mass
balance obtained was then a preliminary mass balance until this check
was completed the following year. Errors in depth measurement are
± 0.05 m and in density determination ± 0.04 g/cm^3. The resulting error
in mass balance assessment for the accumulation zone is ± 0.10 m. It
should be noted that internal accumulation and superimposed ice are
insignificant components in the mass balance,if present at all (Meier
and Tangborn. 1965).

Below the snowline, ablation triangles are used to determine
annual ablation. An ablation triangle consists of three stakes driven
or drilled into the ice at 3 m intervals. Ablation measurements are
made at nine points on the triangle periphery,and again at the
conclusion of the ablation season. Error in annual ablation measurements are ± 0.03 m, due to ice density variations,stake settling and a
low sampling density.

The error in annual mass balance calculation for the entire
glacier is then $\pm 0.17 - 0.22$ m, except during years of extreme
ablation, when the error is higher. The goal of this project was to
obtain a rough mass balance estimate for a large number of glaciers,
not strict accuracy for a few.

Table 1 The annual mass balance of ten North Cascade glaciers for
1984, 1985 and 1986.

Glacier	Area	1984	1985	1986
Columbia	0.9	+0.21	-0.31	-0.20
Daniels	0.5	+0.11	-0.51	-0.36
Eldorado	1.4	+0.25	-0.14	-0.02
Foss	0.7		-0.69	+0.12
Lewis	0.1	+0.67	-1.16	-0.34
Lower Curtis	0.9	+0.39	-0.16	-0.22
Lynch	0.8	+0.33	-0.22	-0.07
Rainbow	1.5	+0.58	+0.04	+0.20
Spider	0.1	+1.12	-0.63	+0.10
Yawning	0.2	+0.09	-0.23	-0.14

A comparison of the measured mass balance for the nine glaciers in
1984, 1985 and 1986 (Table 1) to that estimated for the glacier from
AAR observations and the activity index measured using the techniques
discussed previously for activity index determination, is shown in
Figure 2. The data indicate errors in annual mass balance measurement

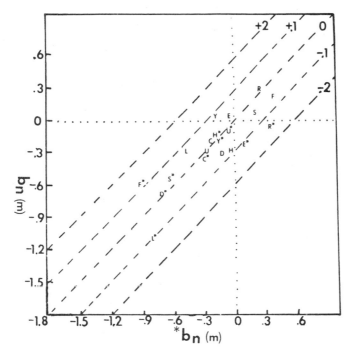

Figure 2   The measured mass balance (bn) of 10 North Cascade
glaciers in 1985 and 1986, compred to the predicted mass
balance (*bn) using the AAR-activity index method. C=Col-
umbia, D=Daniels, E=Eldorado, F=Foss, H=Lynch, L=Lewis,
R=Rainbow, S=Spider, U=Lower Curtis, and Y=Yawning.

(using the latter method) of $\pm 0.25 - 0.30$m, and that the errors are
random.

Climatic analysis

The mass balance of North Cascade glaciers is determined by the
accumulation season cyclonic activity, ablation season temperature
and summer anticyclonic activity. During the accumulation season, 90%
of the precipitation is associated with cyclonic systems, hence as
accumulation season cyclonic astivity increases, so does winter
balance. The number of days during the accumulation season associated
with cyclonic conditions then provides an estimate of balance.
    Ablation season temperature is predominantly controlled by the
persistence of anticyclonic conditions along the Pacific Coast or
over the study region. Ablation season temperatures and ablation
increase with increasing frequency of anticyclonic   conditions.
Ablation is highest during periods of weak air flow and clear skies.
Summer balance can then be estimated from the number of days with
prevailing anticyclonic conditions and the number of days of clear
windless conditions.
    Local climatic conditions were compared to daily synoptic scale

atmospheric circulatic, to relate synoptic climatology to local climate. Daily synoptic scale 500 mb pressure maps produced by NOAA were used to identify daily atmospheric circulation. Nineteen atmospheric circulation types were found accounting for 97% of the observed days. The atmospheric circulation types were correlated with records from eight weather stations in the region (Figure 1),for the period 1974-1986. The effect of each circulation type on local weather mass balance was then identified. This technique and the atmospheric circulation types for this area are best described by Yarnal (1984).

It was found that 86% of the accumulation season cyclonic precipitation was related to five circulation types, each exhibiting a low pressure system over the eastern Pacific between 50 N and 60 N and a low pressure system centred SE of and extending over the North Cascades at the 500 mb level. At the surface low pressure systems embedded in the resulting SW to W airflow caused heavy precipitation in the North Cascades.

During the ablation season it was found that 83% of the cooling degree days for Seattle and Snoqualmie Pass were accounted for by seven high ablation atmospheric circulation types. Five of the circulation types were associated with a high pressure system over the study area and weak SW-NW airflow. Two of the circulation types had a low pressure system north or east of the study area and no significant air flow.

It is evident that high pressure systems over the study area during the summer and NE of the study region during the winter have become increasingly stable. The resulting $1.1^{\circ}C$ rise in ablation season temperature and the 15 % decrease in winter precipitation has led to declining glacier mass balances and extreme climatic swings. In Figure 3 the relation between the number of days annually of these mass balance dominating climatic types is related to measured mass balance. The best fit for this data was obtained by equation (2):

$$bn = 5.0 (AC) - 2.7 (AB) \qquad (2)$$

where AC is the number of days during the accumalation season associated with the five high precipitation circulations types, and AB is the number of days during the ablation season associated with high ablation circulation types. The error in mass balance calculation using the above method is $\pm 0.40 - 0.50$m.

Reservoir capacity

North Cascade glaciers supply 40 to 50 % of the summer runoff for the region. Runoff peaks in a non-glacierized basin during May and June, the spring snowmelt period. For a glacierized basin runoff peaks in July and August (Tangborn and Rasmussen,1976). This two month delay is due to the glaciers ability to store meltwater and the increased ablation during mid-summer (Fountain and Tangborn,1985). The amount of meltwater that can be stored by the glacier is its reservoir capacity. A glacier's reservoir capacity is determined by the depth of annual snow-firnpack and the areal extent of the accumulation area. The reservoir capacity decreases with decreasing snow-firnpack

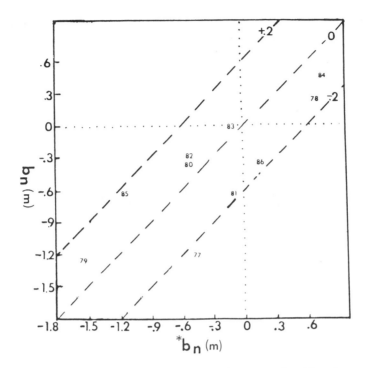

Figure 3   The mass balance of the North Cascade Glaciers, based on
AAR-activity index mass balance calculations for 16-47
glaciers (bn) versus the mass balance predicted by
consideration of atmospheric circulation type using
equation 2 (*bn).

depth and a real extent of the accumulation area.

Tangborn et al. (1975) found that in May 1970 54% of the total
possible May runoff was temporarily stored in the glacier. Meltwater
accumulates in the snow-firnpack until it is saturated, usually in
early to mid June. In the ablation zone meltwater is not effectively
stored after the saturation is reached. Meltwater quickly percolates
down to the rather impermeable glacier ice and rapidly escapes the
system. In the accumulation zone the impermeable layer is at a much
greater depth and an effective drainage network is not established.
Meltwater is stored in transit through the snow-firnpack for approximately 4 to 14 days in July and 2 to 7 days in August (Behrens, 1979;
Krimmel and others, 1973).

From 1977 to 1986 the accumulation area decreased and the snow
depth decreased, hence the reservoir capacity declined, saturation
occurred earlier and the transit time decreased. The result was
higher glacier runoff in June and less in July and August. This
resulted in an increase in the danger of spring flooding. This trend
is evident in the runoff records for the Stehekin drainage basin,
which is 3.4% glacier covered. June runoff had increased 12% relative
to the neighbouring unglaciated Andrews Creek and Twisp River
watershed. In contrast August runoff has decreased by 6% versus a 19%
increase in runoff from the Andrews Creek and Twisp River watershed.

Only changes in the reservoir capacity can be calculated accurately, the actual reservoir capacity is unknown.

Since 1977 the annual snowpack depth is 10% below the 1938-1984 mean, as indicated by Washington State snowpack surveys. The accumulation area is 12% below that neccessary to maintain equilibrium, which had been maintained from 1944-1974 (Tangborn, 1980). The result is a 21% decrease in glacier reservoir capacity.

Glacier mass balance changes

The mass balance of 22 North Cascade glaciers was determined for the 1977-1983 period from photograhic records of the FGER, Seattle Mountaineers, USGS and North Cascades National Park Service. These records were used to determine the annual AAR for the glaciers. Mass balances were then calculated based on a measured activity index, using the AAR-activity index method. There was a complete record for only two glaciers, but there was a four-year record for each glacier and no year had fewer than fourteen suitable AAR observations. This data, in combination with South Cascade Glacier data, was used to calculate the average annual mass balance for North Cascade glaciers. The glacier climatic sensitivity type was ascertained for each North Cascade Glacier. The mass balance for each glacier was assigned from the mass balance of a nearby glacier of the same climatic sensitivity type, for which a record was available. The mean annual mass balance for the North Cascade glaciers between 1977 and 1986 was +0.37 m. The annual error is +0.30m dropping to +0.15m for the entire 1977-1986 period. In Table 2 the mass balance data is shown. Winter balances were obtained from crevasse stratigraphy on the Eldorado, Lower Curtis and Lynch Glaciers and South Cascade Glacier data. The summer balance is then the difference between winter balance and annual mass balance. The resulting reduction in glacier volume in conjunction with the 21% reduction in reservoir capacity will cause below normal summer glacier runoff.

Table 2 The mean annual mass balance(bn), winter balance (bw) and summer balance (bs) of North Cascade glaciers between 1977 and 1986 (Determined using the AAR-activity index method)

Year	bn(m)	bw(m)	bs(m)
1977	−1.20	1.70	2.90
1978	+0.20	2.80	2.60
1979	−1.30	2.40	3.70
1980	−0.40	2.20	2.60
1981	−0.80	2.80	3.60
1982	−0.30	3.10	3.40
1983	−0.20	2.70	2.90
1984	+0.40	3.30	2.90
1985	−0.80	2.80	3.60
1986	−0.40	3.00	3.40
Mean	−0.48	2.68	3.16

The climatic causes for the large negative balances is an increasing
stability of anticyclonic conditions over the eastern Pacific during
the accumulation season and over the study area and Pacific Coast
during the ablation season. At a larger scale, the cause for this
climatic change is unknown. The climatic effects on changing glacier
mass balance are discussed in more detail by Pelto(1987).

References

Behrens, H., Loschorn, U. Ambach, W. & Moser, H. (1977) Studie zum
    shmelzwesser ablfluss uaus dem akkumulationsgebeiteine Alpine
    gletschers.Z. fur Gletscherkund und Glazialeologie. 12(1), 69-74.
Fountain, A.G. & Tangborn, W.V. (1985) The effect of glaciers on
    streamflow variations. Wat. resour. Res. 21(40, 579-586.
Krimmel, R.M., Tangborn, W.V. & Meier, M.F. (1973) Water flow through
    a temperate glacier. In: Role of Snow and Ice in Hydrology(Proc.
    Banff Symposium, September, 1972). IAHS-Aish Publ. No. 104, 401-
    416.
Meier, M.f. (1969) Glaciers and water supply. J. Am. Wat. Wks. Ass.
    61(1), 8-12.
Meier, M.F. & Post, A.S. (1962) Recent variations in mass net budgets
    of Western North America. In: Commission of Snow and Ice(Proc.
    Obergurgl symposium, September, 1962). IAHS Publ. no. 58, 63-77.
Meier M.F. & Tangborn W.V. (1965) Net budget and flow of the South
    Cascade Glacier, Washington. J.Glaciol 5(41), 547-566.
Ostrem G. & Stanley, A (1969) Glacier mass balance measurements,
    Canadian Department of Energy, Mines and Resources and Norwegian
    Water Resources and Electricity Board.
Pelto, M.S. (1987) Spatial variatons in mass balance of North Cascade
    glaciers related to atmospheric circulation. Presented at the IAHS
    Symposium on the Influence of Climatic Change, Vancouver,August.
Post, A., Richardson, D., Tangborn W.V. & Rosselot, F.L. (1971)
    Inventory of glaciers in the North Cascades, Washington U.S.A. J.
    Glaciol. 30(105), 3-21.
Tangborn, W.V., Meier, M.F. & Krimmel R.M. (1975) A comparison of
    glacier mass balance by glaciological, hydrological and mapping
    methods, South Cascade Glacier, Washington. In: Snow and Ice(Proc.
    Moscow, Symposium, August, 1971.
Tangborn W.V. & Rasmussen L.A. (1976) Hydrology of the North Cascade
    Region, Washington. I. Runoff, precipitation and storage characte-
    ristics. Wat. Resour. Res. 12(2), 187-202.
Yarnal, B. (1984) Relationship between synoptic scale atmospheric
    circulation and glacier mass balance in SW Canada during the IHD,
    1965-1974. J.Glaciol. 30(105), 188-198.

The Influence of Climate Change and Climatic Variability on the Hydrologic Regime and Water Resources (Proceedings of the Vancouver Symposium, August 1987). IAHS Publ. no. 168, 1987.

Quantification of long term trends in atmospheric pollution and agricultural eutrophication: A lake-watershed approach

Ian D.L. Foster & John A. Dearing
Geography Department
Coventry Polytechnic
Priory St.
Coventry, CVI 5FB, U.K.

ABSTRACT   Historical data, documenting trends in atmospheric pollution by heavy metal emissions from industry and motor vehicles, and increasing agricultural eutrophication, caused by post war agricultural industrialisation, are difficult to establish. This is primarily because extensive monitoring of various aspects of environmental pollution was generally not carried out until after the major problems were identified. Direct quantitative evaluation of such trends are important for many undocumented environments when judging present levels of sediment associated nutrients and contaminants contained in the fluvial environment in relation to baseline conditions for the same area.
    Information may be derived from analysis of single lake sediment cores collected from basins acting as pollutant sinks but the data derived from such studies is far from satisfactory because they only indicate rates of accumulation at a single point in the lake and do not take account of a number of important considerations. These include the lake catchment area ratio, the sources of accumulating sediments and the volumetric loading of these contaminants to the lake bed.
    A multiple lake sediment coring exercise applied to two lakes is used to derive historical sediment yield information for the English Midlands since 1765. These data are used in conjunction with chemical extraction techniques to budget atmospherically derived contaminants and agricultural fertilisers accumulating in the lake sediments. Comparison of a semi-natural woodland and an intensively farmed landscape shows that carefully selected lake-watersheds can be used to provide data of sufficient reliability to quantify long term response of the environment to pollution from various sources. It is argued that not only can baseline information be rapidly obtained by careful analysis of lake sediment properties, but that sufficiently long records may be obtained in order to quantify the response to and recovery of an environment from different types of environmental change. It is concluded that the lake-catchment should be recognised as an important hydrological framework for quantitative analysis.

Introduction

Since the pioneering work of Mackereth (1966) in the English lake district, limnologists and palaeolimnologists have devoted consider- able time to the study of the physical and chemical properties of lake sediments in an attempt to understand contemporary elemental fluxes and migrations, and to reconstruct environmental histories. These studies have been undertaken in a range of environments but in temperate regions generally relate to the post-glacial period. Many of these studies have relied on an analysis of the changing concentrations of various chemical elements throughout the lake sediment column, but often pay little regard to the influence of sedimentation rate or sedimentation dynamics on the reconstructed record. Palaeohydrological interest in these records is relatively recent and relates in part to the development of both accurate dating methods and means by which total sediment influx over varying time- scales can be calculated (cf Oldfield and Appleby, 1984, Dearing 1986, Dearing and Foster, 1986; Foster et al, 1985; 1986). The potential of lake catchment based frameworks for studying and quantifying hydrological change over timescales of 100 to 1000 years has yet to be fully evaluated and realised, and the present paper attempts such an evaluation on the basis of the chemical stratigra- phies of two contrasting lake catchment systems in the English Midlands.
  Assuming that the lake sediment record can be used to reconstruct sediment yield histories, this paper attempts to answer the following questions. First, are chemical records sufficiently well preserved in the lake sediments for quantitative detection? Secondly, if a variety of elements can be detected, does their presence and/or concentration reflect processes operating in the catchment, in the lake or in both? Thirdly, is it possible to selectively investigate certain properties in order to quantify long term pollution delivered to the lake either from the atmosphere or from accelerated erosional processes reflecting land management practices in the catchment?

Site details

The sites selected for this investigation are small drainage basins (less than 5 sq.km) contributing their outputs to man-made reservoirs (Figure 1) and were chosen to represent contrasting contemporary and historical land uses within the English Midlands. The Merevale Catchment historically and at present is predominantly covered by forest although the relative balance between hardwoods and softwoods has shifted towards the latter in the post-war period. The Seeswood Pool catchment has been dominated since Domesday times by a mixed agricultural economy. Geological, pedological and land use details are given in three recent papers which deal with the rates and patterns of sedimentation in the lakes, the means by which historical changes in catchment erosion is calculated and the development of a mixing model to identify contemporary and historical sediment sources (Foster et al, 1985; 1986; Dearing and Foster, 1987). Summary statis- tics for the two environments are given in Table 1.

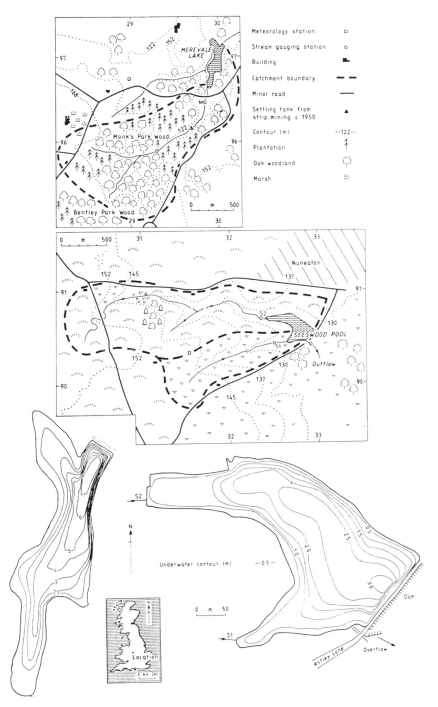

Figure 1   Catchment location and lake characteristics.

Lake sediment retrieval and analysis

Multiple undisturbed sediment cores were retrieved from the beds of Merevale Lake and Seeswood Pool using a pneumatic corer operated from

Table 1   Site details for the present day Merevale and Seeswood
lake-catchments

A. Regional characteristics

Climatic classification - Humid Temperate

Min mean monthly temp (°C) - 3.1 (Jan)

Max mean monthly temp (°C) - 15.9 (Aug)

Mean annual rainfall (1941-70 in mm) - 674

Mean annual Pot. Evap. (mm) - 492

Mean annual Act Evap. (mm) - 385-455

B. Catchment & Lake Properties	Merevale	Seeswood	Land use (% area)	Merevale	Seeswood
O.S. grid ref.	SP300970	SP327905			
Lake area (ha)	6.5	6.7	Deciduous woodland	78.4	4.0
Catchment area (ha)	195	221	Coniferous plantation	7.5	0
Lake catchment ratio	1:30	1:33	Permanent pasture	14.1	55.0
Maximum altitude (m)	175	160	Grass ley	0	11.0
Minimum altitude (m)	118	125	Arable	0	30.0
Relative relief (m)	57	35			
Maximum lake depth (m)	8	3.5	Soil groups (% area)		
Mean lake depth (m)	4	1.5			
Lake volume (m³)	154219	103241			
Date of impoundment	c. 1838	c. 1765	Stagnogleys	c. 50	c. 70
			Argillic brown earths	c. 15	c. 20
			Reclaimed	c. 20	
			Other	c. 15	c. 10

a rubber dinghy. On a grid with intersections at 25m intervals, over
70 sites were sampled at Merevale Lake. On a grid with intersections
at 50m intervals, over 30 sites were sampled at Seeswood Pool. The
sediment cores from both sites were divided into 1 cm slices for
further analysis. From both lakes, two cores were analysed for the
^{137}Cs and ^{210}Pb content, and an absolute chronology was derived by
applying the CRS model recommended by Oldfield and Appleby (1984).
The master dated chronology was transferred to surrounding cores on
the basis of magneto-stratigraphic correlations (cf Dearing, 1986),
and sediment accumulation rates between dated horizons estimated on
the basis of dry sediment density. The lake sediment yield record
was obtained after dividing the period 1860 to 1983 into 9 unequal
intervals of time. Sediment yields ranged from less than 4 to over 9
t km² per yr in the Merevale catchment. At Seeswood Pool, estimates
for 11 unequal time intervals between 1765 and 1983 ranged from a pre
20th century level of around 8 to a most recent estimate of 36 t km²
per yr (Table 2).
    Chemical analysis of all sediment cores based on the 1 cm slices
used in sediment mass calculations was impractical since the combined
lake study provided in excess of 4000 samples from the two lakes.
Two different approaches were adopted for chemical analysis depending
on the nature of the sediment digestion or extraction employed.
First, a total digestion, as described below, was undertaken on
composite samples from 7 cores at Merevale and on 4 cores from
Seeswood Pool, selected to represent different environments of
sedimentation. The composite samples were obtained by physically
mixing equal quantities of sediment derived from the appropriate

Table 2   Summary statistics for sediment yields and acid digestions

Time Zone	Years	Sediment[1] Yield kg ha⁻¹yr⁻¹	Pb	Zn	Ni	Cu	Fe	Mn	P
Merevale Lake									
1	1964–82	114.1	77.7 (14.6)	785.7 (109.4)	143.5 (24.3)	114.3 (44.2)	76200 (5413)	2118 (410)	982.3 (229.5)
2	1953–63	70.5	79.7 (14.7)	600.1 (225.7)	138.9 (25.9)	82.0 (14.9)	80513 (10219)	1740 (591)	680.8 (219.4)
3	1943–52	77.4	70.0 (15.5)	559.5 (148.4)	114.9 (19.5)	78.9 (18.5)	79048 (9023)	1634 (561)	599.6 (175.3)
4	1936–42	143.1	62.6 (20.4)	434.7 (130.1)	93.7 (24.0)	64.0 (15.2)	71067 (15564)	1394 (678)	654.7 (224.2)
5	1922–35	116.7	53.5 (12.5)	316.7 (94.7)	91.8 (8.6)	70.1 (18.1)	77295 (5608)	1453 (608)	579.3 (144.0)
6	1914–21	147.0	55.7 (16.3)	260.8 (69.7)	82.6 (9.7)	82.8 (30.5)	83498 (7371)	1503 (504)	482.9 (139.0)
7	1906–13	142.1	51.1 (15.3)	219.2 (65.7)	78.3 (9.4)	66.8 (19.4)	87251 (6439)	1579 (553)	552.0 (133.9)
8	1879–1905	61.8	44.2 (19.6)	180.9 (31.2)	70.2 (8.1)	68.7 (19.5)	84062 (8500)	1553 (558)	511.6 (183.2)
9	1861–1878	72.7	37.7 (13.4)	152.4 (23.3)	61.8 (9.4)	65.2 (32.6)	79298 (10617)	1292 (389)	567.8 (102.9)

Time Zone	Years	Sediment[1] Yield kg ha⁻¹yr⁻¹	Pb	Zn	Ni	Cu	Fe	Mn	P
Seeswood Pool									
1	1978–83	453.1	61.7 (16.9)	400.0 (101.0)	54.5 (5.2)	80.0 (11.1)	35907 (2678)	1426 (399)	1760.6 (417.7)
2	1973–77	228.4	39.5 (7.8)	443.2 (155.4)	58.6 (5.9)	55.6 (11.2)	36094 (4791)	1063 (209)	1655.3 (106.7)
3	1965–72	173.9	51.6 (26.7)	473.5 (113.0)	56.5 (7.2)	69.1 (21.3)	40813 (6222)	916 (85)	1553.7 (226.7)
4	1948–64	150.0	42.7 (23.3)	491.1 (104.7)	59.1 (3.8)	73.9 (16.4)	40417 (5299)	879 (122)	1387.7 (212.8)
5	1934–47	159.7	56.7 (19.9)	531.4 (45.0)	61.7 (7.7)	67.1 (16.7)	41904 (6728)	986 (294)	1597.8 (430.2)
6	1926–33	201.3	51.9 (13.8)	488.9 (71.7)	62.8 (5.9)	84.2 (24.1)	40525 (6738)	886 (251)	1429.0 (548.6)
7	1920–25	270.2	48.7 (19.8)	388.3 (55.3)	62.0 (5.9)	56.2 (2.6)	39774 (6157)	859 (263)	1361.4 (450.2)
8	1903–19	119.8	57.7 (31.2)	356.8 (41.5)	63.2 (7.6)	60.2 (8.2)	42181 (7334)	838 (305)	1018.4 (440.2)
9	1881–1902	101.8	46.5 (21.4)	288.9 (84.7)	62.7 (8.8)	62.4 (15.0)	42300 (8289)	763 (226)	739.5 (409.2)
10	1854–80	152.6	41.0 (15.6)	204.5 (43.5)	58.4 (8.7)	55.3 (12.2)	43656 (11213)	997 (434)	767.5 (178.6)
11	1765–1853	88.3	18.4 (14.9)	124.0 (12.3)	41.9 (6.3)	33.3 (13.3)	39447 (8656)	603 (289)	602.4 (28.1)

Element concentrations in acid digests[2] in µg g⁻¹

[1] Uncorrected for allochthanous and atmospheric contributions

[2] Standard deviation of 7 and 4 cores at Merevale and Seeswood in brackets

number of 1 cm slices in each of the correlated time zones. This procedure was adopted in order to estimate the variability in chemical composition at the lake bed and at depth in the sediment cores. Secondly organic carbon, diatom silica and the relative proportion of various forms of inorganic P were obtained by analysing the same cores after physically mixing sub-samples together on the basis of respective time zones to provide 1 composite sample for each time zone in the two lakes.

Digestion and extraction techniques

A large literature exists in fields of pedology and limnology which debate the efficiency of various extraction and digestion procedures for selectively removing chemical elements in soluble form for quantitative analysis (cf Mackereth, 1966; Hesse, 1971; Allen et al, 1974; Williams et al, 1976; Hieltjes and Lujklema, 1980; Bostrom et al, 1982; Engstrom and Wright, 1984; Bengtsson and Enell, 1986). Selection of the most suitable method will often be site specific or specific to the element under consideration. The present study utilised both an acid digestion procedure in order to analyse total core chemistry and a range of fractionation-extraction procedures for the study of inorganic phosphorus and "weakly held" elements.

The treatments utilised in this study are summarised in Table 3. Total acid digests were analysed on an atomic absorption spectrophotometer for major cations and heavy metals content. Phosphorus in the total digest and in extractions was analysed colorimetrically on an EEL model 197 spectra or on an LKB Ultrospec II using the ammonium molybdate- stannous chloride colour producing reaction at 640 nm.

Table 3   Extraction/digestion procedures

---

1. Acid digestion:-   nitric - perchloric - sulphuric acid
                            with c 0.5 g sediment. Digested on
                            Kjeldahl heating stand stand.

2. Acetic acid extraction[1]:-   1 m glacial acetic acid with
                            5g sediment shaken for 1 hour.

3. Inorganic Phosphorus fractionation[2]:-
            a) Loosely bound 2 extractions for 2 hours with
                1M Ammonium chloride at pH 7.

            b) Fe + Al bound 17 hours with 0.1N Sodium hydroxide.

            c) Ca bound 24 hours with 0.5N hydrochloric acid.

---

[1]See Foster et al (1987)

[2]See Hieltje and Lijklema (1980)

Organic carbon content of composite samples was determined by the wet oxidation method and diatom silica by a gravimetric determination following an alkali extraction using sodium carbonate which digests diatom silica but only causes minor degradation to clay and other mineral silicates (Engstrom and Wright, 1984).

One of the most important problems relating to the presentation of sediment chemistry data is the selection of appropriate units of

measurement.   It is common in the limnological literature to find
units   quantifying lake bed accumulation   rate (eg. in g $m^{-2}yr^{-1}$)   or
concentration   per unit mass of sediment (mg per g).   Choice of   the
appropriate   unit is important because,   for example,   in a situation
where   accumulation rates in two lakes of similar area are the   same,
no account is taken of varying catchment   area.   Here,   accumulation
rate may well be a function of the catchment:lake area ratio.
Similarly, concentration units will in part be a function of sediment
accumulation   rate especially where the element under   investigation
arrives at the lake bed by a process independent of that   responsible
for the delivery of most sediment (cf Engstrom and Swain, 1985).   The
apparent difficulty in interpreting the lake sediment chemical record
can   be overcome by reporting results in both units of   concentration
and units of accumulation.   For hydrological studies, the latter may
be   usefully converted to more familiar measures of catchment
denudation in kg ha^{-1} yr^{-1}or t $km^{-2}yr^{-1}$.   Choice of the analytical
units are related in this study to the hypothesised principal source
of   the element such that trace metals data are reported in units   of
concentration   and accumulation rate per unit area of lake bed   since
it   is assumed that the principal source of these elements to the two
midland   lakes is atmospheric.   Phosphorus data   are reported   in
concentration   units and units of phosphorus yield per unit   area   of
basin.   The   latter computation requires,   of course,   that sediment
mass   accumulating across the lake bed for each time zone has been
calculated.

Results

Summary   statistics   for the concentration of trace metals,   Fe,   Mn
and   P in the   total acid digests are given in Table 2.   These   data
represent the mean and standard deviation of concentrations in the   7
and 4 cores analysed in Merevale Lake and Seeswood Pool respectively.
Average   concentrations of P from 5 different extraction or digestion
procedures   are presented in Table 4.   Columns 1 and 2   compare   the
efficiency of two commonly used methods for the extraction of "weakly
held" P, the latter generally appearing more consistent and highlighting   greater differencws between the Seeswood Pool and Merevale lake
sediments.   Despite   this difference,   the amount of P extracted   by
either of these methods is at most only 1% of the total contained   in
the   acid digests.   The HCl and NaOH extractions of columns 3 and   4
are   thought   to quantify effectively the relative proportions of   Ca
(apatite) and Fe or Al bound P in the sediments respectively (ie   the
inorganic P fraction),   whereas column 5 includes organic and inorganic   forms.   The relative proportion of organic to inorganic P may be
estimated by subtracting the sum of columns 3 and 4 from column 5.
   The   organic   matter and diatom silica contents of bulked   samples
are   given   in   Figure 2 in addition to the ratio of   average   Fe:Mn
concentrations in the total digestions from both lakes.
   Preliminary   results   describing   the   chemistry of   the   total
digestions   and   the   chemical   characteristics of   the   contributing
streams are presented elsewhere (Foster et al., 1987;   Foster and
Dearing, 1987).

Table 4 Phosphorus concentration ($\mu g \ g^{-1}$) in digests and
extractions

Merevale Lake	Time Zone	Digestion/extraction				
		1	2	3	4	5
	1	12.2	4.5	163.9	166.2	982.3
	2	4.3	4.3	199.5	111.1	680.8
	3	2.2	3.6	148.5	88.7	599.6
	4	0.0	5.0	161.8	93.6	654.7
	5	2.7	4.5	100.1	52.8	579.3
	6	0.0	5.1	110.9	95.9	482.9
	7	0.0	6.7	94.7	81.2	552.0
	8	0.9	6.5	74.2	160.5	511.6
	9	0.7	4.3	23.1	74.0	567.8
Seeswood Pool						
	1	7.3	29.7	396.0	155.7	1760.6
	2	2.9	32.5	319.6	133.8	1655.3
	3	4.9	32.5	448.3	140.5	1553.8
	4	2.5	26.5	361.0	106.2	1387.8
	5	2.6	25.9	375.5	144.8	1597.8
	6	0.8	23.6	268.9	219.0	1429.0
	7	1.4	21.8	312.2	130.2	1361.4
	8	1.0	19.1	374.4	152.2	1018.4
	9	0.0	19.1	168.2	80.2	739.5
	10	0.0	11.1	164.3	88.2	767.5
	11	0.0	12.1	107.2	115.6	602.4

1 = Ammonium chloride

2 = Ammonium acetate

3 = Hydrochloric acid

4 = Sodium hydroxide

5 = Acid digestion (Table 2)

Phosphorus concentrations in lake sediments

Phosphorus, more than nitrogen and potassium, has probably been the
most intensively studied macronutrient. The reasons for this lie in
its association, since the 1960's, with the process of accelerated
eutrophication in water bodies deriving both from fertiliser and
slurry application in the rural environment and with detergent
disposal in the urban environment (Reynolds, 1978). According to
Meybeck(1982) the per capita loadings of P in river waters correlates
directly with the Demophoric Index of development defined as the
ratio between per capita energy consumption over a given period and
the energy required to satisfy basic human needs.

The role of phosphorus is often stressed not simply because it is
an essential nutrient, but often because it is the limiting element
for plant growth and aquatic productivity (Vollenweider, 1968;
O.E.C.D., 1982). Phosphorus is therefore often seen as the single
most important factor in the deleterious fertilisation of lake waters
(Li et al, 1972; Gower, 1980).

The record of phosphorus in lake sediments is complicated by the

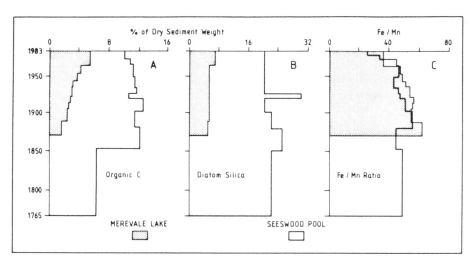

Figure 2   Organic   carbon   (a),   Diatom silica (b) and the Fe :   Mn
ratio (c) of Merevale Lake and Seeswood Pool sediments.

many   processes   responsible   for   its   erosion,   transportation   and
deposition within the sediment column.   Mackereth (1966),   and Wentz
and   Lee,   (1969)   have   postulated six modes by which   phosphorus   is
deposited in lake sediments which include:
   (a) Detrital phosphorus minerals eroded from the catchment.
   (b) Phosphorus co-precipitated with iron and manganese.
   (c) Sorbed phosphate.
   (d) Phosphate associated with carbonates.
   (e) Phosphorus associated with autochthonous organic matter.
   (f) Phosphorus associated with allochthonous organic matter.
   Of the main P minerals,   the most commonly identified in soils and
fresh water sediments is apatite,   occuring in many different phases,
and   predominantly derived in lake sediments from eroded soils (Jones
& Bowser,   1978).   However, many other P minerals have been identified
or   hypothesised to exist in lake sediments (cf Williams   and   Mayer,
1972;   Stumm   and Morgan,   1981) and include iron phosphates such   as
vivianite   and strengite and the aluminium phosphates,   variscite and
wavellite.   Although no mineralogical study has yet been performed on
the Seeswood Pool and Merevale Lake sediments, the diagnostic mineral
vivianite,   which usually occurs in strongly reducing conditions, has
been   observed in the Seeswood   sediments   only.   Palaeolimnological
interpretation   of the trends in the P content of lake sediments   not
only   requires   the   fractionation of this element into   organic   and
inorganic forms,   but also requires analysis of other elements   which
may   affect   its precipitation in the   sediments.   Mackereth (1966)
identified   the   possibility   of   a   close   association   between   the
precipitation   of   phosphorus with both Fe and Mn which in   part   may
reflect   variations in the oxidising/reducing conditions at the   mud-
water interface rather than the loading of P to the lake. In the same
study, Mackereth suggested that the final concentration of phosphorus
residing in the sediment would be a function of the rate of supply to
the   lake basin,   the efficiency of the precipitating mechanism,   the
rate of accumulation of the sediment as a whole and the rate of   loss
of   phosphorus from the recently sedimented material.   Bortleson and

Lee, (1974) for example, examined the phosphorus content of sediments from 5 Wisconsin lakes and concluded that historical patterns of Fe and Mn deposition were closely related to P deposition in most of the cores. Furthermore, Brenner, (1983) in a study of Mayan cultures associated with P loading to Lake Quexil, Guatemala, identified a correlation between deposition rates of P, carbonate and silica and suggested, but did not specify, a chemical link between them. Nevertheless, Brenner's study suggested that the P levels in the Guatemalan sediments was a direct index of cultural development. Similar studies by Engstrom and Swain (1985) based on an analysis of the sediments of Lake Minnetonka, Minnesota, identify an increase in sediment P directly attributable to recent trends in eutrophication.

The analysis of phosphorus in the two Midland Lakes has identified a range of total P concentrations between 500 and 1700 μg per g. The levels vary significantly between the two lakes and between the upper and lower zones of Seeswood Pool sediments reported by Syers et al, (1973) of c 580 – 7000 μg per g of dry sediment. The form in which phosphorus occurs in the sediment was divided into three fractions as shown in Figure 3. The total phosphorus content of both lakes is dominated by organic P despite the fact that the organic matter levels of the Merevale Sediments are considerably lower than the Seeswood Pool sediments. In both lakes, organic P accounts for between c 55 and 80% of the P contained in the sediments, with an average for both lakes of around 65%. This level is almost five times that reported for Lake Erie sediments, which have a much lower organic matter content, by Williams et al (1976) and is slightly higher than the maximum levels reported by Bostrom (1982) in a review

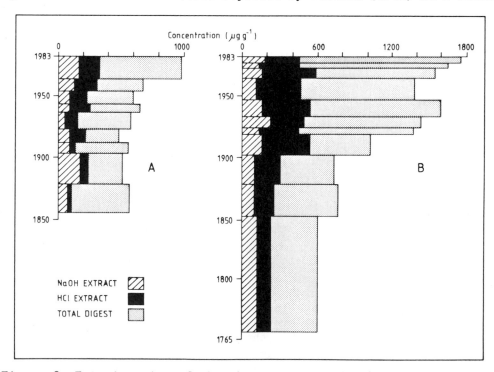

Figure 3   Fractionation of phosphorus content in the
lake sediments.

of 15 lakes throughout the world.   Apatite P,   derived from analysis
of   the acid extraction,   averages 52% of inorganic phosphorus in the
Merevale sediments but is significantly higher,   68%, in the Seeswood
sediments.   This increase in the relative proportion of apatite P may
reflect   the   generally   higher levels of catchment erosion   in   this
basin.   The concentration of Fe and Al bound P extracted with sodium
hydroxide   is similar in both lakes (Table 4) and although   remaining
relatively constant in concentration with depth in the sediment, does
not   appear to reflect total Fe content of the lake sediment which is
considerably   higher   in   Seeswood   Pool   (Table 2).   Despite   the
differences in absolute concentration,   the Fe and Mn levels in   both
lakes, with their associated ratios given in Figure 2c, suggest that
redox   conditions   at   the   lake bed have   not   changed   dramatically
throughout   the   lifespan of either lake   and   that   P   precipitation,
unlike   the   Bortleson and Lee (1974) study,   is not associated   with
this process.

Since   the   precipitation of P in these catchments appears   to   be
independent of co-precipitation with Fe and Al,   it is suggested that
the   concentrations given in Figure 3 and Table 4 reflect loading   to
the lake from the catchment and the dominance of processes a, e and f
listed above.

Trace metals concentration in lake sediments

Analysis   of   lake   sediment   cores   for   trace   metals   can   provide
important   information   regarding   the   fate   and   cycling   of   these
elements   through environmental systems.   Many   investigations   have
demonstrated   that   the   fine   sediments   contained   in   lakes   and
reservoirs   form   an important sink for these metals as   they   settle
through   the   water column.   In some situations,   these   metals   are
directly correlatable with the onset of mining in a particular region
such as in   Clearwater Lake,   Southeast Missouri (Gale et al 1976) or
to   the   discharge of waste from industrial processing (Harding   and
Whitton (1978).   In many cases, however, there is no direct correlation   between heavy metals content and mining or industrial pollution
and an atmospheric origin is hypothesised.   For example, Christensen
and Chien,   (1981), examined the metal content of sediments from Lake
Michigan and from Green Bay and established a clear post 20th century
increase   in   the   concentrations of Pb,   Zn and Cd.   In   the   United
Kingdom, Taylor (1979), found dramatic increases in the concentration
of   Pb,   Zn,   and   Cu over the last 130 years   and   attributed   these
increases   to   the atmospheric supply of pollutants.   The uptake   of
heavy metals by diatoms was seen as an important secondary factor   in
controlling sediment concentrations. Much published evidence suggests
that   the   average concentration of heavy metals in   lake   sediments
reflects proximity to source of emission (cf Rippey,   et al 1982) yet
the   existence   of   heavy metals in remote lakes   such   as   that   in
Ellesmere Island (Muller and Barsch,   1980) and the presence of heavy
metals   in arctic and alpine ice cores (Nriagu,   1979) testify to the
global distribution of atmospheric pollutants.

One of the most comprehensive reviews of anthropogenic and   global
emissions   is provided by Nriagu (1979).   It was calculated in   this
review   that   the   present day annual anthropogenic emission of   Pb

exceeded the natural rate by well over an order of magnitude, whereas the figures for Cu, Ni and Zn were over 300%, 200% and 700% of natural emissions respectively. The global emissions of the 4 metals under consideration in the present study for decadal data for the years from 1850 to 1980 are given in Figure 4. In terms of gross emissions, the metals Pb and Zn dominate the outputs, the former predominantly derived from oil consumption and the latter predominantly derived from non-ferrous metal production but also from wood combustion and iron and steel production (Nriagu, 1979).

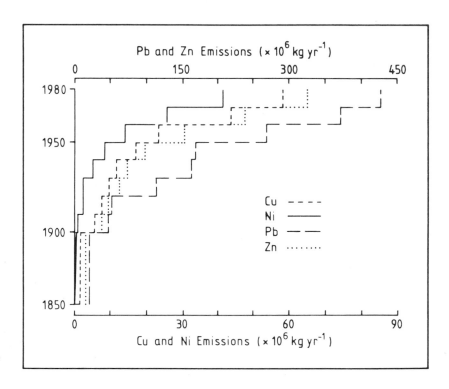

Figure 4 Historical anthropogenic emissions of trace metals
(based on data in Nriagu, 1979).

The summary statistics for the concentration of trace metals in Merevale Lake and Seeswood Pool of Table 2 show the dominance of Zn, Ni and Cu in the sediments, with concentrations ranging from around 785 μg per g for Zn in the upper sections of the Merevale cores to under 40 μg per g for Pb in the basal sequences of Seeswood Pool. Ni levels are similar to those reported in the sediments of Lough Neagh (Rippey et al 1982), whereas Zn, Cu and Pb levels are considerably higher. Differences in concentration, however, will not only reflect trace metal supply but will also relate to the dilution caused by changes in sedimentation rate. This point is demonstrated clearly in the comparative trace metal profiles of Figure 5. The profiles for Merevale Lake show a remarkable degree of similarity with the anthropogenic emission data of Figure 4, and are suggested to accurately reflect these trends since sedimentation rates vary little throughout time at this site (Table 2). In contrast, the Seeswood

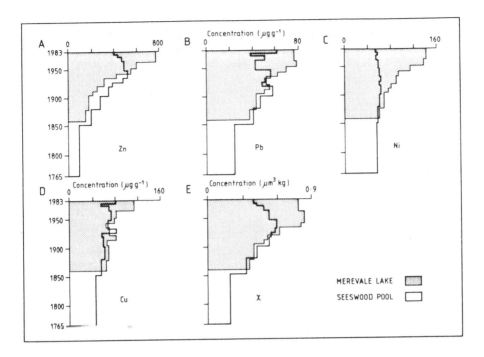

Figure 5   Trace metals concentration (a-d) and the magnetic susceptibility (e) of the lake sediments.

Pool sediments either show little downcore variation or, in the case
of Zn and Pb in particular, a reversal of the increasing trend in the
upper levels.   It is suggested that this reversal is the direct
result of increasing catchment erosion delivering sediment of lower
trace metals concentration to the lake basin. If the entire catchment
receives fallout, and the source of the sediment remains constant,
then it would be expected that the relative proportion of trace metal
concentrations in the sediment would remain constant (with an
increased loading to the lake).   However, a shift in sediment source
from a trace metal enriched topsoil to non enriched channel sources
might be expected to produce the observed trends in the Seeswood Pool
concentration record.

Also included in Figure 5 is a measure of magnetic susceptibility
for the bulked lake sediment samples (cf Thompson and Oldfield,
1986). In this particular instance, this easily determined parameter
appears to be closely related to the average trace metal concentration data in both lake systems.   Correlation of metals concentration
data with magnetic susceptibility values in both lakes indicates that
all relationships, except for Cu, are statistically significant above
the 95% level.

Phosphorus and trace metal loadings

It was suggested earlier that interpretation of sediment chemistry
data would be incomplete unless units of both concentration and
loading were utilised.   These data for total P and Zn in the lake

sediments are given in Figure 6. Direct comparison with other studies
reveals that for total P loading, Merevale Lake and Seeswood Pool lie
midway between upland grazing ecosystems in N Wales (Dearing et al,
1981) and intensive arable systems in S. Sweden (Dearing, unpub
data).

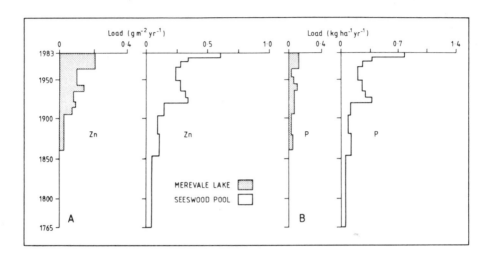

Figure 6   Zinc (a) and the total phosphorus (b) loading
            to the lake sediments.

The Zn profiles can be compared with review data provided by
Rippey et al (1982), which shows Zn loadings varying from 0.593 g m^{-2}
per yr in Lake Erie to 0.065 g m^{-2} per yr in Woodhull Lake in a remote
area of the Adirondack Mountains of New York State. The calculated
flux for Merevale Lake probably reflects atmospheric loading to this
lake basin, whereas the difference between these levels and the
higher values of Seeswood Pool is likely to be a direct consequence
of the erosion of trace metal enriched topsoil and transport to the
lake basin. As suggested in the previous section, the concentration
profiles of Figure 5 would imply a change in sediment source in the
Seeswood catchment on the basis of this argument, yet it still
contributes more enriched topsoil than the Merevale site. This point
is substantiated by the development of mixing models based on the
magnetic characteristics of soils and sediments reported elsewhere
(Dearing and Foster, 1987).

Hydrological implications

As yet, few hydrological studies have considered the general
applicability of data derived from an analysis of lake-catchment
systems, despite the fact that almost a decade has passed since the
publication of Oldfield's (1977) important general review paper
devoted to the methodological framework. Direct links between
catchment processes and limnic sedimentation are as yet imperfectly
understood although some attempts have been made to link paleohydro-
logical interpretations to the conceptual framework in terms of

sediment yield data and palaeodischarge reconstructions based on lake level fluctuations (cf Foster et al, 1985; Dearing and Foster, 1986; 1987).

The analysis of lake sediment chemistry data, such as that presented here, could add an important dimension to palaeohydrological studies over timescales poorly represented in the literature. Phosphorus concentrations and loadings, in association with fertiliser application rates and crop yield information, for example, could provide important quantitative data on the impact of human activity on both terrestrial and aquatic components of the P cycle which could form the basis of simple simulations. Similar analyses based upon the trace metals record contained in lake sediments may permit quantification of the global and local distribution of atmospherically derived pollutants.

Conclusions

In relation to the questions posed in the introduction, the following conclusions may be drawn.

(a) Both heavy metals and phosphorus concentrations and loadings may be accurately determined in lake sediments.

(b) The P profiles in Seeswood Pool appear to reflect the land use history of the region and could be used as a basis for modelling land use impact.

(c) Sites like Merevale lake which demonstrate slow sedimentation and little variation in sediment yield are most appropriate for quantifying trends in atmospheric pollution.

References

Allen, S.E., Grimshaw, H.M., Parkinson, J.A. and Quarmby, O. (1974) Chemical analysis of ecological materials. Blackwell, Oxford.

Bengtsson, L. and Enell, M. (1986) Chemical analysis. In: Berglund, B. (ed.), Handbook of Holocene Palaeoecology and Palaeohydrology. pp.423-454. Wiley, London.

Bortleson, G.C. and Lee, G.F. (1975) Recent sedimentary history of Lake Monona, Wisconsin. Water Air and Soil Pollution 4, 89-98.

Bostrom, B., Jansson, M. and Forsberg, C. (1982) Phosphorus release from lake sediments. Arch. Hydrbiol. beih. Ergeben Limnol. 18, 5-59.

Brenner, M. (1983) Palaeolimnology of the Peten lake District, Guatemala. Hydrobiologia. 103, 205-210.

Christensen, E.R. and Chien, N.K. (1981) Fluxes of arsenic, lead zinc and cadmium to Green Bay and Lake Michigan sediments. Environ. Sci. Technol. 15 (5), 553-558.

Dearing, J.A. (1986) Core correlation and total sediment influx. In Berglund, B. (ed.) Handbook of Holocene Palaeoecology and Palaeohydrology. pp 247-272. Wiley, London.

Dearing, J.A. and Foster, I.D.L. (1986) Lake sediments and palaeohydrological studies. In Berglund, B. (ed) Handbook of Holocene Palaeoecology and palaeohydrology. Wiley. 67-90.

Dearing, J.A. and Foster, I.D.L. (1987) Limnic sediments used to reconstruct sediment yields and sources in the English Midlands since 1765. In Geomorphology '86. Proc 1st Int. Conf. in Geomorph. Wiley, London. Vol. 1., pp. 853–868.

Dearing, J.A., Elner, J.K. & Happey-Wood, C.M. (1981) Recent sediment flux and erosional processes in a Welsh upland lake-cachment based on magnetic susceptibility measurements. Quaternary Research. 16, 356–372.

Engstrom, D.R. and Swain, E.B. (1985) The chemistry of lake sediments in time and space.   Paper presented at 4th Int. Symp. Palaeolimnology. Ossiach, Carinthia. Austria. 2nd–7th Sept.1985.

Engstrom, D.R. and Wright, H.E. (1984) Chemical stratigraphy of lake sediments as a record of environmental change.   In Hawarth, E.Y. and Lund, J.W.G. (eds.) Lake Sediments and Environmental History. pp 11–67.

Foster, I.D.L. and Dearing, J.A. (1987) Lake catchments and environmental chemistry: a comparative study of contemporary and historical catchment processes in Midland England. GeoJournal.

Foster, I.D.L., Dearing, J.A., Simpson, A., Carter, A.D. and Appleby, P.G. (1985) Lake catchment studies of erosion and denudation in the Merevale Catchment, Warwickshife, UK. Earth Surface Processes and Landforms 10, 45–68.

Foster, I.D.L., Dearing, J.A. and Appleby, P.G. (1986) Historical trends in catchment sediment yields: a case study in reconstruction from lake-sediment records in Warwickshire, UK. Hydrological Sciences Journal, 31 427–443.

Foster, I.D.L., Dearing, J.A., Charlesworth, S.M. and Kelly, L.A. (1987) Paired lake-catchment studies: a framework for investigating chemical fluxes in small drainage basins. Applied Geography. 7, 115–133.

Gale, N.L., Bolter, E. and Wixwon, B.G. (1976) Investigation of Clearwater Lake as a potential sink for heavy metals from lead mining in southeast Missouri. In Hemphill, D.D. (ed) Trace Substances in Environmental Health. Univ. Missouri, Columbia, 1–16.

Gower, A.M. (ed) (1980) Water quality in catchment Ecosystems. Wiley, London.

Harding, J.P.C. and Whitton, B.A. (1978) Zinc, cadmium and lead in water sediments and submerged plants of the Derwent Reservoir, Northern England. Water Research. 12, 307–316.

Hesse, P.R.: A Textbook of Soil Chemical Analysis. Murray, London. 1971.

Hieltjes, A.H.M. and Lujklema, L. (1980) Fractionation of inorganic phosphates in calcareous sediments. J. Environ. Qual. 9 (3) 405–407.

Jones, B.F. and Bowser, C.J. (1978) The mineralogy and related chemistry of Lake sedimengs. In Lerman, A. (ed.) Lakes; Chemistry, Geology, Physics. pp 179–235. Springer, New York.

Li, W.C., Armstrong, D.E., Williams, J.D.H., Harris, R.F. and Syers, J.K. (1972) Rage and extent of inorganic phosphate exchange in lake sediments. Soil Sci. Soc. Amer., Proc. 36 (2), 279–285.

Mackereth, F.J.H. (1966) Some chemical observations on post-glacial lake sediments.   Philosophical Transactions of The Royal Society, London B250, 165–213.

Meybeck, M. (1982) Carbon, nitrogen and phosphorus transport by world rivers. Amer. J. Sci: 282, 401–450.

Muller, G. and Barsch, D. (1980) Anthropogenic lead accumulation in the sediments of a high Arctic Lake, Ooblayouh Bay, N. Ellesmere

Island. N.W.T. (Canada). Environmental Technology Letters 1, 131–140.

Nriagu, J. (1979) Global Inventory of natural and anthropogenic emissions of trace metals to the atmosphere. Nature 279, 409–411.

OECD. (1982) Eutrophication of Waters; Monitoring, Assessment and Control. OECD Paris.

Oldfield, F. (1977) Lakes and their Drainage Basins as units of Ecological Study. Progress in Physical Geography 1, 460–504.

Oldfield, F. and Appleby, P.G. (1984) Empirical testing of Pb dating models for lake sediments. In Hawarth, E.Y. and Lund, J.W. (eds.) Lake Sediments and Environmental History. Leicester University Press, Leicester.

Reynolds, C.S. (1978) Phosphorus and the eutrophication of lakes – a personal view. CIBA Fnd. Symp. 13–15 9 1977. Phosphorus in the Environemts: Its Chemistry and Biochemistry. Elsevier. 201–216.

Rippey, B., Murphy, R.J. and Kyle, S.W. (1982) Anthropogenically derived changes in the sedimentary flux of Mg, Cr, Ni, Cu, Zn, Hg, Pb and P in Lough Meagh, Northern Ireland. Environmental Science and Technology 16, 23–30.

Stumm, W. and Morgan,J.J. (1981) Aquatic Chemistry 2nd ed. Wiley.

Syers, J.K., R.F.Harris & Armstrong, D.E. (1973). Phosphate chemistry in lake sediments. J. Environ, Qual. 2 (1), 1–14.

Thompson, R. and Oldfield, F. (1986) Environmental magnetism. Allen and Unwin.

Taylor, J. Hamilton– (1979) Enrichment of Zinc, lead and copper in recent sediments of Windermere, England. Environ. Sci. Technol. 13 (6), 693–697.

Vollenweider, R.A., (1968) Scientific fundamentals of the eutrophication of lakes and flowing waters, with particular reference to nitrogen and phosphorus as factors in eutrophication. OECD Paris. Technical Report DA 5/SCI/68.27.

Wentz, D.A. and Lee, F.G. (1969) Sedimentary phosphorous in lake cores – analytical procedures. Environ. Sci. Technol. 3 (8), 750–754.

Williams, J.D.H. and Mayer, T. (1972) Effects of sediment diagenesis and regeneration of phosphorus with special reference to lakes Erie and Ontario. In Allen, H.T. and Kramer, J. (ed) Nutrients in Natural Waters. Wiley. 281–316.

Williams, J.D.H., Murphy, T.P. & Mayer, T. (1976) Rates of accumulation of phosphorus forms in Lake Erie Sediments. Journal of the Fisheries Research Board, Canada 33, 430–439.

The Influence of Climate Change and Climatic Variability on the Hydrologic Regime and Water Resources (Proceedings of the Vancouver Symposium, August 1987). IAHS Publ. no. 168, 1987.

Critical precipitation conditions for landslide and tree ring responses in the Rokko Mountains, Kobe, Japan

Kenji Kashiwaya
The Graduate school of Science and Technology,
Kobe University, Kobe 657, Japan
Takashi Okimura & Takeshi Kawatani
Faculty of Engineering
Kobe University, Kobe 657, Japan

ABSTRACT   Heavy rainfall in the vicinity of the Rokko Mountains, Kobe, can cause landslides and debris flows. Analysis of the annual summation of heavy rainfall (over 100mm/day) shows that the rainfall has dominant periodicities at about 30,10-13,and 5-7 years, and that the maximum years of a harmonic function with the periods, correspond with severe landslides and debris flows. Analytical results of tree rings in the district indicate that the sequences of the tree ring width have a dominant periodicity of around 30 years. Cross spectral analysis between the rainfall and the ring width shows high coherency in the periodicity of about 30 years.

Conditions de précipitation critiques pour les glissements de terrain et lien avec la largeur des anneaux d'âge des arbres dans les Monts Rokko, Kobe, Japon

RESUME   Les fortes pluies de la région des Monts Rokko, Kobe,   peuvent provoquer des glissements de terrain et des écoulements de gravats. L' analyse du total annuel des précipitation importantes (plus de 100 mm/jour) montre que la pluie suit un cycle avec des périodes principales de 30, 10 à 13, et 5 à 7 ans, et que les ans plus grands extrême de la fonction harmonieuse avec les périods coincident avec d'importants glissements de terrains et écoulements de gravats. L'etude des anneaux d'âge des arbres de la région indique que la largeur de ces anneaux varie selon un cycle avec période d'environ 30 ans. La comparison du cycle de précipitations et de celui de la largeur des anneaux montre un accord au niveau de la période de 30 ans.

Introduction

Heavy rainfalls often cause various changes in the earth surface environment by landslides, slope failure and so on. It is known that different geomorphic processes have influence on vegetational conditions, such as tree ring evolution (cf. Alestalo, 1971; Shroder, 1978). The authors also reported a preliminary study on the relation between rainfall and tree ring width for the area discussed in this paper (Kashiwaya et al., 1986). The changes of tree structure (ring width, density, etc.) have been used to reconstruct the variation of climatic conditions in various parts of the world (cf. Hughes et al., 1982). In this paper, we shall discuss the time variations of characteristic heavy rainfall as it relates to landslides and tree ring width in order to recognize the relationship between them.

Kobe city at the south foot of Rokko Mountains is noted for its severe natural disasters caused by heavy rainfall. There are frequent landslides and debris flows and particularly in 1938, 1961, and 1967, many people were killed or injured by them (cf. Kashiwaya et al., 1984). Therefore, it is necessary to find out the periodicity of those natural disasters not only for recognizing the time variation of erosional force (cf. Kashiwaya, 1986) but for predicting the heavy rainfalls concerning such landslides.

Trees employed for the analysis were all Pinus densiflora of about 100 years old. They were sampled at the sites considered to be in the highest precipitation areas during natural disasters (Figure 1). Data from the Kobe Marine Observatory, near the sampling sites, were used for the precipitation analysis. Sampling areas are composed of granitic rock (Rokko granite) – erosional landforms here are thought to be related to the weathering of the granite (Suzuki et al., 1977).

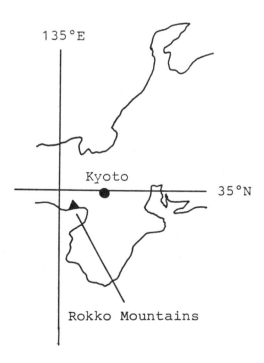

135°E

Kyoto

35°N

Rokko Mountains

Figure 1 Location of sampling site (Rokko Mountains).

Analysis of rainfall

We shall employ here the annual summation of excess rainfall over 100 mm/day (excess rainfall) as the characteristic rainfall concerning

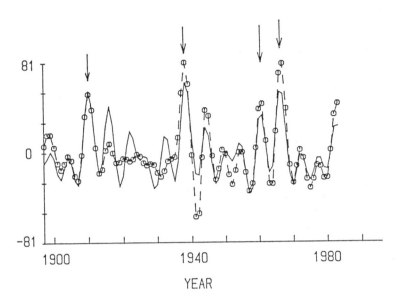

Figure 2 Time variation in the excess rainfall. The solid line
represents a synthesized curve from harmonic analysis.
The circles and dotted line indicate filtered original
data. The arrows indicate years of severe natural
disasters due to heavy rainfalls.

landslides. Circles and the dotted line in Figure 2 indicate the time
variation of filtered original excess rainfall. Spectral analysis
(maximum entropy method) is applied to the variation of the rainfall
to identify dominant periodicities in it. Figure 3 shows an example
of the power spectral density for the variation. It has dominant
periodicities of around 30, 10-13, and 5-7 years. The periods of 30
years and 10-13 years are close to those of the Brückner cycle and
the sunspot cycle respectively.

Next, harmonic analysis was performed using the dominant periods.
The solid line in Figure 2 shows the synthesized excess rainfall on
the basis of the analysis. From this figure, we can see the time
variation of the excess rainfall concerning severe landslides (arrows
in the figure indicate the years of severe natural disasters by heavy
rainfall) is approximately reconstructed from the periods in this
time domain.

Analysis of tree ring

Ring width of trees was measured on cross sections for various
directions and averaged values were used for statistical analysis.
Examples of time variation of the ring width are shown in Figure 4.
These examples show that the tree ring widths are comparatively
narrow around the years with heavy rainfalls, but a few years later,
they become wider again. This may indicate that trees are damaged by
the earth surface process from heavy rainfalls, but the processes, on

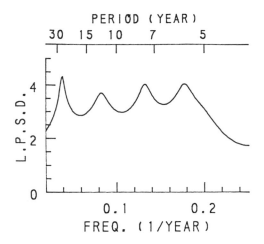

Figure 3   Power spectral density of the excess rainfall
(L.P.S.D. means Log Power Spectral Density).

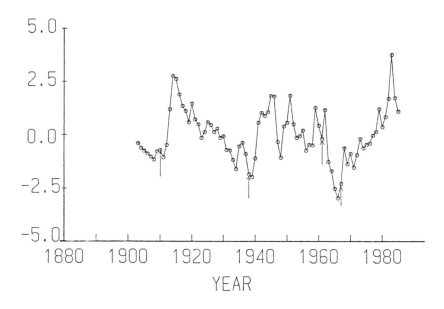

Figure 4a Time variations in filtered ring width (Example 1).

the other hand, provide a convenient earth surface environment for
trees to grow again after a few years in these cases. Some trees
display the same trend in time variations of ring width.

Spectral analysis is also applied to the time series of tree ring
widths in order to identify dominant periodicities in them. Figure 5
shows some examples of the resultant power spectral density for the
series. These spectra are characterized by a definite peak corresponding to around 30 years. We can find this period in all tree rings

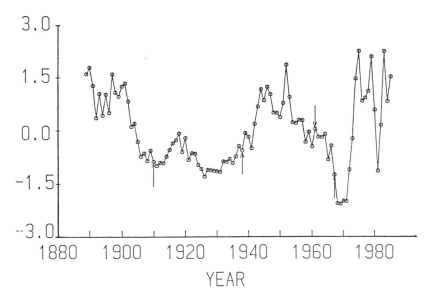

Figure 4b Time variations in filtered ring width (Example 2).

analyzed here. Therefore, it may be said that at that period the connection between the excess rainfall and the tree ring width is very strong, though the reason is unclear. We can also see a rather broad and unclear peak around 10 years close to the period of the sunspot cycle in some tree rings.

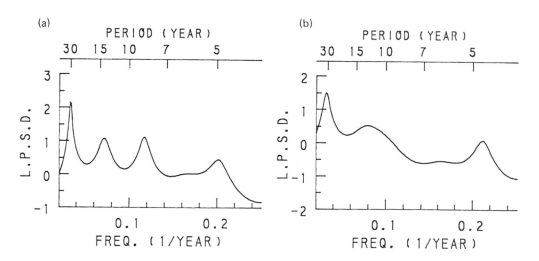

Figure 5   Power spectral density of the changes in ring width.

Next, we performed a cross-spectral analysis to examine in detail the relationship between the rainfall variation and the ring width

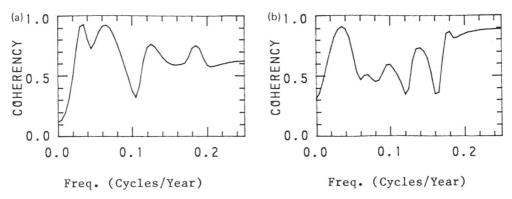

Figure 6 Coherency between the excess rainfall and ring width.

variation. Figure 6 shows the coherency between the rainfall in Figure 2 and the tree ring width in Figure 4(a), (Fig 6(a)) and Figure 4(b), (Fig 6(b)). The coherencies between the variations are surprisingly high for the 30-year period in both cases. This result supports the above inference.

The phase spectra from the analysis are shown in Figure 7. At the period about 30 years, the phase lag between the rainfall and the tree ring width is about a few years in one case (Figure 7a, cf. Figure 4a) and a half cycle of the period in another (Figure 7b cf. Figure 4b). This means that the extrema of the tree ring width appear about a few years after or 10-15 years after the peaks of the rainfall respectively. It is compatible with the inference discussed above the curves in Figure 4.

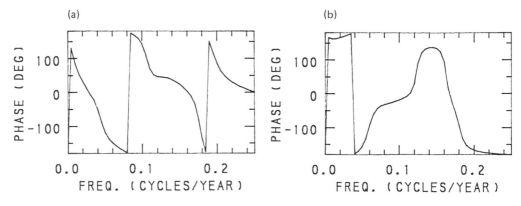

Figure 7 Phase spectra between the excess rainfall and ring width.

Concluding remarks

From the results described above, we can say that there is a definite period of about 30 years in the periodic variations of the tree ring width. It is also said that the excess rainfall has dominant periodicities of 30, 10-13, and 5-7 years. The coherency at the period of 30 years is very high. Therefore, it may be inferred that there is a strong correlation between the excess rainfall relating to the change of the earth surface and the tree ring width.

AKNOWLEDGEMENTS   The authors would like to express their thanks to the staffs of the Department of Parks and Green Land, Public works Bureau of Kobe City Government for their help in sampling trees, and Mr. H. Yoshida for his help in their analysis.

References

Alestalo, J. (1971): Dendrochronological interpretation of geomorphic processes, Fennia, 105, 1-140.
Hughes, M.K., Kelly, P.M., Pilcher and LaMarche, V.C. Jr. (1982): Climate from tree rings, Cambridge University Press Cambridge.
Kashiwaya, K., Okimura, T., Hirano, M. and Okuda, S. (1984): Geomorphological aspects of landslide and time change of rainfall character in the south-west part of Rokko-san Mountains, Annuals, D.P.R.I., Kyoto Univ., 27B, 397-408.
Kashiwaya, K. Okimura, T. and Kawatani, T. (1986): Dendrochronoligical informations and hydrological conditions for landslides in Mt. Futabisan area of Rokko Mountains, Trans. Japan. Geomorph. Union 7, 281-290.
Kashiwaya, K., (1986): A mathematical model of the erosional process of a mountain, Trans. Japan. Geomorph. Union 7, 69-77.
Shrodder, J.F. Jr. (1978): Dendrogeomorphological analysis of mass movement on Table Cliffs Plateau, Utah, Quaternary Research 9, 168-185.
Suzuki, T., Hirano, H., Takahashi, K., and Yatsu, E.(1977): The interaction between the weathering processes of granites and the evolution of landforms in the Rokko Mountains, Japan, Bull. Fac. Sci. Eng. Chuo Univ. 20, 343-389.

**Climatic variability and
stochastic hydrology**

The Influence of Climate Change and Climatic Variability on the Hydrologic Regime and Water Resources (Proceedings of the Vancouver Symposium, August 1987). IAHS Publ. no. 168, 1987.

The role of stochastic hydrology in dealing with climatic variability

Marshall E. Moss & Gary D. Tasker
U.S. Geological Survey
Reston, Virginia 22092 U.S.A.

ABSTRACT   Climate and the weather that it comprises provide the inputs and one of the primary driving forces to the hydrologic cycle. Because the climate and its interaction with hydrology cannot be described with full accuracy and precision, stochastic hydrology has been a useful technology. Three purposes for the use   of stochastic hydrology can be identified: (1) a language for description of the uncertainties of hydrological processes, (2) a vehicle for the development of hydrologic insight, and (3) a tool in water-resources decisionmaking. Each purpose will be reinforced during the next several decades in the light of anticipated climatic reactions   to anthropogenic forces. However, stochastic hydrology must be carried to a higher scientific plane to deal with the transitory expectations of the future.

Introduction - Climate yesterday, today, and tomorrow

To many laymen, climate is a static concept: it is the expected weather at a given time of year. Until the last generation or so, even climatologists considered their field to be at least quasi-static. Thirty-year averages of temperature, precipitation, humidity, insolation, and wind (each expressed seasonally) were thought to be apt descriptors of a location's climate. However, today, next month's weather is a valid topic for inquiry by the climatological community. Thus, it might be stated that the most variable aspect of climate over the recent past has been its definition. Nevertheless, for purposes of this paper, climatic variability is defined as changing patterns of weather of extended duration beyond the time scale at which weather forecasters can demonstrate their skill with significance. Less formally, it could be said that, in general, climate influences the decisions on the type of clothes that we buy, while weather influences the decisions on the type of clothes that we wear.
     Why have climatologists recently begun to concern themselves with the transitory aspects of their realm of science? First and probably foremost, it is because they now can, whereas before the advent of modern measurement and computing technologies, they were limited mostly to speculation on this matter. An added driving force for change in climatology is the increasing time horizons of societal decisions that are being made today. As an extreme example, some strategies for the disposal of nuclear wastes with half-lives well in excess of the longest of continually recorded weather (or climate)

records are very sensitive to the climate of the disposal site over the time scale at which the waste can be considered to be hazardous.

A perhaps less extreme, but more pervasive example is impacts of changing concentrations of greenhouse gases in the atmosphere on the magnitude and distribution of water resources around the world (World Climate Program, 1986). The concensus now is that significant changes in the patterns of temperature and precipitation can occur in the timeframe of 50 to 100 years, which is the relevant time scale for most major water-resources plans and projects. Thus, the hydrologist can ill afford to disregard the new type of information that is being generated in the climatology field.

Stochastic hydrology as a tool

Because weather and climate, which provide the inputs and one of the primary driving forces to the hydrologic cycle, vary, the stores and flows of water on and under the surface of the earth vary in space and time. The composite knowledge of these stores and flows is the science of hydrology, which to date has not been perfected to the state where the water resources can be described with full accuracy and precision. Therefore, the study of the random nature of water resources has come to be known as stochastic hydrology, which is the application to water resources of the classical field of statistics known as stochastic processes. Parzen (1962) states that the purposes for the study of stochastic processes are:

(a) "to provide a language in which assumptions may be stated about observed-time series."

(b) "to provide insight into the most realistic and/or mathematically tractable assumptions to be made concerning the stochastic processes that are adapted as models for time series."
Obviously, the time series to be interpreted in the above light for stochastic hydrology are the time series of observations of water resources at diverse locations above, on, and beneath the surface of the Earth.

In addition to Parzen's two purposes, stochastic hydrology has an added category of purpose that frequently is predominant when stochastic studies of hydrology are undertaken, which is:

(c) to provide a tool for obtaining approximate solutions to the complex problems that arise in making water-resources decisions.
The remainder of this paper will discuss some classical examples of stochastic hydrology from each of these categories and attempt to reach some tentative conclusions about the potential role of stochastic hydrology in light of climate variability, which in all likelihood will entail more or less permanent perturbations of an anthropogenic origin.

The language of stochastic hydrology

Much of the language of stochastic hydrology has evolved from statistical hydrology, which is an earlier branch of hydrologic science that has been practically subsumed by the more modern stochastic branch. Terms like mean-annual flood, 7-day 10-year low flow, and mean-annual sediment load (see Riggs, 1985 for definitions of these terms) are basic statistical descriptors that serve as the underpinnings of stochastic hydrology. However, statistical hydrology does not take into account the time-series aspects of the flows and stores of water that comprise the hydrologic cycle; stochastic hydrology does this and thus its jargon becomes richer than that of its predecessor.

Persistence, which is the temporal interdependence of the measures of a random phenomenon, is the keyword in defining the difference between statistical and stochastic hydrology. Hydrologic phenomena exhibit interdependence through time; consecutive observations of a hydrologic variable usually are not statistically independent like the classical games of chance such as the tossing of coins or dice. The simplest form of serial dependence is caused by storage of water in the hydrologic system (Moss and Bryson, 1974), which can be reasonably well described by autoregressive moving-average (ARMA) time series (Box and Jenkins, 1970). The simplest form of climatic variability, the change of seasons, can be seen to impact not only the statistical properties of streamflow, but also its simple persistence as measured by the first-order lag correlation coefficient as shown in Figure 1.

The simple persistence caused by water storage has been found inadequate, however, in describing more complex patterns of hydrologic occurrences. For example, Hurst (1951) found that the flows of the Nile River had a stronger pattern of persistence than can be explained by ARMA descriptions. This complex form of persistence came to be known as the Hurst phenomenon (Wallis and Matalas, 1970) and stimulated one of the most active areas of research in stochastic hydrology during the late 1960's and 1970's. Bras and Rodriguez-Iturbe (1985) provide a recent review of these activities. Nevertheless, a fully satisfactory explanation of the Hurst phenomenon is yet to be realized.

One new line of attack in exploring the Hurst phenomenon might be that of studying the storage of energy within the hydrologic cycle. Recent studies of the El Nino Southern Oscillation (Rasmusson, 1984) have demonstrated dramatic impacts on terrestial hydrologic processes caused by storage and release of energy in the Pacific Ocean. With the vast energy reservoirs that the oceans comprise, it is not unreasonable to speculate that they could be at least one source of the hydrologic persistence that cannot be explained by the storage of water in the terrestial phase of the hydrologic cycle.

Like the climatologists, hydrologists have been reasonably comfortable with stationarity or quasi-stationarity assumptions in stochastic hydrology. However, all hydrologists recognize that anthropogenic effects cause non-stationarities of varying degrees in most hydrologic time series today. As a drastic example, the mouth of the Colorado River at the head of the Sea of Cortez (Gulf of California) stood essentially dry for decades because of storage and diversion of its water. From hydrologic reconstructions using tree rings in the Colorado River drainage basin, Stockton and Jacoby (1976) have estimated that the long-term average runoff of the upper part of the Colorado River is in excess of 500 cubic meters per second. A stochastic hydrologist would have to be brash indeed to model such a situation as a stationary process.

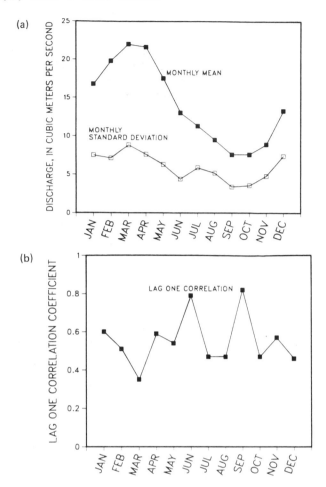

Figure 1   (a) Mean and standard deviation of monthly flows and
           (b) correlation coefficients between flows for indicated
           month and previous month; Toccoa River, near Dial,
           Georgia (1913-1986).

    The   extreme example of non-stationarity just mentioned was caused
primarily by man's development in the downstream part of the basin of
the   water resource derived primarily from the upstream part   of   the
basin.   Development   of   other   natural   resources,   primarily   fossil
fuels, may be on the verge of causing less extreme but more pervasive
changes   in the stochastic patterns of hydrologic phenomena.   General
circulation   models   have   shown   significant   changes   in   regional
patterns   of   precipitation   and temperature occurring   in   the   next
century   if consumption of fossil fuels continues the current   trends
of   building up concentrations of greenhouse gases in the atmosphere.
Thus,   it   would seem that,   if stochastic hydrology is to   remain   a
healthy   and   useful component of hydrologic science in   the   Twenty-
First Century,   its "language" must be expanded to describe processes
that are in transition.

Hydrologic insight

Precipitation causes runoff -- this is a basic tenet of surface-water hydrology. However, knowing when and where the precipitation occurred that culminated in runoff at any given time and place requires a great deal of insight. Studies by Matalas (1963) and by Yevjevich (1963) both propose that annual streamflow can be described as a moving-average process with net annual precipitation (precipitation - evapotranspiration) as the independent variable. They surmised that the lagged coefficients of the moving-average process were the fractions of net precipitation that appeared as streamflow in subsequent years. These were the earlist known attempts at the use of stochastic processes to develop insight into the persistence of hydrologic phenomena. However, a short time later Wemelsfelder (1964) did devise a scheme by which persistence of monthly discharges on the Rhine River could be used to estimate ground-water storage in the Rhine basin.

These studies were first approximations at linking stochastic and physical understanding of hydrologic processes, and it is sometimes difficult to determine whether stochastic modeling provided physical insight as Parzen (1962) seems to state that it should or, if indeed, it was just the opposite. In actuality, there is a great deal of feedback when an investigator is attuned to both stochastic and physical analysis; the state of physical knowledge can help dictate the stochastic approaches to be used, while significant deviations between assumed stochastic processes and observed hydrologic phenomena can indicate fruitful areas for research of a physical nature.

Although the feedback between stochastic and physical understanding probably has been the least utilized of the three purposes for stochastic modeling, it may have the most potential for studying hydrology in the next century. Even though the general circulation models that indicate coming hydrologic changes are very complex and sophisticated relative to most other computer modeling of today, they are still crude and coarse approximations of nature. In the future, stochastic analysis must be superimposed on these models so that a comparison between the model output and the observations can be used to guide further development of the science of global modeling and to provide guidance for improved data-collection networks.

Water-resources decisionmaking

The role of stochastic hydrology in water-resources decisionmaking was first previewed by Thomas and Fiering (1962) in the context of system design. Synthetic sequences of streamflow generated by Monte Carlo simulation were used to test various designs in which the functional relationships among the objective function of the design, the design-decision variables, and the hydrologic parameters were too complex to be derived. By simulating the outcomes of various system designs when subjected to the synthetic streamflows, the best design could be identified. Subsequent to this initiation in the Harvard Water Program, stochastic hydrology has been used for practically all phases of water-resources decisionmaking. Kibler and Hipel (1979) and Sorooshian (1983) reviewed some of the work in this field.

Indeed, even the design of the networks that provide hydrologic data for water-resources decisionmaking have begun to be impacted by stochastic hydrology; many methodologies for the design of these networks depend on stochastic descriptions of the hydrologic phenomena to be monitored. However, Moss (1986) points out that uncertainties in these designs are generated by unrealistic assumptions of stationarity that, in turn, result in non-optimal design decisions. To overcome the non-stationarity problem, Moss (1986) introduced the concept of the maximization of the integral of instantaneous information over an area and over time. Although this concept is yet to be completely developed and implemented, it is one example of how one might deal with ever-changing climate in a practical stochastic-hydrology problem.

Conclusions

The use of modern technology, resulting in dramatically increased use of fossil fuels, for example, creates new types and levels of uncertainty in the world's climate and hydrology. Other types of modern technology, both analytical and computational, provide a means for mankind to cope with these uncertainties; stochastic hydrology is one such technology. One of the greatest challenges for the hydrologist of today is to develop and utilize the one type of technology to fend off the impacts of uncertainty and change caused by use of the others before the world's water resources are unduly degraded.

References

Box, G. E. P., and Jenkins, G. M., 1970, Time series analysis forecasting and control: Holden-Day, Inc., San Francisco, 553p.

Bras, R. L. and Rodriguez-Iturbe, I., 1985, Long-term persistence in hydrologic modeling, chapter 5 in Random Functions and Hydrology, Addison-Wesley Publishing Co., Reading, MA, p.210-265.

Hurst, H. E., 1951, Long-term storage capacity of reservoirs, Transactions of ASCE, v.116, no.776, p.770-799.

Kibler, D. F. and Hipel, K. W., 1979, Surface water hydrology, Reviews of Geophysics and Space Physics, v.17, no.6, p.1186-1209.

Matalas, N. C., Statistics of a runoff-precipitation relation: U.S. Geological Survey Professional Paper 434-D, 1963, 9 p.

Moss, M. E. and Bryson, M. C., 1974, Autocorrelation structure of monthly streamflows: Water Resources Research, v.10, no.4, p.737-744.

Moss, M. E., 1986, Management of water-resources information during changing times, Integrated Design of Hydrological Networks, IAHS Publ. no.158, p.307-317.

Rasmusson, E. M., 1984, El Nino: The ocean/atmospheric connection, Oceanus, v.27, no.2, p.5-12.

Riggs, H. C., 1985, Streamflow characteristics, Elsevier, Amsterdam, 249p.

Sorooshian, S., 1983, Surface water hydrology's on line estimation, Reviews of Geophysics and Space Physics, v.21, no.3, p.706-721.

Stockton, C. W. and Jacoby, G. C., Jr., 1976, Long-term surface-
supply and streamflow trends in the upper Colorado River basin
based on tree-ring analysis, Lake Powell Research Bulletin Number
18, 70p.
Wallis, J. R. and Matalas, N. C., 1970, Small sample properties of
H and K--estimates of the Hurst coefficient h, Water Resources
Resaerch, v.6, no.6, p.1538-1594.
Wemelsfelder, P. J., 1964, The persistency of river discharges and
groundwater storage: Berkeley Symposium Proc., International
Association of Scientific Hydrology Pub., no.63, p.90-106.
World Climate Programme, 1986, Conference statement--international
assessment of the role of carbon dioxide and other greenhouse
gases in climate variations and associated impacts, WCP
Newsletter, no.8, World Meteorological Organization, Geneva,
Switzerland, p.1-3.
Yevjevich, V. M., 1963, Fluctuations of wet and dry years, pt. I,
research data assembly and mathematical models: Hydrology Paper,
no.1, Colorado State University, 55p.

The Influence of Climate Change and Climatic Variability on the Hydrologic Regime and Water Resources (Proceedings of the Vancouver Symposium, August 1987). IAHS Publ. no. 168. 1987.

Long water balance time series in the upper basins of four important rivers in Europe — indicators for climatic changes?

Bruno Schädler
Swiss National Hydrological & Geological Survey
CH-3003 Berne, Switzerland

ABSTRACT   Long term time series of climate and water balance components are often used to identify climatic changes and check on climate models. Eighty-year series of water balance components from four different basins in Switzerland (central Europe) are discussed together with a 175-year series of the alpine basin of the Rhine: overall precipitation and evaporation have increased in accordance with the general rise in temperatures in the last 175 years. The increase in evaporation has been especially marked in this century despite major fluctuations in the precipitation figures. This effect is attributable, on the basis of comparisons with other areas, to human factors involving intensive agriculture. During winter months in recent years there has been a significant increase in the precipitation figures. The runoffs vary relatively little because of compensating effects. Time series of individual components of the water balance are evidently insufficient to allow climatic variations to be recognized or climatic models to be checked.

Longues series chronologiques des éléments du bilan hydrique dans le bassin versant supérieur de quatre fleuves européens: indiqueraient-elles une modification du climat?

RESUME   De longues séries chronologiques de variables climatiques et d'éléments du bilan hydrique sont souvent utilisées pour déceler des changements climatiques et pour vérifier des modèles du climat. Les chronologies des éléments du bilan, longues de 80 ans, pour quatre bassins fluviaux différents de Suisse (Europe centrale) sont comparées avec des séries des mêmes variables, longues de 175 ans, pour le Rhin alpin. Il en ressort qu'à la suite d'un réchauffement général au cours de ces 175 ans, les précipitations, et l'évaporation ont dans l'ensemble augmenté. Malgrè de fortes variations dans les précipita-tions, l'augmentation de l'évaporation est frappante, tout particulièrement au cours de ce siècle. A l'aide de comparaisons avec d'autres bassins, cet effet a pu être attribué aux influences anthropogènes d'une agriculture intensive. Au cours de ces dernières années, les précipi-tations ont augmenté de façon significative, surtout pendant les mois d'hiver. Les débits subissent des fluctuations relativement faibles, en raison d'effets se compensant mutuellement. Il est évident que des séries chronologiques portant sur un seul des éléments du bilan hydrique ne suffiraient pas à mettre en évidence des variations du climat, ni à vérifier des modèles du climat.

Introduction

Long term measurement series of such individual climate components as temperature or precipitation are of interest for two reasons when considering the problem of long term climate variations:

(a) They provide evidence of whether and how the climate has changed.

(b) They permit the use of historical data to check the behaviour of climate models, designed to predict future climatic changes.

Evaporation has a special importance among the different components included in the water balance because it is the common factor linking the energy and the water balances (Figure 1). The interaction between the individual components is highly complex. Thus for example an increased temperature can result from a long dry spell with little evaporation (low consumption of latent energy). Conversely, a high temperature can be responsible for increased precipitation; the time when such precipitation occurs will control whether there is more rather than less evaporation. In addition increased evaporation caused by a rise in temperature can have a feedback effect on the temperature: the rising water vapour content in the atmosphere causes an increased greenhouse effect, so that the temperature can climb even higher.

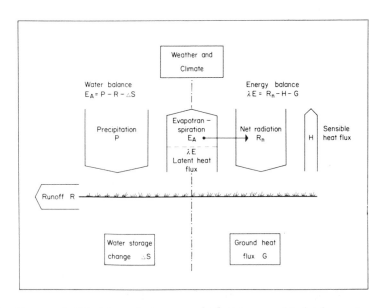

Figure 1   Water balance and energy balance as linked system.

However in each case evaporation, the bearer of latent energy, links the two fundamental cycles of our climate. Conversely the runoff must be seen as the result of all these processes. In view of this one may ask whether discharge or water level data are forces to be reckoned with as far as climatic changes are concerned?

Long term measurement series in Switzerland

Several long term series of hydrologic data have been available in Switzerland whose records commenced near the beginning of this century. The systematic observation of precipitation and temperature began in the middle of the 19th century. The extension of the glaciers has been systematically observed since 1890, and runoff since about 1905. This means that from about 1910 the water balance data is of comparable quality for all areas.

Switzerland is the source of four major rivers: the Rhine, Rhone, Po, and Danube. They flow to all four points of the compass and each one into a different sea (Figure 2). The drainage basins have different climates because the Alps form a weather barrier. However the varying location of the drainage basins also influences the land use (Table 1). These areas are thus ideal for comparative studies.

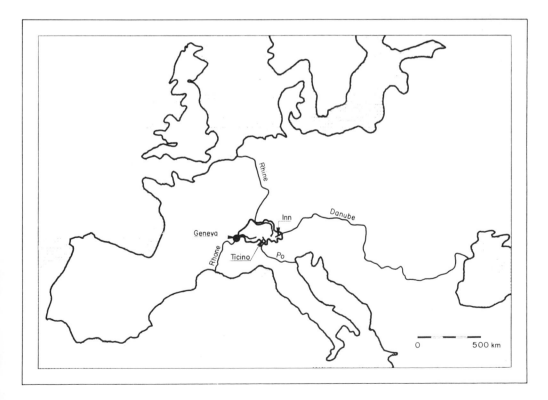

Figure 2   Sketchmap of Europe.   Illustrated are the four rivers which have their sources in Switzerland.

Table 1   Data on drainage basin characteristics and land utiliza-
tion in the Swiss parts of the four river basins

River basin	Area	Mean elevation	Glaciers	Waste land	Lakes and rivers	Wood	Pasture	Cultivated land	Built up area
	(km^2)	(m a.s.l.)	(%)	(%)	(%)	(%)	(%)	(%)	(%)
Rhine	27 970	1110	2	10	4	27	19	33	5
Rhone	7 521	1670	10	21	4	20	22	20	3
Po(Ticino)	3 851	1560	2	21	3	33	25	13	3
Danube(Inn)	1 814	2340	5	37	1	12	34	10	1

The water balance components were calculated for the Swiss part of
these four drainage basins at a monthly time step over the 80 year
period from 1901 to 1980 (Schadler, 1985).

Description of the method of calculation employed

The areal precipitation P was evaluated using uniform precipitation
measurement series provided by the Swiss Meteorological Institute.
The long term characteristic precipitation distribution was taken
into account (Uttinger, 1949) and the stations were weighted using
the polygon-method. The systematic error in precipitation measurement
(about 3 - 25%) was not taken into account because of insufficient
data. The random error should total less than 2%.
    Almost complete measurements for runoff depth were available
from the Swiss National Hydrological and Geological Survey. The
runoffs had to be estimated for some partial areas. Accidental
errors were generally below 1% with the exception of the decade from
1901-10. No major underground imports or exports are known.
    The storage changes ΔS were dealt with in the following manner.
Uninterrupted series of measurements are available from the Federal
Office for Water Economy for the larger lakes· and artificial
reservoirs, permitting very accurate tabulation of the storage
changes. The change in groundwater storage is not known, although it
should be scarcely noticeable over the periods of 10 hydrological
years being considered here. The glacier changes were extrapolated
on the basis of observations of the mass balance of different alpine
glaciers in all the drainage basins by the Glacier Commission of the
Swiss Academy of Natural Sciences. The errors can vary considerably
and should be generally around 10% to 20%.
    The actual evaporation E_A was calculated as the closing term in
the water balance equation

$$E_A = P - R - \Delta S \tag{1}$$

According to compounded error theory the random errors for 10-20 year
periods should be around 4%. The systematic measurement error in the
precipitation figures, which was not taken into consideration, is
carried over to the derived evaporation values. For the time being
its order of magnitude remains unknown.

Results

The precipitation figures present an almost unvarying pattern (Figure 3). Despite major deviations from year to year, the decadal precipitation quantities have remained almost stable since the beginning of the century, no trend is evident and the temporal distribution is about the same in all areas. Precipitation is noticeably heavier in the Ticino area (Po), which is affected by the Mediterranean climate, so that the extreme values are even more extreme. In the Danube source basin, in the Inn area, precipitation quantities are noticeably lower since despite the greater altitude, the drainage basin is located in a high inner-Alpine valley, protected by the lee effect of the surrounding mountains.

As for evaporation, the results present a somewhat different picture (Figure 3). In the Rhine basin, where the most accurate data are available, there is a clear upward trend: reflected not only in the average values, but also in the extreme values. The same observations can be made in the Rhone and Ticino (Po) basins, despite probably greater errors, while conversely no indications of increasing or decreasing evaporation are apparent for the Inn (Danube) basin.

No doubt the increase in temperature which has been recorded throughout the world (Jones et al. 1986) is an important reason for the upward trend in evaporation. However the fact that the precipitation distribution also plays an important role can be seen clearly in the 1940's. At that time evaporation was apparently limited by a major deficit in summer precipitation.

The temperature increased by a similar amount in practically all regions of Switzerland, so why should no increase in evaporation be recorded in the Inn basin (Danube)? The very different land utilisation in this area could provide an explanation, although this cannot really be proved (cf Table 1). From about 1940 in Switzerland generally, there was a great increase in agricultural use: more extensive employment of artificial fertilizers, improvement of crop varieties, planting of new crops (e.g. maize), reduction of the fallow period, and intensification of irrigation. The combined result of these factors was a major increase in the crop yield per unit area. Since the dry-matter production of plants is directly related to transpiration (Slayter & Mabutt, 1964), it is clear that evaporation must have increased in areas used for agriculture. On the other hand since 1929 there has been a 12% reduction in the area of land used for agriculture.

However, the high-altitude Inn basin (Danube) was hardly affected by these national trends. This fact could explain the differences in evaporation between this drainage basin and the other areas.

This example shows that the evaporation and the related energy balance are not influenced purely by other climate and water balance components but also by man-made changes.

The changes in the glacier reserves should receive special mention, in addition to minor changes in reserves due to the construction of artificial reservoirs for hydroelectric power production (Figure 4). Here one can see the situation which prevails almost world wide: a long and intensive melting period, which was interrupted in the second decade by a period of positive mass

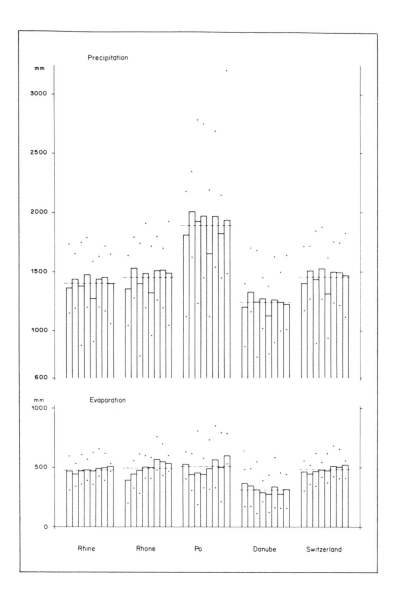

Figure 3   Precipitation and evaporation for the Swiss parts of the 4
river basins and for the whole of Switzerland. The 8
columns show the average values for each of the 10 year
periods (1901-1910 etc.). The points represent the maximum
and minimum values in each ten year period.

balance, and came to an end in 1965. At this time some 30% of the
glacier ice in existence around 1900 had melted and had thus
increased the discharge quantities. The glaciers have rebuilt
reserves since 1965 and thus withdrawn some water from the rivers.
    The glaciers' behaviour is very dependent on changes in climate to
which they react with great sensitivity. However in this context it
should be noted that the connections between precipitation quantity,
temperature and changes in glacier mass are highly complex and very

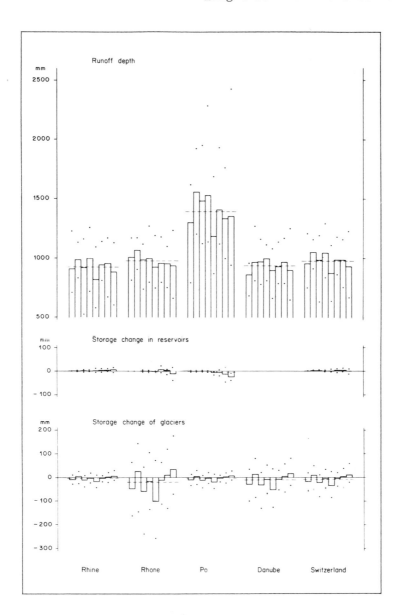

Figure 4 Runoff depth and storage changes for the Swiss parts of 4
river basins and for the whole of Switzerland. the 8
columns show the average values for each 10 year periods
(1901-1910 etc.). the points represent the maximum and
minimum values for each ten year period.

dependent on the seasonal distribution of these influencing factors.
For example the interpretation of glacier variations is complicated
by the fact that a rise in temperature, which would lead to a melting
of the glacier, is at least partly compensated by a simultaneous
related increase in precipitation.

The runoffs show a slight downward trend. However this trend is
the result of the increasing evaporation and the glacier mass

balance, negative at the beginning of the period under review, but later positive.

The temporal distribution of precipitation within the year is of decisive importance as far as the behaviour of the water resource system is concerned. For this reason the data were analyzed by season. Figure 5 for example shows the results for a partial area of the Rhine. The significant variations from year to year can be seen clearly. Wet and dry periods appear to change in an unsystematic way. There is no discernible trend. On the other hand from about 1976 there has been a singular and conspicuous increase in the winter months to above average precipitation quantities.

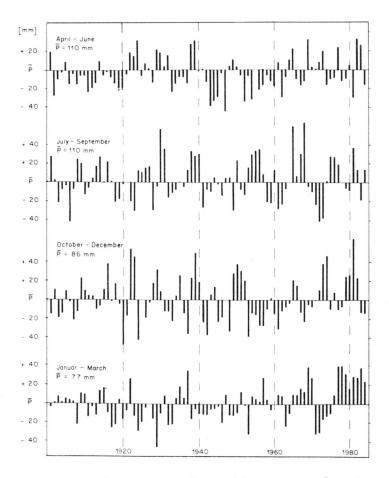

Figure 5   Deviation from seasonal monthly average for the whole of the review period 1901–84 for the part of the Rhine drainage basin.

A 175-year series of hydrological measurements

Although the series of measurements discussed thus far covers more than 80 years or so, this time period is very short when one is trying to deduce climatic changes and connections between individual measurement series and climate changes. Fortunately, we have a series of measurements of water balance components for the Rhine basin since 1810 at our disposal, involving very reliable measurements, which are fully comparable with the data which we are considering.

Different sources of information and methods of evaluation must be employed for the years preceding 1900. Details of the runoff volume were described and published by Ghezzi (1926) on the basis of systematic water level observations and occasional discharge measurements. The precipitation data were obtained from individual series of measurements together with proxy data compiled by Pfister (1984) using written sources. These data were recalculated using 1901-1960 as a reference period to provide comparable area precipitation data.

No figures are available for changes in reserves, especially for the changes in glacier reserves prior to 1900. However, it can be assumed following Zumbühl (1980) that the glaciers grew considerably between 1810-20, remained stable until 1860, then melted rapidly until 1880 when they reached their 1900 level.

Evaporation was calculated as the simple difference between precipitation and runoff. The lack of data on changes in glacier reserves is likely to have resulted in too high values for the years from 1860-80 and too low values for the years from 1860-80. These errors might amount to some 10-20 mm per year.

The average annual temperature in Basel (Schüepp, 1961), which is representative of the whole drainage area, is illustrated to complement the runoff components (Figure 6). The upward trend in the evaporation series and its relationship with temperature have been confirmed. At the same time it has become clear that it is not just the yearly annual temperature but also the seasonal distribution of the temperature, and especially the seasonal distribution of the precipitation, which are decisive. The first two decades were both cold and dry, so that evaporation was very limited. However the evaporation was above average in the years 1850-60 despite similar low temperatures and average precipitation, thanks to very warm, wet summers.

The unclear relationship of precipitation with temperature also becomes evident. The very cold decade from 1830-40 and the warmest decade from 1940-50 exhibit practically the same precipitation deficit as the average decade from 1860-70. However this should come as no surprise given the complex connection between the energy balance and water balance outlined above.

Finally the runoff data show practically no connection with the temperature or the other components. The patterns of the precipitation and evaporation compensated each other to some extent (1810-1830) or had a reinforcing effect (1880-1900; 1910-20).

Conclusions

Series of discharge or water level measurements, considered on their own or together with average temperatures, do not provide evidence of the changes in hydrologic characteristics of a drainage basin or even about changes in the climate. They are too heavily influenced by the

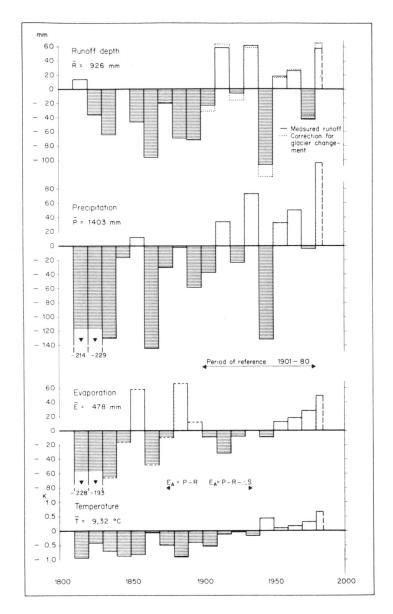

Figure 6   175-year   measurement   series   of   the   water   economy
components   of   the   Swiss   Rhine   drainage   basin.   The
deviations   are   shown   in the decade   averages   from   the
average of the 1901–1980 reference period.

opposing factors of precipitation and evaporation,   as well   as   by
changes   in   the glacier reserves and in more recent times   by   human
behaviour.

Series   of   annual precipitation measurements cannot be   employed,
even when considered with average annual temperatures, as evidence of
the   actual   evaporation   and thus of   the   influence   on   the   energy
balance.   This   requires data on the seasonal distribution of   these
factors.   The   annual precipitation quantities are not linked to the

average annual temperatures. What happens seasonally is decisive here as well.

Only the verification of climate models on the basis of long-term historical series of measurements, can lead to the correct conclusions and then only if either several components can be compared simultaneously or if the comparative values available are at least broken down by seasons.

Since the beginning of the 19th century the components of the water balance in the alpine and pre-alpine area of the Rhine drainage basin have developed as follows: The precipitation and evaporation have in general increased.   The upward trend in evaporation is especially notable in the 20th century despite major variations in precipitation quantities.   In the most recent period especially, precipitation has exhibited the first increase noted for a very long time.   The runoffs in the 20th century are generally higher than in the previous century.   However they are again showing more of a downward trend.

ACKNOWLEDGEMENTS   The illustrations were prepared by R. Bigler. Translation was carried out by F. de Montmollin and A.N. Other. Thanks are due to them and all the bodies who provided valuable data for this study.

References

Ghezzi, F. (1926) Die Abflussverhaltnisse des Rheins in Basel (Discharge conditions in the Rhine in Basel). Mitteilungen   des Amtes fur Wasserwirtschaft Nr. 19, Bern.

Jones, P.D., Wigley, T.M.L. & Wright, P.B. (1986) Global temperature variations between 1861 and 1984. Nature 322,   430-434.

Pfister, Ch. (1984) Klimageschichte der Schweiz 1525-1860   (History of climate in Switzerland 1525-1860). Academica   Helvetica(6), Haupt, Bern, Switzerland.

Schadler, B. (1985) Der Wasserhaushalt der Schweiz (Water balance of Switzerland). Mitt. Landeshydrologie (6), Bern.

Schuepp, M. (1961) Klimatologie der Schweiz (Climatology of Switzerland), Vol.2, Zurich.

Slayter, R.O. & Mabutt, J.A. (1964) Hydrology of arid and   semiarid regions. In: Handbook of applied hydrology (ed. Ven Te Chow),24, 34-38. McGraw Hill, New York, USA.

Uttinger, H. (1949) Die Niederschlagsmengen in der Schweiz (Precipitation in Switzerland). Fuhrer durch die   Schweizerische Wasser- und Elekrozitatswirtschaft, Zurich,   Switzerland.

Zumbuhl, H.J. (1980 Die Schwankungen des Grindelwaldgletschers in den historischen Bild- und Schriftquellen des 12, bis 19. Jahrhunderts (Variations of the Grindelwald-glaciers in historical sources from the 12th to the 19th century).   Denkschrift d. Schweiz. Naturf.Ges. 92, Basel, Switzerland.

The Influence of Climate Change and Climatic Variability on the Hydrologic Regime and Water Resources (Proceedings of the Vancouver Symposium, August 1987). IAHS Publ. no. 168, 1987.

Analysis of patterns in a precipitation time sequence by ordinary Kalman filter and adaptive Kalman filter

Akira Kawamura, Kenji Jinno,
Toshihiko Ueda & Reynaldo Real Medina
Department of Civil Engineering Hydraulics
Faculty of Engineering
Kyushu University
6-10-1 Hakozaki
Higashi-ku, Fukuoka 812 Japan

ABSTRACT   This paper proposes a methodology of investigating the patterns in a precipitation time sequence by an ordinary Kalman filter (OKF) and an adaptive Kalman Filter (AKF). Specifically this research aims to investigate the occurrence and characteristics of drought precipitation patterns. The methodology is applied to the 92-year long monthly precipitation time sequence at Fukuoka City in Japan. OKF identifies the periods (in the time sequence) with abnormal precipitation by comparing the observed and average precipitation patterns. AKF detects the changes of the long term precipitation patterns in the time sequence. These changes are associated with the abrupt changes in the parameters of the periodic-stochastic model. The time sequence is divided into several precipitation epochs, where an epoch is uniquely represented by one set of parameter values. The precipitation pattern in each epoch is appropriately characterised and reveals whether the risk of drought occurrence is high or not in an epoch.

Analyse des Courbes de Précipitation Séquentielle dans le Temps par le Filtre Ordinaire de Kalman et le Filtre Adaptable de Kalman

RESUME   Cet article propose une méthodologie pour investiguer les courbes d'une précipitation séquentielle dans le temps par le filtre ordinaire de Kalman (OKF) et le filtre adaptable de Kalman (AKF). Plus précisement, cette étude à pour but d'investiguer la survenue et les caractéristiques des courbes de précipitation de sécheresse. Cette méthodologie est appliquée à la précipitation à séquence mensuelle d'une longue période de 92 ans de la ville Japonaise de Fukuoka. OKF identifie les périodes (dans le temps) avec précipitation anormale en comparant les courbes de précipitation observées avec la courbe moyenne. Et AKF détecte les changements à long terme des courbes de précipitation. Ces changements sont associés avec des variations brusques dans les parametres du modèle périodique avec des pointes en crochets. La séquence dans le temps est divisée en plusieurs époques de précipitation, une époque étant uniquement représentée par un ensemble de valeurs des parametres. La courbe de précipitation dans chaque époque est bien caracterisée et révèle si le risque de survenue d'une sécheresse est grande ou non dans une époque.

Notation

A_i, B_i	periodic coefficients
f_i	frequency component
G	unknown $(n \times 1)$ abnormality vector
H	known $(m \times n)$ observation matrix
H_o	the hypothesis that no change in precipitation pattern has occurred
H_1	the hypothesis that precipitation pattern has changed at $k = \theta$
I	$(n \times n)$ identity matrix
k	time instant
l	innovation cumulative number
m	dimension of observation vector
n	dimension of system state vector; number of system parameters
q	number of significant frequency components
s	length of moving average (months)
T	return period (years)
u	independent, zero mean, white Gaussian $(p \times 1)$ system noise vector
w	independent, zero mean, white Gaussian $(m \times 1)$ observation noise vector
x	$(n \times 1)$ system state vector
y	$(m \times 1)$ observation vector $(m \leq n)$; smoothed log-transformed monthly average precipitation $(\log (mm/day))$
z	monthly average precipitation (mm/day)
α	smoothing coefficient
Γ	known $(n \times p)$ system matrix
$\delta_{k\theta}$	Kronecker's delta ($\delta_{k\theta} = 1$ if $k = \theta$ and $\delta_{k\theta} = 0$ if $k \neq \theta$)
η	threshold value
θ	unknown time instant when abrupt change occurred
Φ	known $(n \times n)$ state transition matrix
ϕ_*	abnormality detection index

Introduction

Droughts are often caused by adverse weather (Nemoto, 1974), and their impact on the economy, society, and political situation can be disastrous. The need to limit the undesirable consequences of drought has increased society's interest in long term prediction of adverse weather. However we believe that the evaluation of the possibility of occurrence of adverse weather that would eventually result to drought can be improved if variations in meteorologic and climatologic parameters are fully understood. In this report, the long term variations of precipitation are studied.

Specifically, this report aims to investigate the occurrence of drought by analyzing the dynamic characteristics of the monthly precipitation time sequence using an ordinary Kalman filter (OKF) and an adaptive Kalman filter (AKF). OKF is used to detect periods in the sequence with abnormal precipitation and to determine quantitatively the period's magnitude of abnormality. We classify the periods into three types and evaluate the degree of possibility of drought occurrence in each type. AKF is used to detect changes in the long term precipitation pattern of the time sequence. The shifts in the precipitation pattern divide the sequence into several precipitation epochs, where an epoch represents one precipitation pattern. Each epochs precipitation is characterized, and the risk of drought occurrence is revealed high or not in the epoch. Finally, drought duration curve analysis is performed to evaluate the robustness of the water resources system during the different epochs.

Modeling of the precipitation sequence

The 92 year long (1890-1981) precipitation sequence at Fukuoka City forms the basis for the analysis in this investigation. The use of monthly average precipitation provides a suitable approach to the investigation of long term variations in precipitation pattern.

The 92 annual totals are plotted on normal paper using the Weibull plotting method as shown in Figure 1. In this figure, the annual precipitation appears normally distributed, however the abnormally wet and dry years deviate from the normal line. From this result, one may think that the years with abnormally high and low precipitation belong to a population different from those on or close to the normal line.

Logarithmic transformation of the 1104(92 years) monthly average data (in mm/day) was done to normalize the data. The log-transformed precipitation data was smoothed using the recursive low-pass filter shown in eq.1 (Bendat and Piersol, 1976).

$$y(k) = (1-\alpha)\log\{z(k)\} + \alpha y(k-1) \tag{1}$$

We chose $\alpha = 0.6$. The smoothed log-transformed monthly average precipitation $y(k)$ at time instant k is modeled by a periodic function as follows.

$$y(k) = M_y + \sum_{i=1}^{q}(A_i\sin2\pi f_i k + B_i\cos2\pi f_i k) + w(k) \tag{2}$$

The purpose of smoothing and the reasons for choosing and for modeling the sequence by a periodic function are mentioned in detail by Kawamura et al. (1985). We set q=5 in eq.2. The dominant frequency components used in this study are $f_1=1/48$, $f_2=5/72$, $f_3=1/12$, $f_4=1/3$ and $f_5=5/12$ (cycles/month), having periods of 4 years, 1.2 years, one year, 3-months and 2.4 months respectively, which were obtained by MEM spectral analysis of the smoothed transformed data (Kawamura et al. 1985). Comparison of the power spectra of z(k) and y(k) shows similar dominant peaks, indicating that the periodic properties of z(k) are retained in y(k).

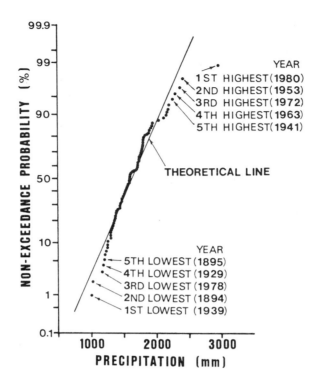

Figure 1　Plot of the annual precipitation data
on normal distribution paper.

Kalman filter formulation

Now we consider the problem of identifying M_y and A_i and B_i by OKF and AKF, assuming f is known. The system equations for OKF and AKF are given respectively in eqs. 3 and 4, while the observation equation for both OKF and AKF is given in eq. 5.

$$x(k+1) = \Phi(k)x(k) + \Gamma(k)u(k) \tag{3}$$

$$x(k+1) = \Phi(k)x(k) + \Gamma(k)u(k) + \delta_{k\theta} G(k) \tag{4}$$

$$y(k) = H(k)x(k) + w(k) \tag{5}$$

Here n is 11 and m is one; $x=[M_y \; A_1 \; B_1 \ldots A_5 \; B_5]$; $\Phi(k)=I$; the observation eq.5 corresponds to eq.2; and $H(k) = (1 \; \sin2\pi f_1 k \; \cos2\pi f_1 k \ldots \sin2\pi f_5 k \; \cos2\pi f_5 k)$. Here the state variables are the system parameters M, A, and B. Details of parameter estimations by OKF and AKF and the various properties of the two filters for the analysis of time series expressed by a periodic function and the definition of $\phi_*(k,1)$ which expresses quantitatively the system abnormality at time instant k are given by Ueda, et al. (1984), Kawamura, et al. (1984) and Kawamura et al. (1986).

Analysis of abnormal precipitation

Since the plot of the one-step ahead predictions by OKF are almost the same as the average precipitation pattern (Ueda et al. 1984; and Kawamura et al. 1984), we can calculate recursively $\phi_*(k,l)$ using the l-month series of residuals between the predicted and observed values. Thus we can determine the magnitude of abnormality of an l-month observed precipitation period from the magnitude of ϕ_*. Here we set l = 15. From the information on how a 15-month observed precipitation differs from the average precipitation pattern, we can decide whether the observed precipitation of the period is abnormal or not.

Using the above procedure, the numbers of peak ϕ_* above $\phi_* = 6.0$, 5.0, 4.0 and 3.0 are determined to be 6,10, 19 and 44 respectively. The respective recurrence intervals of the periods with abnormal precipitation identified by these peak ϕ_* are 15,9,5 and 2 years on the average. In this report, we analyze the 19 abnormal precipitation periods arranged in descending order of magnitude of peak ϕ_* in Table 1. Figure 2 illustrates the first five most abnormal precipitation periods (ranked 1-5 in Table 1).

Table 1 The abnormal precipitation periods detected by OKF

	Peak ϕ_*		
Rank	Magnitude	Time of occurrence	Type
1	7.90	Apr. 1894	A
2	7.85	July 1904	C
3	6.14	July 1956	C
4	6.09	May 1980	B
5	6.08	Feb. 1939	A
6	6.02	Oct. 1944	C
7	5.68	Oct. 1971	C
8	5.16	Aug. 1933	C
9	5.15	Oct. 1962	B
10	5.08	June 1902	C
11	4.93	July 1899	C
12	4.90	June 1948	B
13	4.72	May 1953	B
14	4.52	Jan. 1951	C
15	4.45	July 1977	A
16	4.21	Apr. 1913	C
17	4.19	Aug. 1936	C
18	4.19	Apr. 1941	B
19	4.15	July 1976	C

We illustrate the characteristics of the abnormal precipitation in the 19 periods detected by OKF. Each period whose length is 15 months, starts from the time of occurrence of peak ϕ_* . In Figure 3(a), the abnormal precipitation period ranked I is characterized by abnormally low precipitation in May, July, August and October of 1894 and April, May, August and September of 1895 and below average

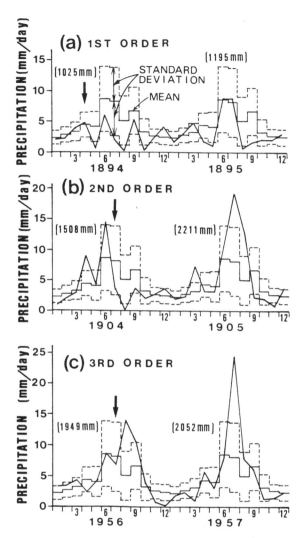

Figure 2　Abnormal　precipitation periods (up to rank 5) detected by OKF.　The figures in parentheses are annual　precipitation amounts;　downward-arrow　indicates the time of occurrence of peak ϕ_* as detected by OKF.

precipitation in the rest of the months.　As a result, the second and fifth　lowest annual precipitation totals were recorded in　1894　and 1895 respectively.

The detection of abnormal precipitation periods by OKF, through ϕ_*, results　in the identification of three types of abnormal　precipitation periods:　Type A,　Type B and Type C. Type A is characterised by months with below average precipitation depths. Type B is typified by months with above average precipitation amounts.　Type C is characterised　by　both extremely high and low precipitation　amounts,　which occur　alternately at an interval of two or more months.　These three types　of　abnormal precipitation periods could span for one　or　two years. The 19 abnormal precipitation periods are classified according to these types, as shown in Table 1. Of the 19 abnormal precipitation

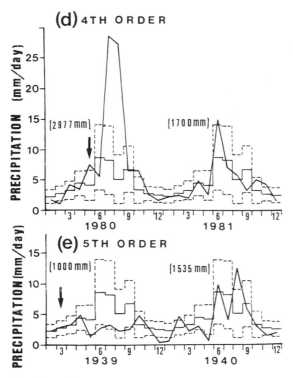

Figure 2   Continued.

periods, three, five and eleven are Type A, Type B, and Type C respectively (as shown in Figure 3), suggesting that Type C is most likely to occur whenever there is an increase of ϕ_*. This increase of ϕ_* should warn us of the possible occurrence of both abnormally high and low precipitation.

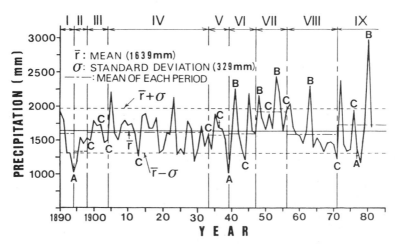

Figure 3   The annual precipitation sequence divided by AKF into nine different epochs.   A, B and C are the types of the abnormal precipitation detected by OKF.

As shown in the above discussion, the risk of drought occurrence in each period can be evaluated using OKF.

Analysis of long-term precipitation patterns

AKF detects whether or not an abnormality is present in x by evaluating the variation of the innovations using a generalized likelihood ratio test (GLRT). If such abnormality is detected as an abrupt change in the system parameters, the time instant θ and its magnitude $G(\theta)$ are estimated quantitatively, and the state variables and the error covariance matrix are compensated according to the magnitude of abnormality at the time instant when the abnormality is detected (Ueda et al. 1984 and Kawamura et al. 1986). In this report, θ is the time when the change in precipitation pattern occurs. An equivalent GLRT is performed to detect whether or not a change in precipitation pattern has taken place, as follows:

$$\phi_*(\theta,1) \underset{H_0}{\overset{H_1}{\gtrless}} \eta \qquad (6)$$

The detection of the time occurrences of the changes of precipitation patterns leads to the division of the precipitation sequence into several epochs, where each epoch represent one precipitation pattern.
   According to the statistical divisions made by the World Meteorological Organization (WMO), the interpretation of climatic change differs by time scale. A phenomenon regarded as abnormal in one time scale may be a normal event in a longer time scale. Since the 92 year precipitation sequence is analyzed using a monthly step, the changes in the precipitation pattern correspond to time scales from several years to several decades, which conform to the modern time scale defined by WMO (Nemoto, 1974).
   In this study we aim to detect the change of precipitation pattern which recurs at an interval of about 10 years on the average; thus we set $\eta = 5.0$. Ten peak ϕ_* corresponding to ten abnormal precipitation periods (ranked 1-10 in Table 1) are above this level of η . AKF yields finally eight changes in precipitation pattern in the 92 year sequence. This divides the sequence into nine precipitation epochs. Table 2 lists the parameters identified at the last time instant k by AKF for each epoch, and by OKF for the whole sequence. Figure 3 exhibits the nine precipitation epochs identified by AKF and the time of occurrences of the 19 abnormal precipitation periods identified by OKF.
   The drought duration curve (DDC) of the monthly precipitation is drawn for the 92 year record (Figure 4) and epochs II, IV, VII and IX (Figure 5). Each curve is prepared by first taking the s-month moving average of the precipitation sequence, abstracting the annual minima for that s-month interval and estimating the return period for each annual minimum. The DDC is the plot of the annual minima of the different s-month intervals (s=1 to 24), having the same return period (Takeuchi 1986).

Table 2 Identified system parameters by AKF and OKF

| | Mean | Frequencies(cycle/month) and their amplitudes | | | | | | | | | |
| | | $f_1=1/48$ | | $f_2=5/72$ | | $f_3=1/12$ | | $f_4=1/3$ | | $f_5=5/12$ | |
Epoch	M_y	A_1	B_1	A_2	B_2	A_3	B_3	A_4	B_4	A_5	B_5
I	0.53	.06	.02	.06	−.00	−.10	−.09	.01	.00	−.01	.00
II	0.43	−.15	−.01	.00	−.12	.01	−.06	.01	.04	−.04	.02
III	0.55	.03	−.02	−.02	.02	−.12	−.13	.00	−.01	−.01	.01
IV	0.53	.03	−.01	.03	.02	−.10	−.10	.03	.02	−.03	−.00
V	0.50	−.03	−.01	−.01	−.02	−.18	−.04	−.00	.01	−.02	.01
VI	0.51	.03	.13	−.05	.02	−.19	−.09	.01	.03	−.03	−.00
VII	0.61	−.01	.03	−.03	.04	−.11	−.12	.02	.02	−.04	.00
VIII	0.53	.02	.01	.01	−.00	−.14	−.10	.01	.00	−.02	−.01
IX	0.54	−.12	−.04	−.02	−.04	−.14	−.11	.01	.01	−.01	−.01

AKF — rows I–IX

	Mean										
OKF Total	o.54	−.00	.00	.01	.01	−.12	−.10	.02	.01	−.02	−.00

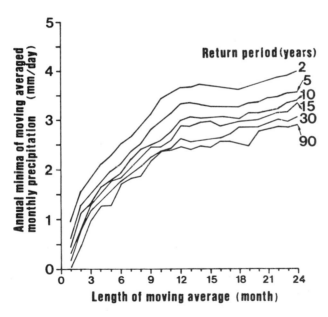

Figure 4 DDC of the monthly precipitation for the 92-year record.

Discussion

Characterization of the long term precipitation patterns

We have confirmed that the parameters identified by AKF for each epoch were almost the same as those by the least-square method for the same epoch, and the parameters identified by OKF and by the least

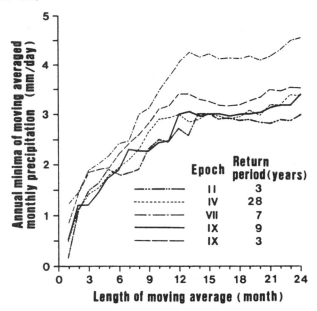

Figure 5   DDC of the monthly precipitation for epochs II, IV, VII
and IX.

square method for the whole sequence are equal. The average precipitation pattern defined by the parameters estimated by OKF (shown in
Table 2) has a very dominant one year cycle, as indicated by
amplitudes A_3 and B_3; these amplitudes A_3 and B_3 are far higher
than the rest of the amplitudes which are almost equal to zero. The
change from one epoch to another is mainly caused by the mean M_y and
frequency components lower than or equal to one year (f_1, f_2, and f_3).
As shown in Figure 1, the second and the fifth lowest years
occurred in epoch II. This epoch is typified by less precipitation,
as indicated by the mean of the epoch being very much below the mean
of the whole sequence (Figure 3). As shown in Table 2, the estimate
of M_y of epoch II is the lowest in the nine epochs. Amplitudes of the
one year cycle are relatively dominant, and the amplitudes A_1 and B_2
are very dominant, suggesting that the epoch's precipitation pattern
differs much from the average precipitation pattern. Hence the risk
of drought occurrence is certainly the highest among the nine epochs.
The   precipitation patterns in epochs III, IV, V and VIII   are
similar to the average precipitation pattern, so that the risk of
drought occurrence is not high.
The highest value of M_y in Table 2 occurred in epoch VII which is
characterized by much precipitation, as indicated by the very high
mean of the epoch, well above the mean of the sequence (Figure 3).
However the amplitudes of the five frequency components in this epoch
are practically the same as those identified by OKF. Thus the epoch's
precipitation pattern is very similar to the average precipitation
pattern, except for the abrupt change in the mean level. Hence we can
safely say that drought is least expected in this kind of precipitation pattern.
In epochs VI and IX (present epoch), the amplitudes A_1 and B_1 for
the 4-year cycle are very dominant (shown in Table 2). The three
types of abnormal precipitation periods (A, B, and C) tend to happen
at the same epoch characterized by a very dominant 4-year cycle

(Figure 3). The first lowest (1939) and the fifth highest (1941)
annual precipitation totals appeared in epoch VI, whereas the third
highest (1972), the third lowest (1978), and the first highest (1980)
yearly precipitation totals were observed in epoch IX. Although
abnormal precipitation type B appeared in these epochs, droughts also
happened in these epochs. Hence the risk of drought occurrence is
regarded high in these epochs.

Drought duration curve analysis

In this section, the robustness of the water resources system in the
different epochs is evaluated using the drought duration curves shown
in Figure 4 and 5. These figures show that the ascending slope of
each DDC suddenly flattens at s equal to 12 months. In Figure 5, the
DDC of epoch II for T=3 years is lower than the DDC of epoch IV for
T=28 years. (As explained in the previous section, epoch II has a
drought precipitation pattern, epoch IV has a normal precipitation
pattern, epoch VII has a flood precipitation pattern and epoch IX a
drought-flood precipitation pattern.) This means that the drought
with a return period of three years in epoch II is more severe than
the drought with a 28-year recurrence interval in epoch IV. In
addition, this DDC of epoch II is lower than the DDC of the 92-year
record for T=15 years and the former is also less than or equal to
those portions between s=6 and 9 months and between s=20 and 24
months of the DDC of the record for T=30 years and T=90 years.
    Also, the DDC of epoch VII for T=7 years is higher than the DDC of
the 92 year record for T=2 years. This includes that, if a water
resource system is planned using data belonging to epoch VII, its
robustness is diminished if that epoch is terminated abruptly by
another epoch characterized by a lower DDC. Moreover, the DDC of
epoch IX for T=9 years is nearly equal to the DDC of epoch IV for
T=28 years. It indicates that a drought in an epoch characterized by
epoch IX is more severe than the drought in an epoch having a normal
precipitation pattern. As presented here, the DDC which expresses the
robustness of the water resources system, differs significantly
among epochs, and the DDCs of some of the epochs are more critical
than those of the 92-year record. Hence in order to have a more
accurate evaluation of its robustness, a planned water resources
system should be analysed not only by record but also by epoch.

Conclusions

In this report, we have used two filtering techniques (OKF and AKF)
to study the patterns of a precipitation time series. OKF has been
used successfully to detect periods with abnormal precipitation,
which are usually indiscernable by visual inspection of the precipitation sequence. AKF has been applied effectively to divide the
precipitation sequence into several epochs, where each epoch is
characterised by one precipitation pattern. The characteristics of
the precipitation pattern in each epoch can be used as a basis for
predicting the future behaviour of the precipitation sequence and
evaluating the possibility of the occurrence of a drought.

Finally the DDC analysis has shown that the robustness of water resources system would differ significantly among epochs.

References

Bendat, J.S. & Piersol, A.G. (1971) Random Data: Analysis and Measurement Procedures. John Wiley & Sons, Inc., New York.

Kawamura, A., Jinno, K. & Ueda, T. (1984) On characteristics of the adaptive Kalman filter for detecting abnormality in a periodic time sequence. Proc. 28th Japan Conf. Hydraul. 28, 383-390 (in Japanese).

Kawamura, A., Ueda, T. & Jinno, K. (1985) Analysis of long-term pattern fluctuations in a precipitation sequence. Proc. Japan Soc. Civ. Engrs. no 363/II-4, 155-164(in Japanese).

Kawamura. A., Jinno, K., Ueda, T. & Medina, R.R.(1986) Detection of abrupt changes in water quality time series by the adaptive Kalman filter. In: Monitoring to Detect Changes in Water Quality Series (Proc. Budapest Symp., July 1986), 285-296. IAHS publ. no. 157.

Nemoto, J. (1974). Search for abnormal Weather, Chuokoron Book Co., Tokyo (in Japanese).

Takeuchi, K. (1986) Chance-constrained model for real-time reservoir operation using drought duration curve. Wat. Resour. Res., vol. 22, no. 4, 551-558.

Ueda, T., Kawamura, A. & Jinno, K. (1984) Detection of abnormality by the adaptive Kalman filter. Proc. Japan Soc. Civ. Engrs. No. 345/II-1, pp. 111-121 (in Japanese).

The Influence of Climate Change and Climatic Variability on the Hydrologic Regime and Water Resources (Proceedings of the Vancouver Symposium, August 1987). IAHS Publ. no. 168, 1987.

The identification of recent rainfall fluctuations in the Philippines

Reynaldo Real Medina, Kenji Jinno
Toshihiko Ueda & Akira Kawamura
Department of Civil Engineering Hydraulics
Faculty of Engineering
Kyushu University, 6-10-1 Hakozaki, Higashi-ku
Fukuoka 812, Japan

ABSTRACT   Our approach to the identification of rainfall fluctuations is to determine whether or not the occurrence of a short-duration rainfall pattern anomaly will cause an abrupt change in rainfall characteristics, i.e., the rainfall characteristics within a period will be terminated by a rather instantaneous shift to another period with different rainfall characteristics. We apply this approach to the recent (1951-1983) monthly rainfall data of four stations representative of the four types of climates in the Philippines. We use the adaptive Kalman filter to identify such kinds of fluctuations in a rainfall sequence by directly linking them with the abrupt changes in the parameters of the periodic-stochastic model of the rainfall time sequence. Fluctuations of 10% or more in the mean between non-overlapping adjacent periods have been detected. However, the Chow test for parameter change has shown that the mean is still not sufficient to describe the rainfall fluctuations.

Identification des fluctuations récentes de pluviosité aux Philippines

RESUME   Notre approche pour identifier les fluctuations de pluviosité est de déterminer si oui ou non la survenue d'une courbe de pluviosité anormale de courte durée peut causer un changement brusque dans les caractéristiques de pluviosité, c'est-à-dire que les caractéristiques d'une pluviosité dans une période seraient plutôt terminée par un virement instantané à une autre période avec des caractéristiques différentes. Nous appliquons cette approche aux données récentes de pluviosité mensuelle (1953-1983) aux 4 stations représentatives des 4 types de climat aux Philippines. Nous utilisons le filtre adaptable de Kalman pour identifier de pareilles fluctuations dans une sequence de pluviosite en les liant directement aux changements brusques des paramètres du modèle périodique avec des pointes en crochet de la pluviosité séquentielle dans le temps. Des fluctuatious de 10% ou plus de la moyenne entre 2 periodes adjacentes et ne se chevauchant pas ont été observées. Toutefois, le

> test de Chow pour changement des paramètres a montrè que
> la moyenne n'est pas encore suffisante pour décrire les
> fluctuations de pluviosité.

Introduction

The biggest problem besetting the water resource management planners
in the Philippines is the scarcity of hydrologic data. According to
the report prepared by the Philippine National Water Resources
Council (1976), there are 369 rain gaging stations in the country as
of 1976; 66% of this number have less than ten years of record. To
complement the rain gaging network, there are 496 stream gaging
stations, and some 20% of this number have been abandoned or
discontinued. Less than 20% of the 496 stations have more than 20
years of record. Also the density of the existing data collection
network is insufficient to meet the needs for long-range development
planning.
    Potential users of the short records of data must be warned that
careless use of these data may lead to wrong assumptions on some
parameters, which may cause some undesirable consequences in the
future. This is so because of the inherent fluctuations in rainfall
characteristics; a reservoir designed using data belonging to a high
rainfall period will be vulnerable if that period is terminated
abruptly by another period characterized by less rainfall. In order
to minimize the risk involved in using short records of data in the
design and operation of water resource systems, the rainfall fluctuations in the country must be identified and their characteristics
fully understood.
    Our approach to the identification of recent rainfall fluctuations
is to determine whether or not the occurrence of a short-duration
rainfall pattern anomaly will appear as an abrupt change in rainfall
characteristics, i.e., the rainfall characteristics within a period
will be terminated by a rather instantaneous shift to another period
with different rainfall characteristics. We design the identification
of fluctuations in this way because it is believed that it is often
important not only to detect when a rainfall fluctuation has occurred,
but also to know when rainfall is about to change or is in the
process of changing its characteristics. In this way, we explore the
possibility of early detection of moderate climate change (Nemec,
1985). We apply this approach to the recent (1951–1983) monthly
rainfall data of four stations representative of the four types of
climates in the Philippines.
    In this approach, we use the methodology explained by Kawamura, et
al. (1985) to identify and characterize the rainfall fluctuations.
This methodology utilizes the ordinary Kalman filter (OKF) to detect
short intervals with abnormal rainfall patterns. It also applies the
adaptive Kalman filter (AKF) to identify abrupt changes in the
parameters of the periodic-stochastic model of the rainfall time
sequence. These abrupt changes in the model parameters, which occur
in the short intervals with abnormal rainfall patterns, divide the
time sequence into several rainfall periods.
    Since rainfall fluctuations have always been associated with
changes in the mean (Karl & Riebsame 1984, & Nemec 1985), we identify

fluctuations of   10%   or   more in the   mean   between   two   adjacent
rainfall periods. This value (namely changes in precipitation of 10%)
has   been considered in defining moderate climate variations   (Nemec,
1985).   To   give   prominence to the differences   in   characteristics
between   two rainfall periods,   statistically significant changes   in
parameter   structures are detected using the Chow test.   Results   of
this   test   have   indicated the inability of the mean   by   itself   to
define   how   severe   and   how frequent intervals of   abnormally   low
rainfall may occur;   the occurrence of such adverse rainfall has been
directly linked with the behaviour of the model parameters.

Identification of rainfall fluctuations by the adaptive Kalman filter

Each   monthly   rainfall time series is transformed and modeled   as   a
periodic-stochastic process in the form:

$$y(k) = M_y + \sum_{i=1}^{q} (A_i \sin 2\pi f_i k + B_i \cos 2\pi f_i k) + w(k) \qquad (1)$$

where   y(k) is the transformed monthly mean rainfall at time step   k;
M_y is   the   mean   of the transformed series;   q is   the   number   of
significant frequency components; f_i is the frequency component; A_i
and B_i are the periodic coefficients;   and w(k) is   the   stochastic
component which is assumed to be white Gaussian noise with zero   mean
and   variance   W(k).   The dominant harmonics f_i (in cycles/month)   in
each series are obtained using MEM spectral analysis. The identification   of rainfall fluctuations is associated with the abrupt   changes
in the model parameters M_y, A_i and B_i.
    The   ordinary Kalman filter is used to detect the   intervals   with
abnormal rainfall patterns in the four rainfall records. It estimates
recursively   the   abnormality detection index $\phi_*(k,l)$ (Ueda   et   al.,
1984)   at   each   time   step k from   a   finite   innovations   (step-one
prediction   residuals)   sequence.   The   occurrence   of   a   peak ϕ_*
identifies   an   interval with an abnormal rainfall pattern,   and   its
value measures the size of the abnormality.
    The adaptive Kalman filter detects whether an abrupt change in the
system state variables M_y, A_i and B_i occurs by evaluating the finite
innovations   sequence   using   the generalized likelihood   ratio   test
(GLRT).   This   test   compares the value of $\phi_*(\theta,l)$ with   a   threshold
value η ,   where θ is   the unknown time step   when   the   abnormality
occurred and l is the length of the innovations sequence.   If $\phi_*(,l)$
is   greater   than η ,   the hypothesis (H_1) that   an   abrupt   change
occurred   at time k=θ is accepted;   otherwise,   the hypothesis (H_0)
that   no   abrupt   change has occurred is accepted.   Once   an   abrupt
change is detected,   its time of occurrence and magnitude   are   estimated   quantitatively,   and   the   state variables   are   appropriately
corrected   according to the magnitude of this abrupt change to   allow
the filter to adjust to the new rainfall characteristics.   The abrupt
changes   in model parameters divide the time sequence into   parameter
regimes,   where   each   parameter regime corresponds to   one   rainfall
period.   Moreover,   the estimates of the parameters in a period   describe   the   rainfall characteristics of that period   and   the   occurrences of abnormal rainfall patterns. The occurrences of the abnormal
patterns in a period characterize the rainfall of that period.

In the absence of prior information on parameter change, it is only possible to make valid inferences, if great care is taken in the interpretation of any parameter change (Bennett,1979). Chosen for reason of its simplicity, the Chow test (Chow,1960) is used for testing each pair of nonoverlapping adjacent rainfall periods for the presence of two parameter regimes.

Short-duration rainfall pattern abnormalities do not happen at regular intervals and, like droughts and floods, can be expressed in terms of return periods. We feel that the most appropriate return period for such abnormalities would be one related to the design and operation of water resource systems. We consider the period in years on the average of about a decade during which rainfall fluctuations can be expected to recur. In designing reservoir capacity, Japan's Ministry of Construction recommends the use of the same drought return period (Hori, 1978). Also, this return period is considered in choosing η.

Rainfall in the Philippines

The Philippine archipelago is comprised of approximately 7000 islands having an aggregate area of 300,000 sq km. It is located in the tropics and the climate prevailing in any particular place in the country is influenced by its geographical position and wind system prevalent at certain times of the year. The prevailing wind systems over the country are as follows: the northeastern monsoon (December to January), trade wind (April), and southwestern monsoon (July, August and September). The driest of these is the trade wind season, while the wettest is the southwestern monsoon season. The classification of Philippine climatic conditions is based on the characteristics of the distribution of rainfall received in a locality during the different months of the year. On the basis of this classification, four types of climate are adopted. Figure 1 shows the climatologic map of the country. It should be noted in this figure that the dividing lines between different climatic types occur along mountain ranges which are high enough to cause variations in rainfall distribution. Figure 1 also shows the locations of the four stations selected for analysis: Vigan, Legaspi, Zamboanga and Davao. Figure 2 presents the mean monthly rainfall at each station (see also Medina et al., 1985).

Results and discussion

Figure 3 illustrates the time series plots of $\phi_*(k,1)$ calculated by OKF (broken line), assuming no abrupt change in system parameters, and by AKF (full line), implementing GLRT. The numbered peaks identify the time and magnitudes of the abnormal rainfall patterns. These abnormal rainfall patterns are classified into three types: Type A which is characterized by a dominance of monthly rainfall depths of below the mean values and presence of abnormally dry months, Type B which is typified by a dominance of rainfall depths of above the monthly mean values and existence of abnormally wet months, and Type C which is characterized by both abnormally dry and wet months

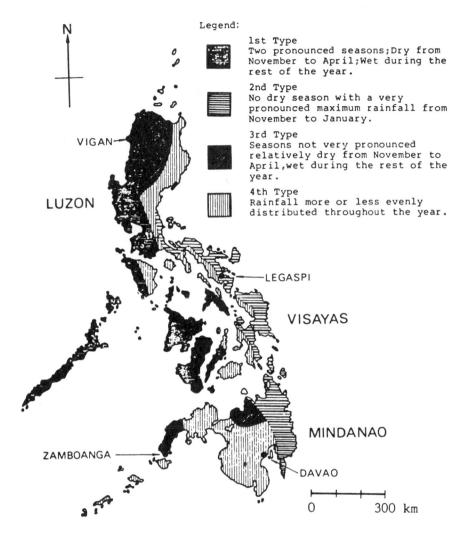

Figure 1   Climatological map (after "Philippines Water Resources",
           1976). Vigan, Legaspi, Zamboanga and Davao are the
           stations selected for analysis.

occurring more or less alternately.   These abnormal rainfall patterns
may last for one or two years.
    In the same figure, the arrowed peaks above the threshold
identify   the   time   of   occurrences   of   abrupt   changes   in   model
parameters. As can be observed in this figure, these changes coincide
with   occurrences of abnormal rainfall patterns,   indicating that the
occurrence of an abnormal rainfall pattern may induce significant
rainfall fluctuation. For each sequence, the abrupt changes in model
parameters   identify three rainfall periods with average duration of
11 years.   The estimates of the parameters by AKF are shown   to   be

Table 1   Parameters slected by AKF

Station	f_i	Parameters	Period I	II	III
Vigan		M_y	1.23	1.29	1.24
	1/12	A_1	-0.85	-0.97	-0.91
		B_1	-0.81	-0.95	-0.88
	1/6	A_2	0.08	0.21	0.23
		B_2	0.02	0.04	0.12
	70/396	A_3	-0.03	-0.10	-0.02
		B_3	-0.02	-0.18	-0.01
	132/396	A_4	-0.17	-0.11	-0.14
		B_4	0.00	0.15	0.00
Legaspi		M_y	2.05	1.99	2.04
	7/396	A_1	-0.02	0.14	0.05
		B_1	0.07	0.00	-0.02
	1/12	A_2	-0.10	-0.22	-0.26
		B_2	0.35	0.20	0.21
	1/6	A_3	-0.09	0.02	0.01
		B_3	0.18	0.12	0.17
	99/396	A_4	-0.12	-0.07	-0.06
		B_4	0.15	0.03	0.05
	160/396	A_5	-0.11	-0.02	0.00
		B_5	0.03	0.08	0.00
Zamboanga		M_y	1.40	1.38	1.36
	21/384	A_1	-0.01	0.05	0.09
		B_1	-0.04	0.03	0.03
	1/12	A_2	-0.36	-0.29	-0.36
		B_2	-0.12	-0.11	-0.14
	1/6	A_3	-0.11	-0.05	-0.08
		B_3	0.09	0.02	0.09
	96/384	A_4	-0.01	0.01	-0.01
		B_4	0.08	-0.11	-0.14
	178/384	A_5	-0.05	-0.09	-0.11
		B_5	0.01	0.00	0.00
Davao		M_y	2.16	2.16	2.10
	7/396	A_1	0.08	0.07	0.20
		B_1	0.10	-0.10	-0.06
	1/12	A_2	-0.22	-0.29	-0.39
		B_2	-0.23	-0.24	-0.31
	1/6	A_3	-0.14	-0.13	-0.07
		B_3	0.10	-0.09	0.26
	99/396	A_4	0.14	0.14	0.16
		B_4	-0.03	0.06	-0.22

effectively equal to those (not shown in this paper) by least squares method, which verifies the validity of the estimates in Table 1.

Table 2 shows that only the I vs II pair at Vigan and Legaspi can be interpreted as obeying two different parameter structures at the 5% level of significance, which suggests that serious changes in rainfall characteristics took place between these periods. Figure 4 presents in a convenient form the three rainfall periods and the distribution of the abnormal patterns identified by OKF.

We propose the following discussion and conclusions regarding rainfall fluctuations on the basis of the behaviour of M_y and A_i and B_i which correspond to the periodicities (longer than one year, one year, and six months) accepted as real (Medina et al.,1985). For these, we refer the reader to Figure 4 and Tables 1 and 2.

The occurrence of Type B abnormal pattern in Vigan initiates a period (II) with an abundance of abnormally high rainfalls as indicated by the presence of Type B and Type C abnormal patterns. This shift from period I to period II is accompanied by an increase in the estimated value of M_y, as well as in coefficients A_1, B_1 and A_2 of the one-year and six-month harmonics. The abrupt change in M_y

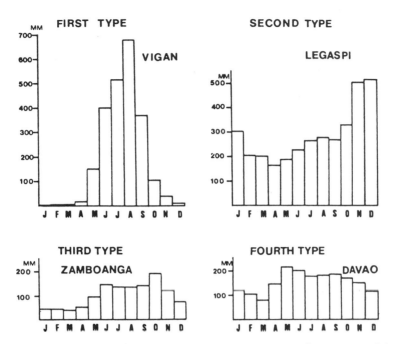

Figure 2 Mean monthly rainfall. First type, second type, third type
and fourth type are the types of climates in the Philippines.

Table 2   Results of the Chow test for parameter change.
(p is number of parameters, n and m are the
numbers of observations in the two periods.)

Station	Pair of periods	F-statistic (critical value)	Degrees of freedom p, n + m – 2p
Vigan	I vs II	2.202 (1.92)	9, 239
	II vs III	1.110 (1.92)	9, 260
Legaspi	I vs II	2.218 (1.83)	11, 251
	II vs III	0.605 (1.82)	11, 293
Zamboanga	I vs II	1.487 (1.83)	11, 195
	II vs III	0.456 (1.83)	11, 283
Davao	I vs II	0.823 (1.91)	9, 296
	II vs III	1.266 (1.95)	9, 125

amounts to an increase of 32.7% in mean rainfall corresponding to an
increase of 854 mm (45.6%) in the annual rainfall during 1961-72 over
that of the previous ten years (1951-60).   In contrast, the shift
from period II to period III, which is also initiated by a Type B
abnormal pattern, is characterized by a decrease in M_y, A_1 and B_1 and
an increase in A_2 and B_2.   Although the F-statistic finds these

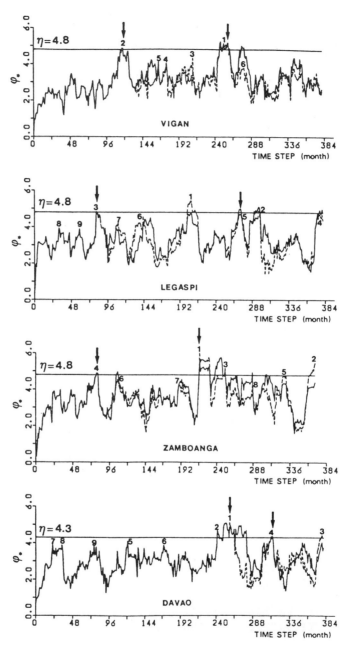

Figure 3   Time series plots of the abnormality detection index ϕ_* as calculated by OKF (broken line) and by AKF (full line). Numbers indicate periods with abnormal rainfall patterns. Downward arrows indicate rainfall fluctuations.

changes to be not significant, it corresponds to a decrease of almost 700 mm in the annual total rainfall that occurred from 1961-72 to 1973-83.

At Legaspi station, the fluctuation from period I to period II is initiated by a Type C abnormal pattern. The decrease in both M_y and

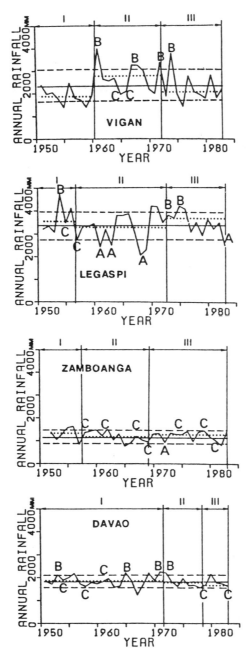

Figure 4   Time series plot of the annual rainfall of each station,
depicting two rainfall fluctuations and distribution of
abnormal rainfall patterns (A, B and C).

amplitudes of one-year and six-month periodic components plus a
remarkable increase in A_1 of the 4.7-year oscillation accompany this
fluctuation.   The decrease in M_y reflects a drop of only 8.2% in
actual monthly mean rainfall during period II from period I.   In this
shift, the presence of Type B and Type C abnormal patterns typifies
period I and accounts for the above-average rainfall in this period,

whereas Type A, abnormally dry pattern, prevails in period II. On the other hand, the change from period II to period III is initiated and characterized by Type B occurrences. This shift is described by an increase in M_y of 11.2%, a decrease in the amplitude of the 4.7-year harmonic, and increases in the amplitudes of the one-year and six-month components. Period III experiences Type A and Type B abnormal patterns.

Using drought duration curve analysis, Kawamura et al. (1987) have demonstrated that droughts could be very severe in periods when the rainfall characteristics are similar to those displayed in period II at Legaspi station. This period is characterized by frequent occurrences of Type A abnormal patterns, presence of intervals of above-average rainfall, and particularly, a remarkable 4.7-year oscillation. One of the three Type A abnormal patterns in this period is the 1969-drought episode (Medina et al., 1985) which caused crop failure in the region where Legaspi is situated. It is also the most abnormal rainfall pattern at this station in the sense that the value of its peak θ_* is the highest. Also, similar characteristics emerged in period III of Davao, where the two significant droughts in the country in 1982 and 1983 appear as the occurrence of Type C abnormal pattern.

In the Zamboanga and Davao rainfall sequences, the occurrences of abnormal rainfall patterns, which resulted in the detection of abrupt changes in model parameters, cause neither fluctuation of 10% in the mean nor statistically significant changes in parameter structures. This is understandable, since rainfalls in both Zamboanga and Davao have much lower variances (therefore have much less fluctuations) than those in Vigan and Legaspi. Also, with this result, it appears that rainfall fluctuations would take place more readily at places with climates similar to those in Vigan and Legaspi than at localities with climates like those in Zamboanga and Davao.

Although the shift in the mean is useful for the identification of significant rainfall fluctuations, it is clear from this study that it is not a sufficient parameter. This observation is illustrated by the fluctuations from period I to period II in Legaspi, where the shift in M_y exhibits less than a 10% drop in actual monthly mean rainfall while the F-statistic shows significant changes in the parameter structures between these periods. In particular, the mean describes the average level. This parameter provides a vague description of just how poorly a water resource system might behave in the infrequent situation when flood or drought does occur. Although rainfall may be satisfactory in a decade, our concern must extend to the short intervals of several months when water resources systems might be seriously 'depleted' (at least temporarily). For example, our attention should not be focused exclusively on the ten-year low rainfall as things can be worse in critical parts of the system during several intervals of adverse rainfall. Period II of Legaspi, as shown in Figure 4, illustrates the inability of the mean by itself to define how severe and how frequent intervals of abnormally low rainfall may occur.

However, in cases where rainfall is highly variable, or if the consequences of intervals of abnormal rainfall are severe, then it is appropriate and desirable to consider also the changes in the parameter structures which (unlike the shifts in the mean) describe in a clear and meaningful way what the character of fluctuations might be.

Conclusions

The AKF has proved to be a useful procedure for the identification of the time of occurrence and magnitude of rainfall fluctuations. The AKF approach applied to periodic-stochastic rainfall time series model has not only provided information on shifts in the mean but also automatically exposed hidden periodicities in a particular period, which have been found indispensable in probing for occurrences of adverse rainfall and rainfall fluctuations.

The rainfall fluctuations have been described not only in terms of rainfall amounts but also in terms of the occurrences of adverse rainfall. The magnitude of changes in the mean exhibited in the Vigan record, and the high incidence of adverse rainfall in period II of both Vigan and Legaspi, have demonstrated that the occurrence of a short interval with rainfall pattern anomaly may induce serious changes in rainfall characteristics. These kinds of severities of rainfall fluctuations may serve as a basis for evaluation as to whether or not the existing and proposed water resources systems are robust, vulnerable or resilient. Nevertheless, the occurrence of an abnormal rainfall pattern, which led to the detection of an abrupt change in model parameters, does not always mean that significant rainfall fluctuation is happening or is going to happen as shown by the Zamboanga and Davao rainfall sequences.

While the existence of rainfall fluctuations has been ascertained, it is obvious that the four stations involved are not sufficient to derive general conclusions applicable to other stations of similar or different climatic conditions. With this number of stations, it is also difficult to examine the interrelationships among the four types of climates in the country regarding the spatial occurrence of rainfall fluctuations. Moreover, the shortness of records limits the interpretation of the results reported above. Also, three periods are not enough to draw complete characterization of the temporal occurrence of rainfall fluctuations.

ACKNOWLEDGMENT   We express our gratitude to the Philippine Atmospheric Geophysical and Astronomical Services Administration (PAGASA) for providing rainfall data.

References

Bennett, R. J. (1979) Spatial Time Series. Pion Limited, London.
Chow, G. C. (1960) A test of equality between sets of observations in two linear regressions. Econometrica. 28, 591-605.
Hori, K. (1978) Water resources. In: Encyclopedia of Civ. Engng. 88. vol. 24. Shokoku Press. In Japanese.
Karl, T. R. and Riebsame, W. E. (1984) The identification of 10- and 20-year temperature and precipitation fluctuations in the contiguous United States. J. of Climate and Applied Meteor., 23,

950-966.

Kawamura, A., Ueda, T. and Jinno, K. (1985) Analysis of long-term pattern fluctuations in a precipitation sequence. Proc. Japan Soc. Civ. Engrs., 363, 155-164. In Japanese.

Kawamura, A., Jinno, K., Ueda, T. and Medina, R. R. (1987) Analysis of patterns in a precipitation time sequence by ordinary Kalman filter and adaptive Kalman filter. In: this publication.

Medina, R. R., Jinno, K., Ueda, T. and Kawamura, A. (1985) Study on the statistical and dynamic characteristics of rainfall in the Philippines. Memoirs of the Faculty of Engng., Kyushu University, 45(1) Fukuoka, Japan.

Nemec, J. (1985) Water resource systems and climate change. In: Facets of Hydrology II (ed. by J. Rodda). John Wiley & Sons, New York.

Philippine National Water Resources Council (1976) Philippine Water Resources. Quezon City, Philippines.

Ueda, T., Kawamura, A. and Jinno, K. (1985) Detection of abnormality by the adaptive Kalman filter. Proc. Japan Soc. Civ. Engrs., 345, 111-121. In Japanese.

The Influence of Climate Change and Climatic Variability on the Hydrologic Regime and Water Resources (Proceedings of the Vancouver Symposium, August 1987). IAHS Publ. no. 168, 1987.

Consistent parameter estimation in a family of long-memory time series models

A. Ramachandra Rao
School of Civil Engineering
Purdue University
W. Lafayette, IN 47907 USA
Gwo-Hsing Yu
Department of Hydraulic Engineering
Tamkang University
Tamsui, Taipei Hsien, Taiwan, ROC

ABSTRACT Models with short correlation characteristics such as AR or ARMA models do not offer much flexibility in simultaneous modeling of short and long term characteristics of time series with a small number of parameters. Fractional difference models, on the other hand, have recently been proposed to overcome this deficiency in short correlation or Markovian models. Both the long and short term correlation characteristics may be satisfactorily modelled by using these fractional difference models.

   The characteristics of fractional difference models depend on the difference parameter d and the noise variance. It is therefore important to estimate these parameters accurately. Recently two approaches have been proposed for estimation of parameters in these fractional difference models. The first estimator is based on the least squares method in the frequency domain. This estimator is unbiased and consistent in the mean square sense. The second estimator is based on the spectral characteristics of the model. These two estimators are compared in this study.

Introduction

A model based on fractional Brownian motion has been proposed by Bickel and Doksum (1977) who have also investigated many of its desirable properties. This model is represented by (1) where $x(t)$ is the time series to be modelled, Z^{-1} is the unit delay operator and $w(t)$ is the zero mean white Gaussian random sequence with variance ρ.

$$x(t) = (1 - Z^{-1})^{-d} w(t) \tag{1}$$

   If the sequence (1) is differenced d times then it will result in a zero mean Gaussian white noise sequence. This model may also be represented as an infinite order moving average sequence. By expanding $(1 - Z^{-1})^{-d}$ as a binomial expansion, Equation 2 may be derived.

$$x(t) = (1 + \sum_{k=1}^{\infty} b_k Z^{-k}) w(t) \tag{2}$$

245

$$b_k = \frac{\Gamma(k+d)}{\Gamma(d)\Gamma(k+1)} \tag{2a}$$

By using the well known relationships between the AR and MA processes, the infinite order MA model in (2) may be written as the AR model in (3).

$$(1 - \sum_{k=1}^{\infty} a_k Z^{-k})x(t) = w(t) \tag{3}$$

$$a_k = \frac{\Gamma(k-d)}{\Gamma(-d)\Gamma(k+1)} \tag{3a}$$

These equivalences between the fractional difference and AR (∞) and MA (∞) models cannot be easily exploited for modelling because the coefficients a_k and b_k decay very slowly. For example the 10th AR coefficient for d=0.4 is approximately equal to 0.01 and it may not be possible to ignore it. Also, if we have only a limited number of observations, which is generally the case for annual hydrologic time series, it would not be possible to obtain good parameter estimates when the number of parameter estimates is large. Consequently, it is preferable to use the model in its original form (1) and estimate the parameter d and the variance ρ. Estimation of parameters d and ρ by the usual methods is very difficult. Hosking (1981) has given a parameter estimation method which was tested by Rao and Yu (1984).

There have been other approaches to estimating the parameter d. Granger and Joyeux (1980) approximated the fractional difference model by a 100th order AR model and estimated the difference parameter by comparing variances for different choices of d. This method requires a lot of computing time, is arbitrary and not optimal. The maximum likelihood estimators in the frequency domain have been demonstrated to be consistent. However, the maximum likelihood method requires the solution of nonlinear simultaneous equations by numerical methods.

In view of these developments, Kashyap and Eom (1984) developed a parameter estimation method based on the least squares principle in the frequency domain. This estimator is very easy to use and is discussed in the next section. Janacek (1982) has also developed another parameter estimation method which is based on the power spectral characteristics of the model (1). These two methods are applied to annual meteorologic and hydrologic time series.

Parameter estimation

(a) Kashyap and Eom's (1984) method

We have a sequence x(t), t=0, 1, ..., N-1, for which the fractional difference model in (1) must be fitted. In order to estimate the parameter d, the discrete Fourier transform of 1, given in (4) is used.

$$X(k) = F[x(t)] = [1 - \exp(-j2\pi k/N)]^{-d} W(k) \tag{4}$$

or
$$X(k) = \exp(-j\pi kd/N)[2j\ \sin(\pi k/N)]^{-d}W(k) \tag{5}$$

$$k = 0, 1, \ldots N$$

$X(k)$ and $W(k)$ are discrete Fourier transforms of $x(t)$ and $w(t)$ respectively. The magnitude of $X(k)$ is given by (6). The sequence $|w(k)|$ may be proved to be a white sequence but is distributed as the Rayleigh density in (7).

$$|x(k)| = [2|\sin(\pi k/N)|]^{-d}|w(k)| \tag{6}$$

$$\frac{f(w)}{|W(k)|} = \begin{vmatrix} (2w/\rho N)\ \exp(-w^2/\rho N) & W \geq 0 \\ 0 & \text{otherwise} \end{vmatrix} \tag{7}$$

Logarithmic transformation of (6) yields (8). In (8), $X(k)$ and $\sin(\pi k/N)$ are known and $W(k)$ is a transform of white Gaussian noise sequence $w(t)$. Consequently (8) is in the well known form of a deterministic signal plus noise. The mean and variance of log $w(k)$ are given by (9) and (10) respectively, where γ is the Euler's constant (0.5722157) and ρ is the variance of $w(t)$.

$$\log|X(k)| = -d\ \log|2\sin(\pi k/N)| + \log W(k) \tag{8}$$

$$E[\log|w(k)|] = [-\gamma + \log(\rho N)]/2 \tag{9}$$

$$\text{var}[\log|w(k)|] = \pi^2/24 \tag{10}$$

Now let

$$\alpha = -E[\log|w(k)|] = [\gamma - \log(\rho N)]/2 \tag{11}$$

and
$$V(k) = \log|w(k)| + \alpha \tag{12}$$

$V(k)$ is a zero mean white noise sequence. With the definitions in (11) and (12), (7) may be written as in (13).

$$\log|X(k)| = -d\ \log|2\sin(\pi k/N)| - \alpha + V(k) \tag{13}$$

In (13), $\log|X(k)|$ is given in terms of deterministic trend terms and additive zero mean white noise. Consequently the least squares estimation algorithm can be applied to (13) to obtain the estimate $\hat{\theta}$.

Let
$$\underset{\sim}{Z}(k) = \begin{vmatrix} \log|2\sin(\pi k/N)| \\ -1 \end{vmatrix}, \quad \underset{\sim}{\theta} = \begin{vmatrix} d \\ \alpha \end{vmatrix} \tag{14}$$

Then the least squares estimate of $\underset{\sim}{\theta}$ is given by (15).

$$\underset{\sim}{\hat{\theta}} = [\sum_{k=1}^{\frac{N}{2}} \underset{\sim}{Z}(k)\underset{\sim}{Z}^T(k)]^{-1}[\sum_{k=1}^{\frac{N}{2}} \underset{\sim}{Z}(k)\ \log|x(k)|] = \begin{vmatrix} \hat{d} \\ \hat{\alpha} \end{vmatrix} \tag{15}$$

The estimators $\hat{d}$ and $\hat{\alpha}$ in (15) are unbiased and consistent in the mean square sense (Kashyap and Eom (1984)). The variances of $\hat{d}$ and $\hat{\alpha}$ were proved by Kashyap and Eom to be $1/N$ and $\pi^2/12N$ respectively. The noise variance ρ must also be estimated. A consistent estimator of ρ, with α defined as in (14) is given by (16). $\hat{\rho}$ is unbiased and consistent in the mean square sense with variance $\hat{\rho}^2\pi^2/3N$.

$$\hat{\rho} = \frac{1}{N} \exp[\gamma - 2\hat{\alpha} - \frac{\pi^2}{6N}] \tag{16}$$

(b) Janacek's method

An alternative method for estimating the parameter d has been proposed by Janacek (1982). Janacek starts with (4) which can be written as (17) where $\omega = 2\pi k/N$

$$f_X(\omega) = [|1 - e^{j\omega}|^2]^{-d} f_\Omega(\omega) \tag{17}$$

Taking natural logarithms of (17) we can write (18)

$$\ln[f_X(\omega)] = -d \ln[2(1 - \cos\omega)] + \ln[f_\Omega(\omega)] \tag{18}$$

or

$$\ln[f_X(\omega)] = -d H(\omega) + \ln[f_\Omega(\omega)] \tag{19}$$

where $H(\omega)$ is a weight function given by (20).

$$H(\omega) = -0.5[\log 2 - \sum_{k=1}^{\infty} \frac{\cos^k W}{k}] \tag{20}$$

The similarity between (19) and (8) is apparent. Instead of utilizing the signal plus noise structure of (8) to estimate the parameter d in (19), Janacek uses the property of the log spectra of $x(t)$ and $w(t)$ to estimate it.

Let

$$S = \frac{1}{\pi} \int_0^\pi \ln[f_X(\omega)] H(\omega) \, d\omega \tag{21}$$

$$Q = \frac{2}{\pi} \int_0^\pi H^2(\omega) \, d\omega \tag{22}$$

$$P = \frac{1}{2\pi} \sum_{k=1}^{\infty} \int_0^\pi \frac{\cos^k\omega}{k} \ln[f_W(\omega)] \, d\omega \tag{23}$$

$$\ln \sigma^2 = \frac{1}{\pi} \int_0^\pi \ln[f_X(\omega)] \, d\omega \tag{24}$$

The estimator d is then given by

$$\hat{d} = \frac{1}{Q} [S + \frac{1}{2}\ln(2 + \sigma^2) - P] \tag{25}$$

The estimator $\hat{d}$ in Janacek's method is thus based entirely on the spectral properties of the data. Given the ad-hoc nature of many of the assumptions made in estimating the spectra, such as choice of the maximum lag, and window functions in the Blackman-Tukey method for example, the spectral properties vary. Consequently, the estimates $\hat{d}$ would also vary. Furthermore, the consistency and variance of Janacek's estimates are unknown. The details of Janacek's method are found in Janacek (1982) and in Rao and Yu (1984). These two estimators are used to estimate the parameters of meteorologic time series and the results are discussed below.

Data used in the study

Six meteorologic and hydrologic time series are used in the present study. The two meteorologic series are the central England mean annual temperature series which have been compiled by Manley (1974) and the Eastern American mean annual temperature series (Lamb (1977)). The central England temperature series starts in 1659 and ends in 1976 with 318 observations. The mean value of this series is 9.15°C and its variance is 0.38. The Eastern American mean annual temperature series covers the period 1738 to 1967 with some missing data (for years 1741, 1764, 1778, 1779 and 1780). The average value of the series is used for these five missing values. This series has 230 observations with a mean value of 12.4°C and a variance of 0.29. The hydrologic data consists of four annual flow series from St. Lawrence, Gota, Blacksmith and Gunpowder rivers. Some information about these riverflow data is given in Table 1.

Table 1   Statistical characteristics of the Hydrologic
           Data Used in the Study

River	Location	Period	$N^{(1)}$	$AVG^{(2)}$	$VAR^{(2)}$
St.Lawrence	Ogdensburg, New York	1860-1957	98	1.0	0.00753
Gota	Sjotorp-Vanersburg,Sweden	1807-1957	150	1.0	0.03276
Blacksmith	Hyrum, Utah	1913-1979	66	1.0	0.11200
Gunpowder	Lock Raver, Maryland	1883-1978	95	1.0	0.10895

1. Number of data points (series is given as the modular coefficients of observed and computed annual flows for water year).
2. AVG, VAR are the mean value and variance of the series.

Results and conclusions

The parameter d estimated by the algorithm developed by Kashyap and Eom (1984) and by Janacek (1982) are given in Table 2. For the algorithm by Kashyap and Eom the standard errors for d and ρ are also given in Table 2.

The results in Table 2, for the Janacek's method, indicate that these are close to the estimates by Kashyap and Eom when $\hat{d}$ values are small. The deviations are quite substantial for larger values of $\hat{d}$ such as those corresponding to St. Lawrence and Blacksmith rivers. Whether this tendency is valid in general may be tested by using simulated data.

Table 2 Estimates $\hat{d}$ and $\hat{\rho}$

Series	Kashyap and Eom		Janacek
	$\hat{d}$	$\hat{\rho}$	$\hat{d}$
Central England	0.23	0.3706	0.270
Annual temperature	(0.06)	(0.0377)	
Eastern U.S.	0.38	0.2352	0.350
Annual temperature	(0.07)	(0.0281)	
Gota: Annual Flows	0.45	0.0265	0.359
	(0.08)	(0.0039)	
Gunpowder: Annual Flows	0.28	0.1277	0.305
	(0.10)	(0.0237)	
St.Lawrence: Annual Flows	0.72	0.0039	0.535
	(0.10)	(0.0007)	
Blacksmith: Annual Flows	0.61	0.0921	0.429
	(.12)	(0.0206)	

The correlation function of the X(t) series is given by (26) and the partial autocorrelations are given by (27).

$$\rho_X(k) = \frac{d(1+d)(2+d)\ldots(k-1+d)}{(1-d)(2-d)\ldots(k-d)} = \frac{\Gamma(1-d)\,\Gamma(k+d)}{\Gamma(d)\,\Gamma(k+1-d)} \qquad (26)$$

$$\phi_X(k) = \frac{d}{k-d} \qquad (27)$$

The power spectrum of the x(t) series may be written as in (28).

$$f_X(\omega) = \left| \begin{array}{ll} [2\sin(\omega/2)]^{-2d} & 0 < \omega \leq \pi \\ \omega^{-2d} & \omega \to 0 \end{array} \right. \qquad (28)$$

The correlograms and partial autocorrelations for Gunpowder and St. Lawrence river flows and the Central England temperature series are shown in Figures 1 through 3. The correlogram and partial autocorrelation function of Central England annual temperature are shown in Figure 1. By definition the theoretical power spectrum approaches infinity as $\omega \to 0$. Consequently it is not possible to compare the empirical and theoretical spectra near $\omega = 0$. The correlogram and partial autocorrelation function of the data match the corresponding theoretical estimates for the Central England annual temperature data.

The theoretical and observed autocorrelation and partial autocorrelation values for the Gunpowder River annual flows match each other closely. Consequently, it may be concluded that the fractional difference model is a good candidate model for these data. For the Gota riverflow data, maybe because the d value is very high (0.45) and approaching 0.5, the persistence indicated by the model is much higher than that indicated by the data. This has resulted in values of correlation coefficients which are much larger than the

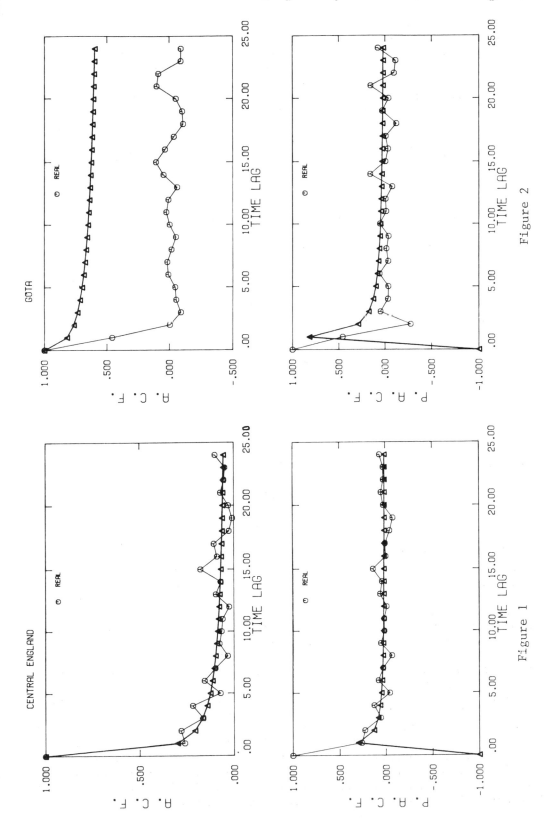

Figure 1

Figure 2

Figure 1   Autocorrelation coefficients (top), and partial autocorre-
            lation coefficients (bottom) of Central England Tempera-
            ture Series. Circles represent the results from observed
            data and triangles from the model.

Figure 2   Autocorrelation coefficients (top), and partial autocorre-
            lation coefficients (bottom), of Gota River flow series.
            Circles represent the results from observed data and
            triangles from the model.

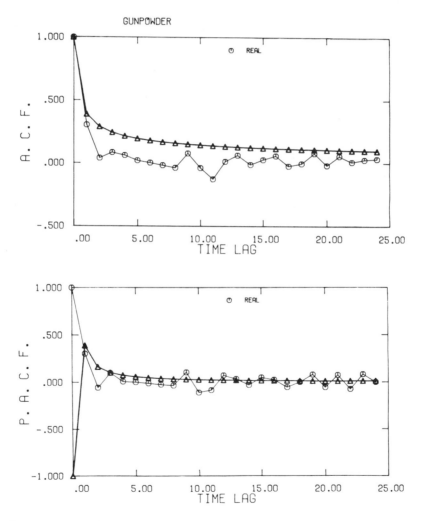

Figure 3   Autocorrelation coefficients (top), and partial autocorre-
            lation coefficients (bottom), of Gunpowder River flow
            series. Circles represent the results from the observed
            data and triangles from the model.

empirical values. The simple fractional difference model is not appropriate in this case. The fit between theoretical and empirical partial autocorrelation coefficients is better for the Gunpowder data, especially for large lags.

Based on the preceding discussion of the two methods of parameter estimation the parameter estimation method by Kashyap and Eom is recommended. The reason for this choice is that it avoids the use of ad-hoc assumptions inherent in estimation of spectra and is much simpler to use.

References

Bickel, P.J. & K.A. Doksum (1977) Mathematical Statistics, Holden-Day Inc., San Francisco.

Granger, C.W.J. & R. Joyeux (1980) An Introduction to Long Memory Time Series Models and Fractional Differencing, Jour. Time Series Analysis, 1(1), 15-30.

Hosking, J.R.M. (1981) Fractional Differencing, Biometrika, 68, 1.

Janacek, G.J. (1982) Determining the Degree of Differencing for Time Series Via the Log Spectrum, J. Time Series Analysis, 3, 177-183.

Kashyap, R.L. & K.B.Eom (1984) A consistent estimation of parameters in a family of long-memory time series models, Personal Communication.

Lamb, H.H. (1977) Climate: Present, Past and Future, Vol. 2, Climatic History and the Future, London, Methuen, 572-579.

Manley, G. (1974) Central England Temperatures: Monthly Means 1659-1973, Quart. J., Royal Met. Soc., 100, 389-405.

Rao, A.R. & G.H. Yu (1984) Investigation of Fractional Difference Models of Hydrologic and Climatologic Time Series, Tech. Rept. CE-HSE-84-2, School of Civil Engineering, Purdue University, W. Lafayette, IN 47907, USA, p. 151.

The Influence of Climate Change and Climatic Variability on the Hydrologic Regime and Water Resources (Proceedings of the Vancouver Symposium, August 1987). IAHS Publ. no. 168. 1987.

Temporal disaggregation of monthly rainfall data for water balance modelling

Thomas W. Giambelluca
Department of Geography
University of Hawaii at Manoa
Honolulu, Hawaii 96822, USA
Delwyn S. Oki
Water Resources Research Center
University Of Hawaii at Manoa
Honolulu, Hawaii 96822, USA

ABSTRACT   Water balance computations are often done using a monthly time step because of data or computation time constraints. However, because of the variable intensity of rainfall during a month, evapotranspiration is generally overestimated and groundwater recharge underestimated when balancing is done using a monthly interval. Water balance calculations using hourly, daily, and monthly intervals for a site in Hawaii result in recharge estimates of 40%, 37%, and 19% of precipitation, respectively. For locations where daily rainfall data are unavailable, but for which daily rainfall characteristics may be estimated, a method is developed for simulating sequences of daily rainfall which equate with a known monthly total. Using this method for the Hawaii sample site, soil moisture, evapotranspiration, and recharge estimates were significantly improved over monthly based estimates. However, using the simulated data, recharge estimates were consistently about 23% lower than estimates based on hourly data.

Désaggrégation temporelle des données
sur la pluie mensuelle afin de modeler
le bilan de l'eau

RESUME   En calculant le bilan de l'eau, on emploie généralement des intervalles d'un mois à cause des données ou des contraintes du calcul. Mais, à cause de l'intensité variable de la pluie pendant un mois, ordairement l'évapotranspiration est surestimée et la recharge de l'eau phréatique est sous-estimée quand on calcule le bilan sur une base mensuelle. Pour les besoins de cet article, des calculs du bilan de l'eau, effectués aux intervalles d'une heure, d'un jour et d'un mois dans un site à Hawai ont donné des estimations de recharge de 40%, 37%, et 19% de la précipitation, respectivement. Pour les endroits où les données quotidiennes ne sont pas disponibles, mais où l'on peut estimer caractéristiques de la précipitation journalière, une méthode est developpée pour simuler des séquences de précipitation journalière, lesquelles sont égales à un total mensuel connu. En employant cette méthode pour le site d'Hawai, il apparaît que les estimations du contenu de l'eau dans le sol, de l'évapotranpiration, et de la recharge sont meilleures que les estimations fondées sur les intervalles d'un mois. Cependant, en utilisant les données simulées, les estimations de recharge sont restées à un niveau environ 23% inférieur aux estimations fondées sur un intervalle horaire.

Introduction

Water balances are applied not only to practical water resource problems such as streamflow prediction, groundwater recharge estimation, evaluation of land use effects on the hydrologic cycle, determination of irrigation requirements and crop yield modeling, but also to problems related to climatic variation such as drought assessment (Palmer 1965).

Because of data or computation time constraints, water balances are often carried out using a monthly time step. Several authors (Rushton & Ward 1979; Howard & Lloyd 1979; Alley 1984) have noted that use of a monthly interval tends to result in overestimation of evapotranspiration and underestimation of groundwater recharge. However, the monthly water balance remains popular, perhaps because alternatives are lacking. In this paper, the effect of different time intervals on the results obtained from a simple water balance model are compared for a sample site in Hawaii. Subsequently, a method is suggested for disaggregating monthly rainfall into daily sequences for water balance computations.

Water balance model

The water balance model used in this study is a variant of the Thornthwaite and Mather (1955) bookkeeping procedure. Water fluxes through the plant-soil system are determined for each time interval in the following manner. A state variable X_i is first computed as,

$$X_i = S_{i-1} + P_i - R_i - E_i \qquad (1)$$

where S_{i-1} = ending available soil moisture (difference between total soil moisture content and the wilting point) for previous time interval, P_i = precipitation during time interval i, R_i = surface runoff during time interval i, and E_i = evapotranspiration during time interval i. (All variables are expressed as equivalent water depths.) On the basis of X_i, groundwater recharge (drainage flux) and end-of-interval soil moisture are determined according to the following drainage rules,

$$S_i = 0$$

$$Q_i = 0 \qquad\qquad\qquad \text{for } X_i \leq 0 \qquad (2)$$

$$E_i = S_{i-1} + P_i - R_i$$

$$S_i = X_i$$

$$Q_i = 0 \qquad\qquad \text{for } 0 < X_i \leq \Phi \qquad\qquad (3)$$

$$S_i = \Phi$$

$$Q_i = X_i - \Phi \qquad\qquad \text{for } X_i > \Phi \qquad\qquad (4)$$

where S_i = available soil moisture content at the end of time interval i, Q_i = groundwater recharge during time interval i, and Φ = available soil moisture capacity.

Direct runoff, under the assumptions of this model occurs as an instantaneous response to rainfall. Runoff may be estimated from streamflow measurements or computed using a rainfall-runoff model such as the Soil Conservation Service (1972) runoff curve number method. Runoff for this study was estimated using a regression-type rainfall-runoff model calibrated with field data measured locally by the U.S. Agricultural Research Service (El-Swaify & Cooley, 1980).

Evapotranspiration is determined as a function of environmental demand, potential evapotranspiration (PE_i), and soil moisture availability during the interval. Unlike the usual Thornthwaite procedure, depression of E_i below PE_i is strictly determined by soil moisture availability and no differentiation is made according to whether P_i exceeds PE_i. Soil moisture availability at the beginning of an interval (Z_i) is determined as,

$$Z_i = S_{i-1} + P_i - R_i \qquad\qquad (5)$$

The instantaneous rate of evapotranspiration (E) is assumed to vary as a function of instantaneous soil moisture (S) according to the following rules:

$$E = PE_i \qquad\qquad \text{for } S \geq C_i \qquad\qquad (6)$$

$$E = SC_i^{-1} PE_i \qquad\qquad \text{for } S < C_i \qquad\qquad (7)$$

The quantity C_i, sometimes called the root constant (Penman 1949), may be interpreted as the available soil moisture content below which E is depressed below the potential rate. A model was developed to estimate C_i (Giambelluca 1983) having the form,

$$C_i = \min[a + b(ROOT) + c(PE_i), 1] \Phi \qquad\qquad (8)$$

where a,b,c = calibration coefficients, ROOT = root depth (mm), and where PE_i is in units of mm d^{-1}. Data from lysimeter studies by Ekern (1966) were used to calibrate the model for conditions in Hawaii: a = 1.25, b = -1.87×10^{-3}, and c = 5.20×10^{-2} for PE $\leq$ 6 mm d^{-1}; a = 1.41, b = -1.87×10^{-3}, and c = 2.20×10^{-2} for PE > 6 mm d^{-1}.

Based on this model E_i is determined as,

$$E_i = PE_i T_i + C_i\{1 - \exp[-\alpha_i(1 - T_i)]\} \qquad \text{for } Z_i > C_i \qquad (9)$$

$$E_i = Z_i[1 - \exp(-\alpha_i)] \qquad\qquad \text{for } Z_i \leq C_i \qquad (10)$$

where T_i is the fraction of the current time interval during which soil moisture is above C_i,

$$T_i = \min[(Z_i - C_i)PE_i^{-1}, 1] \tag{11}$$

and where,

$$\alpha_i = PE_i C_i^{-1} \tag{12}$$

PE is estimated from pan evaporation measurements (Ekern & Chang 1985) adjusted by a crop factor to represent the surface of interest.

Effect of time interval

The assumptions of the above model differ somewhat from the many Thornthwaite-type water balance models. However, as with all such models, there is a tendency to overestimate E_i and underestimate Q_i. This bias results from the failure of the model to account for within-interval variability in soil moisture input. For longer time intervals, the importance of the rainfall variability and the resulting bias increases.

To examine the effect of time interval on soil moisture, evapotranspiration, and recharge estimation, the water balance model was applied to 3 sequences of input data derived from a single set of measurements and differing only in degree of aggregation. Specifically, an 8-year record of measured hourly data from an autographic gage, U.S. Weather Bureau (USWB) Station 0300 (Camp 84, maintained by Del Monte Corporation) in Central Oahu, Hawaii was aggregated into daily and monthly sequences. Daily runoff was computed using the previously mentioned regression model. Hourly and monthly runoff were derived from the daily sequence. Monthly pan evaporation records from a nearby station were used to estimate potential evapotranspiration. Daily potential evapotranspiration was assumed to be uniformly distributed, while the hourly sequence was computed as a function of the seasonally-adjusted diurnal cycle of incident solar radiation.

The intention here is to focus on the effect of the temporal resolution of rainfall. Non-periodic variability in potential evapotranspiration is low and direct runoff (a small portion of the total water balance in this region) variability is substantially accounted for by the variability of rainfall.

The water balance model was run with each set of input data and with the following parameter settings: Φ = 33.0 mm, ROOT = 300 mm, S_0 (initial soil moisture content) = 24.75 mm, crop factor =1.00 (PE = 100% pan evaporation).

In Table 1, the 8-year water balance totals are presented. The degree of bias in the monthly-based estimates of evapotranspiration and recharge are evident. Recharge as a percentage of precipitation is 19% for the monthly run, 37% for the daily run, and 40% for the hourly run. Figure 1 shows the monthly recharge time series for each run. Hourly and daily run values are aggregated into 96 monthly totals for comparison. Using the hourly run as a standard for comparison, it is evident that the daily interval accurately

TABLE 1 Eight-year (1966-1973) water balance totals using hourly,
         daily, and monthly intervals for USWB Station 0300, Oahu,
         Hawaii

Interval	P(mm)	R(mm)	ET(mm)	Q(mm)
Hourly	6585	354	3582	2641
Daily	6585	354	3807	2420
Monthly	6585	354	4968	1269

Note: P=precipitation, R=runoff, ET=evapotranspiration,
       Q=groundwater recharge.
Note: Because beginning and ending soil moistures differ, the
       output terms do not sum to the precipitation.

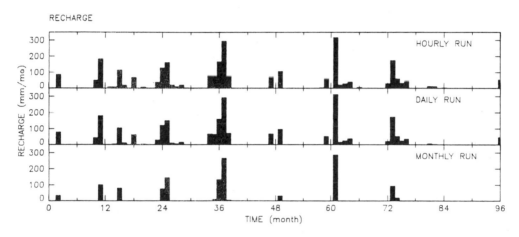

Figure 1 Monthly recharge time series for hourly, daily, and
          monthly-interval water balance runs for USWB station 0300,
          1966-1973.

estimates the occurrence of recharge events and only slightly
underestimates their magnitude. On the other hand, use of the monthly
interval results in a failure to recognize many recharge events as
well as a consistent and substantial underestimation of the magnitude
of the recharge. Daily and monthly (versus hourly) results are shown
in Figure 2 in the form of scattergrams. The corresponding
coefficients of determination and standard errors of the estimate are
listed in Table 2. For soil moisture, end-of-month values are used.
Average soil moisture is obtained for each month by averaging end-of-
day values from hourly and daily runs. This, of course, was not
possible for the monthly-interval run.
   The foregoing comparison indicates that use of the monthly
interval in water balance computations, at least for the central Oahu
sample site, results in unacceptably large error variance and bias
when compared with hourly-interval computations. The daily interval,
however, produces results which are very close to the hourly
calculation, though still slightly biased.

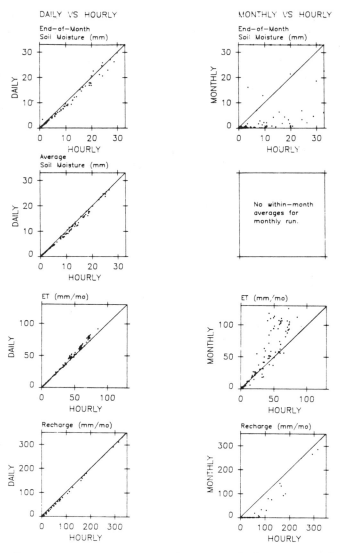

Figure 2 Scattergrams of monthly water balance results for daily and monthly vs. hourly runs.

Table 2 Coefficients of determination (R^2) and standard errors of the estimate (SEE) for daily and monthly vs hourly intervals(N=96 months)

	Daily vs Hourly		Monthly vs Hourly	
	R^2	SEE (mm)	R^2	SEE (mm)
Soil Moisture				
End-of-Month	0.989	1.01	0.340	9.26
Average	0.995	0.52	-----	----
Evapotranspiration	0.993	2.06	0.789	17.62
Recharge	0.998	2.75	0.853	17.78

Rainfall disaggregation model

Water balance models based on a daily time step require daily rainfall as input. Accurate and continuous daily rainfall records, however, are scarce relative to monthly records. A first-order Markov
chain model can be employed to disaggregate monthly rainfall at a
particular location into daily rainfall. To do so, some daily rainfall data must be available to facilitate the estimation of rainfall
distribution functions and transitional probability matrices, the
basic elements of this model. The presumption here is that although
the particular location and period of interest may lack continuous
daily rainfall records, measurements at the site during another
period or measurements at a nearby site may be available.

   In order to simulate daily rainfall and at the same time utilize
the information provided by the known monthly rainfall totals at the
location of interest, values of the non-dimensionalized parameter Y,
expressed as,

$$Y = P_d P_m^{-1} \tag{13}$$

where P_d= daily rainfall (mm d ^{1}) and P_m= monthly rainfall(mm d^{-1}),
are simulated. The chain of dimensionless Y values can subsequently
be multiplied by the known monthly rainfall (mm d^{-1}) to obtain a
simulated daily rainfall chain.

   Using historical data transformed according to equation 13, a
probability histogram of Y can be constructed. The range of non-
dimensionalized daily rainfall can be divided into a finite number(n)
of discrete classes or states. A rainfall distribution (probability
density function, PDF) must be fit over each of the non-zero rainfall
states of the histogram.

   Transitional probability matrices are also based on the available
historical data. The matrix elements, $p_{i,j}$, represent the probabilities of transition of Y from state i to state j on successive days
(i=1,n; j=1,n). With n states defined, an nxn transitional probability matrix PM can be expressed as

$$PM = [p_{i,j}] \qquad \text{for } i,j=1,2,\ldots,n. \tag{14}$$

   The year can be divided into a finite number of seasons assuming
that the transition probabilities for each season remain constant
throughout that season. Using this criterion, it would certainly be
acceptable to form 12 seasons corresponding to the months of the
year. However, this may require the estimation of parameters beyond
the capacity of the data (Allen & Haan, 1975). By grouping the
months to form a fewer number of seasons, the number of transition
probabilities which must be estimated can be reduced. Each season has
its associated state definitions, probability density functions, and
PM.

Daily rainfall simulation

In order to apply a Markov chain model, there must exist a dependence
between successive daily values of Y. Such a dependence is a

distinguishing characteristic of a Markov process. In addition, the historical time series of Y must be stationary within each season. Thus, a Markov property and stationarity test must be performed in order to validate statistically the use of a Markov chain model.

A Markov chain model was employed to disaggregate monthly rainfall at Station 0300. A 34-year record of daily rainfall from a nearby rain gage at USWB Station 8945 (Wahiawa Dam, Oahu, Hawaii) was used to define the Markov transitional probabilities. Markov property and stationarity tests were successfully passed. The key steps in the model application are briefly described below:

(a) Based on median monthly rainfall values at Station 8945, two seasons were defined for the area of interest. A dry season was found to occur during the months of April to October while a wet season was found to occur during the months of November to March.

(b) Seasonal probability histograms of Y were formed based on the historical data at Station 8945.

(c) For each of the seasons, the range of Y values was found to be adequately described by 10 discrete states or classes.

(d) In order to describe the nine non-zero rainfall states, a Weibull distribution function was fitted to each of the seasonal histograms (Figure 3).

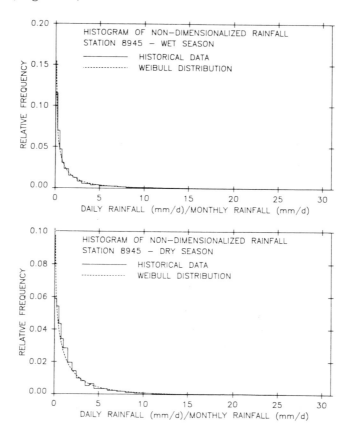

Figure 3   Histograms of non-dimensionalized wet season (Nov.-Mar.) and dry season (Apr.-Oct.) rainfall for USWB station 8945 and the fitted Weibull distributions.

(e) Two 10 x 10 transitional probability matrices (one for each season) were formed based on the historical Y sequence.

(f) A random number was generated to determine the next day's rainfall state (k) given the current state (i).

(g) A second random number was generated to obtain a simulated value of Y with state k.

(h) The daily rainfall value was obtained by multiplying the simulated Y value by the monthly rainfall.

The cumulative distribution of simulated daily rainfall at Station 0300 compares favorably with the historical distribution at that station(Figure 4). It should be noted, however, that the Markov model based on the transition probabilities and distribution functions of Station 8945 seems to underestimate the number of extreme daily rainfall events at Station 0300. It is likely that the distribution

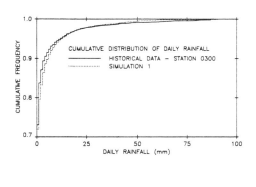

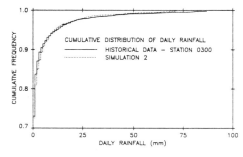

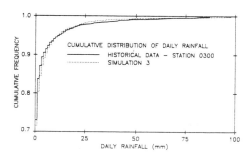

Figure 4 Cumulative daily rainfall distributions for simulations 1, 2 and 3 in comparison with historical data for USWB Station 0300.

functions based on Station 8945 data (Figure 3) fail to account fully for extreme rainfall events at Station 0300.

Water balance with simulated rainfall data

Using the Markov-Weibull disaggregation technique described above, three daily rainfall sequences were generated from monthly data for Station 0300 during 1966-1973. Using these data, daily runoff sequences were generated and adjusted to be equivalent to the runoff sequence computed for the actual daily data. Monthly pan evaporation was assumed to be uniformly distributed over each month. Water balance calculations were done for each daily rainfall simulation, using the same model parameters as in the real data runs previously described.

TABLE 3 Eight-year (1966-1973) water balance totals using 3 daily rainfall simulations and using monthly and hourly data for USWB Station 0300, Oahu, Hawaii

Run	P(mm)	R(mm)	ET(mm)	Q(mm)
Simulation 1	6585	354	4234	2020
Simulation 2	6585	354	4205	2021
Simulation 3	6585	354	4195	2038
Simul. Ave.	6585	354	4211	2026
Monthly Data	6585	354	4968	1269
Hourly Data	6585	354	3582	2641

Note: P = precipitation, R = runoff,   ET = evapotranspiration, Q = groundwater recharge.
Note: Because beginning and ending soil moistures differ, the output terms do not sum to the precipitation.

Eight-year water balance totals for each simulation are given in Table 3. Also shown for comparison are the results using monthly and hourly data. Estimates based on the daily rainfall simulations consistently overestimate evapotranspiration and underestimate recharge, but to a much lesser degree than the monthly-interval run. Scattergrams of soil moisture, evapotranspiration, and recharge for each daily rainfall simulation run vs the hourly data run are given in Figure 5. Table 4 lists corresponding coefficients of determination and standard errors of the estimate.
As the scattergrams and R^2 values indicate, water balances using simulated daily rainfall result in evapotranspiration and recharge estimates which are in reasonable agreement with results of the hourly data calculations. While significant bias remains, the use of simulated data greatly improves the model accuracy over monthly data.

Discussions and conclusions

The consistent underestimation of the recharge estimate based on

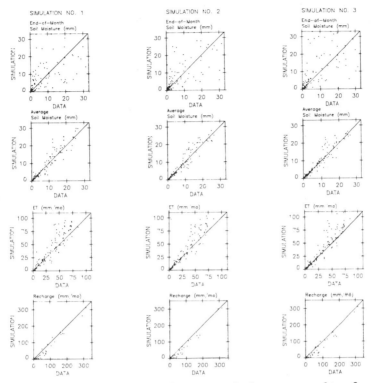

Figure 5   Scattergrams of monthly water balance results for simulated daily rainfall runs vs hourly rainfall data runs.

Table 4 Coefficients of determination (R^2) and standard errors of
the estimate(SEE) for water balance results based on simulated daily rainfall vs hourly rainfall data (N=96months)

	S1 vs Hourly R^2	SEE(mm)	S2 vs Hourly R^2	SEE(mm)	S3 vs Hourly R^2	SEE(mm)
Soil Moisture						
End-of-Month	0.319	7.66	0.485	7.10	0.405	7.98
Average	0.881	2.82	0.895	2.58	0.882	2.74
Evapotranspiration	0.819	12.38	0.842	11.79	0.838	11.91
Recharge	0.957	11.08	0.956	11.26	0.951	11.61

Note: S1=Simulation No.1, S2=Simulation No.2, S3= Simulation No.3

simulated daily rainfall, averaging about 23% compared with the
hourly data results, is higher than expected. Using real data, the
daily recharge estimate was biased by only 8%. The discrepancy here
seems to be the result of the imperfect fit of the rainfall
simulation model. While comparison with actual rainfall data (Figure
4) showed the fit to be reasonably good, presumably, the water
balance is sensitive to even minor differences in the frequency

distribution or persistence of daily rainfall.

The results of this study confirm that the use of a monthly interval in water balance models results in a consistent overestimation of evapotranspiration and underestimation of recharge. The bias is much less pronounced for the daily interval. If daily rainfall data are available at the site for some period, or if daily data from a nearby site exist, a Markov-Weibull model calibrated to non-dimensionalized daily rainfall can be used to disaggregate monthly data. Use of simulated sequences of daily rainfall improves estimates of evapotranspiration and recharge but still results in a significant bias.

It is likely that the assumptions of the water balance model can be modified in such a way as to reduce the interval biasing effect. Specifically, the bias stems from the assumption that all water input is available at the beginning of an interval. It is possible that by shifting the time of input to some point within the interval (to the mid-interval point for example), or by distributing the input within the interval according to a fixed pattern, the interval bias can be alleviated. It is doubtful that such a solution will have much effect on the error variance produced by long interval data, however, it may offer a simple method of reducing the bias error.

The results presented here are of a preliminary nature, having been based on one sample site. Subsequent testing will be done using data from contrasting climates. Additionally, various alternative assumptions regarding the within-interval timing of water input will be tested.

References

Allen, D. M. & Haan, C.T. (1975) Stochastic simulation of daily rainfall. Res. Rep. no. 82, Water Resources Research Institute, University of Kentucky, Lexington, Kentucky, USA.

Alley, W.M. (1984) On the treatment of evapotranspiration, soil moisture accounting and aquifier recharge in monthly water balance models. Water Resour. Res. 20(8), 1137-1149.

Ekern, P.C. (1966) Evaporation by Bermuda Grass sod, Cynodon dactyon L. Pers., in Hawaii. Agron. J. 58(4), 387-390.

Ekern, P.C. & Chang, J.H. (1985) Pan evaporation: State of Hawaii, 1894-1983. Rep. R74, Department of Land and Natural Resources, Division of Water and Land Development, State of Hawaii, Honolulu, Hawaii, USA.

El-Swaify, S.A. & Cooley, K.R. (1980) Sediment losses from small agricultural watersheds in Hawaii (1972-1977). A.R.M.-W.-17 Agricultural Research, Science & Education Administration, U.S. Department of Agriculture, Oakland, CA, USA.

Giambelluca, T.W. (1983) Water balance of the Pearl Harbor-Honolulu basin, Hawai'i, 1946-1975. Tech. Rep. no.151, Water Resources Research Center, Univ. of Hawaii at Mamoa, Honolulu, Hawaii, USA.

Howard, K.W.F. & Lloyd, J.W. (1979) The sensitivity of parameters in the Penman evaporation equations and direct recharge balance. J. Hydrol. 41, 329-344.

Palmer, W.C. (1965) Meteorologic drought. Res. Pap. 45, U.S. Weather Bureau. U.S. Government Printing Office.

Penman, H.L. (1949) The dependence of transpiration on weather and soil conditions. J. Soil Sci. 1(1), 74-89.

Rushton, K.R. & Ward, C. (1979) The estimation of groundwater recharge. J. Hydrol. 41, 345-361.

Thornthwaite, C.W. & Mather, J.R. (1955) THe water balance. Publ. Climatology. 8(1), 1-104.

U.S. Department of Agriculture (1972) Hydrology. National Engineering Handbook. Soil Conservation Service, Government Printing Office.

The Influence of Climate Change and Climatic Variability on the Hydrologic Regime and Water Resources (Proceedings of the Vancouver Symposium, August 1987). IAHS Publ. no. 168, 1987.

Space-time modelling of rainfall-runoff process

Kaz Adamowski & Fadil B. Mohamed
Department of Civil Engineering
University of Ottawa
Ottawa, Ontario   K1N 6N5
Nicolas R. Dalezios
INTERA Technologies Ltd
785 Carling Avenue
Ottawa, Ontario   K1S 5H4

ABSTRACT   An explanatory model belonging to the class of space-time   transfer function-noise (STTFN) processes is presented.   The paper develops a three-stage iterative procedure for building a STTFN model of the rainfall-runoff process.   Four precipitation and runoff stations located in a watershed in Southern Ontario, Canada, sampled at 15-day intervals are used for the numerical analysis.   The identified model is STTF (2,1,0,0,1,0,0). The model parameters are estimated by the polytope method, the optimal-step steepest descent method, a conjugate gradient method, and a quasi-Newton method.   The developed space-time   model proved to be adequate in describing the spatio-temporal   characteristics   of precipitation   and runoff time series.
Note: A French abstract of this paper is in the back of the text.

Introduction

In recent years a large number of stochastic models have been developed to represent different aspects of the rainfall-runoff process.   The most used approach has been the Box-Jenkins (Box and Jenkins, 1976) transfer function-noise (TFN) modelling of hydrologic time series.   This method of time series analysis relates the output (runoff) of a hydrologic system to the input (rainfall) of the system by adding a noise series.

There is an increasing interest in hydrology to develop empirical spatio-temporal models of rainfall-runoff in the context of regional hydrologic analysis (Salas et al., 1980). Since rainfall and runoff series are correlated in space and time, the Box-Jenkins TFN modeling procedure is extended to a multivariate input-output hydrologic system (Cooper and Wood, 1982; Adamowski and Hamory, 1983; Mohamed, 1985) resulting in a general model class of space-time transfer function-noise (STTFN) models.

The main purpose of this paper is to develop an input-output STTFN model of the rainfall-runoff process for regional hydrologic analysis and to examine different nonlinear optimization techniques for parameter estimation of the identified space-time models.

The model building procedure

In STTFN modelling, the output y_{it} from i=1, 2, ..., N zones over t=1, 2, ..., T time periods is assumed to be linearly dependent upon the input series X_1, X_2, ..., etc. in time and space. The STTFN model may take the form

$$y_{it} = \frac{\sum\limits_{s=0}^{\ell} \sum\limits_{k=1}^{p} \omega_{sk} B^k L_s}{(1 - \sum\limits_{s=0}^{m} \sum\limits_{k=1}^{q} \delta_{sk} B^k L_s)} B^b L_j x_{it} + a_{it} \tag{1}$$

where ℓ and m are the spatial orders, p and q are the temporal orders, b and j define an initial period of pure delay or dead time before the response to a given input change begins to take effect, a_{it} is the output noise series independent of x_{it}, B is the backward shift operator in time defined as $B^k Y_{it} = Y_{i(t-k)}$, L_s is the spatial lag operator and ω_{sk} and δ_{sk} are parameters.
   The spatial lag operator L_s is defined such that

$$L_s y_{it} = \sum\limits_{j=1}^{N} w_{ijs} y_{it} \qquad\qquad \text{for } s > 0 \tag{2}$$

where w_{ijs} are a set of weights scaled so that

$$\sum\limits_{j=1}^{N} w_{ijs} = 1 \tag{3}$$

for all i and w_{ijs} nonzero only for i and j sites being sth order neighbours. For s=0, equation (2) becomes $L_0 Y_{it} = Y_{it}$. The weights follow a hierarchical ordering of spatial neighbours based on distances between the observation sites in the watershed and may reflect physical characteristics of the observed time series.

Identification

The space-time cross-correlation function (STCCF) between Y_{it} and x_{it} series at spatial lag s and time lag k is given by

$$r_{sk}(\ ) = \frac{\sum\limits_{t=k+1}^{T} \sum\limits_{i=1}^{N} [v_{it} (\sum\limits_{j=1}^{N} (w_{ijs} z_{j(t-k)}))]}{(\sum\limits_{t=1}^{T} \sum\limits_{i=1}^{N} v_{it}^2)^{1/2} (\sum\limits_{t=1}^{T} \sum\limits_{i=1}^{N} (\sum\limits_{j=1}^{N} (w_{ijs} z_{j(t-k)}))^2)^{1/2}} \tag{4}$$

where $v_{it} = y_{it} - \bar{y}$ and $z_{it} = x_{it} - \bar{x}$ with $\bar{y}$ and $\bar{x}$ being respectively estimates of the space-time grand means given by

$$\bar{y} = \frac{1}{NT} \sum\limits_{i=1}^{N} \sum\limits_{t=1}^{T} y_{it} \tag{5a}$$

and $\qquad \bar{x} = \dfrac{1}{NT} \sum\limits_{t=1}^{T} \sum\limits_{i=1}^{N} x_{it}$ $\qquad\qquad\qquad\qquad\qquad\qquad\qquad$ (5b)

The identification of STTFN models is based on the estimation of the STCCF between the rainfall series x_{it} and the runoff series y_{it}. From a physical understanding of the hydrologic cycle there should be at least one value of the STCCF significantly different from zero. These values can explain the lagging of runoff with respect to rainfall referred to as the delay parameter.

A model from the general family of space-time autoregressive moving average (STARMA) models (Martin and Oeppen, 1975; Pfeifer and Deutsch, 1980; Mohamed, 1985) can be identified for the input rainfall series x_{it}. This model is used to transform the correlated input series x_{it} to the prewhitened z_{it} series. The same model is used to transform the output y_{it} series to the transformed v_{it} series. The STCCF is then computed between the new z_{it} and v_{it} series (Box and Jenkins, 1976).

The general family of STARMA models is given by

$$y_{it} = a_{it} + \sum_{s=0}^{\varrho} \sum_{k=1}^{p} \phi_{sk} \, L_s \, y_{i(t-k)} - \sum_{s=0}^{m} \sum_{k=1}^{q} \theta_{sk} \, L_s \, a_{i(t-k)} \qquad (6)$$

where p is the autoregressive (AR) order, q is the moving average (MA) order, ϱ and m are the spatial orders of AR and MA, respectively and ϕ_{sk} and θ_{sk} are parameters to be estimated. The identification of STARMA model for the rainfall series x_{it} is based on the inspection of the space-time autocorrelation function (STACF) given at spatial lag s and time lag k by

$$r_{sk} = \frac{\sum\limits_{t=k+1}^{T} \sum\limits_{i=1}^{N} \left[z_{it} \left(\sum\limits_{j=1}^{N} (w_{ijs} \, z_{j(t-k)}) \right) \right]}{\left[\sum\limits_{t=1}^{T} \sum\limits_{i=1}^{N} z_{it}^2 \right]^{1/2} \left[\sum\limits_{t=1}^{T} \sum\limits_{i=1}^{N} \left(\sum\limits_{j=1}^{N} (w_{ijs} \, z_{jt}) \right)^2 \right]^{1/2}} \qquad (7)$$

Following the prewhitening of x_{it} to z_{it} series and transforming y_{it} to v_{it} series using the identified STARMA model for the rainfall series x_{it}, the STTFN model (ϱ, p, m, q, b) may be written as

$$v_{it} = \frac{\sum\limits_{s=0}^{\varrho} \sum\limits_{k=1}^{p} \omega_{sk} \, B^k L_s}{\left(1 - \sum\limits_{s=0}^{m} \sum\limits_{k=1}^{q} \delta_{sk} \, B^k L_s \right)} \, B^b \, L_j \, z_{it} + N_{it} \qquad (8)$$

where N_{it} is the transformed noise series defined by

$$N_{it} = \frac{\sum\limits_{s=0}^{\varrho} \sum\limits_{k=1}^{p} \phi_{sk} \, B^k L_s}{\sum\limits_{s=0}^{m} \sum\limits_{k=1}^{q} \theta_{sk} \, B^k L_s} \, a_{it} \qquad (9)$$

The impulse response functions are given by

$$\nu_{sk} = r_{sk}(v,z) \frac{SD_v}{SD_z} \qquad (10)$$

where SD_z and SD_v are the standard deviations of the z_{it} and v_{it} series, respectively.

The STCCF $r_{sk}(v,z)$ and the coefficients ν_{sk} are used to identify the orders of ℓ, p, m, q and the pure delay parameter b and j using the following rules (Martin and Oeppen, 1975): zero or near zero correlation values up to spatial lag j-1 and time lag b-1 followed by irregular or rising values up to spatial lag $j+\ell-m+1$ and time lag b+p-q and correlation $r_{sk}(v,z)$, $s>j+\ell-m+1$, $k>b+p-q+1$, which decay exponentially in time and space. It should be mentioned that no such rising correlation values occur if $\ell<m$ and $b<q$. Once the values of these parameters are determined the initial values of the coefficients ω_{sk} and δ_{sk} can then be estimated. In this way the STTF model is identified for the space-time system.

Parameter estimation methods

Estimates of the parameters ω_{sk}, δ_{sk}, ϕ_{sk} and θ_{sk} of the tentative STTFN model can be obtained by minimizing the residual sum of squares:

$$s(\omega,\delta,\phi,\theta) = \sum_{i=1}^{N} \sum_{t=1}^{T} a_{it}^2 \qquad (11)$$

Given any initial values for the parameters ω_{sk} and δ_{sk} the errors N_{it} can be estimated from

$$N_{it} = y_{it} - \frac{\sum_{s=0}^{\ell} \sum_{k=1}^{p} \omega_{sk} B^k L_s}{(1 - \sum_{s=0}^{m} \sum_{k=1}^{q} \delta_{sk} B^k L_s)} L_j X_{i(t-b)} \qquad (12)$$

and the transformed noise series N_{it} is then defined by

$$N_{it} = \frac{\sum_{s=0}^{\ell} \sum_{k=1}^{p} \phi_{sk} B^k L_s}{\sum_{s=0}^{m} \sum_{k=0}^{q} \theta_{sk} B^k L_s} a_{it} \qquad (13)$$

Equation (12) is a modified version of the Box-Jenkins (1976) transfer function equation given by:

$$y_{it} = \frac{(\omega_0 - \omega_1 B^1 - \omega_n B^n)}{(1 - \delta_1 B - \delta_2 B^2 - .. \delta_r B^r)} x_{i(t-b)} \qquad (14)$$

where n and r are orders to be determined.

Since the STMA and STTFN models are nonlinear in form, four different unconstrained function minimization procedures were applied to determine optimal values for the parameters in the identified models: namely, the polytope method, the optimal-step steepest descent method, a conjugate gradient method, and a quasi-Newton method. In this section, we briefly describe the basis for each of these minimization procedures in the context of the general unconstrained function minimization problem; i.e., the problem of determining the n-vector x* which yields the least value for a given scalar (objective) function, f(x).

The polytope method (frequently referred to as the simplex method of Nelder and Mead, 1965) is an example of a heuristic procedure for solving the unconstrained function minimization problem. The process begins with the specification of a regular simplex which is defined in terms of (n+1) points in n-space (hence in 2-space the simplex is simply a triangle). Through a sequence of operations referred to as reflection, expansion and contraction, the simplex changes shape and moves through the parameter space until it (hopefully) encompasses, and then contracts upon, the minimizing argument x*. Each basic step begins with a particular simplex characterized by its vertices x_1, x_2, ..., x_n, x_{n+1} and ends with a new simplex whose shape and location have been altered in response to the local topology of the function. (Note: throughout this discussion, if v is a vector variable, then the notation v_j is used to denote a particular occurrence of the vector v rather than its jth component). The process can be terminated either when the vertices of the simplex become sufficiently clustered or when the function values at the vertices are all within a prescribed tolerance.

The optimal-step steepest descent method is the most fundamental of the gradient dependent function minimization procedures. Such procedures rely on information about the local topology of the objective function as reflected by its first partial derivatives which are contained in the gradient vector $f_x(x)$. The statement of the optimal-step steepest descent method is as follows:

(a)   choose x_0 (the priming guess), let $g_0 = f_x(x_0)$ and set k=0
(b)   let $x_{k+1} = x_k - \alpha * g_k$ where $f(x_k - \alpha * g_0) = \min f(x_k - \alpha\, g_0)$
(c)   compute $g_{k+1} = f_x(x_{k+1})$
(d)   check a termination criterion and if not satisfied then repeat from step (b) with k replaced with k+1.

The implementation of the above procedure has two prerequisites; namely, a means for computing the gradient vector and a mechanism for solving the scalar (one-dimensional) minimization problem with step (b). Frequently in practical problems, the determination of an analytic expression for the partial derivatives contained within the gradient vector is either awkward or unfeasible. In such cases a simple finite difference approach can be used; to compute an approximation to each of the components of the gradient vector; e.g.,

$$g_j(\overline{x}) = \frac{f(\overline{x} + \varepsilon\, e_j) - f(\overline{x})}{\varepsilon} \tag{15}$$

where $\varepsilon > 0$ is a small scalar and e_js the jth column of the n x n identity matrix. With suitable care in selecting an appropriate value for ε, such an approach can yield reasonable performance for gradient dependent methods in general (Birta, 1976).   This was, in fact, the procedure used in the experiments reported in this study.

A wide variety of methods has been proposed in the literature for solving the scalar (linear) minimization problem of step (b).   The specific approach used in the experiments was based on a sequence of quadratic polynomial approximations.   Data for establishing each polynomial in this sequence is conditioned by the minimum point predicted by its predecessor.   This process is terminated when the estimated solution from two successive polynomials is within a prescribed tolerance.

The original procedure in the class of conjugate gradient algorithms was proposed by Fletcher & Reeves, 1964, and is as follows:

  (a)   choose x_0 (the priming guess), let $p_0 = g_0$ and set k=0
  (b)   let $x_{k+1} = x_k + \alpha p_k$ where $f(x_k + \alpha p_k) = \min f(x_k + \alpha p_k)$
  (c)   set $p_{k+1} = -g_{k+1} + \beta_k p_k$ where $\beta_k = g_{k+1}g_{k+1}/g_k g_k$
  (d)   check a termination condition and if not satisfied, repeat from step (b) with k replaced with k+1

The significant feature of conjugate gradient methods is that when applied to a quadratic objective function, the search directions, p_k, that are generated have the property of A - conjugacy.   This, in particular, implies that the minimizing argument of a quadratic function will be located in at most n steps (i.e. linear searches). Variations on the original method are possible by selecting different values for the parameter β_k.   The value $\beta_k = g_{k+1}^T y_k / p_k^T y_k$ (with y_k = $g_{k+1} - g_k$) as proposed by Sorenson 1969, was used in the study.

The quasi-Newton (or variable metric) methods can be viewed as an evolution of the classical Newton-Raphson method.   The generic representation for methods of this class is as follows:

  (a)   choose x_0 (the priming guess) and a value for the nxn symmetric positive definite matrix H_0 (normally the identity matrix); set k=0
  (b)   set $x_{k+1} = x_k + \alpha_k d_k$ where $d_k = -H_k g_k$
  (c)   update the value of H_k by replacing it with H_{k+1}
  (d)   check a termination condition and if not satisfied, repeat from step (b) with k replaced with k+1

Two implicit subproblems are contained in this procedure: namely the matter of choosing the scalar α_k and the specification of the update formula which generates H_{k+1}.   Much of the original work with the quasi-Newton methods was based on an optimal choice for α^k in the sense that $\alpha_k = \alpha^*$ where $f(x_k + \alpha^* d_k) = \min_\alpha f(x_k + \alpha d_k)$

The original quasi-Newton update formula is the DFP formula (Davidon, 1959; Fletcher and Powell, 1963) which was used in this study; namely,

$$H_{k+1} = H_k + \frac{s_k s_k^T}{s_k^T y_k} - \frac{(H_k y_k)(H_k y_k)^T}{y_k^T H_k y_k} \tag{16}$$

Efforts at generalization have given rise to various families of such up-dates which contain the DFP formula as a special case; e.g., Broyden, 1967.   When used with an optimal step-size strategy (i.e. α_k chosen optimally), these formulae generate A-conjugate search directions when the process is applied to a quadratic function.

Diagnostic checking

Diagnostic checking is performed to examine the adequacy of the selected STTFN model (Box and Jenkins, 1976). The residuals STACF and the Portmanteau tests are used to examine the whiteness of the residuals. The cumulative periodogram test is also used to investigate the presence of any periodicities in the residuals.

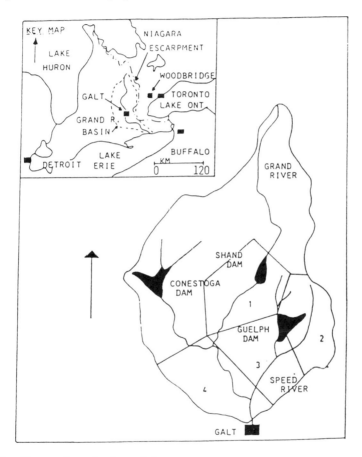

Figure 1   The watershed and key map.

Space-time modeling of the rainfall-runoff process

The data used in this study consist of precipitation and runoff time series from four gauging stations located within the Grand River basin in southern Ontario, Canada (Figure 1). Data are available for the period of July 1966 to January 1974. A time step of 15-day precipitation totals is used in this study, which allows the time series to preserve the characteristics of the storm events. Similarly, the runoff series are sampled every 15 days. There were a few missing values in the precipitation time series and the normal-ratio method (Linsley et al., 1982) was used to estimate these data points. The watershed is divided into four polygonal subareas using the Thiessen method. The four runoff gauge stations correspond to the same

subareas where the four rain gauge stations are located (Figure 1).
Equal weights are selected for this hierarchical weighting scheme of
the space system with a maximum spatial order of 2 (Table 1).

Table 1   Neighbors of each site for each spatial order

Spatial Order	0	1	2
Site 1	1	2, 3	4
2	2	1, 3	4
3	3	1, 2, 4	
4	4	3	1,2

The STACF and STPACF of the original precipitation series
suggested that the space-time system is nonstationary. First
differencing in time is applied to the precipitation system to
achieve stationarity. The identification of the differenced precipitation series resulted in an STMA(1_2) model of order one in time and
two in space. The parameters of the STMA (1_2) precipitation model
are estimated using the polytope algorithm. Diagnostic checking of
the tentative model indicates that the STMA (1_2) model adequately
describes the spatio-temporal characteristics of the precipitation
time series. The STMA (1_2) model takes the following form:

$$x_{it} = x_{i(t-1)} + a_{it} - 0.9448a_{i(t-1)} + 0.0672W_1a_{i(t-1)} - 0.004W_2a_{i(t-1)} \quad (17)$$

The STMA (1_2) model is used in the STTFN model-building procedure of
the rainfall-runoff process to determine the prewhitened series z_{it}
and the transformed series v_{it} .
The STACFs of the original runoff suggested that the space-time
system is nonstationary since the STACF failed to tail off quickly at
all times and spatial lags. The seasonal components are removed from
the time series to achieve stationarity. The STCCF between the z_{it}
and v_{it} series given in Table 2 and the estimated impulse response

Table 2   STCCF between z_{it} and the v_{it} series

Spatial lag (s) Time lag (k)	0	1	2
0	-0.0186	-0.0236	-0.0076
1	0.1357	0.1000	0.1312
2	0.0275	0.0030	0.0237
3	0.0220	0.0078	0.0490
4	0.0015	-0.0073	0.0029
5	-0.0044	0.0041	-0.0058
6	0.0522	0.04121	0.0857

functions ν_{sk} (Table 3) indicate that there are values significantly different from zero only at time lag one (k=1) for all the spatial lags s=0, 1 and 2.   In other words, runoff is lagged 15 days or one time step behind precipitation. Therefore, the memory of the rainfall runoff process is 15 days or one time lag.

The identification of the STTF (ℓ,p,m,q,b,n,r) model of equation (8) based on the Box-Jenkins procedure suggests that the rainfall-

Table 3   Impulse response functions

Spatial lag (s) Time lag (k)	0	1	2
0	-0.014	-0.016	-0.005
1	0.095	0.070	0.092
2	0.019	0.002	0.016
3	0.015	0.005	0.034
4	0.001	-0.005	0.002
5	-0.003	0.003	-0.004
6	0.036	0.029	0.060

runoff process could be identified as an STTF (2,1,0,0,1,0,0) model by

$$y_{it} = \sum_{s=0}^{2} \omega_{os} W_s x_{i,t-1} + a_{it} \tag{18}$$

The parameters of the STTF (2,1,0,0,1,0,0) model estimated by the polytope method, the optimal-step steepest descent method (STPSDS), a conjugate gradient method (SOREN) and a quasi-Newton method (DFP) are shown in Table 4.   Each of these procedures was initialized from the same starting (priming) point and the final value obtained for the objective function was the same in all cases. The results obtained in these experiments are shown in Figure 2.   The STACF (Table 5) and the Port Manteau test (Table 6) indicate that the residuals are uncorrelated and consequently white noise. The cumulative periodogram (Figure 2) of the residuals at the 0.05 confidence level suggest that the residuals contain no periodicities. Therefore, the STTF (2,1,0,0, 1,0,0) model is accepted.

Conclusions

The comprehensive procedure presented in this paper for building a space-time transfer function noise (STTFN) model is useful in detecting dynamic relationships between hydrologic time series.   The developed STTF (2,1,0,0,1,0,0) model for the selected watershed is proved adequate in describing the spatio-temporal characteristics of the precipitation and runoff time series. For the selected watershed, the output (runoff) lags behind the input (rainfall) and the delay parameter is 15 days or 1 lag in time.   The unconstrained function minimization procedures, namely, the polytope method and a quasi-Newton method performed well in estimating the selected model

Table 4   Parameter estimates of the STTF (2, 1, 0, 0, 1, 0, 0) model

Method	Parameter	Initial Guess	Final Guess	Initial S	Final S
STP$DS	ω_{00}	0.0	−0.0457		
	ω_{01}	0.0	0.0155		
	ω_{02}	0.0	−0.0243	0.238×10^6	0.235×10^6
SOREN	ω_{00}	0.0	−0.0457		
	ω_{01}	0.0	0.0157		
	ω_{02}	0.0	−0.0243	0.238×10^6	0.235×10^6
DFP	ω_{00}	0.0	−0.0457		
	ω_{01}	0.0	0.0157		
	ω_{02}	0.0	−0.0243	0.238×10^6	0.235×10^6
POLYTOPE	ω_{20}	0.0	−0.0457		
	ω_{01}	0.0	0.0157		
	ω_{02}	0.0	−0.0243	0.238×10^6	0.235×10^6

S is the residual sum of square.

Table 5   STACF of the residuals of the STTF (2, 1, 0, 0, 1, 0, 0) model

Spatial lag (s) Time lag (k)	0	1	2
1	0.0646	0.0088	−0.0040
2	−0.0553	−0.0286	−0.0014
3	−0.0277	−0.0098	0.0026
4	0.0778	0.0134	0.0083
5	0.0512	0.0120	0.0009
6	0.1057	0.0214	−0.0074

Table 6   Results of Port Manteau test on residuals

Spatial lag (s)	Port Manteau test	Chi-square statistics $\alpha = 0.05$	Decision
0	11.59	27.6	accepted
1	8.97	27.6	accepted
2	6.68	27.6	accepted

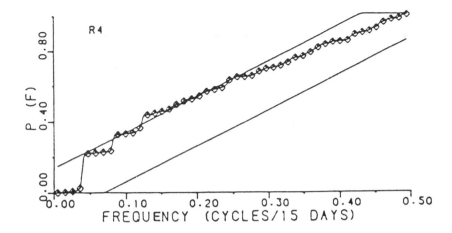

Figure 2   Behaviour of different minimization algorithms.

parameters based on the residual sum of squares. The polytope
algorithm is recommended.

ACKNOWLEDGMENTS   The research was supported by a grant provided by
NSERC Canada. The data were provided by Environment Canada. Computing
expenses were covered by the University of Ottawa.

References

Adamowski, K. & T. Hamory, 1983; Time series analysis of groundwater
    level fluctuations. J. Hydrology, 62.
Birta, L.G. (1976).   Some investigations in function minimization.
    IEEE Trans. on System, Man and Cybernetics, SMC-6, 186-197.
Box, G.E.P. & G.M. Jenkins, 1976. Time Series Analysis: Forecasting
    and Control. Holden-Day, San Francisco, 2nd edition, 553.
Broyden, C.G. (1967). Quasi-Newton methods and their application to
    function minimization. Math. of Comp. 21, 368-381.
Cooper, D.M. and E.F. Wood, 1982. Identification of Multivariate
    Time Series and Multivariate Input-Output Models. Water Resour.
    Res. 18(4), 937-946.
Davidon, W.C. (1959).   Variable metric method for minimization.
    Atomic Energy Commission, Research and Development Report No. ANL-
    5990, (Revised).
Fletcher, R. and M.J.D. Powell. (1963). A rapidly convergent descent
    method for minimization. Computer J., 6, 163-168.
Fletcher, R. and C.M. Reeves. (1964). Function minimization by
    conjugate gradients. Computer J., 7, 149-154.
Linsley, R.K., M.A. Kohler and J.L.H. Paulhus, 1982. Hydrology for
    Engineers, McGraw Hill Co., 2nd Edition, 482.
Martin, R.L. and J.E. Oeppen, 1975. The identification of regional
    forecasting models using space-time correlation functions.
    Trans. Institute of British Geographers. 66, 95- 118.

Mohamed, F.B. 1985. Space-time ARIMA and Transfer Function Noise Modelling of Rainfall-Runoff Process. M.A.Sc. Thesis, University of Ottawa, Ottawa, Canada, 170.

Nelder, J.A. and A. Mead. 1965. A Simplex Method for Function Minimization, Comp. Journal, 7, 308-313.

Pfeiffer, P.F. and S.J. Deutsch, 1980. A Three-stage iterative procedure for space-time modelling. Technometrics, Vol. 22, No. 1, February, 35-47.

Salas, J.D., J.W. Delleur, V. Yevjevich and W.L. Lane, 1980. Applied Modelling of Hydrologic Time Series. Water Resources Publications, 484

Sorenson, H.W. (1969). Comparison of some conjugate directions procedures for function minimization. J. Franklin Inst. 288, 421-441.

The Influence of Climate Change and Climatic Variability on the Hydrologic Regime and Water Resources (Proceedings of the Vancouver Symposium, August 1987). IAHS Publ. no. 168, 1987.

Fourier-ARIMA modelling of the multiannual flow variation

I. Haidu
Department of Geography
University of Cluj-Mapoca,
Str. Tăṣnad, Nr. 20, Bl. E3, Sc. IV, Ap.
39,3400, Cluj-Napoca, Romania
P. Serbrn & M. Simota
Institute of Meteorology and Hydrology,
Ṣos. Bucuresti-Ploiesti, Nr. 97
Bucharest, Romania

ABSTRACT   In order to analyse multiannual flow variations, Fourier, ARIMA and combined Fourier-ARIMA models were used. The combined Fourier-ARIMA model takes into account the following components: the general trends, the periodical component, the moving average and the random component. Eleven chronological series of annual mean discharges, extending over periods of 55-145 years, were analysed by means of the models. Relative error analysis of the mean annual computed hydrograph and the mean annual measured hydrograph emphasized the superiority of the combined Fourier-ARIMA model over the Fourier and ARIMA models. This combined model provides for a substantially more accurate eleboration of a flow prognosis for groups of years with different qualitative hydrologic characteristics.

Modélisation Fourier-ARIMA de la variation
multiannuelle de l'écoulement

RESUME   Modèles du type Fourier, ARIMA et une combinaison des modèles Fourier et ARIMA ont été utilisés pour analyser la variation multiannuelle de l'écoulement. Le modèle combiné Fourier+ARIMA tient compte des composantes suivanles: la tendance générale, la composante périodique en forme des séries Fourier. La composante autorégressive et moyen glise, et le component aléatoire. En utilisant ces modèles on a analysé onze séries chronologiques des débits moyens annuels ayant une durée de 55-145 ans, representant des unités climatiques homogènes du basin du Danube. L'analyse des erreurs relatives entre les débits moyens annuels calculés et mesurés a mis en évidence la supériorité du modèle combiné Fourier-ARIMA par comparison aux modèles Fourier ou ARIMA. Le modèle Fourie-ARIMA permet d élaborer avec une précision suffisante de la prognose de l'écoulement pour des groupes d années ayant des caractéristiques qualitativement différentes.

Introduction

The nature of the factors which condition flow variability can be described according to the interval and time unit of the series under study. In hydrology one considers flow variability within the year separately to the flow variations from one year to another. In the study of flow variation with the year the seasonal cycle is dominant and hence most simulation models explain the flow changes from one month to another with a seasonal component.

In the case of annual discharge variation seasonal influences cannot be modelled explicitly despite the fact that hydroclimatic elements during the year can have a significant impact upon the flow variation from one year to another. An important part is played by the influence of the time distribution of precipitation within the various phases of the annual thermal regime, together with the carry over due to water reserves in the hydrologic basin.

The runoff quantities contain information regarding all the factors influencing flow. Annual mean discharge integrates all the causative factors: time varying factors such as precipitation, temperature and situation of the previous year's water reserves; and quasi constant factors such as the morphometrical and physico-geographical conditions of the basin.

The models used for the study of flow variation from one year to another have not been, up to now, based on the analysis of causal factors, but on the decomposition of the time series into its basic components by means of the dynamic series model:

$$y(t) = u(t) + v(t) + s(t) + e(t) \tag{1}$$

where:

$u(t)$ = trend due to the action of permanent factors
$v(t)$ = cyclic variations due to the action of rhythmical factors
$s(t)$ = stochastic components assumed to follow an autoregressive moving average process
$e(t)$ = random component, due to the action of random factors

Obvious connections exist between these elements of the series and the actual physical situation. Thus, the general trend reflects the general direction of the dynamics of the phenomenon; the cyclic variations give the periodical fluctuations about the trend; and the non-rhythmical variations (stochastic variations) and the random variations give the result of the impact of random factors on the natural development of the phenomenon.

Trend

The general trend of the hydrologic series can be ascribed to climatic change or to the influence of man's activities such as agriculture, forestry, and urbanization.

The extraction of the trend from the raw data series can be achieved by means of a polynomial fit:

$$u(t) = a_1 + a_2t + a_3t^2 + \ldots + a_nt^{n-1} \tag{2}$$

After having used the least square method with a number of hydroclimatic dynamic series, we can state that, with regard to long-term variations, the trend term presents a linear shape in some cases, and parabolic shape in other cases.

After the elimination of the trend a residual component $r(t)$ results:

$$r(t) = v(t) + s(t) + e(t) \tag{3}$$

In order to establish the general structure of the residual series $r(t)$ the correlogram method is used. The autocorrelation coefficient CA_L is determined:

$$CA_L = \frac{\frac{1}{N-L} \sum_{t=1}^{N-L}(r_t - \bar{r})(r_{t+L} - \bar{r})}{\frac{1}{N} \sum_{t=1}^{N} (r_t - \bar{r})^2} \tag{4}$$

where:
  N = the series length,
  L = the difference between two values of the series expressed in time units,
  $\bar{r}$ = the mean of the series.

The correlogram is obtained by the graphical representation of the variation of CA_L with L. If the data series is random, the computed value of CA_L is other than only due to internal (random) variations of the series, the points of the correlogram being very close to the horizontal axis.

The confidence limits corresponding to a given significance level is expressed by the Anderson test (Yevjevich 1972)

$$CL\ (CA_L) = [-1 \pm z_\alpha\ (N-L-2)^{1/2}]/(N-L-1) \tag{5}$$

where:
  z_α = the standard normal deviate corresponding to probability α. For α = 5% the significance level is z_α = 1.645,
  N = the series length
  L = the difference between two values of the series L =1,2....N/4.

Periodic component

The determination of the periodic component was achieved by the representation of series (3) by a Fourier series of the shape:

$$v(t) = \frac{1}{2} a\ + \sum_{n=1}^{\infty} (a_n\cos\frac{2\pi}{T}nt + b_n\sin\frac{2\pi}{T}nt) \tag{6}$$

where:
  n = the order of the harmonic corresponding to the period established by the analysis of the correlogram,
  $2\bar{\pi}/T$ = the frequency of the oscillations,
  a_o, a_n, and b_n = Fourier coefficients.
The residual component is isolated by the difference

$$r(t) - v(t) = s(t) + e(t) \tag{7}$$

Autoregressive and moving average component

After eliminating the trend and periodic components from the original
data the remainder is the stochastic component. The stochastic
component is modelled by means of parametric models of an autoregressive and moving average type. (ARIMA, Box and Jenkins, 1976).
    The general structure of a model of a nonstationary ARIMA type
(p,d,q) is the following:

$$W_t = \phi_1 W_{t-1} + \ldots + \phi_p W_{t-p} + a_t - \theta_1 a_{t-1} - \ldots - \theta_p a_{t-q} \qquad (8)$$

where:

W $= \nabla^d S_t$,
∇ = the difference operator ($\nabla S_t = S_t - S_{t-1}$),
d = the difference order,
S_t = the current value of the modelled series at moment t,
$\phi_1 \ldots \phi_p$ = parameters of the autoregressive model,
$\theta_1 \ldots \theta_p$ = parameters of the moving average model,
p,q = order of the autoregressive model and moving average
         model, respectively,
a_t = current value of an independent random variable mean zero
         and constant square mean deviation σ_a^2 (of the "white
         noice" type).

    When $d = 0$, $W_t = S_t - \mu$, where μ is a parameter determining the
"level of the process" (in the case of a stationary process, μ is the
series average).
    The steps necessary for establishing the ARIMA type model to
describe optimally the data series are:
    (a) estimation of order p,q and d, by the analysis of the autocor-
relation and partial autocorrelation functions,
    (b) determination of parameters ϕ and θ by means of the maximum
veri-similarity criterion,
    (c) analysis of the structure of the residuals obtained after the
application of the model in order to verify the appropriateness of
the chosen model. If the residuals obtained do not comply with the
above mentioned conditions indicated for a_t, the model should be
reanalyzed.

Application

The models presented herein were applied to the analysis of eleven
annual mean discharge series of between 55 and 145 years duration.
These data represented homogeneous climatic units within the Danube
basin. The results of the analysis of the deterministic component of
the study series are summarised below:
    (a) A slightly rising trend, due mainly to human influence-defore-
station and expansion of urban and industrial areas.
    (b) The cyclic component exhibited:
    (i) periods of 55-60 and 30-35 years which reflect the general
climatic characteristics of the entire Danubian basin, established on
the basis of series of more than 100 years (Figure 1),
    (ii) periods common to wide geographical areas, which lead to the
consideration of the circulation factor; 10-13 and 19-21 years in

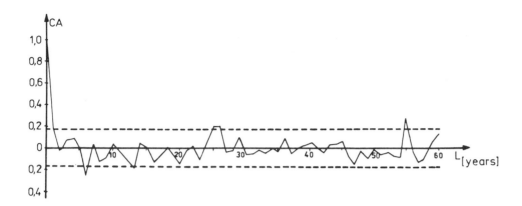

Figure 1   The autocorrelogram of data services No. 1.

Transylvania and 15-17 years in Moldovia and Muntenia,
    (iii) periods of 3-6 years, not common to wide regions and
probably imposed by the time expressions of each basin's regulating
capacity.
    The analysis of the periods and phases of the periods of the
harmonics shows substantial synchrony in the multiannual flow
variation in the areas with homogeneous general atmospheric circulation imposed by the natural conditions caused by the Carpathian arc.
The periods of 10-13 and 19-21 years characteristic of the western
side of the Carpathian arc can be traced throughout the entire
Danubian basin, but not in the areas south and east of the Carpathian
arc. Thus, it is possible to agree that these periods are brought
about by the western and north-western component of the general
circulation.
    Thus, the multiannual flow of the rivers in the north-west of
Romania accords with that specific to central and south-eastern
Europe, and is typified by the Fourier parameters of the Danube at
hydrometric station No. 1 (Orşova). For the rivers in the south and
east of the country, the dominant characteristic appears to be a
multiannual flow periodicity of 15-17 years, due mainly to the
Mediterranean atmospheric circulation.
    The eleven series under study were simulated by means of models of
the type: Fourier, ARIMA, Fourier-ARIMA. The relative classification
between measured annual mean discharges and computed mean discharges
demonstrated (Table 1) the superiority of the combined Fourier+ARIMA
model over the Fourier and ARIMA model considered separately.

Table 1   Relative error classification for the annual
           discharges at hydrometric station No. 1

Error class % / Model	0 - 5	5 -10	10-15	15-20	20-25	25-30	30-35	35-40	40-45	Maximum error %	Average error %
FOURIER	30,4	20,3	24,6	5,9	11,6	2,9	-	4,3	-	11,4	38,6
ARIMA	34,8	21,7	23,2	11,6	4,4	4,3	-	-	-	9,6	28,0
FOURIER+ARIMA	39,1	29,0	23,3	4,3	-	4,3	-	-	-	9,0	26,0

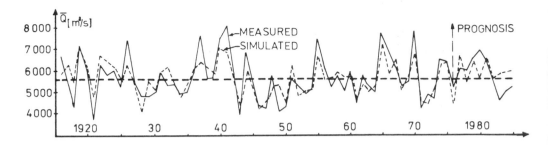

Figure 2   Simulation and prognosis of the mean annual discharges.

The combined Fourier+ARIMA model provides for a substantially accurate prediction of mean annual flow distribution (Figure 2), for groups of years with different qualitative hydrologic characteristics. The deviations occurring in the simulation are due mainly to the small volume of the selections utilised, as well as to human impact on the hydrologic and climatic factors on the local and continental scale.

More details on the analysis presented in this paper are presented in Haidu and Farcas (1986).

References

Box, G.E.P. & Jenkins, G.M. (1976) Time Series Analysis. Forecasting and Control. Revised Edition. San Francisco, Holden Bay.
Haidu, I. & Farcas, I. (1986) Studial variatiei de lunga durata a parametrilor hidroclimatici in scopul elaboraii prognozei prin extrapolare analitica(L'etude de la variation de longue duree des parameters hydroclimaique en vue de 1 elaboration de la prognose par extrapolation analytique). In: Probleme de Geografie Aplicata, 53-63. Universitatea Cluj-Mapoca.
Yevjevich, V. (1972) Stochastic Processes in Hydrology. Fort Collins.

The Influence of Climate Change and Climatic Variability on the Hydrologic Regime and Water Resources (Proceedings of the Vancouver Symposium, August 1987). IAHS Publ. no. 168. 1987.

Modélisation d'un processus non-stationnaire — Application à la pluviométrie en zone semi-aride

A. Musy & P. Meylan
Institut de Génie Rural
Hydrologie & Aménagements
Ecole Polytechnique Fédérale de Lausanne
CH-1015 Lausanne (Suisse)

RESUME   La persistance d'un déficit pluviométrique dans les régions sahéliennes durant les quinze dernières années conduit à mettre en doute la légitimité des méthodes classiques de dimensionnement des ouvrages. Les séries chronologiques présentent en effet un caractère non stationnaire. Aussi, les auteurs proposent-ils une technique de modélisation d'un tel processus, et tentent de l'appliquer aux observations pluviométriques du Burkina-Faso. L'utilisation de ce modèle pour le dimensionnement nécéssite la connaissance de la tendance climatique régionale, dont l'évolution future est incertaine. Quelques éléments d'approche de ce problème ardu sont esquissées.

Modelisation of a non-stationary
process - Application to the rainfall
data in semi-arid areas

ABSTRACT   The persistence of a rainfall deficit in the Sahelian area for fifteen years lead us to be doubtful about the legitimacy of the classical methods for the design of construction works. Indeed, time series show an unstationary behaviour. The authors propose a technique for the modelisation of a nonstationary process, which is applied to the rainfall data of Burkina-Faso. The use of this model for the design of construction works requires the knowledge of the regional climatic trend, whose future evolution is uncertain. Some possibilities to face that difficult problem are outlined.

Introduction

Les ouvrages d'aménagement des eaux, tant en milieu rural qu'urbain, sont en général dimensionnés pour satisfaire à des besoins divers, compte tenu d'un risque calculé. Ce dernier dépend d'un événement climatique extrême, pas encore connu mais dont l'apparition est fort probable durant la période de vie de l'ouvrage considéré. Ainsi par exemple, le déversoir de crue d'une digue en terre, tout comme l'exutoire recevant des eaux pluviales d'un bassin-versant sont calculés pour résister à des crues provoquées par une pluie exceptionnelle, d'intensité critique fixée, et de fréquence d'apparition supposée connue. Les notions de seuil critique, de probabilité d'apparition et de temps de retour se déterminent alors à partir d'une analyse statistique spécifique, basée sur des données de qualité. Or précisément ces données sont souvent fort fragmentaires et de surcroit peu homogènes. Il convient alors de s'assurer au préalable de la qualité de celles-ci avant d'entreprendre une statistique, aussi élaborée soit-elle.

Deux contrôles préalables au moins doivent être effectués: l'un concerne l'homogénéité spatiale des données acquises, l'autre l'homogénéité temporelle. Le premier de ceux-ci est nécessaire car il permettra par la suite d'interpoler les résultats: le site à aménager ne disposant trop souvent que de très peu d'information. Le deuxième contrôle demeure indispensable, car la stationnarité des séries d'observations est l'hypothèse fondamentale de toute analyse hydrologique statistique communément utilisée.

La régionalisation des données hydro-pluviométriques a déjà fait l'objet de nombreuses études. Plutôt que d'en citer une longue liste qui serait encore lacunaire, nous illustrerons l'évolution des techniques utilisées par deux travaux des auteurs. Ainsi Musy (1973) effectue la "spatialisation" des paramètres d'une loi statistique, représentative du phénomène dans la région considérée, par une méthode de triangulation. Plus récemment, Jordan & Meylan (1986) ont étudié la répartition spatiale de la pluviométrie par la méthode du krigeage, modifiée par Meylan (1986) pour permettre la prise en compte des erreurs de mesure. Dans les deux cas, l'efficacité et la fiabilité d'une telle régionalisation sont surtout fonctions du nombre et de la répartition des stations d'observation. Ce problème majeur est du ressort des services hydrologiques concernés. Cependant, dès que celui-ci est résolu, les questions subséquentes trouvent réponses de manière généralement satisfaisante. C'est la raison pour laquelle, dans cette étude, l'accent ne sera pas porté sur ce point.

L'homogénéité temporelle des séries chronologiques est moins évidente à cerner. Plusieurs chercheurs ont tenté, avec plus ou moins de succès, de mettre en évidence des cycles liés à l'apparition de phénomènes divers (cycles sur les sécheresses notamment). Mais plus rares sont les personnes qui preconisent des méthodes d'analyse efficaces, tenant compte d'un effet de non-stationnarité de ces phénomenès. Or, dans des conditions climatiques particulières, il est nécessaire de prendre en compte ce type d'effets. Il en résulte une meilleure projection dans le temps des divers éléments de calcul indispensables au dimensionnement des ouvrages.

Le cas qui nous intéresse est celui du Sahel et nous allons tenter de répondre à la question suivante: "Comment procéder correctement à l'ajustement statistique d'une série d'observations sachant que, qualitativement du moins, celle-ci dérive dans le temps"?

Au Sahel, on assiste depuis 1969 environ à une diminution systématique du module pluviométrique annuel. Dès lors, si ce facteur nous intéresse pour le remplissage d'une retenue par exemple, il est faux d'appliquer sans précaution les méthodes statistiques "classiques". Le phénomène n'est en effet pas stationnaire. Il convient alors de voir comment on peut contourner cette difficulté. La présente étude en expose quelques principes.

Figure 1   Situation géographique des stations pluviométriques.

L'application se porte sur les modules pluviométriques de la
saison des pluies d'avril à octobre du Burkina-Faso (Afrique
occidentale). Nous n'avons retenu que les observations des stations
principales de ce pays qui ont un certain historique (cf figure 1).
Leur nombre limité ne permet pas d'établir une régionalisation
spatiale sérieuse. Toute fois, l'effet régional est tout de même pris
en compte ne serait-ce que pour vérifier si la dérive temporelle que
l'on constate se comporte de la même manière, quelle que soit la
position géographique des stations. L'étude porte donc principalement
sur la mise en évidence de la non-stationnarité des séries observées
et sur la modélisation de ces séries non-stationnaires.

Recherche d'une structure temporelle

De façon classique, l'analyse fréquentielle d'une série chronologique
suppose que l'échantillon étudié est aléatoire et simple: toutes les
valeurs proviennent d'une population-mère de caractéristiques
stationnaires, par tirages aléatoires et indépendants. Brunet-Moret
(1979) classe les causes de non-respect de cette hypothèse en 4
catégories:
    (a) effet de persistance: la population est stationnaire, mais les
valeurs successives ne sont pas indépendantes
    (b) effet de dérive: la population n'est pas stationnaire, en
particulier son espérance mathématique croît ou decroît dans le temps
    (c) effet cyclique: l'espérance mathématique est fonction du
temps, mais varie cycliquement, de sorte que sa valeur lissée peut
être considérée comme stationnaire
    (d) effet d'erreurs systématiques d'observation
    Les deux seuls cas envisages ici sont les causes de violation (a)
dues à la persistance et (b) dues à une dérive.

Etude de la dérive

Matheron (1962) définit la dérive comme l'espérance mathématique, dépendante du temps, de la variable étudiée.

L'observation des séries brutes des modules d'avril à octobre (cf par exemple figure 2) révèle la présence d'une dérive du processus. Pour permettre la comparaison inter-stations de ces modules, Triboulet (1983) utilise la normalisation de Lamb: $p_{ij} = (P_{ij} - \overline{P}_i)/s_i$ Dans cette relation p_{ij} est le module normé pour la station i et l'année j, alors que P_{ij} représente le module d'avril à octobre de l'année j pour la station i, $\overline{P}_i$ et s_i sont respectivement la normale et l'écart-type de ces modules pour la station i. Une telle normalisation permet le calcul d'une moyenne régionale des précipitations normées (cf figure 3) qui fait apparaître clairement une tendance à la baisse dans la période de 1950 à 1985.

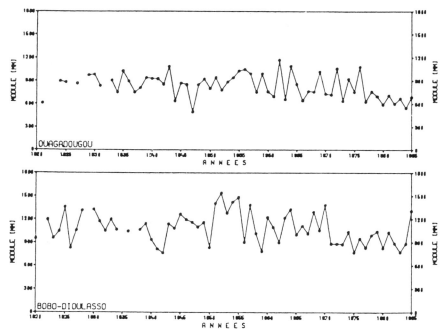

Figure 2   Exemples de séries chronologiques
des modules d'avril à octobre.

La non-stationnarité des séries chronologiques peut encore être vérifiée par des tests statistiques. Si l'on considère, comme Albergel et al (1985), la période "excédentaire" de 1950 à 1969 et la période "déficitaire" de 1970 à 1985, on constate, par un test de la différence des moyennes des modules des deux périodes, que la probabilité critique du test (probabilité de conclure de façon erronée en rejetant l'hypothèse d'une différence nulle des moyennes) reste inférieure à 0.5% pour les sept stations. La diminution des moyennes est donc statistiquement significative.

L'examen des diminutions relatives des moyennes est instructif par

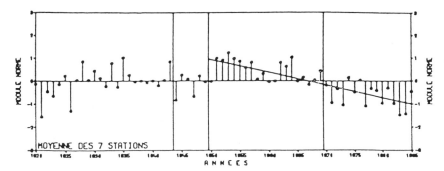

Figure 3   Série des moyennes régionales des modules normés.

lui-même.   L'écart le plus faible est obtenu pour la station de Fada
N'Gourma (-12%),   les plus forts pour Dori et Ouahigouya (-24%),   les
autres   stations   présentent   un écart   voisin de   -15%.   Les   deux
stations   septentrionales (Dori et Ouahigouya) sont   donc   nettement
plus affectées par la sécheresse que les autres.
    Une   autre   constatation   intéressante   concerne   la   remarquable
stabilité des coefficients de variation: Leur moyenne vaut 0.18 avec,
toutefois,   un   écart-type différent entre la première période   (0.02
pour 1950 à 1969) et la seconde période considérée (0.05 pour 1970   à
1985).

Etude de la persistance

L'effet de persistance, ou mieux la structure temporelle, évoquée par
Brunet-Moret   (1979),   peut   être   mise   en   évidence   par   plusieurs
techniques   dont   l'une   des plus connues est basée sur   l'étude   des
corrélogrammes (voir par exemple Bouvier (1983)).   Pour identifier ce
phénomène,   Brunet-Moret & Roche (1975) proposent un "coefficient   de
persistance",   basé   sur une modélisation du processus par chaîne   de
Markov d'ordre 1.   Sans vouloir mettre en cause   cette méthode, nous
préférons   cependant utiliser les techniques de l'analyse structurale
basées   sur   l'étude du variogramme.   Celles-ci permettent en   effet,
selon   nous,   une   approche   plus rigoureuse du   problème   (voir   par
exemple Serra (1967) et Meylan (1986b)).
    Ainsi,   par   exemple,   l'étude du comportement du   variogramme   au
voisinage   de   l'origine   renseigne   sur   la   nature   de   la   variable
étudiée.   Les cas typiques sont représentés à la figure 4, empruntée
à Meylan (1986) et Haas & Viallix (1976).
    Si   l'on   étudie   les modules annuels normés p_{ij} ,   le   variogramme
expérimental   peut être calculé,   pour une station i donnée,   par   la
relation:

$$g_i(h) = 1/2 \ N_h \sum_{}^{N_h} (p_{ij} - p_{ij+h})^2 \qquad\qquad (1)$$

avec $g_i(h)$ comme la valeur du variogramme pour une distance   de   h
années, p_{ij} le module normé de l'année j, p_{ij+h} celui de l'année j+h,

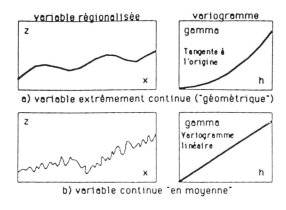

a) variable extrêmement continue ("géométrique")

b) variable continue "en moyenne"

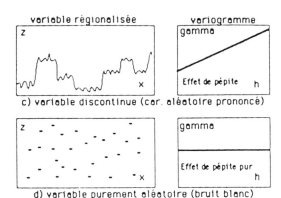

c) variable discontinue (cor. aléatoire prononcé)

d) variable purement aléatoire (bruit blanc)

Figure 4   Types de continuité d'une variable régionalisée
et comportement correspondant du variogramme
au voisinage de l'origine.

la somme étant étendue aux N_h couples de valeurs séparés de h années.
La figure 5 présente les variogrammes obtenus pour les 7 stations,
ainsi que le variogramme régional moyen.

Avant d'interpréter ces variogrammes il convient de noter que g(h)
n'est un estimateur non biaisé du vrai variogramme que si la variable
étudiée ne présente pas de dérive. Dans notre cas, ou une dérive est
manifeste, on se bornera à n'utiliser ces variogrammes qu'au
voisinage de l'origine, en supposant que l'effet d'une dérive est
négligeable en cette zone. De plus, il conviendra de faire la part
de la fluctuation d'échantillonnage dans l'interprétation des
variogrammes obtenus. L'examen de la figure 5 conduit à la
conclusion que les modules d'avril à octobre se comportent comme des
variables purement aléatoires, ce qui se traduit par un variogramme
de niveau constant. La seule station qui semble présenter une
structure est celle de Ouahigouya. Il est toutefois difficile de
l'affirmer.

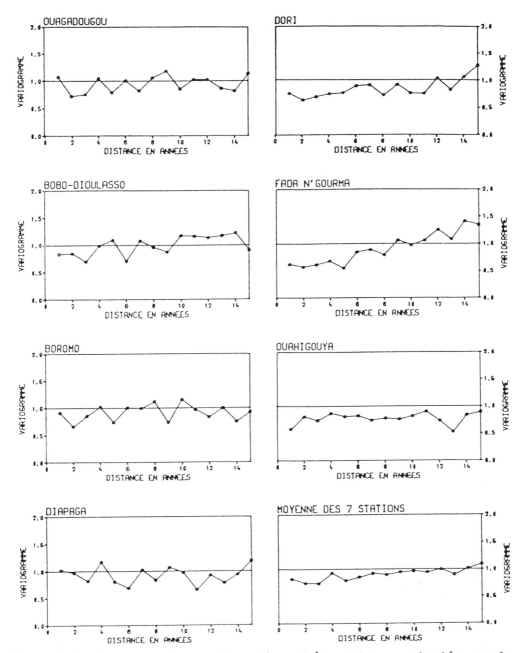

Figure 5 Variogrammes des modules d'avril à octobre,   calculés sur la
période de 1920 à 1985.

Au vu des résultats ci-dessus on ne peut que conclure à
l'inhomogénéité des séries pluviométriques, déjà constatée et relevée
par exemple par Triboulet (1983), Albergel et al. (1984) ou Sircoulon
(1976).   La cause de cette inhomogénéité doit toutefois être

recherchée uniquement dans la non-stationnarité des séries
pluviométriques. Les données disponibles ne permettent pas de
conclure à un effet de persistance inter-annuelle.

Recherche d'une structure spatiale

Les données traitées ne concernent que 7 stations du Burkina-Faso
et ne permettent évidemment pas de résoudre le problème de la
régionalisation des données.   Il est toutefois intéressant d'étudier
quelques caractéristiques élémentaires, afin de mieux cerner le
problème.

Coefficient de variation

Les coefficients de variation, calculés pour chacune des 7
stations sur les données disponibles de la période de 1920 à 1985
sont pratiquement tous constants, comme en témoigne la figure 6.
Leur moyenne s'établit à 0.2, avec un écart-type de 0.02. Toutefois,
sur la base de 7 stations uniquement, il n'est guère possible
d'envisager une régionalisation de ce paramètre.   Tout au plus nous
noterons que les deux valeurs les plus fortes concernent Ouahigouya
(Cv=0.215) et Dori (Cv=0.231).

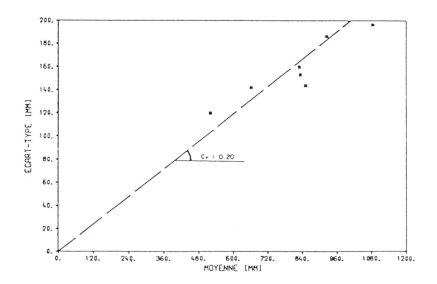

Figure 6   Relation entre moyenne et écart-type
           pour les sept stations du Burkina-Faso.

Corrélation des modules bruts

La forte variabilité spatiale des modules a été relevée par de
nombreux auteurs, en particulier par Sircoulon (1976).   La matrice

des coefficients de corrélation a été calculée sur la base des 45 années d'observation simultanée disponibles aux 7 stations (de 1920 à 1985). Ceux-ci sont très faibles, ce qui confirme la très forte variabilité spatiale. Sur les 21 couples de stations, seuls quatre présentent un coefficient de corrélation statistiquement différent de zéro au seuil usuel de 5%: (a) Ouagadougou – Bobo-Dioulasso, (b) Boromo – Diapaga, (c) Boromo – Fada N'Gourma et (d) Boromo – Ouahigouya. Cette constatation permet de conclure à l'absence de structure spatiale des modules.

Corrélation des modules lissés

Une dernière analyse a été tentée, basée sur les modules d'avril à octobre, lissés par la relation d'Albergel et al. (1984):

$$PL_j = \sum_{k=0}^{6} 0.5^k P_{j-k} / \sum_{k=0}^{6} 0.5^k \qquad (2)$$

ou PL_j est le module lissé correspondant à l'année j, P_j étant le module de l'année j. Dans ce cas, les coefficients de corrélation sont tous significativement différents de zéro au seuil usuel de 5%.

Modélisation d'une série non-stationnaire

Considérons la série chronologique des modules d'avril à octobre P_{ij} (où j est l'indice de l'année), pour une station i donnée. Forts des résultats précédents nous pouvons décomposer le module P_{ij} en deux termes:

$$P_{ij} = m_{ij} + A_{ij} \qquad (3)$$

Dans cette relation, m_{ij} est la valeur de la dérive pour l'année j, définie par Matheron (1962) comme l'espérance mathématique (non-stationnaire) des modules, alors que A_{ij} est un résidu aléatoire de moyenne nulle.

Posé en ces termes, et sans hypothèses supplémentaires, le problème de la modélisation d'une série non-stationnaire reste insoluble. En effet:

(a) la dérive m_{ij} ne peut pas être déterminée sur la base de l'unique réalisation P_{ij} disponible pour chaque année j.

(b) s'il est raisonnable de supposer, dans notre cas, que le terme aléatoire A_{ij} suit une distribution connue de moyenne nulle (par exemple la distribution normale ou log-normale), il convient encore d'admettre que la variance de cette distribution est stationnaire. De plus, avant de réaliser un ajustement des résidus A_{ij}, il faut pouvoir les calculer, ce qui suppose connue la dérive m_{ij}.

Pour sortir de cette impasse, nous considérerons les éléments suivants:

(1) La tendance climatique, caractérisée par une dérive des modules, est une réalité physique à caractère régional. Il est donc légitime de supposer que les dérives m_{ij} varient de la même façon dans une région "climatologiquement homogène". Il est dès lors

possible d'envisager l'évaluation de la dérive, par exemple par le biais d'un indice climatique régional annuel C_j, conduisant à une relation du type $m_{ij} = C_j k_i$, ou k_i est un paramètre dépendant de la station i considérée.

(2) Le concept même de tendance climatique n'a de sens physique ou pratique que si la dérive qui lui correspond varie de façon régulière dans le temps. Pratiquement, cela signifie que l'on peut obtenir une approximation de la dérive par un lissage inter-annuel des modules P_{ij} ou des indices climatiques C_j.

(3) Comme nous l'avons constaté expérimentalement, les coefficients de variation restent pratiquement constants au cours du temps et entre stations. En conséquence, il est légitime d'admettre la stationnarité du coefficient de variation des résidus A_{ij}.

Méthode des indices annuels de précipitation

Il est de pratique quotidienne d'effectuer la vérification de l'homogénéité des données pluviométriques, pour l'ensemble des stations d'une zone climatique. Une méthode très couramment utilisée à cet effet est celle du double cumul. Une amélioration substantielle a été apportée par Brunet-Moret, consistant non pas à comparer les stations entre elles mais par rapport à une station fictive, représentative de l'ensemble de la région. A cet effet Brunet-Moret (1979) définit un indice régional annuel des précipitations par la relation $P_{ij}/\overline{P}_i = z_j + e_{ij}$, avec P_{ij}: module pour l'année j à la station i, $\overline{P}_i$: normale à la station i, z_j: indice régional annuel des précipitations pour l'année j et, finalement, e_{ij}: résidu aléatoire. Les hypothèses posées sont donc les suivantes:

(a) l'espérance mathématique des z_j égale un,
(b) l'espérance mathématique des e_{ij} vaut zéro,
(c) les variances des e_{ij}, pour chaque station i, sont égales entre elles.

La méthode de Brunet-Moret consiste alors à déterminer simultanément les valeurs de l'indice régional z_j pour chaque année j et les valeurs des normales $\overline{P}_i$ pour chaque station i en minimisant la somme des carrés des écarts $(P_{ij}/\overline{P}_i) - z_j$.

Modèle du processus non-stationnaire

Nous nous sommes intéressés à l'approche de Brunet-Moret, car elle répond aux conditions (1) et (3) examinées plus haut.

Pour la première, l'indice annuel z_j est équivalent à l'indice C_j, le coefficient de proportionalité k_i étant tout simplement la normale $\overline{P}_i$: on peut en effet écrire le modèle de Brunet-Moret sous la forme $P_{ij} = (z_j + e_{ij}) \overline{P}_i$.

Pour la troisième de ces conditions, il nous suffit de constater que l'hypothèse d'homocédacticité des e_{ij} est équivalente à celle d'un coefficient de variation constant des résidus A_{ij}.

A ce stade, il est déjà possible de proposer un modèle du processus non-stationnaire des précipitations, puisque soit les indices z_j, soit les normales $\overline{P}_i$ peuvent être déterminés.

Cependant, conformément à la deuxième condition, il est souhaitbale de lisser les valeurs z_j de l'indice annuel des précipitations, pour obtenir une approximation plus "régulière" de la dérive régionale. Nous avons choisi la relation:

$$z1_j = 1/45 \sum_{k=0}^{9} (10-k) \, z_{j-k} \qquad (4)$$

Ce choix arbitraire n'affecte toutefois pas la rigueur du modèle. Les indices lissés peuvent être considérés comme une approximation de la dérive régionale (figure 7).

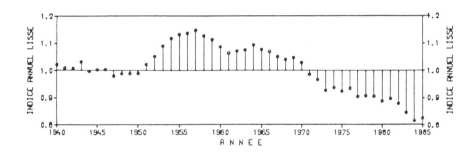

Figure 7 Série chronologique de l'indice régional lissé

Il est alors possible de calculer les écarts lissés correspondants par $el_{ij} = (P_{ij} / \overline{P_i}) - z1_j$. La figure 8 réprésente, à titre d'exemple, la distribution des valeurs obtenue pour la station de Ouagadougaou, ainsi que la distribution de l'ensemble des valeurs el_{ij}, reportées sur un papier probabiliste de Gauss. On constate, à postériori, que l'hypothèse d'homocédasticité est plausible, puisque la valeur moyenne de l'écart-type des différentes distributions des el_{ij} est de 0.171, avec une erreur-type de 0.015.

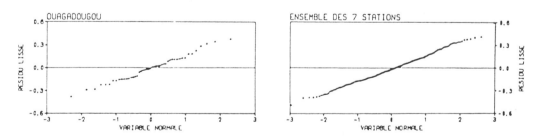

Figure 8 Exemples de distributions des résidus annuels lissés

Utilisation pratique du modèle

Pour une station i donnée, la normale $\overline{P_i}$ peut être déterminée tout comme la distribution des écarts lissés el_{ij} (distribution normale, de moyenne nulle et d'écart-type = 0.171). Si, de plus, on connait la valeur de l'indice régional lissé $z1$, la distribution des modules P_{ij} est donnée par la relation: $P_{ij} = \overline{P}_i (zl + el_{ij})$. Avec ce résultat, il est donc possible d'effectuer une analyse fréquentielle classique.

Il faut cependant convenir qu'à ce stade le modèle proposé n'a pas grande utilité pratique, hormis peut-être l'expertise d'un dimensionnement d'ouvrage effectué il y a 50 ans! Pour une application opérationelle, telle qu'un dimensionnement de retenue à construire pour les 30 prochaines années, il est nécessaire de supputer la valeur la plus probable de l'indice régional lissé zl. Le problème de la projection dans le futur de ce paramètre reste entier. Nous tentons cependant d'indiquer, ci-après, quelques possibilités d'appréhension de ce problème.

Perspectives

La projection dans le temps d'une tendance est toujours risquée. Plusieurs considérations peuvent être prises en compte lors d'une telle extrapolation, allant de l'analyse phénoménologique directe à l'approche purement statistique.

Dans notre cas, l'évaluation de la tendance climatique régionale ne peut guère s'appuyer sur une analyse climatique sérieuse, faute de données historiques permettant de mettre éventuellement en évidence un cycle météorologique réel. L'approche intuitive de l'ingénieur, consistant à faire un choix selon des critères souvent peu rigoureux mais conduisant selon lui à la solution la plus vraisemblable, peut être envisagée tout comme celle qui s'appuie sur une analyse strictement statistique. Mais, prises de manière separée, toutes deux peuvent conduire à des résultats erronés. Un compromis entre ces deux méthodes permettrait certainement de mieux cerner ce problème ardu.

La décroissance des indices climatiques régionaux ne peut continuer indéfiniment. Les valeurs de ceux-ci vont en effet varier à l'intérieur d'un domaine qu'il est difficile de borner. Une analyse fréquentielle de ces indices permettrait toutefois d'évaluer les probabilités d'apparition des valeurs extrêmes. Cette information, situant en probabilité la dernière valeur z_i réellement calculée, pourrait alors être utilisée en second lieu dans une analyse baysienne basée sur la chronologie de ces mêmes indices régionaux. Ceci permettrait enfin de chiffrer la probabilité d'apparition d'un événement futur, compte tenu de l'occurrence d'un événement passé. Projeté ainsi pas à pas, une extrapolation temporelle des indices climatiques régionaux deviendrait alors possible.

Une autre approche statistique, utilisant un processus autorégressif ou markovien peut également être discutée. Celle-ci devrait cependant s'établir à partir des indices régionaux bruts, pris sur une courte période, mais impliquant une extrapolation à partir de la dernière valeur de l'indice lissé; le problème du domaine de variation des z_i restant entier. L'une et l'autre de ces méthodes sont actuellement en étude; chacune d'entre elles comporte des avantages et des inconvénients. Les auteurs ne manqueront pas d'informer le lecteur intéressé de l'avancement de leurs travaux dans ce domaine.

REMERCIEMENTS   Le service national de la météorologie du Burkina-Faso a bien voulu mettre à notre disposition les données pluviométriques nécéssaires à cette étude. Qu'il en soit chaleureusement remercié.

Références

Albergel, J., Carbonnel, J.P. & Grouzis, M. (1984) Péjoration climatique au Burkina-Faso. Incidences sur les ressources en eau et les productions végétales. In: Cah. Orstom, sér. Hydrol., vol XX, no 1, 3-19.

Brunet-Moret, Y. & Roche, M. (1975) Persistance dans les suites chronologiques de précipitations annuelles. In: Cah. Orstom, sér. Hydrol., vol XII, no 3, 147-165.

Brunet-Moret, Y. (1979) Homogénéisation des précipitations In: Cah. Orstom, sér. Hydrol., vol XVI, NO 3&4, 147-170.

Bouvier, Ch. (1983) Etude des effets de dépendance dans une série chronologique. Application à l'étude des séquences de jours de pluies. In: Cah. Orstom sér. Hydrol., vol XX, no 2, 79-116.

Haas, A.-G & Viallix, J.-R. (1976) Krigeage applied to geophysics. The answer to the problem of estimates and contouring. In: Geophys. Prosp., vol 24, 49-68.

Jordan, J.-P & Meylan, P. (1986) Répartition spatiale des précipitations dans l'ouest de la Suisse par la méthode du krigeage. In: IAS, no 12 & 13, 157-162 & 187-189.

Matheron, G. (1962) Traité de géostatistique appliquée. Technip, Paris.

Meylan, P. (1986) Computerized data sets. Unep/Unitar & Epfl Training Programme 1986, Lausanne.

Meylan, P.   (1986b) Régionalisation de données entachées d'erreurs de mesure par krigeage. Application à la pluviométrie. In: Hydrol. Contin., vol 1, 25-34.

Musy, A. (1973) Répartition spatiale et prévision en temps de retour de facteurs climatiques en vue d'un dimensionnement des ouvrages d'aménagement des eaux. In: BTSR, no 15, 319 339.

Serra, J. (1967) Un critère nouveau de découverte des structures: le variogramme. In: Science de la Terre, Vol 12, no 4, 277-299.

Sircoulon, J. (1976) Les données hydro-pluviométriques de la sécheresse récente en Afrique intertropicale. Comparaison avec les sécheresses 1913 et 1940. In: Cah. Orstom, sér. Hydrol., vol XIII, no 2, 75-174.

Triboulet, J.-P. (1983) Persistance du déficit pluviométrique en Haute-Volta de 1965 à 1982. Agrhymet, Niamey.

The Influence of Climate Change and Climatic Variability on the Hydrologic
Regime and Water Resources (Proceedings of the Vancouver Symposium,
August 1987). IAHS Publ. no. 168. 1987.

Bayesian forecasting of hydrologic variables under changing climatology

Lucien Duckstein
Systems and Industrial Engineering,
University of Arizona, Tucson, Arizona 85721
Bernard Bobee
Universite du Quebec, INRS-Eau
Complexe Scientifique, 2700 Rue Einstein
Case Postale 7500, Ste-Foy, Quebec
Canada G1V 4C7
Istvan Bogardi
W-348 Nebraska Hall, Civil Engineering Dept.
University of Nebraska, Lincoln, NE 68588-0531

ABSTRACT   A Bayesian framework is developed to provide a
credible set (or Bayesian confidence interval) to forecast
hydrologic parameters under uncertainty of climatologic
changes. The methodology is illustrated by developing the
forecast for the credible set of parameters of the
distribution of flow exceedance $\underline{x}$ of a Canadian river.
Let the probability distribution of $\underline{x}$ be $f(x \mid \theta)$, where
θ is a vector of parameters (mean u, variance v^2 location
parameter c...): $\theta = (u,v^2,c)$. In the case study, only
the mean is assumed to be unknown so as to simplify the
presentation. This parameter u, a random variable, is
taken as a function of the climatologic regime.
While such a function is difficult to develop, it
appears that a prior distribution of the mean u of future
flood $\pi(u)$ can be estimated using one or a combination of
various techniques. For example, in the hierarchical
Bayes approach, the prior $\pi(u \mid \mu, \tau^2)$ is selected as a
normal distribution with mean μ and standard deviation τ .
The climatologic change is encoded in $\pi_{21}(\mu \mid \tau^2)$ while the
prior $\pi_{22}(\mu \mid \tau^2)$ is chosen initially as a non-informative
prior. Knowing the posterior $\pi(u \mid \underline{x})$, credible sets $C(\alpha)$
at the 100 $(1-\alpha)\%$ confidence level can be calculated.
Furthermore, as new observations on climatology and floods
become available, the posterior distribution and credible
sets can routinely be updated.

Introduction

The purpose of this paper is to develop a Bayesian statistical
framework for the forecasting of hydrologic variables under uncertain
climatologic changes. The essential advantages of such a framework
are (Berger, 1985):
    (a) The ability to incorporate information from various sources
into the forecasts.
    (b) The possibility of defining Bayesian confidence intervals,

also called credible sets, under the form of actual probability statements.

In addition, a complete Bayesian analysis (Davis et al., 1972); Duckstein et al., 1978; Bernier, 1987a, 1987b; Bogardi et al., 1987) yields two more extremely useful concepts.

(c) The definition of the expected worth of perfect information (or expected opportunity loss XOL), which measures the average cost of uncertainty.

(d) The definition of the expected worth of sample information EVSI, which makes it possible to design a sampling strategy.

The elements of the Bayesian framework may be defined as follows:

x: random variable, taken here as the annual peak flow (as an example).

$f(x|\theta)$: the probability density function (pdf) of x given θ; this is the stochastic model of the process.

θ : the parameter vector of the pdf of flow assumed to be uncertain and thus taken as a random variable.

$\pi(\theta)$: the prior pdf of θ estimated by means of regional data, climatologic information, or even subjective procedures if necessary.

$\pi(\theta|x)$: the posterior pdf of given a random sample $x=(x_1,\ldots,x_n)$

$m(x)$: the marginal, predictive or Bayesian pdf of x.

Next, the analysis is illustrated by means of the example of the river Harricana in Quebec (Canada).

Statistical analysis

Model choice and standard analysis

Because of U.S. Water Resources Council recommendations (WRC, 1976), most North American river annual peak flows have been fitted to the Log Pearson type 3 pdf. However, in the case considered, as in many other examples, (Davis et al, 1972; Tecle et al, 1987) a two-parameter log-normal pdf seems to provide a statistically acceptable fit. Let the parameter vector θ be:

$$\underline{\theta} = (u,v^2)$$

with

$$y = \ln x; \text{ and } u = E(y); \ v^2 = Var(y)$$

then the log-normal pdf becomes

$$f(y|u,v^2) = (2\pi v^2)^{-\frac{1}{2}} \exp[-(y-u)^2/2v^2] \tag{1}$$

Using n = 61 years of data at the gaging station Trois-Pistoles, the maximum likelihood estimator of u and v are estimated to be $\hat{u}$ = 5.3309, $\hat{v}$ = 0.3481. Then $\hat{x}$ and Var $\hat{x}$ can be calculated by the relations:

$$\hat{x}_T = \exp[\mu_T + k_T v_T]$$

$$\hat{k}_T = z^{-1}(\exp[(\ln(1+z^2))^{\frac{1}{2}} U_T - 0.5\ln(1+z^2)] - 1)$$

$$z = \hat{v} \cdot \hat{u}^{-1} \tag{2}$$

U_T = standard normal pdf corresponding to an exceedance probability
p = $1/T$ where T is the return period
and

$$Var(\underline{x}_T) = var(\hat{y}_T)(\hat{x}_T)^2$$

$$Var(\hat{y}_T) = \frac{v^2}{n} \delta_T$$

$$\delta_T = z^{-2}[\ln(z+1)][1+k_T z]^2[1+U_T z] \tag{3}$$

The standard 80% bounds of the confidence interval are then
calculated by the equation

$$x_T = x_T + U(\alpha/2)[VAR \ x_T]^{1/2} \tag{4}$$

where $U(\alpha/2)$ is the standard normal value with exceedance probability
$\alpha/2$ (here =0.20). The results are sketched in Figure 1. In particular, one finds:

for $\frac{1}{T} = 10^{-4}$: $\quad 711 < X_T < 1079$

for $\frac{1}{T} = 10^{-3}$: $\quad 559 < X_T < 784$

for $\frac{1}{T} = 10^{-2}$: $\quad 428 < X_T < 553$

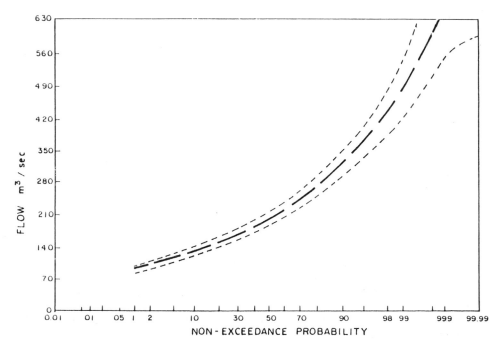

Figure 1   Lognormal fit of annual peaks and 0.80 confidence limits.

Note that the basic premise to calculate a classical confidence interval is that the parameter x_T is deterministic but unknown.

Credible set calculation

The calculation of the credible set is based on a probability statement made from the posterior pdf $\pi(\theta|y)$ given data y. Bayes theorem provides the posterior pdf as

$$\pi(\theta|y) = [m(y)]^{-1} f(y|\theta)\pi(\theta) \tag{5}$$

with

$$m(y) = \int_H f(y|\theta)dF^\pi(\theta) \tag{6}$$

The symbolism in the right-hand of equation (6) represents an integral if θ is continuous and a summation if θ is discrete, in the domain H.

The $100(1-\alpha)\%$ highest posterior density (HPD) credible set for θ is the subset C of H defined as follows (Berger, 1985:

$$C = [\theta \epsilon H , \pi(\theta|y) \geq k(\alpha)] \tag{7}$$

where $k(\alpha)$ is the largest constant such that $Pr(C|y) \geq 1-\alpha$.

In particular, let the uncertainty in equation (1) be a only, with v^2 known, and let the prior of u be normal:

$$\pi(u) = N(\mu,\tau^2;u) \tag{8}$$

Then it can be shown that the posterior pdf is also normal, as follows:

$$\pi(u|y) = N(\mu(y),\rho^{-1};u)$$

with posterior mean

$$\mu(y) = \frac{v^2\mu + \tau^2 y}{v^2 + \tau^2}$$

and posterior variance

$$\rho^{-1} = (v^{-2} + \tau^{-2})^{-1} \tag{9}$$

In this case, the $100(1-\alpha)\%$ HPD credible set on the posterior mean $\mu(y)$ may be calculated as:

$$C = \mu(y) \pm U(\alpha/2) \rho^{-\frac{1}{2}} \tag{10}$$

As a numerical example designed only to illustrate the principle, let

$$f(y|u) = N(u, 1; y), \pi(u) = N(5,1:u)$$

Then the posterior mean and variance of equation (9) become, respectively:

$$\mu(y) = \frac{5 + y}{1 + 1} = \frac{5 + y}{2} \tag{11}$$

$$\rho^{-1} = [1 + 1]^{-1} + 0.5$$

The 80% HPD credible set on the posterior mean is:

$$C = 2.5 + 0.5y \pm (1.28)0.707$$

or else:

$$C = (1.595 + 0.5y, \; 3.405 + 0.5y)$$

where y represents an observation. In our real example y would be the logarithm of the peak annual flood x.

To transform back to x, equations (2), (3), and (4) can still be used but x_T must be replaced by values of the posterior mean within the bounds of equation (11).

Estimation of the prior based on climatology

Climatologic considerations will now be used to estimate the hyperparameters μ and τ^2 of the prior pdf of equation (8).

According to Rosenberg (1986) who reviews recent studies on the effect of man on CO_2 production and ensuing possible climatologic changes, the mid-latitude zones of the northern hemisphere, which are drought-prone, are likely to be strongly affected. The changes in plant production will be the result of two conflicting factors: an increase of CO_2 causing an increase of the photosynthesis rate and a decrease of evapotranspiration because of an increase in stomatal closure. These two effects vary with plant type. In turn, the combination of changing precipitation patterns and vegetation cover influence mean and peak runoff rates, hence floods. Still in Rosenberg (1986), comparisons are provided between the predictions of precipitation and soil moisture changes given by three general circulation models: the Geophysical Fluid Dynamics Laboratory (GFDL), the Goddard Institute of Space Studies (GISS) and the National Center for Atmospheric Research (NCAR). While the predictions given by the three models certainly do not agree everywhere, they may be used to provide the basis for prior pdf assessment.

First, qualitatively, effects of climatologic changes may be expected to influence the mean μ and variance τ^2 of the mean u of peak annual flood as follows:

$$\mu = a_1 R + a_2 S$$

$$\tau^2 = b_1 \Delta R + b_2 \Delta S \tag{12}$$

where

$\quad a_1$, a_2, b_1, b_2 = positive coefficients

R = mean precipitation

S = mean soil moisture

ΔR = a measure of mean precipitation variation, such as number of events per year or rainfall depth per event (Fogel and Duckstein, 1969; Duckstein, et al, 1972; Bogardi et al, 1987)

ΔS = a measure of mean soil moisture variation

For example, Rosenberg (1986) observes that the climatologic models yield variations of mean temperatures between 2 and 8° C, and of precipitation between +2 and -1 mm/day. For a given plant and soil types, ΔR and ΔS can then be calculated.

As an alternative to equation (12), a rainfall-runoff model may be used, such as the one of the U.S. Soil Conservation Service (1979). In any case, the climatologic models derived to date do not make it possible to determine precisely or even statistically the coefficien- ts a_1, a_2, b_1, b_2 in equation (12). However, a subjective determina- tion may be possible, using either a technique such as relative likelihood (Berger, 1985) or even fuzzy sets, as in Bardossy, et al. (1987).   In such a determination, information on likely trends, such as an increase of mean temperatures (Jones, 1986), can be used.

Application

In the case study of the Trois-Pistoles gauging station, assume that a climatologic change leaves the mean (log) flow constant, so that the prior mean is:

$$\mu = u = 5.33$$

but the prior variance doubles:

$$\text{from } \tau_0^2 = 0.15 \text{ to } \tau_1^2 = 0.30$$

Let the variance of the (log) flow be known as $v^2 = 0.10$. Using equation (9), the posterior parameters are found to be, respectively:

$$\mu_0(y) = 2.132 + 0.6y \qquad\qquad \mu_1(y) = 1.333 + 0.75y$$

$$\rho_0^{-1} = 0.060 \qquad\qquad\qquad \rho_1^{-1} = 0.075$$

The corresponding 95% HPD credible sets can be calculated by use of equation (11) as:

$$C_0 = (1.652 + 0.6y, \ 2.162 + 0.6y)$$

$$C_1 = (0.796 + 0.75y, \ 1.870 + 0.75y)$$

The difference between C_0 and C_1 is substantial, as sketched in Figure 2.   In fact, if an observation y is large enough, the two credible sets become disjoint.   Also, note that both width and slope of the credible sets change although only the variance of the prior pdf of u was changed by a factor of 2.   This change is quite modest in view of the high uncertainty of climatologic forecasts.   In the next section, a brief investigation of the effect of uncertainty of

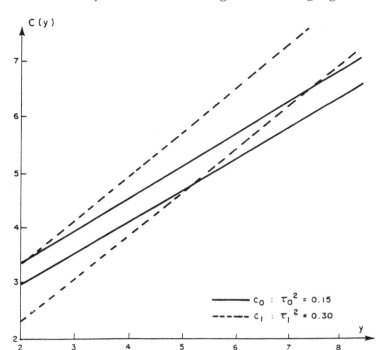

Figure 2   95% HPD credible sets for the case example with common
prior pdf mean μ and different prior pdf variances τ^2.

not only the mean u but also the variance v^2 of x is investigated.
At the same time, the concept of hierarchical Bayes analysis is
introduced to cope with the difficulties encountered in estimating μ
and mainly τ^2 on the basis of climatologic considerations.

Hierarchical Bayes analysis

Principle

Since both u and v^2 are assumed to be uncertain, the prior pdf will
be bivariate, and the hyperparameter vector may be three-dimensional
(Benjamin and Cornell, 1970). In this case, a two-stage approach is
very appropriate and may be illustrated as follows. The prior is
decomposed as:

$$\pi(\underline{\theta}|\underline{\lambda}) = \int_\Omega \pi_1(\underline{\theta}|\mu,\tau^2,n)\pi_{21}(\mu|\tau^2,n)\pi_{22}(\tau^2|n)d\mu d\tau^2 \qquad (13)$$

In this equation, $\underline{\theta}$ = (u,v^2) as before, the parameter vector of the
prior pdf is $\underline{\lambda}$ = ($\overline{\mu}$, τ^2, n), but it will be assumed that n is given
to simplify the presentation; this parameter n will be deleted from
here on. In fact, if past prior pdf selected from the appropriate
conjugate family, then n represents the number of observation points,
which is known. Otherwise, the concept of "equivalent sample size"

(Benjamin and Cornell, 1970) may be used. Using Result 7 in Section 4.6 of Berger (1985) the posterior pdf of $\underline{\theta} = (u,v)$ is found to be

$$\pi(u,v^2|y) = \int_\Omega \pi_1(u,v^2|y,\mu,\tau^2)\pi_{21}(\mu|y,\tau^2)\pi_{22}(\tau^2|y)d\mu d\tau^2 \tag{14}$$

with

$$\pi_1(u,v^2|y,\mu,\tau^2) = [m_1(y|\mu,\tau^2)]^{-1} f(y|u,v^2)\pi_1(u,v^2|\mu,\tau^2)$$

$$m_1(y|\mu,\tau^2) = \int f(y|u,v^2)\pi_1(u,v^2|\mu,\tau^2)d\mu \, dv^2 \tag{15}$$

$$\pi_{21}(\mu|y,\tau^2) = [m_2(y|\tau^2)]^{-1} m_1(y|\mu,\tau^2)\pi_{21}(\mu|\tau^2)$$

$$m_2(y|\tau^2) = \int m_1(y|\mu,\tau^2)\pi_{21}(\mu|\tau^2)d\mu \tag{16}$$

$$\pi_{22}(\tau^2|y) = [m(y)]^{-1} m_2(y|\tau^2)\pi_{22}(\tau^2)$$

$$m(y) = \int m_2(y|\tau^2)\pi_{22}(\tau^2)d\tau^2 \tag{17}$$

Consider the following example. Several new extreme observations of the Harricana river flow at Amos are available:

$$\underline{y} = (5.572, 5.802, 5.568, 5.759, 5.820, 5.568, 5.521)$$

with

$$\overline{y} = 5.6585 \text{ and } s^2 = 0.0232, s = 0.1523$$

We assume that these seven observations are independently distributed $Y_i \sim N(u_i, 0.04)$, where the u_i were drawn independently from a common pdf $N(\mu, \tau^2)$. A two-stage hyperprior is assigned to this pdf. First it is assumed that μ follows the population distribution, that is, here, the one observed at Trois-Pistoles over 61 years:

$$\pi_{21}(\mu) = N(5.3309, 0.1212; \mu) \tag{18}$$

Next, since our knowledge about τ^2 is quite vague, let the second-stage prior be the (improper) constant density

$$\pi_{22}(\tau^2) = 1 \tag{19}$$

Thus assuming independence of μ and τ^2

$$\pi_2(\mu,\tau^2) = \pi_{21}(\mu)\pi_{22}(\tau^2) \tag{20}$$

It can be shown that, in this normal pdf case $\pi_1(u,v|\underline{y}, \underline{\mu}, \tau^2)$ is:

$$N_7 ([\underline{y} - \frac{0.04}{0.04+\tau^2} (\underline{y} - (\mu,\dots,\mu))], \frac{0.04\,\tau^2}{0.04+\tau^2} I) \tag{21}$$

and

$$m_1(y \mid \tau^2) \text{ is } N_7[(\mu, \dots, \mu)', \ (0.04 + \tau^2) I] \tag{22}$$

(N_7 means a 7-variate normal pdf).

Since m_1 is normal and $\pi_{21}(\mu)$ is also normal as given by equation (18), it can be shown that $\pi_{21}(\mu \mid y, \ \tau^2)$ is

$$N(\bar{y} - \frac{0.04 + \tau^2}{0.884 + \tau^2} (\bar{y} - 5.3309), \ \frac{(0.04 + \tau^2)(0.1212)}{0.884 + \tau^2} \tag{23}$$

and

$$m_2(y \mid \tau^2) \text{ is } N_7((5.3309, \dots, 5.3309), \ (0.04 + \tau^2) \ I + 0.1212(\underline{1}) \tag{24}$$

where $(\underline{1})$ is the matrix of all ones.
Hence, all terms on the right hand of equation (14) are known, and the posterior pdf $\pi(u, v^2 \mid y)$ can be calculated using equations (21) to (24) up to a consgant $m(y)$.

Again, once the posterior has been calculated, credible sets can be found without particular difficulty.

Discussion and conclusions

As pointed out in Lamb (1986), long-term results may not provide a reliable basis to predict future climatology. Thus, long-term records may not be usable to estimate future extreme floods; one must then resort to global climatologic models to obtain a forecast of solar energy reaching the atmosphere and the land. The production of CO , ozone, heat, and the changes in land use are altering the energy balance hence the precipitation and the streamflow. This is the type of information that must be used to assess the hydrologic effect of climatologic changes. Bayesian models with subjective assessment of the prior pdf seem to be well suited for this purpose.

In the case of design or operation of water resources systems design under uncertainty of climatologic change, the posterior pdf $\mu (\theta \mid y)$ can be used in an extensive Bayesian analysis. In this case, a loss function $L(y,a)$ is estimated where an action or decision a is to be taken (de Groot, 1970). As pointed out in Davis et al (1972) or Duckstein et al (1978), a Bayesian decision-theoretical analysis yields not only an optimum decision a*, but also the expected worth of perfect information (XOL) and the expected value of sample information (EVSI). One may thus ascertain if it is economically worth pursuing a given path of information gathering.

Concluding points of this brief overview of possible uses of a Bayes viewpoint to assess the effect of climatologic changes on hydrologic systems may be made as follows:

(a) A Bayes analysis makes it possible to study the combined effect of randomness (peak flood) and uncertainty (parameter θ of peak flood pdf), even when few reliable observations are available.

(b) Information from various sources, especially regional infor-mation, global circulation models and forecasts of CO_2 or ozone production may be incorporated in a natural way into the assessment of the prior pdf $\pi(\theta)$.

(c) Using the posterior pdf $\pi(\theta|y)$ calculated by means of Bayes theorem, credible sets on θ can be calculated; such sets represent intervals that correspond to actual probability statements, in contrast with classical confidence intervals.

(d) When difficulties arise in selecting a joint prior pdf to represent the variation of mean and variance of θ, a hierarchical Bayes analysis may be performed.

(e) Both design and operation decision may be evaluated by means of Bayesian extensive analysis.

References

Bardossy, A., Bogardi, I., Duckstein, L. and Nachtnebel, H.P. (1987) Fuzzy decision-making models for regional management. Available as working paper 85-17,Systems and Industrial Engineering Department, University of Arizona.

Benjamin, J.R. and Cornell, C.A. (1970) Probability, Statistics and Decision for Civil Engineers, McGraw-Hill, New York, 684 pp.

Berger, James O. (1985) Statistical Decision Theory and Bayesian Analysis, Springer-Verlag, New York, New York, 2nd ed., 617 pp.

Bernier, J. (1967) Les methodes Bayesiennes en hydrologie statistique, Proceedings, International Hydrology Symposium, pp. 439-470, Colorado State University, Fort Collins.

Bernier, J. (1987a) Elements of Bayesian analysis of uncertainty in hydrological reliability and risk models, (in) Engineering Reliability and Risk in Water Resources, L. Duckstein and E. Plate (Eds.) NATO ASI Series M. Nijhoff, Dordrecht, The Netherlands.

Bernier, J. (1987b) Bayesian analysis: further advances and applications, (in) Engineering Reliability and Risk in Water Resources, Duckstein L. and E. Plate (Eds.) NATO ASI Series M. Nijhoff, Dordrecht, The Netherlands.

Bogardi, J.J., Duckstein, L., Rumambo, O.H., (1987) A simplified event-based rainfall analysis for semi-arid regions in developing countries, to be presented at IAHS General Assembly, Vancouver, B,C., August (1987), available as working paper 86-7, Systems and Industrial Engineering Department, University of Arizona.

Davis, D.R., Duckstein, L. and Kisiel, C.C., (1972) Uncertainty in the return period of maximum events: A Bayesian approach, Proceedings, International Symposium on Uncertainties in Hydrologic and Water Resource Systems, Tucson, Arizona.

Duckstein, L., Fogel, M.M. and Kisiel, C.C. (1972) A stochastic model of runoff-producing rainfall for summer type storms, Water Res. Res.. Vol. 8, No. 2, pp. 410-421, April.

Duckstein, L., Krzysztofowics, R. and Davis, D.R. (1978) To build or not to build: A Bayesian analysis, Journal of Hydrological Sciences, Vol. 5, No. 1, pp. 55-68.

Fogel, M.M. and Duckstein, L. (1969) Point rainfall frequencies in convective storms, Water Res. Res, Vol. 5, No. 6, pp. 1229-1237, December.

Jones, P.D., Wigley, T.M.L. and Wright, P.B. (1986) Global temperature variations between 1861 and 1984. Nature 322:430-434

Lamb, P.J. (1986) A state-of-the-art review of the development of climactic scenarios. Paper presented at the Task Force Meeting on Policy-oriented Assessment of Impact of Climatic Variations, IIASA, Laxenburg, Austria, June 30-July 2.

Musy, A. and Duckstein, L. (1976) Bayesian approach to tile drain design, Journal of the Irrigation and Drainage Division, ASCE Vol. 102, No. IR3, Proc. Paper 12390, pp. 317-334, September.

Rosenberg, N.J. and Holmes, G. (1986) Proceedings Symposium on Climatology and Drought, University of Nebraska, Lincoln, November.

Tecle, A. and Duckstein, L. (1987) Multicriterion forest watershed management, To be presented at International Association of Hydrological Sciences Symposium S2: Forest Hydrology and Watershed Management, Vancouver B.C., August.

W.R.C. (U.S. Water Resources Council) (1976) Guidelines for determining flood flow frequency. Bull. No. 17, The Hydrology Committee, Washington, D.C.

Climatic change and river basin characteristics

The Influence of Climate Change and Climatic Variability on the Hydrologic Regime and Water Resources (Proceedings of the Vancouver Symposium, August 1987). IAHS Publ. no. 168, 1987.

Impact of climate change on the morphology of river basins

F.H. Verhoog
Division of Water Science
Unesco

Introduction

There is sufficient evidence to show that climate is variable. There are also indications that climate can vary considerably over a relatively short time span. If relatively short means 100 or 200 years, if we are presently undergoing such a change it is important for us to recognise the changes and to adapt our hydrologic means.

Within the framework of the IAHS, we are particularly interested in changes in the morphology of river basins related to the hydrologic regime.

When we want to estimate the impact of climatic change on the morphology of river basins, we first have to estimate the impact of climate change on precipitation and evaporation, secondly on natural ecosystems, thirdly on runoff and lastly on the morphology of river basins.

The following discussion, unreasonably, tries to be quantitative as possible. Due to a lack of data and a lack of verified and tested methodologies the tables given in the text, although based on ideas expressed in scientific literature, can only be regarded as hypothetical.

Much use has been made of the listed consulted literature. This use has been extensive and without restraint. As parts of the consulted literature is used outside its context it does not seem correct to quote it.

Impact of global warming on precipitation and evapotranspiration

The surface of the earth is warm largely due to radiation from the sun in the visible part of the spectrum. This energy is partly absorbed by the surface of the earth which warms up as a result. Like all warm objects, the earth itself radiates energy, but in the infrared part of the spectrum. This re-radiated energy is partly absorbed by water vapor and CO_2 in the atmosphere and partly radiated back to the earth, the so called greenhouse effect. The mean temperature of the earth is about 15 degrees centigrade. Without water vapour and CO_2, the mean global temperature would be −23 degrees.

When the earth was formed it was very warm and the sun was relatively cool. Due to volcanic activity, water vapour and CO_2 were put into the atmosphere. The earth cooled and the water vapour precipitated to form oceans. Then about 1 billion years ago, plants began to use the CO_2 to produce oxygen. The CO_2 content began to fall. But as the warming influence of the green house effect declined the sun got warmer.

Hundreds of millions of years ago the atmosphere still contained much more carbon dioxide than today. The carbon dioxide concentration fell until about 1 million years ago, when series of ice ages started to occur. During the ice ages the carbon dioxide content of the atmosphere was about 200 ppm and during the interglacials about 270 ppm. Over the past 100 years natural variations have ranged from 250 to 310 ppm.

In about 1850 the CO_2 concentration was around 280 ppm. At the beginning of this century the concentration was about 300 ppm. In 1958 when systematic measurements were started the average concentration was 316, and at present it is around 350 ppm. The increase in CO_2 is mainly caused by the burning of fossil fuels. About 58% of the CO_2 produced seems to remain in the atmosphere.

In geological terms, an increase in CO_2 concentration is correlated with an increase of the global mean temperatures. It is thus to be expected that the increase in CO_2 since the start of the industrial revolution about 1850, will have had an impact on temperatures. There was little warming in the nineteenth century, marked warming until 1940, relatively steady conditions until the mid 1970s and a rapid warming until now. Five out of nine of the warmest years since 1860 have occurred since 1978. However, the steady conditions maintained from the late 1930s to the mid 1970s still needs an explanation. Carbon dioxide levels are now expected to reach twice the value of 1850 by the end of the next century.

Future warming brought about an increase in the CO_2 concentration is predicted through the use of climatic models -- the so-called general circulation models.

The predicted changes are not the same all over the globe. When the world as a whole warms up, the higher latitudes warm up more than the lower ones which reduces the difference in temperature between the equator and the poles creating a profound effect on the circulation of the atmosphere. The equatorial regions may warm up to 0.5 degrees and the higher latitudes up to 6 degrees.

In the first instance, general circulation models (GCMs) try to predict the temperatures of the globe on the basis of a calibration of the present situation and a diminution of outgoing radiation. Some of these models can also give indications of changes in precipitation in the northern hemisphere.

The latest GCMs predict a global warming up of 3.5 to 4.0 degrees celcius brought about by a doubling of CO_2. One of the features of these GCMs is that they can simulate an exchange of heat between the oceans and the atmosphere creating the formation of sea ice and variations in cloud formation. Warming of tropical surface air ranges from 2 to 4 degrees. The greatest precipitation changes occur between 30 degrees north and 30 degrees south. One of the existing models predicts that the soil in almost all of Europe, Asia and North America will become drier during the summer, while others foresee wetter soil during the summer in most of these continents.

Some of the GCMs predict an increase in temperature and a decrease in precipitation for the areas around the Mediterranean.

Climate model results all point to an increase in the intensity of the hydrologic cycle and an overall small increase (3 to 5%, depending on what models are used) in global mean precipitation. The effects of precipitation change and vegetation changes are expected to reinforce each other. In some regions very large increases in annual mean runoff can be expected.

Those computer models which predict a temperature rise of 4.5 degrees brought about by a doubling of the CO_2 concentration also predict an increase in precipitation of 7-11%. Rainfall will increase at the ocean borders of the USA, over most of Canada, at the mouth of the Mississipi, around the Baltic, Egypt and Sudan, in the Middle East, India and Pakistan and in the middle of China. Precipitation over the rest of the Northern Hemisphere will decline. (source : University of East Anglia, as reported in the Economist of 28 November 1986).

An increase in evaporation is mainly likely to occur in equatorial regions but could also occur at high latitudes.

An approximation of the above information can be found in Table 1.

Table 1   Possible changes in temperatures and precipitation due to a doubling of the CO_2 concentration. The evapotranspiration changes only take into account the increase in temperature. All data are hypothetical

Possible changes in		TEMPERATURE (^{O}C)	PRECIPITATION (%)	EVAPOTRANSPI-RATION(%)
TROPICAL	Arid	+(0.5)	+	+(2)
	Humid	+(0.5)	+/-	+(2)
SUB-TROPICAL	Arid	+(2)	-	+(8)
	Humid	+(2)	-	+(8)
TEMPERATE	Warm	+(3)	-	+(12)
	Cold	+(4)	+/-	+(18)
COLD		+(6)	+	+(74)

Impact of runoff

Static hydrologic changes

Climate change means changes in temperature and precipitation and changes in evaporation caused by changes in temperature, air humidity, windspeed and cloudiness. As hydrologists we are particularly interested in the resulting changes in runoff.

The relation of existing regional climate to runoff is not always straightforward as can be seen from the following table of runoff ratios (the annual volume of discharge divided by the annual volume of precipitation) of large rivers in the world: From

(a) 0.01 to 0.20   Parana, Zambezi, Nile, Niger, Sao Francisco, Murray.
(b) 0.21 to 0.30   Congo, Mississipi, Ob, Danube, Vistula, Volga
(c) 0.31 to 0.40   St. Lawerence, Amur, Mckenzie, Indus

(d) 0.41 to 0.50   Amazon, Yang-tse, Ganges, Yenissei, Orinocco, Lena.
(e) 0.51 to 0.60   Columbia, Irrawady
(f) 0.61 to 0.70   Brahmaputra

The lower ratios clearly belong to arid and semi-arid climates. But the combination in one group of the Congo, Danube and Ob rivers is not obvious. It means that the usual climate classifications are not directly applicable to hydrology.

Global circulation models are used to predict the impact of a doubling of CO_2 on climate and likewise hydrologic mathematical models are used to study changes in runoff caused by changes in climate.

Like the GCM models, rainfall runoff mathematical models are calibrated on existing data and situations. The existing data are mostly daily precipitation, temperature and discharges for periods of 20 to 60 years. The most used model technologies are based on the assumption that physical basin boundary conditions do not change, thus vegetation, soil properties and channel morphology are kept constant.

For hydrologic climate impact modelling, the ideal model would be a distributed deterministic model with the following sub-models: an unsaturated zone model, a root zone model, a saturated flow model, a snowmelt model, a canopy interception model, an evapotranspiration model, an overland and channel flow model. An example of such a model is SHE (Systems Hidrologique Europeen). With SHE it is possible to model changes in vegetation and land use, but it is not possible to model secondary details such as soil macro-pores and an undergrowth of vegetation below the major vegetation.

For the purpose of rainfall-runoff modelling the modelling of evapotranspiration is crucial. Evapotranspiration, unlike runoff, disappears from the basin system and depends very much on ecological factors. Evapotranspiration depends on net radiation, rate of increase of the saturated vapour pressure of water at air temperature, density of air, specific heat of air at constant pressure, vapour pressure, deficit of air, aerodynamic resistance to water vapour transport, latent heat of vaporization of water, psychrometric constant, canopy resistance to water transport.

Nemec and Schaake used the Sacramento soil moisture accounting model to predict changes in streamflow in three existing river basins by changing both precipitation and evapotranspiration. The evapotranspiration change was based on changes in temperature.(1 degree celsius corresponds to a 4 % change in evapotranspiration). The basin as such, including the seasonal distribution of precipitation and evapotranspiration, remained the same.

The runoff ratios for the basins tested were 0.02, 0.13 and 0.31. For the dry basin and decrease of 10% in precipitation and an increase of 1 degree C decreased the runoff by 50%. For the humid one, the decrease was 25 and for the medium basin a decrease in runoff of 40%.

A dry basin is more sensitive to changes in precipitation than to evapotranspiration and is more sensitive to precipitation increases than a more humid basin (humid and dry as indicated by the runoff coefficient). For reductions in precipitation all basins are less sensitive to changes in evaporation than an increase in precipitation. For low runoff ratios, small changes in precipitation may cause large changes in runoff.

An approximation of the above infrastructure is listed in Table 2.

Table 2   Changes   in percentage   of annual runoff   if precipitation
and evaporation   change as   indicated. It is   assumed that
basin characteristics such as vegetation have not changed.
All data are hypothetical

Possible changes in		PRECIPITATION (%)	EVAPOTRANSPI- RATION (%)	RUNOFF (%)
TROPICAL	Arid	+(10)	+(2)	+20
	Humid	+/-(10)	+(2)	+40/-30
SUB- TROPICAL	Arid	-(10)	+8	-50
	Humid	-(10)	+8	-30
TEMPERATE	Warm	-(10)	+(12)	-50
	Humid	+/-(10)	(12)	+10/-35
	Cold	+(10)	+(18)	+5
COLD		+(10)	+(24)	nc

Ecological changes

The importance of vegetation

Runoff ratios for most of the rivers of the world are less than 0.50.
The   water   "lost"   is   lost   through   evapotranspiration.   The   term
evapotranspiration   in   the   water   balance   is   therefore   important.
Taking weather as a boundary condition, evapotranspiration is mainly
dependent   on the soil-vegetation complex. Vegetation,   in addition,
influences the base and rise times of the hydrograph.

While precipitation is the input, the soil cover complex plays the
role   of   the discriminating   element of   the   precipitation - runoff
relations.

Climate influences the microclimate,   the soil and the vegetation.
The   microclimate influences the vegetation and the   soil.   The   soil
influences   the   microclimate   and   the   vegetation.   The   vegetation
influences   the microclimate and the soil.   The fauna influences   the
vegetation.These   circular   relations   are difficult to introduce   into
mathematical model.

In   addition,   because the vegetation depends on both climate   and
soil,   it   is   not   always   a   good   indicator   of   present   climate.
Vegetation   is   also   dependent   on   previous   climates   through   the
historical   diversity   in plant species and   indirectly   through   the
soils that were formed during previous climatic conditions.   Only the
climate   (the   weather   over   a long period)   reacts   immediately   in
temperature and changes in air circulation.

What is the sensitivity of ecosystems to climate change?   First of
all   we   should   be   aware   that   a   change   in   temperature   and/or

precipitation will be accompanied by a change in probabilities of extremes. In certain cases ecosystems may be unexpectedly sensitive. For example, there is a theory that the reason forests are dying in certain parts of Europe is caused by the combination of increased acid precipitation and the drought of 1976. In some regions in the Alps 40% of the fir trees have died or are dying. This will increase as will the probability of avalanches. Thus a single drought period may have very serious consequences.

In general, when plant and animal generation times are short relative to the timescale of environmental change plant, and animal populations tend to react quickly to environmental processes. Environmentally bad times reduce the population but the return of favourable conditions sees a rapid increase in population growth.

The Northern Hemisphere warmed up from the mid-nineteenth century to about 1940. The predicted warming up and historical warming up are of a comparable characteristic timescale, that is, about a century. The characteristic timescale for animal population growth is from a month to 10 years, for vegetation biomass growth from a year to 10 years, for vegetation biomass growth from a year to 15 years, for soil accumulation from 10 to 800 years and for vegetation range extension from 1000 to 8000 years.

This means that biomass in general will adapt gradually to climate change brought about by global warming up. This may not necessarily be the case for forests. Large scale expanses of trees such as those following the last glaciation operated on a timescale of thousands of years. The reverse will also be true, forests and trees will survive longer if the conditions for tree growth and in particular, reproduction deteriorate.

According to some studies if a doubling of CO_2 occurs we may expect the following vegetation range changes:

Forest types	present	predicted	change
boreal forest	23%	1%	–
cold temperate	15%	20%	+
warm temperate	21%	25%	+
subtropical	16%	14%	–
tropical	25%	40%	++

Ecosystems			
tundra	3	–	– –
woodland	58	47	–
grassland	18	29	++
desert	21	24	+

According to some studies increasing atmospheric carbon dioxide concentrations will have a direct effect on plants. There stomata will close down, reducing evaporation and increasing water use efficiency. Reduced evapotranspiration would increase runoff and thus could offset the effects of precipitation reduction or enhance the effects of precipitation increases.

Changes in evapotranspiration will in this case occur due to changes in climate, changes in the area of vegetation cover (due to either climate change and/or the direct effect if CO_2 on plant growth) and the direct effect of CO_2 on evapotranspiration. If the runoff coefficient is high then runoff is more sensitive to precipitation changes than to evapotranspiration changes. Precipitation changes have an amplified effect on runoff, particularly in arid regions where the runoff coefficient is small. Evapotranspiration changes only have an amplified effect on small runoff coefficients.

For a catchment with a present runoff ratio of 0.20, the effect of a 10% reduction in precipitation may range from a 50% reduction in runoff with no direct CO_2 effect, to a 70% increase in runoff with a maximum direct CO_2 effect. For higher runoff ratios the ranges of possible runoff changes is much less.

The increases in evaporation due to global warming up would be much less than the maximum direct CO_2 effects on a global scale. However, there are many unknowns. There will, of course, be large seasonal and regional departures from the global mean.There are also other factors, for example, the seasonal distribution of precipitation may be different from that of evapotranspiration. Direct CO_2 effects may also be reduced due to increased leaf temperatures, in individual plant leaf areas or in the total vegetated area of a catchment.

When the CO_2 content is less than 280 ppm it is probably the limiting factor on plant growth. For every increase of 10 ppm above that level plant growth is stimulated by between 0.5% and 2% depending on the species. Also when more CO_2 is present, plants use water more efficiently in photosynthesis and do not require so much rainfall. More efficient use of water can also cause problems, for example, increased runoff, increased soil erosion, etc.

For rainfall runoff relationships the important geophysical factors are: the nature of geological formation (crystalline or sedimentry); the density of the natural vegetation, the nature of the drainage pattern and the formation of floodplains. Floodplains have high infiltration losses and evapotranspiration losses, which reduce streamflow.

Besides the drainage area and the slopes of the catchment, the influence of the soil vegetation complex, affecting the rainfall runoff transformation, is preponderant. For soils, the clay content seems the most important and as regards the vegetation the percentage of crops in the catchment.

The role played by geophysical factors in determining the runoff coefficient shows that a separate analysis is needed for the arid and semi-arid regions which later are affected by impervious soil surface crusts and hydrographic degradation.

The base and rise times of the hydrographs are much longer in the tropical and forest zones than in the semi and arid zones by:

(a) the attenuating effect of vegetation;

(b) in the semi-arid zones floods are produced by surface runoff alone while in tropical zones floods are, to a considerable extent, generated by sub-surface runoff. In forest zones sub-surface runoff predominates.

The influence of cultivation following the clearing of natural vegetation is as follows:

(1) where mean annual rainfall is less than 650-700 mm, crops reduce mean annual runoff, probably because they consume more water

than the herbaceous stratum.

(2) where rainfall exceeds 650-700 mm, the reverse is true, either because crops consume less water than the tree stratum or because soil exposure promotes runoff.

As variations in climate change the vegetational possibilities in a particular geographic area, changes occur in agricultural methods and extensions or dimunition of cultivated areas. These changes should also be accounted for.

In a forest, any cultivation leads to a considerable increase in runoff. For natural grasslands, subjected to burning every two years and consequently lacking dense plant cover, any modification leads to a decrease in the annual runoff. Finally the installation of anti-erosive structures followed by contour cultivation reduces runoff to a greater extent than traditional cultivation or reafforestation.

Impact of ecological change on hydrology

Table 2 was established assuming that the soil-vegetation complex, the relative distribution of wet and dry seasons, water use and cultivation patterns would not change when the earth warmed up. This, assumption, in the long run, will certainly not be correct.

The following table is an attempt to bring together in a comprehensive form the information described above. The table will certainly be proved wrong as the present knowledge is incomplete and probably partly eroneous. In order to simplify the table, the cold regions and the truly arid regions are left out.

Table 3 Possible relative impacts of vegetation and cultivation changes on the impact figures for runoff of Table 2. The indications in this figure have no real scientific basis

Present situation	Possible future situation	Changes in vegetation cover	Relative impact on runoff of vegetation	Relative impact on runoff of cultivation changes	Corrected runoff changes in %
Humid	more humid	nc	nc	nc	+20
	more dry	-	+	-	less than - 30%
Sub-humid	more humid	+	-	+	more than + 10%
	more dry	-	+	-	less than -40%
Semi-arid	more humid	+	-	+	less than +10%
	more dry	-	+	+	more than -50%

The information in this table, however wrong it may be, is based on the assumption that an increase in CO_2 will not have a direct effect on the water use efficiency of the vegetation. If this is the case evapotranspiration will be reduced. The direct effect of doubling of CO_2 would be most pronounced in semi-arid regions and to a lesser extent in sub-humid regions. The effect would be to increase annual runoff whether the precipitation increases or not. The maximum effect would occur in areas with increased humidity, the runoff may double.

Impacts on river morphology

The variables determining river morphology are: geology, paleoclimatology, relief, valley dimensions, climate, vegetation, hydrology, channel morphology, water discharge, sediment discharge and flow hydraulics. These variables are not, of course, independent of each other they are listed in this manner to facilitate description.

For engineering purposes, for example for hydrologic engineering, we are interested in the water discharge and the sediment discharge. The dependent variables are the observed water and sediment discharges and the hydraulics of flow. All the other variables are independent variables.

For engineering purposes we look at a timescale of weeks to tens of years and, in this case, we do not usually take into account possible changes in vegetation cover. When looking at the impact of climate change on river morphology we must consider larger timescales, say around 100 to 200 years.

When we want to consider the impact of the climate change on river morphology, the channel morphology becomes the independent variable. The observed water and sediment discharges and the hydraulics of flow become indeterminate variables, and all the others independent variables.

Eventual climate change directly influences the vegetation and the hydrology of the basin, which in turn influences the channel morphology, which influences valley dimensions, which influences relief. Relief again influences the hydrology of the basin, etc.

In a natural stream, over longer periods of time, mean water and sediment discharge are independent variables which determine the morphologic characteristics of the stream and, therefore, the flow characteristics.

A major change in the hydrologic regime would trigger a response that would completely change channel morphology. Channel morphology reflects a complex series of independent variables, but the discharge of water and sediment integrates most of the other independent variables; it is the nature and quantity of sediment and water moving through the channel that largely determines the morphology of stable alluvial channels.

A decrease in precipitation in the head waters will not only cause a decrease in annual discharge, but through reduction of vegetation density, it will increase peak discharge and greatly increase the amount of sand load with less water, the channel will become wider and shallower. Many of the wide "unstable" rivers of the world could be transformed into the stable channels by reducing flood peaks and bedload transport.

The shape of the channels is closely related to the percentage of silt and clay in the sediments forming the banks and the bed of the channel. Relatively wide and shallow channels contain only small percentages of silt and clay. Narrow and deep channels contain large percentages of silt and clay.

For semi-arid to sub-humid regions around 90% of the variability of the channel width and 80% of channel depth can be explained by mean annual discharge and type of sediment load. Only about 40% of the variability of channel dimensions can be accounted for by discharge alone.

There does not seem to be much relation between size of sediment and channel dimensions but when an index of the type of sediment load is combined with discharge, good correlations with width and depth were often obtained.

There is a correlation between water and sediment discharge on the one hand and channel width, depth, meander wave length, the width-depth ratio, sinuosity and the slope of the bed on the other hand. Climate change will either change discharge of water and sediment or not. If there are no changes, there will be no changes in morphology. If there are changes, then we may find four cases: the case where both water and sediment discharge increases or decreases and the two cases where one increases and the other one decreases.

Professor Schumm prepared a table giving the consequences of such changes (on a long time scale, a few hundred years);

Q	Qs //	b	d	la	S	P	F
+	+	+	+/−	+	+/−	−	+
−	−	−	+/−	−	+/−	+	−
+	−	+/−	+	+/−	+	−	−
−	+	+/−	−	+/−	+	−	+

b= width of the channel
d= depth of the channel
la=meander wave length
S= gradient
P= sinuosity expressed as ratio of channel length to valley length
f= width-depth ratio

Earlier, Table 3 took a hypothetical look at possible changes in runoff due to global warming. We did not discuss possible changes in erosion.

Table 4 hereunder gives possible changes in both water and sediment discharges.

When we combine the information in professor Schumm's table and the information in Table 4 we get Table 5.

The conclusions we can draw from this table are:

We have seen from Table 4 that the greatest changes in water and sediment discharge are likely to occur in the regions between the, at present, humid and semi arid regions. If these regions become more humid, water discharge will increase considerably and sediment discharge will in the long run become less than at present. The result will be that the depth of channels will increase, the gradient

Table 4 Possible hypothetical changes in both water
and sediment discharge due to global warming up

Present situation	Possible future situation	Possible changes in runoff(Qw) (n%)	Possible changes in sediment discharge (Qs)
Humid	more humid	++	+
	more dry	-	++
Sub-humid	more humid	++	-
	more dry	-	++
Semi-arid	more humid	+	+
	more dry	-	++

Table 5 Possible hypothetical changes in river
basin morphology due to global warming

Possible future situation	b width of channel	d depth of channel	la meander wave length	S gradient (geol)	P sinu- osity	F width/ depth ratio
Humid more humid	+	+/-	+	+/-	-	+
more dry	+/-	-	+/-	+	-	+
Sub-humid more humid	+/-	+	+/-	-	+	-
more dry	+/-	-	+/-	+	+	+
Semi-arid more humid	+	+/-	+	+/-	-	+
more dry	+/-	-	+/-	+	-	+

will decrease and the sinuosity will increase.

On the other hand, if precipitation in the present sub-humid
region decreases, water discharge will decrease and sediment
discharge will increase considerably. The result will be that the
depth of channels will decrease, the gradient will increase and the
sinuosity decrease.

The warming up of the earth due to CO_2 doubling and the increase
in concentration of other man-made gases, will come about gradually
and the impacts will also become visible gradually. Table 5 gives the
directions of the change between the present and the final situation.
In between there will be other important changes not reflected in the
table. For example, when a semi-arid basin becomes more humid storms
may increase in intensity before the vegetation has time to develop.

During this period the erosion and sediment discharge will be greater
than at present although as indicated above in the long run it will
be less than at present.

References

Abbott, M.B., Bathurst, J.C., O'Conell, P.E., and Rasmussen,
    J.(1986) an introduction to the European Hydrological System-
    Systeme Hidrologique Europeen, "SHE", Structure of physically-
    based, distributed modelling system; Journal of Hydrology, 87:61-
    77.
Dubreuil, P.; review of relationships between geophysical factors and
    hydrological characteristics in the tropics; Journal of Hydrology
    87 (1986) 201-222.
Hare, F. Kenneth (1985) Climate variations, drought and desertifica-
    tion WMO publication no. 653.
Manabe, S. and Wetheral, R.T.; Reduction in summer soil wetness
    induced by an increase in atmospheric carbon dioxide; Science Vol.
    232, 1986.
Martin L. Parry (Editor): the sensitivity of natural ecosystems and
    agriculture change. IIASA and UNEP, 1985, reprinted from climatic
    Change, Vol. 7.
Nemec, J. and Schaake, J.; Sensitivity of water resources systems to
    climate variation; Hydrological Sciences Journal, 27, 3, 9/1982.
Schumm, S.A. (1971) Fluvial Geomorphology: Historical Perspective,
    and Channel Adjustment and River Metamorphosis, in "River Mehanics"
    edited and published by Hseieh Wen Shen.
Unesco (1978) World Water Balance and Water Resources of the Earth,
    prepared by the USSR Committee for the Intergovernmental Hydrolo-
    gical Programme, Unesco series Studies and Reports in Hydrology,
    no.25.
Wigley, T.M.L. and Jones, P.D.J.; influences of precipitation changes
    and direct CO_2 effects on streamflow; Nature, vol. 314 of 14 March
    1985.

*The author is employed by Unesco and active in the division of Water
Sciences. Unesco takes no responsibility for the opinions expressed
in this discussion.

The Influence of Climate Change and Climatic Variability on the Hydrologic Regime and Water Resources (Proceedings of the Vancouver Symposium, August 1987). IAHS Publ. no. 168, 1987.

The impacts of alternating flood- and drought-dominated regimes on channel morphology at Penrith, New South Wales, Australia

Robin F. Warner
University of Sydney
N.S.W., Australia, 2006

ABSTRACT   Long flood records for the Hawkesbury-Nepean River have been used to define alternating flood-(FDRs) and drought-dominated regimes (DDRs).   In the former, flood magnitudes and frequencies are higher with mean annual flood discharges ($Q_{2.33}$) from 2 to 4 times greater than for the latter.   If channel size is assumed to be related to some level(s) of discharge, then such regime variations create the potential for channel change or adjustments.   Surveys since 1863 at Penrith provide some insight into adjustments related to regime changes.   In FDRs channel widths generally increase and depths decrease, while in DDRs, these changes are reversed.   Human impacts in and beyond the channel modify the nature of adjustment or response.

Impact des régimes dominés par l'alternance d'inondation et de sécheresse sur la morphologie des chenaux à Penrith, Nouvelle-Galles du sud Australie

RESUME   Les dossiers concernant les longues inondations du fleuve Hawkesbury-Nepean ont été utilisés afin de définir les régimes dominés par l'alternance d'inondation (FDRs) et de sécheresse (DDRs).   Dans le premier cas, les amplitudes et les fréquences d'inondation sont plus élévées, la moyenne annuelle d'inondation ($Q_{2.33}$) etant de 2 à 4 fois plus élevée que dans le second cas. Si l'on assume que la largeur du fleuve est reliée à certains, niveaux de décharge, alors de telles variations dans le régime provoquent la potentiel de changements et d'adaptations fluviaux. Les relevés à Penrith depuis 1863 fournissent un moyen de comprendre les adaptations reliées aux changements de régime. En général dans le cas des FDRs en general, la largeur du fleuve augmente et sa profondeur diminue, alors que dans le cas des DDRs, ces changements sont inversés. La nature de l'adaptation ou de la réaction est modifiée par l'impact humain a l'intérieur et au-dela du chenal.

Introduction

Coastal southeast Australia receives precipitation at any time of the
year.   Winter rains of the south merge with summer-dominant rains of
the north. However, totals are variable and irregular, and consequently flow regimes are very flashy.   Superimposed on these highly
variable annual conditions are periods of higher and lower mean
rainfall lasting up to five decades.   In wetter times, flood
magnitudes and frequencies are very much higher than in drier periods.
The former have been called flood-dominated regimes (FDRs) (Hickin
1983) and the latter, drought-dominated regimes (DDRs) (Erskine
pers. comm.).
    On the Hawkesbury River, west of Sydney, flood-stage records have
been collected at Windsor since 1799.   Upstream, where this river is
called the Nepean, it has been gauged at Penrith since 1891 (Figure
1).   From these records it has been possible to define alternating
FDRs and DDRs at least since European settlement in 1788. Additionally, since this was one of the first areas settled outside of Sydney,
there are several old surveys of the river which allow some insight
into channel changes over the last 120 years.   The major aim of this
paper is to examine these changes and their possible relations with
flood- and drought-dominated regimes.
    After describing the study area, evidence for regime changes is
reviewed and likely impacts on channels are discussed.   There follows
a description of the changes derived from old and new surveys in the
study reach.   These are then discussed in terms both of adjustments
to regime variations and of human impacts in the catchment and
channel.

The area

The Nepean River at Penrith drains 11,000 sq.km of mainly sandstone
uplands southwest of Sydney (Figure 1c).   At Windsor, the drainage-
basin area has increased to about 13,000 sq.km. In this paper flood
records have been used from both sites but most of the observations
on channel changes have been made on the 5 km reach upstream of the
Penrith weir (Figure 1b).   Much of this channel is inset in alluvium
and a Pleistocene terrace, whilst the upper part towards Glenbrook
delta is in a deep sandstone gorge.   The modern alluvia are narrow
zones flanking the older terrace and these have been subject to many
changes in the last 120 years.   These recent materials form narrow
benches in places (incipient flood plains) but the main bank tops in
this "alluvial gorge" are above the 100 year flood level. Consequently, in spite of ponding by the weir, and the flattening of gradients,
most high-event energy is expended within the channel, rather than
overbank.   Net changes associated with intervening events can be
established by cross-section surveys through time.   These began at
the railway bridge in 1863.
    Human impacts affecting this reach include: some clearance
mainly of flood plains and adjacent shalelands of the Cumberland
Plains, some urbanization (including the western parts of the Sydney
metropolitan area), four small water-supply dams on the upper Nepean
(affecting 1,700 sq.km), the large nearby Warragamba Dam (8,500

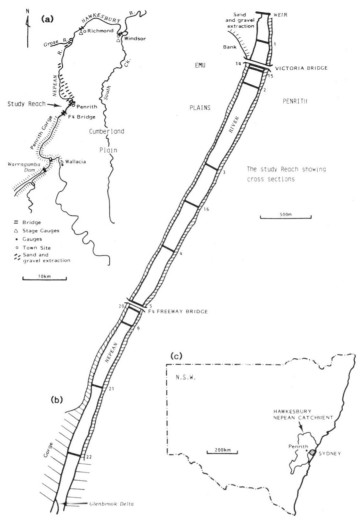

Figure 1

sq.km), the weir (1910), some recent channel-bank improvement works, and large-scale sand and gravel extraction from the channel, flood plains and terraces for 20 km downstream of Penrith. The Glenbrook delta (Figure 1b) now occupies more than half the gorge floor. A May 1944 flood removed about 250,000 m^3 of sandstone blocks from the lower Glenbrook gorge and deposited them in the Nepean. Much of the material had been derived from a railway cutting blasted out from the north side of the gorge. The delta now behaves like a partial weir interrupting the movement of bed load. Boulders from upstream have been trapped and scour has occurred downstream (Warner 1984).

The flood record

Flows above the 6 m stage A.H.D. (Australian Height Datum) are available for Windsor from 1799 (Riley 1980), the first big flood

(15.4 m) after colonization of N.S.W. in 1788. In the next 21 years
(part of a FDR?), there were 11 floods above the 10 m stage (a large
flood) (Table 1) and three above 6 m (a low flood). These caused
much damage and some farming activities were relocated to South Creek
(Jeans 1972).

Table 1 Windsor: floods over 6m and 10m and frequencies

Period	Regime	6m	10m	Years	6m yr^{-1}	10m yr^{-1}
1799–1819	FDR	14	11	21	0.7	0.5
1820–1863	DDR	15	7	44	0.3	0.2
1864–1900	FDR	43	14	37	1.2	0.4
1901–1948	DDR	24	4	48	0.5	0.1
1949–1978	FDR	48	9	30	1.6	0.3
TOTAL		143	45	180	0.8	<0.3

Source: Riley (1980: Table 1)

A DDR followed from 1820 to 1863 when there were only seven 10m
floods in 44 years (Table 1) and the mean stage dropped from 13.0 to
9.6m. In the next 37 years (FDR), there were fourteen 10m and twenty-
nine 6m floods, including the highest at 19.42m in 1867. The large
number of 6m floods helped to reduce the mean stage to 9.5m.
    In the following DDR (1901–1948), when the mean stage was only
8.1m, only four 10m and twenty 6m floods were recorded in 49 years
(Riley 1980). Since 1949, another FDR has had a mean stage of 9.1m,
with nine 10m and thirty-nine 6m floods in only 30 years (Table 1).
    The high mean stages up to 1863 were thought by Riley (1980) to be
associated with a different climate. However it is conceivable that
not all 6m floods were recorded, that the main channel had smaller
dimensions and that higher overbank roughness (before total deforestation of the extensive flood plains) had something to do with higher
stages. A pre-1872 map at Windsor ferry supports the idea of a
smaller channel.
    Based on annual series analysis, the most probable ($Q_{1.58}$), the
mean annual flood ($Q_{2.33}$) (Dury 1969) and the five-year flood (Q_5)
levels do not show such a wide variation (Table 2). In FDRs, the

Table 2 Windsor: stages for $Q_{1.58}$, $Q_{2.33}$ and Q_5 (m)

Period	Annual Series			Partial Duration Series		
	$Q_{1.58}$	$Q_{2.33}$	Q_5	$Q_{1.58}$	$Q_{2.33}$	Q_5
1799–1819	N.A.	9.3	14.6	6.3	14.0	14.5
1820–1863	N.A.	<6.3	6.3	N.A.	<6.3	8.9
1864–1900	7.1	9.1	11.8	8.3	9.5	11.9
1901–1948	N.A.	6.3	7.8	<6.3	6.6	7.8
1949–1978	6.3	9.4	11.7	9.4	9.7	11.6

$Q_{1.58}$ most probable mean annual flood; $Q_{2.33}$ mean annual
flood; Q_5 one in five-year flood. N.A. not available.

Source: Riley (1980: Table 1)

$Q_{2.33}$ stage is 9.1 to 9.4m, while in DDRs they are at least 3m lower. The Q_5 in the first FDR was 14.6m, repeating perhaps the conditions described above, while in later FDRs it was 11.8 and 11.7m.

Partial duration series analysis increases $Q_{2.33}$ in FDR regimes by 4.7, 0.4 and 0.3m respectively. $Q_{1.58}$ values are also greatly increased (Table 2).

The flood record at Penrith began in 1891. In this case, the three later regimes can be confirmed with discharge data (Table 3). However, the effects of the 2000 mill m^3 storage at Warragamba have attenuated flood peaks since 1960.

Table 3 Penrith: discharges for $Q_{1.58}$, $Q_{2.33}$ and Q_5 from annual series (m^{3}s^{-1})

Period	$Q_{1.58}$	$Q_{2.33}$	Q_5
1891-1900	2800	4900	6500
1901-1948	400	1100	2600
1949-1959	900	2900	7000 (pre Warragamba Dam)
1949-1978	700	2000	6000

Source: Water Resources Commission Data.

The same pattern is evident for Wallacia (Pickup 1976a) and for the Warragamba gauge (Figure 1a). However backwater effects of the Nepean have made this difficult to rate. Windsor stages are also influenced by backwater effects from the Colo River, by tidal conditions, as well as by variable inputs from Penrith and from the Grose River (Figure 1a).

Potential effects of regime variations on channel dimensions

If channel-forming discharges are related to some measure of discharge such as most probable ($Q_{1.58}$) or mean annual flood ($Q_{2.33}$) (Dury 1969, Leopold et al 1964), or even to both most effective (Q_{me}) and bankfull discharges (Q_{4-10}) (Pickup and Warner 1976), then changes in these discharges must have some impact on the channel. From the FDRs to DDRs there is a marked reduction in frequency (Table 1) and a big increase in the other direction. Magnitude changes for $Q_{1.58}$, $Q_{2.33}$ and Q_5 have also been considerable (Tables 2 and 3).

Increases in both water (Q^+) and sediment discharge (Q_s^+) should according to Schumm (1971) cause width increases (w^+) and depth decreases or increases ($d^\pm$), but normally the former. Decreases in water and sediment discharges would involve w^- $d^\pm$. Where sediment load is decreased ($Q^+Q_s^-$), as with conservation, interruption to sediment movement by dams and weirs, and the later stages of urbanization, depths would be expected to increase (Schumm 1971).

No details of the 1799 channel have been found. It may have been a mixed-load stream (Schumm 1977) in 1872 because gravels and sand were recorded in the first Windsor bridge survey. In an undated earlier ferry map, seven cross sections over 300m revealed maximum depth variations of 3.7 to 10.7m, hardly characteristic of a wholly sand-bed river. By 1891, well into a FDR, the maximum depth here was only 3m and the channel was wider (w^+d^-).

FDRs are characterised by frequent and larger floods, where bank erosion and bed accretion seem to be the normal responses. DDRs are quieter where bank recovery and depth increases are more common. So far most observations have been based on accelerated bank erosion from the late 1940's onwards (DDR-FDR) (see later discussion). Bank erosion is common, pools have filled in and channels are very different to those depicted in 1940s air photographs.

In DDRs, which have yet to be observed at close hand, channels are subject to lower channel-forming discharges. Bank recovery takes place, low benches are added to the channel floor and the confined flow helps to increase depths. At the end of the FDR regime in 1900, the Nepean was very wide and shallow. By 1949 most widths had decreased, some by up to 70m.

The surveys

It is evident from frequent surveys that there are fluctuations in dimensions rather than single trends, for example, recent surveys at Penrith and bridge-profile data at Windsor (Table 4). Here widths were increasing, while depths decreased and then increased by 1891. Data for that date may be anomalous because an 1891 hydrographic survey shows very shallow depths in that area.

Table 4 Cross section dimension data for the Windsor bridge

Date	Ref level	w	A	d_m	d	Remarks
<1872?	LWM	94	388	6.4	4.1	CS4 undated survey
<1872?	mid tide	96	450	7.1	4.7	CS4 undated survey
1872	AHD	98	375	4.9	3.8	(if deck = 4.3m)
1875?	AHD	107	230	3.8	2.2	
1891	AHD	128	350	4.7	2.7	

LWM low water mark; AHD Australian Height Datum; w width; A cross-section area; d_m maximum depth and d mean depth; CS cross section.

Source: N.S.W. Department of Main Roads maps and plans.

For the low flood stage (6m) only the data shown in Table 5 are available. All these changes were in a FDR (1864-1900). The date of the first map is unknown; it may have predated the 1867 flood. The 1872 survey was 9 years into the regime after seven 10m and eight 6m floods. The post-1875 survey is also undated but it shows the 1875 flood level (11.8m) but not the 1879 (13.1m). Thus it is assumed to predate the latter event. In the 7 year interval, there were ten 6m and only two 10m floods. Prior to 1891, the final survey in Tables 4 and 5, there were another thirteen 6m and six 10m floods.

This evidence shows that in-channel changes did occur within this regime. Changes above 6m are unknown but may be inferred from land

Table 5 Windsor bridge dimensions at 6m stage

Date	w	A	d_m	d	Remarks
1872	183	1130	10.9	6.2	some extrapolation of left bank data
>1875	181	1040	9.8	5.2	-8% area
1891	181	1217	10.7	6.7	+ 17% area

Source: N.S.W. Department of Main Roads plans.

clearance of flood-plain levels. Thus variations in stage have not only involved discharge, as suggested by Riley (1980), but also channel and flood-plain changes.

At Penrith, in the 5.5 km reach from weir to delta, many more surveys are available for comparison (Table 6). Originally this reach had a fairly flat gradient below the Warragamba-Nepean conflue-nce, now enhanced even more by the weir. Deep holes in the sandstone gorge gave way to a more uniform bed downstream. Overlaps in surveys provide the best data on time-based changes but these are not common except at the two bridge sites (Tables 7 and 8).

Table 6 Penrith surveys (Figure 1b for locations)

Location	FDR	DDR	FDR	DDR	FDR	
Rail br = 14 = gauge	-	1863	-	-	1976 1980	1985
46 cross sections	-	-	1900	1949(w)	1982/3	
Cross Sections 1-6	-	-	-	1938(4)	1964 1984	1986
Cross Sections 14-16 20-22	-	-	-	-	1980 1985	

1900 cross sections not shown on Figure 1b

Sources: State Government plans from Public Works Department, Metropolitan Water, Sewerage and Drainage Board and Water Resources Commission. Sydney University Geography Department: 1949 and 1970 (Widths only), 1982/83, 1984, 1985 and 1986.

Table 7 Victoria bridge - channel changes

		At CTF			At BF (24.7m)			Cross section	
Date	w	A	d_m	d	w	A	d_m	d	
1863	156	183	2.7	1.2	239	1971	12.3	8.3	Rail Survey
1900	201				-				Map 1900
1949	160								Air Photo
1976	201	586	4.6	2.9	242	2708	14.2	11.2	Gauge
1980	205	665	5.2	3.2	250	2666	14.8	10.7	14 WC+IC
1985	191	723	5.5	3.7	249	2832	15.2	11.4	14 SUGD

CTF cease to flow at the weir; BF near bankfull at 24.7 m AHD. WC + IC Water Conservation and Irrigation Commission.

Source: State Authorities, air photo analysis and survey.

Table 8   F4 freeway bridge – channel changes

Date	w	at CTF A	d_m	d	w	A	BF d_m	d	Cross Section
1900	173	348	3.5	2.0					42
1949	122	–	–	–					42
1964	144	476	3.8	3.5	244	2278	13.4	9.3	5 at 24.7m
1980	146	675	6.6	4.6	290	2808	17.5	9.7	20 at 26.0m
1984	146	733	6.0	5.0					5 at 24.7m
1985	143	743	7.0	5.2	245	2710	17.9	11.1	20 at 26.0m
1986	146	727	6.0	5.0	241	2462	15.6	10.2	5 at 24.7m

The first survey for the Victoria Bridge was at the end of a DDR in 1863. By the end of the following FDR, the width had been increased by 45m (1900) and then reduced by about 40m in the next DDR (Table 7). By 1976 in the present FDR, this loss had been recovered. Between 1863 and 1976, there had been an increase in capacity of 220% to weir crest level and of 37% to bankfull. Thereafter increases in capacity probably reflect the sediment deficiencies imposed by Warragamba Dam. Recent width reduction is due to bank-protection works.

Cross sections from 1900 (42), 1964 (20) and 1980 (5) are within 80m of each other and near the F4 bridge, built in the early 1970's (Table 8). In 1900 the channel was wide and shallow but at the end of the last DDR (1949), the width had been reduced by about 50m. In the present regime it has been stabilized at 144-146 m by bank works. The big increase in capacity after 1964 is probably due to sediment deficiencies imposed by the dam and delta (Table 9: Figure 1).

The other cross sections are of limited value with only 4 extending back to 1938. They tend to indicate "noise" induced by other factors, as is shown by width and mean depth changes in Table 9.

Table 9   Changes in width and mean depth

Cross Section	1938-1964	1964-1984	1984-1986
1	–	w-(14) d-(0.4)	w-(9) d+(0.1)
2	w+(5) d+(0.7)	w-(6) d-(0.1)	w-(4) d+(0.1)
3	w-(5) d+(0.6)	w+(8) d-(0.3)	w-(1) d+(0.2)
4	w-(5) d+(0.2)	w-(2) d+(0.6)	w-(1) d-(0.1)
6	w-(5) d+(0.2)	w-(26) d+(1.0)	w-(1) d-(0.2)
	1980-1985		
15	w-(10) d+(0.2)		
16	w+(3) d+(0.2)		
21	w+(13) d+(0.4)		
22	w-(3) d-(0.2)		

w width; d mean depth. (4) (0.2) width and depth changes in m.

Between 1938 and 1964, width should have been increasing and depths decreasing. The reverse indicates that sediment deficiencies

were already apparent.   In the next 21 years widths continued mainly
to decrease with depths nearest the weir decreasing and those
upstream increasing greatly (1.0m at C.S.6).   After the 1986 flood,
changes were minor except near the weir, where widths decreased.   In
the period 1980 to 1985, width changes were variable, while depths
increased downstream from the delta accretion (C.S.22). These changes
indicate that general tendencies are masked by other impacts, like
closeness to weir (1), bank protection works (2, 4 and 6), scour
below the delta (6 and 21), and accretion near delta (22), as well as
the general effects of the dam.

Discussion

The secular change in climate in the late 1940's has been well
documented (Pittock 1975; Cornish 1977), and the geomorphic effects
of this change to a FDR were initially studied in the Upper Nepean by
Pickup (1976a and 1976b).   Its impacts elsewhere have received much
attention (Henry 1977; Bell and Erskine 1981; Erskine and Bell 1982;
Erskine and Melville 1983; Riley 1981; Warner 1983 and 1985; Warner
and Paterson 1985).
     The longer records have only been used more recently to define
variable regimes and their consequences (Warner 1984 and 1986). These
have imposed longer-term adjustments, which have been superimposed on
the effects of short-term variations. Built into these longer changes
are lagged responses based on variable thresholds in both bed and
bank materials.
     Where rivers are in gorges or source zones (Pickup 1984 and 1980),
adjustments are minimal. Where they flow in armoured bed zones, major
responses are infrequent even in FDRs, if thresholds for movement are
related to low frequency events (Graf 1983).   However, in transition
and mobile zones, some or much of the bed load is moved more frequently and bed adjustments may be rapid.   In backwater zones or sediment
sinks (Pickup 1984 and 1986), low gradients and more cohesive
perimeters again slow down changes (Warner 1986; Warner and Paterson
1985).   Where beds are armoured and banks cohesive or well protected
by vegetation, dimensional changes may also be slow, or in some cases
they may occur overbank, involving flood-plain stripping (Nanson
1985).   Then there is essentially a two-stage channel: an inner,
smaller form conforming to lower flows (possibly the $Q_{2.33}$ of a DDR),
and a larger channel flanked by high eroded alluvial banks, relating
to lower frequency flows of the FDR.   Where bank vegetation fails,
there are rapid increases in width, especially where much of the
energy is confined to the channel. This is in marked contrast to the
smaller cohesive channels flanked by wide flood plains in backwater
zones.   Here flood-water ponding is common in low energy environments
and there is no stripping in FDRs, only a greater incidence of
flooding and slow adjustments in the main channel (Nanson and Young
1981; Warner and Paterson 1985).
     In the study reach, the banks are fairly cohesive while the bed
varies from armoured to mobile.   At the end of the FDR in 1900 the
channel was wide, shallow with numerous sand shoals and mobile.   By
1949, after a DDR, it was much narrower. When surveys began in 1982,
it was evident that a boulder bed existed in several parts, especial-

ly approaching the weir. Armouring continues below the weir where large volumes of sand and gravel have been removed.

The banks are a combination of several types of alluvium. There is even sandstone in the bed and left bank upstream of the F4 bridge. Well weathered Pleistocene gravels outcrop on the right bank near the Victoria bridge. In general the banks are fairly cohesive, with failure in amphitheatre-like scars, seemingly related to basal weakness in layers of wet, sandy alluvium. Benches often occur below these forms. Water level erosion involves tree falls and sandy bench trimming. Dense native riparian and exotic vegetation offers some protection to banks which extend up to 15m above low-water stage. Some bench surfaces equate to Q_{2-3} levels and might be viewed as incipient flood plains; others "relate" to less frequent flow stages, while the bank tops are at or above Q_{100} levels. Banks are being progressively engineered with grassed, graded slopes, rock-fill bases and gabions. This is to provide protection for residential areas on the bank tops, safer viewing areas for spectators at rowing competitions, and recreational parklands.

Conclusions

One hundred and eighty years of flood stage records at Windsor and 90 years of discharge data at Penrith have allowed alternating FDRs and DDRs to be defined. Two to fourfold shifts in the magnitude of assumed channel-forming discharges have great potential impacts on channel perimeters.

Differences in $Q_{2.33}$ stages at Windsor involve about 3m between FDRs and DDRs. However, frequencies and therefore energy to affect changes are much higher in FDRs. For instance, 6m floods occur 0.7 to 1.6 times per year and 10m floods every 2 to 3 years. In DDRs, their incidence is 2 to 3 years and 5 to 10 years respectively.

Their impacts on channels can be confirmed in part by surveys and air photograph analysis, particularly in the Penrith reach where there are several former surveys. In FDRs channels adjust normally by increasing widths and decreasing depths, while in DDRs, these trends are reversed. However, added complications can occur. These are due to drainage-basin modification, like land-use changes and urbanization, and to the effects of dams, weirs, bank-protection works, sand and gravel extraction, and the Glenbrook delta.

Recognition of these changes, their relations to alternating regimes, and the role of banks (even flood-plain surfaces in some instances) as dynamic buffer zones are important in the management of this river, particularly when contemplating future sand and gravel extraction and costly bank rehabilitation and stabilization works.

ACKNOWLEDGEMENTS   The help of Geography Students and technical staff from 1981 to 1986 is gratefully acknowledged. Staff of the Water Resources Commission, Metropolitan Water, Sewerage and Drainage Board, Department of Public Works and Department of Main Roads were very helpful in the search for old surveys.

References

Bell, F.C. and Erskine, W.D. (1981) Effects of recent increased rainfall on floods and runoff in the upper Hunter Valley. Search, 12, 82-83.

Cornish, P.M. (1977) Changes in seasonal and annual rainfall in New South Wales. Search, 8, 38-40.

Dury, G.H. (1969) Hydraulic geometry. In Water Earth and Man (ed. by R.J. Chorley), 319-330. Methuen, London, U.K.

Erskine, W.D. and Bell, F.C. (1982) Rainfall, floods and river channel changes in the upper Hunter. Aust. Geog. Studies, 20, 183-196.

Erskine, W.D. and Melville, M.D. (1983) Impact of the 1978 flood on the channel and flood plain of the Lower MacDonald River, N.S.W. Aust. Geogr., 15, 284-292.

Graf, W.L. (1983) Flood-related channel change in an arid-region river. Earth Surf. Processes and Landforms, 8, 125-139.

Henry, H.M. (1977) Catastrophic channel changes in the MacDonald Valley, New South Wales. J. Roy, Soc. N.S.W., 11, 1-16.

Hickin, E.J. (1983) River channel changes: retrospect and prospect. Spec. Publ. Inter. Assoc. Sedimentol., 6, 61-83.

Jeans, D.N. (1972) An Historical Geography of New South Wales, to 1901. Reed Educ., Sydney, Aust.

Leopold, L.B., Wolman, M.G. and Miller, J.P. (1964) Fluvial Processes in Geomorphology. Freeman, San Francisco, California, U.S.A.

Nanson, G.C. (1985) Cycles of floodplain stripping and reconstruction along coastal rivers of New South Wales. Geol. Soc. Aust. Abs., 13, 17-19.

Nanson, G.C. and Young R.W. (1981) Overbank deposition and floodplain formation on small coastal streams of New South Wales. Z. Geomorphl, 25, 332-347.

Pickup, G. (1976a) Geomorphic effects of changes in runoff, Cumberland Basin, N.S.W. Aust. Geogr., 13, 188-193.

Pickup, G. (1976b) Adjustment of stream channel shape to hydrologic regime. J. Hydrol., 30, 365-373.

Pickup, G. (1984) Geomorphology of tropical rivers: I landforms, hydrology and sedimentation in the Fly and lower Purari, Papua New Guinea. Catena Supp., 5, 1-17.

Pickup, G. (1986) Fluvial landforms. In. Australia - a Geography (ed. by D.N. Jeans), 148-179. Syd. U. Press, Sydney, Aust.

Pickup, G. and Warner, R.F. (1976) Effects of hydrologic regime on magnitude and frequency of dominant discharge. J. Hydrology., 29, 51-75.

Pittock, A.B. (1975) Climatic change and patterns of variation in Australian rainfall. Search, 6, 498-504.

Riley, S.J. (1980) Aspects of the flood record at Windsor. Proc. 16th Inst. Aust. Geogr. Conf. Newcastle, 325-340.

Riley, S.H. (1981) The relative influence of dams and secular climatic change on downstream flooding, Australia. Wat. Resour. Bull., 17, 361-366.

Schumm, S.A. (1971) Fluvial geomorphology: channel adjustment and river metamorphosis. In. River Mechanics Vol. 1, (ed. by H.W. Shen), 5.1-5.22. H.W. Shen, Fort Collins, Colorado, U.S.A.

Schumm, S.A. (1977) The Fluvial System. Wiley, New York, U.S.A.

Warner, R.F. (1983) Channel changes in the sandstone and shale reaches of the Nepean River, New South Wales. In Aspects of Australian Sandstone Landscapes (ed. by R.W. Young & G.C. Nanson) 106-119. ANZ. Geom. Group. Spec. Publ. No.1.

Warner, R.F. (1984) Impacts of dams, weirs, dredging and climatic change: an example from the Nepean River, NSW, Australia. Paper at 25th IGU Congress, Paris.

Warner, R.F. (1985) Downstream variations in channel morphology in the Bellinger valley, New South Wales. Geol. Soc. Aust. Abs., 13, 22-24.

Warner, R.F. (1986) Spatial adjustments to temporal variations in flood regime in some Australian rivers. In: Rivers: Environment Process and Morphology, (ed. by K.S. Richards), (in press). Inst. Brit. Geog. Spec. Publ.

Warner, R.F. and Paterson, K.W. (1985) Bank Erosion in the Bellinger Valley, NSW: definition and management. Paper at IAG Conf., Brisbane.

The Influence of Climate Change and Climatic Variability on the Hydrologic Regime and Water Resources (Proceedings of the Vancouver Symposium, August 1987). IAHS Publ. no. 168 1987

Is the largest North American sub-arctic sand dune disappearing?

Jeff Whiting & Elaine Wheaton
Sask. Research Council
Saskatoon, Canada
S7N 2X8

ABSTRACT   The Athabasca Sand Dunes in northern Saskatche-wan are the largest active dune complex in Arctic North America and the largest area of open sand in the world at this northern latitude.   The aeolian environment is controlled by climatic, geologic, morphologic, hydrologic, geohydrologic, pedologic, botanic, zoologic, anthropogenic and temporal factors.   Each of these factors act through a number of attributes which interact with one another. While such interaction of various factors are complex, a number of elements dominate in this area.   Parabolic dunes develop in this area from moist sand.   The oldest dunes are orientated to the south and are stabilized with a northwesterly orientation.   This orientation is counter to the presently occurring westerly winds.   The newer dunes are in an easterly alignment.   The difference in orienta-tion and level of groundwater have been related to the horst and graben features of the underlying Sandstone bedrock.   In response to lowering of the level of Lake Athabasca, the rivers have become entrenched and angulated. The present level of Lake Athabasca straddles a fault block step which is not horizontal.   The present active dunes are located against the rise of the fault step and are some 70 to 100 m thick.   The sand dune complex is in an uneasy equilibrium between the combined erosive action of the wind and water on one hand and the stabilizing actions of groundwater and vegetation on the other hand. Man has also induced changes with the introduction of hydro-electric dams hundreds of kilometres away, starting forest fires on and near the dunes and driving vehicles over the dunes.

Le plus grand champ de dunes sub-arctique nord-americain est-il en voie de disparition?

RESUME   Les dunes de l'Athabasca dans le nord de la province de   Saskatchewan forment le plus grand champ de dunes vives en Amerique du Nord arctique, et la plus grande surface de sable à nu au monde à se trouver à une latitude aussi septentrionale.   L'environnement éolien y est contrôlé par les facteurs suivants: climat, géologie, morphologie, hydrologie, pédologie, végétation, faune, occupation humaine, et temps.   Chacun de ces facteurs

influence l'environnement par l'intermediaire de plusieurs
variables qui interagissent entre elles. Bien que ces
interactions soient complexes, un nombre relativement
restreint de variables ont un effet dominant dans cette
région. Par exemple, les dunes paraboliques se forment là
où le sable est humide. Les dunes les plus anciennes se
sont formées sous l'influence d'un vent dominant du sud,
et furent ensuite stabilisées sous l'influence d'un vent
dominant du sud-est. Cette orientation est très différente
de celle des vents dominants actuels, qui proviennent de
l'ouest et qui orientent les dunes plus récentes vers
l'est. La localisation des dunes ainsi que le niveau des
nappes d'eau sonterraines peuvent être corrélés avec la
structure en horst et graben du socle gréseux sous-jacent.
En réponse à un abaissement du niveau du lac Athabasca,
les rivières se sont encaissées et sont devenues anguleu-
ses. La surface actuelle du lac Athabasca chevauche un
gradin de bloc faillé qui a une position proche de
l'horizontale. Les dunes vives sont situées contre le
talus formé par ce gradin et ont une épaisseur comprise
entre 70 et 100 mètres. Le complexe de dunes vives est le
résultat d'un équilibre précaire entre d'une part,
l'érosion combinée du vent et de l'eau, et d'autre part,
l'action stabilisatrice des eaux souterraines et de la
végétation. L'influence humaine est également notable et
résulte de la construction de barrages hydro-électriques à
quelques centaines de kilomètres en amont, de feux de
forêt déclenchés sur ou près des dunes, ainsi que de la
conduite de véhicules sur les dunes.

Introduction

The Athabasca Sand Dunes in Northern Saskatchewan are the largest
active sand dunes complex in Arctic North America and the largest
area of open sand in the world at this northern latitude (59.5°N)
(Abouguendia et al. 1981). Most of the active dunes in the complex
lie within a 1200 km^2 area along the south shore of Lake Athabasca
(Figure 1). The sand dunes activity started as soon as the last
continental ice sheet withdrew from the area. But these dunes
quickly stabilized when climatic conditions changed around 8,000
years B.P. Dune activity resumed when glacial Lake McConnell receded
and exposed the glaciofluvial and glaciolacustrine deposits to
westerly winds.
    South of Lake Athabasca, the PreCambrian plain rises gradually
with a relief of 160 m. The underlying bedrock, the Athabasca
Sandstone Formation, surfaces in only a few places near the lake
shore and on the inland rises (Figure 2). The Formation covers an
oval area of about 100,000 km^2. The sandstone has a shallow dip
towards the centre of the basin south of the dune complex. The
Formation is 1.5 km thick.
    The sandstone is the chief source of the sand dune material.
Seismic studies (Abouguendia et al., 1981) suggest depths to bedrock
range from 70-100 m in the sand sheets near the Williams River and

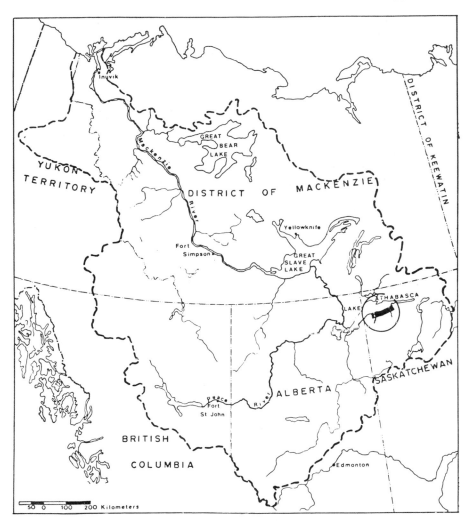

Figure 1   Location of the study area
within the Mackenzie River basin.

Cantara Lake to 20 m near Pederson and Gull Lakes. Because of the
permeable nature of the surficial materials the entire area is an
aquifer. The sand sheets and dunes store about one half (175 mm) of
the annual precipitation, the remainder evaporates from the land,
water or plants (Abouguendia et al., 1981). To some extent the
rivers are "fed" by aquifers through which they flow; by late summer,
the William River is discharging colder water and twice as much in
the lower section than the upstream discharge volume. Periodic
discharge of glaciofluvial deposits into lakes has also occurred (for
example in Kettle Lake near Campbell Lake) (Abouguendia et al.,
1981).

The precursor to Lake Athabasca, proglacial Lake McConnell,
inundated the land surface below 305 m above sea level (a.s.l.).

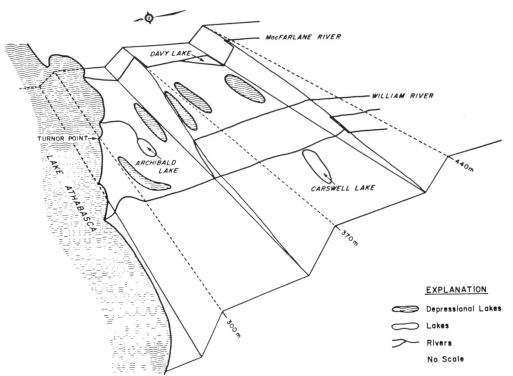

Figure 2 A schematic representation of changes in
river patterns due to fault block ridges.

Below 305 m, glaciolacustrine and glaciofluvial deposits prevail,
covering the till which otherwise appears as morainic material at
higher elevations.   Early post-glacial winds blew sand into the area
from sources to the southeast, forming hairpin-shaped (windrift dunes
(long since stabilized).   Subsequently the prevailing winds have
changed to the west and northwest, winnowing sand from the near-shore
deltaic, glaciolacustrine and glaciofluvial sediments to form large
dunes and dune fields.   Many of these are active today and may well
have been continuously so for the last 6000 years.   Further evidence
of a long aeolian history is provided by the polished and ventiface-
ted lag gravel peripheral to the dunes – the desert pavements.
    Under natural conditions, the level of Lake Athabasca rises and
falls two to three metres per year (presently at 209 m a.s.l. (WSC
1984)).   During low water level years, winds and wind driven waves
work the newly exposed sand into beach ridges.   A series of 35 such
ridges at the William delta are apparently related to periodic
lowering of the lake every 30 years or so.   The entire sandstone
formation drains into Lake Athabasca.   The main rivers are the
William, and MacFarlane Rivers which drain the east and west parts,
respectively.   The major river draining the centre is the Archibald
River.
    The   prevalent vegetation of the upland, well-drained sand is an

open-crowned pine forest with a light coloured carpet of lichens and blueberries. Where soil moisture increases as on the lower slopes, Labrador tea and mosses replace the lichens and blueberries under a closer crown of pine. In low wet areas, peat covers the mineral soil with intermittent zones of permafrost. Black spruce dominates acid bogs that are uninfluenced by groundwater, while larch predominates where groundwater flows through fens. Fire is a common disturbance, especially in the uplands and creates a mosaic of jack-pines. In some dune areas there is evidence that fire has contributed to the infiltration of wind erosion and to expansion of sand sheets. The wetter low lands are less likely to burn. The only fireproof areas are the wet peat-lands and uncommon mineral soil islands within them. On the south shore of Lake Athabasca large white pine have survived in this manner.

Archaeological investigations (Abouguendia et al., 1981) have shown that most human inhabitation was on old raised beaches and within two kilometres of Lake Athabasca. Intermittent occupation has occurred at least as far back as 7000 to 8000 years ago (Paleo-Indian Tradition) (Abouguendia et al., 1981).

The aeolian environment

The Dunes of the Athabasca region are parabolic. Parabolic dunes develop from moist sands, the moisture originating from and being recharged by precipitation. Two groups of dunes, differing in orientation and representing two different environments, are recognized in the region (Abouguendia et al., 1981). The first group comprises the oldest, northwesterly orientated parabolic dunes all of which are stabilized today. They all occur to the south of the area which was not inundated by Glacial Lake McConnell. These dunes were formed by strong southeasterly winds which originated from the continental ice sheet, in the form of katadiabatic air masses and in the major climate event about 8,000 years ago (Berger, 1986).

The second group of dunes are easterly oriented active and stabilized dunes that formed of glaciofluvial and glaciolacustrine sediments soon after the draining of Lake McConnell. These dunes are grouped into sectors, several of which have their own particular style. The most useful parameter in distinguishing one group from another is by the level of groundwater. The active sectors can be divided into the following areas: William River, Yakow Lake, Pederson Lake and Archibald Lake sectors. A rise in groundwater tends to stabilize the dunes and a lowering would de-stabilize the dunes. The stabilized dunes gradually loose sand resulting in development of narrow dunes (Pederson sector) and low dune ridges or dune tracks (Archibald sector), and which are finally covered by vegetation. These vegetated dunes become active again as a result of fire or similar destructive agents.

Methodology

Hobson and McAulay (1969) first suggested that the Athabasca Formation has been subjected to uplifting forces which have created a

series of block faults.   These blocks trend generally from northeast
to southwest (Figure 2).   Evidence from maps, air photography and
satellite imagery indicates that the blocks may be tilted.   Further
evidence for the existence of these faults is the presence of rapids
and orthogonal changes in river channel direction as the channels
cross the faults.

Fieldwork in the region was done by a team of scientists in 1979
and 1980.   The glacial and post-glacial landforms were mapped and
described and a tentative history of ice retreat and lake levels was
outlined. A number of active and stabilized dunes were examined in
detail to determine their rates of movement and rates at which the
soil profiles were developed.   Seismic studies revealed the depth of
unconsolidated overburden and subsurface hydrology, while instrument
stations on streams and lakes provided information on flow regimes,
sediment loads and erosion-deposition along the shoreline.   Weather
records were used to calculate stream water budgets, shoreline water
levels, beach erosion, and dune activity.   The results from this
fieldwork was published in 1981 (Abouguendia et al., 1981).   This
present paper interprets the findings differently and with a
different emphasis.

With new technology available in remote sensing, the available
information was re-examined. Of particular significance was the
location of strandlines, meltwater channels, spillways and deltas.
The vegetational patterns were also examined to determine whether the
species or alignment was related to lineament.

Satellite technology now allows large areas to be examined in an
objective manner with the application of rigorous statistics.
Directional lineament analysis is a standard subroutine of the Dipix
Aries II system.   The computer system is on loan from the Canada
Centre for Remote Sensing to the Saskatchewan Research Council under
the Saskatchewan Technology Enhancement Program.   The Aries II
program analyzes for lineament in the following manner -- applying
linear filters to digital image data to enhance or suppress spatial
features as opposed to spectral or intensity features.   Spatial
features include edges or lines, speckle, homogeneous areas, etc. and
by appropriate selection of the convolution matrix, different spatial
features can be enhanced and/or other features suppressed.
Directional filters suppress low frequencies, highlighting rapid
transitions such as edges (lines) in chosen directions (Dipix, 1984).

The convolution is done in the following manner. Let P(I,J) be an
image. The values of P() are between 0 and 255 in Landsat data. The
indices I and J are the line and pixel (column) number of the image.
Let C(K,L) be a matrix of dimension 2M+1 rows and 2N+1 columns (i.e.,
number of rows and columns is odd).   An output image is computed
O(I,J) by convolving the input image with the matrix C. That is:

$$O(I,J) = P \text{ "CONVOLVED WITH" } C$$
$$= \text{SUM } (K=-M \text{ TO } M) \text{ OF THE SUM } (L=-N \text{ TO } N)$$
$$\text{OF } (P(I+K, J+L) * C(K+M+1, L+N+1))$$

Conceptually, the matrix C is overlaid onto the image P so that
the centre point of C, C(M+1, N+1), lies on top of the image point
(I,J).   Each point in matrix C is then multiplied by the image point
under it. Finally, all these products are summed.

This definition of the convolution differs from the normal one. Generally, the convolving matrix C is reflected through its central point before being multiplied to the image. (i.e. elements (1,1) and (3,3) are swapped in C in the 3 by 3 example, as are (1,3) and (3,1) etc.) Using the above definition, it is easier to see how the matrix overlays the image.

Spatial filtering is often discussed in terms of spatial frequencies. If a two dimensional Fourier transform is performed upon an image, the resultant transform is given in terms of spatial frequencies. High spatial frequencies give rapid intensity transitions over a small range of pixels. For example, the sequence 0,255,0,255,0,255,...along an image line would be the highest spatial frequency an image can have (0.5 cycles per pixel). Alternatively, low frequencies cause very small intensity changes across an image.

If the convolution matrix is asymmetric or anti-symmetric about an axis, then a directional enhancement can be developed. For example, the matrix will detect edges of pixels that are vertical.

$$C = \begin{array}{ccc} -1 & 0 & 1 \\ -1 & 0 & 1 \\ -1 & 0 & 1 \end{array} \quad = 3 \text{ LEVEL, EAST MASK}$$

If an edge is lying vertically (north-south), with the bright side of the edge to the east (right), then this filter produces maximum output. If the edge has the bright side to the west (left), a minimum output is produced. No vertical edge should produce an output near 0.

In this task, matrix or kernels whose elements sum to 0 are differential or difference kernels, while those that sum to 1 are enhancement kernels. In general, an enhancement kernel can be derived by adding the input image to the differential image or by adding 1 to the central element of the differential kernel. Differential kernels are used for edge detection, while enhancement kernels are used for edge enhancement.

When a kernel has been specified, the elements of the kernel are summed. If the sum is non-zero, the user is asked if the kernel should be normalized to a given value, normally 1. If so, each element is multiplied by a constant to the required sum.

Directional filters enhance features with a particular orientation. For the system masks, the 8 major compass directions (N, NE, E, SE, S, SW, W, NW) were used. The direction of the filter will cause a dark to bright transition along that direction to produce a high value at output. If the transition is bright to dark, then a low output is produced. No transition results in a middle value output. Thus, a west filter will produce a reverse image from an east filter at transition areas.

The user may select any size of filter kernel, providing that the kernel is square (in this study a 3 X 3 was used). The user selects the required compass direction. The central pixel of the kernel may be weighted.

Another factor in the future of the Athabasca Sand Dunes is the activities of man. These effects are examined by performing a sensitivity analysis on the parameters discussed above.

Discussion and results

Division of aeolian features into sectors

David (1981) separated the Athabasca Sandstone Formation dunes depending upon whether they are developed on moist areas. Two groups of dunes, differing in orientation and representing two different environments, are recognized in the region. The first group comprises the oldest, northwesterly oriented parabolic dunes all of which are stabilized today. They all occur to the south of the area which were not inundated by Glacial Lake McConnell. The second group of dunes are easterly oriented active and stabilized dunes that formed of glaciofluvial and glaciolacustrine sediments soon after the draining of Lake McConnell. These dunes are grouped into sectors, several of which have their own particular style. The stabilized dunes gradually loose sand resulting in development of narrow dunes (Pederson sector) and low dune ridges or dune tracks (Archibald sector), and which are finally covered by vegetation.

The stabilized parabolic dunes are located on the step. The active dunes are located on the riser of the fault block. Also being on the riser makes sand more susceptible to wind movement from the northern sector.

The most useful parameter in distinguishing one group from another is by the level of groundwater. Figure 3 shows the relationship of the fault locations to the highest areas of groundwater. The parabolic dunes have an average moisture content of 4-6%. The threshold wind velocity to move the sand must be double the threshold velocity for dry sand (Abouguendia et al., 1981).

The dunes can be further divided according to their orientation. The dunes between Ennuyeuse$_o$ Creek (west of William River) are oriented to the northwest (313° $\pm$ 5). These type of dunes run along the step. For example the same orientation can be found near Archibald Lake. These types of dunes also run along an orthogonal fault line along which the William River runs (Figure 4) (David, 1977). These dunes start at 245 m a.s.l. in the west and rise to 360 m – a gradient of 1 m/1000 m. The active life of these dunes is 100-1000 years about 8800 to 10,000 years B.P.

The second division by orientation are those at 114° $\pm$ 14. The wider deviation shows that these have been acted upon by wind from many directions. The majority are located within 10-15 km of the shores of Lake Athabasca. These dunes are the most active and show signs of being active since de-glaciation.

Forces affecting the dunes

Lake Athabasca is one of a ring of lakes which lie along the outer edge of an elliptical trough consisting of: St. Lawrence River, Great Lakes, Lake Winnipeg, Lake Athabasca, Great Slave Lake, Great Bear Lake and Melville Sound (some are shown in Figure 1). This trough occurs at the outermost ring of zero uplift. The west end of Lake Athabasca is assumed to have reached the end of its uplift (Walcott, 1972). Based upon the series of abandoned beach west of Lake Claire, it was calculated that the area is tilted up towards the

Figure 3 Fault lines as given by analysis of Landsat digital data.

northeast at the rate of 0.4 m/1000 m and at the rate of 0.8 m/100
years. The tilt axis lies at about 40° west of north. The elevation
of these beaches above the present lake elevation is 92 m.   The Cree
Lake Moraine on the very southern part of the Formation has been
dated at 10,700 years B.P.   (Rutherford et al., 1974).   Due to the
tilting and uplifting, the MacFarlane River will continue to entrench
itself while the William River is less likely to do so.
    The recession of the continental glacier led to isostatic
readjustment of the land.   During the period of maximum development
of the Wisconsin Glacier, the glacier was about 3,300 m thick over

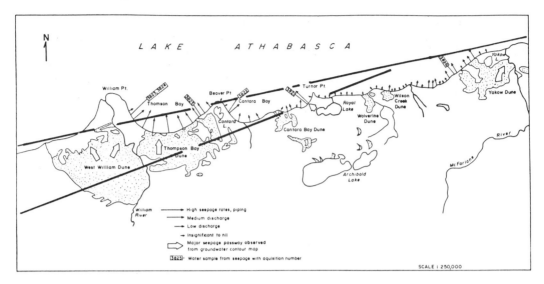

Figure 4   Groundwater seepage at the lake shore.

the study area (Walcott, 1970). Ice cover of this thickness would
depress the continent isostatically by about one-third of the
thickness of the glacier.   The glacier was thickest towards its
centre.   Therefore the rebound is greatest towards the east of the
study area and least to the west. This differential movement results
in tilting of the continental surface during rebound (Environment
Canada, 1973).

The ice left the area around 10,000 years ago retreating in an
east-northeast direction (Prest, 1969). Since the present northerly
outflow river (MacKenzie River) was blocked by the Keewatin Glacier
(Cameron, 1972; Taylor, 1960), the resulting lake reached an
elevation of 500 m a.s.l. before an outlet formed through two gaps at
the southern edge of the Formation (Taylor, 1960).   These gaps may
have been lower than the present elevation of 500 m a.s.l. because of
differential uplift.   As de-glaciation progress, another outflow
channel emerged from the ice at 420 m (Cameron, 1972; Taylor, 1960).
This channel would be located on the back of one of the steps and
starting at about the present location of Davy Lake (Figure 2).   Not
too long after, the 340 m elevation was reached and outflow through
the MacKenzie River started.   The lake quickly drained to 250 m
(5,000 years ago).   The lowering of Lake Athabasca to its pre-1968
level of 218 m was caused by differential uplifting and downcutting
of the MacKenzie River.   The construction of the Bennett Dam and the
filling of the Williston Reservoir dropped Lake Athabasca by 0.76 m
(1968-1972) (Environment Canada, 1973).   Further decreases have
occurred and the Lake is presently at 209 m $\pm$ 0.3 m (WSC, 1984).

The lowering of Lake Athabasca resulted in the emergence of a
series of shoreline deposits and of the surface drainage system.   In
response to the receding lake level, rivers have become entrenched
and groundwater levels in the small closed lakes have fallen (Royal

Lake which has steep sides; Pederson where outflow channel has filled
with sand). The William River is entrenched 30 m while the Archibald
and the MacFarlane Rivers are entrenched 2-10 m and 5-10 m. The
present Lake position straddles a fault block step (Figure 2). Both
the mouths of the MacFarlane and the William Rivers are on steps
close to the riser. Therefore a small lake elevation change will
create a large horizontal displacement of the shoreline. Beaches
with offshore bars are found only along the step. The present active
dunes are found along the rise of the step. The stabilized dunes are
found against other steps further inland south of Carswell Lake and
Cree Lake. The fault blocks act as controlling structures and limit
the dynamics of the sand supply by restricting the landward supply of
sand.

Strandlines of Glacial Lake Athabasca are found at several
locations. The highest strandlines within the dune area are marked by
terraces cut into the drift on highland areas south of Pederson and
Royal Lakes at 305 m. Other strandlines were developed at 290 m, 260
m, and 230 m. These levels correspond with the elevations of the
steps and changes in drainage ways. Between more well-drained levels,
especially below 230 m, are successively lower strandlines. These
lines were developed at regular intervals, commonly about one metre
difference in elevation. These likely correspond to the raising of
the land surface due to isostatic rebound.

The William and MacFarlane Rivers functioned as meltwater channels
carrying water from the melting ice to Glacial Lake Athabasca. The
rivers are braided and meander and have wide sand plains bordering
the present water course, especially below 300 m. A number of
terraces are cut along the William River at about 290 m, 280 m, 260
m, as well as several lower level terraces. Similarly, terraces
along the MacFarlane have elevations of about 300 m, 280 m, and 260 m
also with lower level terraces. Again notice the elevation
difference between the east and west side (about a 10 m difference).
These terraces apparently coincide with successively lower levels of
the glacial lake and with the step faulting. The terraces are cut at
the bottom of the riser or where there is an abrupt rise in
elevation. Sediment carried from the ice was deposited as prograding
deltas into the glacial lake. These deltas continue to grow today
especially the William Delta. The William Delta is growing fastest
because it is closest to a riser (see Figure 2).

Many of the meltwater channels acted as spillways through which
glacial lakes drained: MacFarlane, Archibald, and William Rivers.
Terraces cut at about 350 m along the MacFarlane River just north of
Davy Lake at about 5 m above the present level of 345 m. This
coincides again with the transition from the riser to the step.

Parallel channels from Davy Lake disappear into the large active
sand area (Figure 5). These abandoned channels served as successive
spillways down the riser but were successively blocked by the large
amounts as sand (deltaic material). The Archibald River may also
have acted as a spillway draining a higher level of Archibald Lake.
The east tributary of the William River may once have used the
Archibald River channel. The west tributary of the William River may
have drained waters ponded around Carswell Lake at around 330 m.

The satellite imagery analysis (Figure 3) shows many more faults
than the simplified map in Figure 2. The location of the fault lines

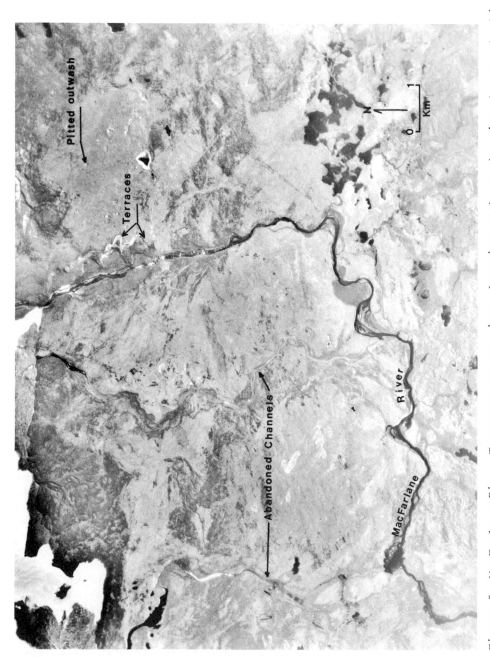

Figure 5   MacFarlane River Terraces are cut into deposits such as the pitted outwash which flanks the river. Two abandoned channels west of the MacFarlane River may have served as temporary spillways or meltwater channels.

and the position of rivers and dunes shows very dramatically the influence of the bedrock faulting on the dune landscape.

Evaluation of man's activities on the dunes

The present average water level of Lake Athabasca is 209 m a.s.l. (1984). The probability that the mean lake level will equal or exceed one metre change in any one year is 10%. However, during any one day in each year there is a 50% chance that the lake level will be one metre different from the previous day due to wave action (Abouguendia et al., 1981). Furthermore, for every one year in five years the lake level change will exceed one metre. The probability that the lake will change by more than 3 m in any one year is 2%. The Bennett Dam and subsequent filling of the Williston Reservoir caused a drop of 0.76 m.

The two areas with major raised beaches are the William and MacFarlane Deltas. A lake level drop of 8.5 m would be required to produce the raised beaches at the MacFarlane Delta. This would require about a million years in today's climatic conditions (Abouguendia et al., 1981). The William River would require a 5 m drop in lake level which has a 1:600 year probability of occurring (return period), Abouguendia et al., 1981). In comparison using 30 years as the average number of years between major low elevations, it would take 1050 years to create the 35 raised beaches in the William Delta. Therefore additional factors must be involved.

Water level changes have three effects on shorelines: a) they alter the angle at which the waves attack; b) they alter the stability of the beach by exposing unstable slopes; and c) they alter the distribution of sediment in the shoreline zone. Human actions have produced lowering of the lake level which can cause the above alterations. The effect is the cutting of beach slopes greater than 0.03 and forming barrier islands on slopes less than 0.03. Evidence of this can be seen along Thomson Bay on the back shore. Lowering of the lake levels leads to movement of material inshore and an increase in supply of sediments to the back shore dunes. Construction of harbours and jetties result in longshore drift of sediments into deeper water. This construction would cut down the drift beaches and cause build up of sediments in front of the risers (Hails, 1977).

The construction of upstream or downstream dams, river diversions, or large scale land use changes within the MacKenzie River Basin can cause large scale changes. Lowering of Lake levels could be equivalent to isostatic changes and cause abandonment of shorelines, exposure of coastal sediment, river entrenchment and lowering of the water table in the present dune areas. The effect would be to drown present dune areas and erosion of surfaces with slopes greater than 0.03 (the present beach gradient). The latter would reduce the sand supply to the present dunes. Upstream changes would cut off the supply of sand and accelerate shoreline erosion. Examples of this have been seen in Egypt (Aswan High Dam – Nielson, 1973) and in California (Hales, 1975) and Florida (Scott and Rawls, 1972). The sediment load in MacFarlane River can change by 60% during a natural flood but will only decrease by 50% from normal in a drought (Abouguendia et al., 1981). On the William River, the effect is even

larger (6 times normal sediment capacity). There is a marked
difference on the William River between the sediment stations on the
step as compared to on the riser.

Conclusion

There are many interconnected processes which are affected by a
change in the sand dunes. The location of the induced change is
important; i.e. whether on the upthrown or downthrown surfaces of the
fault block or on the escarpment is important. For example, a small
increase in sand supply could result from the decrease in lake level
on an escarpment. However, the increase in sand supply would be
considerably larger on a gently sloping block surface. The sand
supply and other related processes are more sensitive to lake level
changes over a block surface than to the same change over an
escarpment.
   Processes which should be monitored, especially if affected by
man-made events, include: changes in water table, instability of
dunes, and erosion of beaches. Significant ecological events which
could occur in the dune areas are: flooding of the area between the
dunes, sand drowning of rivers, rapid drainage of lakes, and dune
formation from turbidites (high density sand flows which occur under
water). These processes have repercussions on other ecosystems such
as soil, flora and fauna.
   Part or all of the Athabasca Sand Dunes should be designated an
ecological reserve, as it contains a large number of rare and endemic
species as well as unique habitats. The area is very important from
the botanical and evolutionary point of view. Strict environmental
measures are necessary to safeguard the sensitive aeolian and sand-
based ecosystem, the gravel plains, and the cultural (archaeological)
resources of lake and river shores. No development affecting the
dune areas should be allowed without adequate environmental impact
studies. Although the dunes are in a delicate balance, the present
dunes are expected to be active in the present location for the next
500 years.

ACKNOWLEDGEMENT   The field program was carried out by the Saskatchewan Research Council under contract to Saskatchewan Environment under
the MacKenzie River Basin Committee.

References

Abouguendia, Z. Whiting, J., FitzGibbon, J., Wheaton, E., Schreiner,
   B. Acton, D., David, P., Schneider, A., Godwin, R., Harms, V.,
   Hudson, J., Rowe, S., Wright, R., Polson, J., Barclay, R.,
   Paquin, B., Wilson, J., and Sawchyn, W. (1981) Athabasca Sand
   Dunes in Saskatchewan (MacKenzie River Basin Study Report
   Supplement #7) Environment Canada, Regina, Canada.
Berger, A. (1986) Nuclear winter, or nuclear fall EOS American
   Geophysics Union.
Cameron, A.E. (1972) Post glacial lakes in the MacKenzie River

Basin, N.W.T. Can. J. Geol. 30:353-377.

David, P. (1977) Dunes types and their use in interpreting past dune environments in Canada. Intern. Quat. Assoc. (10th Congress, Burmingham, England) Abstracts, p109.

(1981) Stabilized dune ridges in northern Saskatchewan. Can. J. of Earth Sci. V. 18 N. 2 pp 286-310.

Dipix (1984) Operations manual Dipix Ltd., Ottawa.

Environment Can. (1973) The Peace-Athabasca Delta Project Peace-Athabasca Group, Queen's Printer, Ottawa.

Hails, J.R. (1977) Applied Geomorphology Elsevier Sc. Pub., N.Y.

Hale, J.S. (1975) Models of the ocean shoreline. J. of Shore Beach 43:35-41.

Hobson, G.D., and MacAuley (1969) A seismic reconnaissance survey of the Athabasca Formation. Alberta and Sask. Survey Can. Paper 69-18.

Nielson, E.V. (1973) Coastal erosion in the Nile Delta. Nat. Res. N. 9, 14-8.

Prest, V.K. (1969) Retreat of Wisconsin and recent ice in N.A. Can. Dept. E.M.R. Map 1257A.

Rawls, O.J. (1972) A case history of shoreline effects, of jetties, and channel improvements at the mouth of St. John's River. J. of Shore Beach 40:33-5.

Rutherford, A. McCallum, K.J., Wittenberg, J. (1974) Radiocarbon 17 #3 Sask. Research Council, Saskatoon, Canada.

Scott, H.A. (1972) Design versus ecology in St. John's and Indian Rivers. Proc. ASCE, J. Workways, Habours and Coastal Eng. Div. 98: 425-432.

Taylor, R.S. (1960) Some Pleistene lakes of northern Alberta and adjacent areas Alta. Soc. Petr. Geol. J. 8:167-178.

Walcott, R.I. (1970) Isostatic response to loading of the crust in Canada Can. J. of Earth Sci. 7:716-734.

(1972) Lake Quaternary vertical movements in eastern North America. Quaternary evidence of glacio-isostatic rebound Rev. Geophys. 10:849-884.

Water Survey of Canada (WSC) (1984) Water Flow Data Queen's Printer, Ottawa.

The Influence of Climate Change and Climatic Variability on the Hydrologic Regime and Water Resources (Proceedings of the Vancouver Symposium, August 1987). IAHS Publ. no. 168, 1987.

Sécheresse, désertification et ressources en eau de surface — Application aux petits bassins du Burkina Faso

Jean Albergel
Hydrologue UR B12
ORSTOM, Laboratoire d'Hydrologie
Miniparc Bât.2, Rue de la Croix Verte
34100 Montpellier, France

RESUME L'étude des conséquences de la sécheresse sur les écoulements des petits bassins versants d'Afrique soudano-sahélienne nécessite la constitution de chroniques de ruissellement. A cet effet, un modèle pluie-lame ruisselée a été mis au point et appliqué sur deux bassins de quelques dizaines de km^2. Ce modèle est basé sur la cartographie des etats de surface qui ont évolué sous l'effet conjugué de la modification climatique et de l'extension des surfaces cultivées. Les séries ainsi obtenues pour la période humide 1950–1968 et la période sèche 1969–1984 ont pu être comparées. Il en ressort que les conditions de ruissellement favorisées par une dégradation des bassins dans la seconde période compensent globalement le déficit pluviométrique. Ponctuellement un evenement pluviométrique fort qui a la meme probabilité d'occurrence avant et après 1969 engendre une crue plus forte dans la période actuelle.

Mots clefs:

Sahel, Bassin Versant Experimental, sécheresse, états de surface, ruissellement.

ABSTRACT A chronological succession of runoff events was built in order to study the effect of the drought on the runoff of small watersheds in sahelian Africa. To fulfill that aim, a specific rainfall-runoff model was derived. This model is based on the mapping of superficial soil features and on the related rainfall-runoff relationships derived from rainfall simulation experiments. It was applied to two small catchments (22 and 54 sq.km) in Burkina Faso. Superficial soil features have evolved along with natural climatic changes and also with changes in land use. The runoff time series obtained for the humid period 1950–1968 and for the dry period 1969–1984 were compared. In the dry period, more favorable conditions for runoff make up for the rainfall deficit, so that globally the runoff distribution remains unchanged. A strong rainfall which has the same probability to occur in the two periods, produces a larger flood at the present time.

Key Words

Sahel, experimental catchment, drought, superficial soil
features, runoff.

Introduction

Les très forts déficits pluviométriques des années 1983 et 1984 sur
l'ensemble de la zone sud sahélienne de l'Afrique ont révélé qu'après
les sécheresses de 1972-1973 la région soudano-sahélienne n'a pas
retrouvé la pluviosité qu'elle connaissait dans les années antérieures. De nombreuses études concordent pour dire que la période
1969-1984 se caractérise par un affaiblissement des totaux pluviométriques jamais encore observés, tant par son intensité, sa persistence
et son extension géographique (Nicholson 1984, Albergel et al.1984,
Snidjers,1986). Les saisons des pluies 1985, 1986 "normales" pour
une majorité des stations ont été déficitaires dans de vastes regions
et ne permettent pas d'affirmer le retour à une période plus
clémente.

Les conséquences sur les grands systèmes hydrologiques ont été
impressionnantes et relatées tant par des scientifiques que par les
médias:

- Effondrement des débits des grands fleuves comme le Senegal
(Olivry,1983), le Niger (Billon,1985) ou l'ensemble des grands
cours d'eau tropicaux (Sircoulon,1986).

- Le bouleversement des systèmes lacustres; lac Tchad, delta
intérieur du Niger (Sircoulon,1985).

- La baisse généralisée de tous les aquifères importants (Leusink
et Tyano,1985).

L'analyse des régimes hydrologiques de cours d'eau moins importants (bassins versants de 1000 à 5000 km^2) a démontré que la
sécheresse climatique a eu des répercussions bien moins importantes
sur le fonctionnement de ces systèmes: pas de modifications
significatives dans les chroniques de modules annuels ou de modules
journaliers maximum, (Pouyaud,1985).

Pour l'ensemble des petits bassins, taille inférieure à 1000 km^2,
le manque de données sur les débits ne permet pas de faire le même
genre d'étude. Rappelons cependant que des crues ponctuelles
causaient des dégâts dans une ville, comme Gorom-Gorom le 29 et 30
septembre 1984 ou emportaient des ouvrages sur des bassins versants
de surface inférieure à 100 km^2, barrage de Zamse au sud de
Ouagadougou à un moment où l'attention était plutôt polarisée sur des
problèmes de manque d'eau. Après avoir rappelé les principales
caractéristiques pluviométriques de la période 1969-1984, cet article
tente d'analyser les répercussions de cette sécheresse sur le fonctionnement des bassins versants de petites dimensions: de quelques
km^2 à 200 km^2.

Principales caracteristiques des pluies durant la periode 1969-1984

Snidjers (1986 op.cit) démontre la non stationarité de la série des
totaux pluviométriques sur l'ensemble des stations du Burkina-Faso.

Carbonnel et Hubert (1985) confirment que la probabilité la plus forte de "rupture" dans ces séries se situe en 1969 ou 1970. Les moyennes interannuelles de la seconde période sont globalement inférieures de 20% à celles de la période précédente (1920-1968).

L'analyse des hauteurs pluviométriques journalières (Albergel, 1986) a conduit aux conclusions suivantes:

- La somme des pluies journalières supérieures à 40 mm sur l'année est significativement plus faible dans la période sèche. Les distributions statistiques de ces pluies ont des coefficients d'asymétrie et d'aplatissement plus forts sur les années 1969-1984 que sur les observations antérieures.

- Les pluies les plus fortes ont une égale probabilité d'apparition dans les deux périodes; en particulier, le calcul de la précipitation journalière de reccurrence décennale donne un résultat équivalent sur les deux séries (résultat vérifié sur 25 stations retenues pour la qualité et la longueur des observations).

Sur les petits bassins versants où l'essentiel de l'écoulement est dû au ruissellement quasi immédiat des plus fortes pluies, nous étudierons comment se sont traduites les modifications du régime des pluies. Dans ce but, nous reconstituerons des chroniques de ruissellement sur deux bassins versants expérimentaux sur lesquels nous disposons de quelques mesures dans les deux périodes:

Le bassin versant de Kazanga (54 km^2) en zone soudanienne (11 40N isohyète 900 mm) a été suivi en 1961-1963 et en 1983.

- Le bassin versant de Kognere dans la région de transition soudan-sahel (12°22N isohyète 700 mm) a été suivi en 1960-1962 et 1984.

Constitution d'une chronique de ruissellement
sur un petit bassin versant

L'état actuel des connaissances sur l'influence des différentes composantes de l'environnement sur le ruissellement met en évidence qu'en zone soudano-sahélienne l'hydrodynamique superficielle est contrôlée essentiellement par le couvert végétal et les organisations pédologiques de surfaces (Albergel et al.,1985). Ce résultat a permis la construction d'un modèle simple de constitution d'une chronique de lames ruisselées à partir de la cartographie des états de surfaces (Albergel et al.,1985). La méthode cartographique développée à cette occasion distingue deux niveaux d'organisation (Valentin,1985):

- La surface élémentaire, caractéristique d'un état de surface et considérée comme homogène quant à son comportement hydrodynamique sous pluie.

L'unite cartographique qui correspond soit à une seule surface élémentaire soit à l'association de plusieurs (généralement interdépendantes au sein de "systèmes de surface") et dont les limites peuvent être tracées à partir des relevés de terrain et de photographies aériennes (figure 1). Sur chaque surface élémentaire, des mesures sous pluies simulées, sur parcelles de 1 m^2, permettent de tester un comportement hydrodynamique et de déterminer une fonction de production dépendante de la pluie et de l'état d'humectation du sol. La fonction de production à l'échelle du bassin est obtenue par la composition des différentes fonctions de production au prorata de

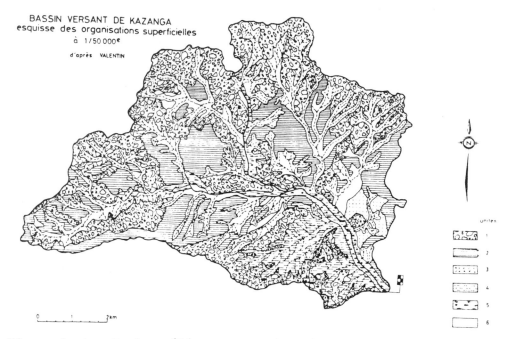

Figure 1   1. Surface éléments grossiers 2. Surface sans éléments
            grossiers 3. Association surface hydromorphe-surface à
            recouvrement sableux 4. Association surface vertique-
            surface a recouvrement sableux 5. Surface hydromorphe
            alluviale claire 6. Surface hydromorphes de bas-fond.

la surface qu'elles représentent et pondérée par un facteur de calage.
    La comparaison de lames ruisselées calculées et observées pour des
memes evenements pluvieux montre une dispersion plus importante pour
les valeurs observees (figure 2).   Sur le bassin de Kazanga, pris en
exemple, le modèle expliquerait 60% de la variance de l'échantillon
"lames ruisselées fonction de la pluie journalière".   Ce résultat
peut être considéré comme satisfaisant vue l'hypothèse de base
implicite: les pluies journalières ont une forme et une répartition
en intensité égales à celles du protocole de pluies simulées qui ne
sont que l'image de pluies statistiquement moyennes pour la région.
    Le tableau n°1 qui compare les paramètres statistiques de la série
de lames ruisselées calculés sur la période d'observation d'un poste
pluviométrique voisin, Manga (34 ans) et celles observées pendant la
periode de fonctionnement du bassin (4 annees) permet de valider ce
modèle.
    Les observations de pluies journalières sont multipliées par un
coefficient d'abattement calculé par régression entre les pluies
moyennes du bassin et les pluies du poste pendant la période commune
de fonctionnement (c = 0.85).

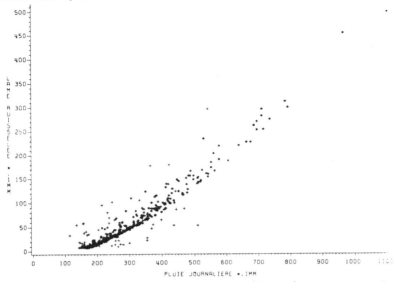

Figure 2   Lames ruisselees observees + (4 ans) et reconstituees
           * (37 ans) sur le bassin versant de Kazanga.

Comparaison des lames ruisselles reconstituees
pendant les periodes seches et humides

Une cartographie des états de surfaces a été realisée pour les deux
periodes à l'aide d'un relevé de terrain et des photos aériennes des
missions IGN (Institut Géographique National (Paris)) 1956 et IGHV
(Institut Géographique de Haute Volta (Ouagadougou)) 1980 sur les
bassins pris en exemple. Le tableau n°2 résume les différences
d'occupation des sols entre ces deux périodes.

Tableau 1   Comparaison des séries: lames ruisselées calculées et
            observées sur le bassin de Kazanga

* Seules les lames ruisselées ≥ 0,7 mm ont été prises en compte

Variable	Nombre d'observat.	Moyenne × 0.1 mm	Ecart type	Valeur max mm	CV
Lames ruisselées reconstituées sur 34 ans	453	56,1	62,2	50,1	110,9
Lames ruisselées observées sur 4 saisons	54	53,3	51,1	29,7	95,9

Tableau 2   Occupation des sols, comparaison des états en 1956 et 1980

Photos	Champs %	Jachère %	Savane Arborée %
KAZANGA 1956	16,2	51,3	32,5
KAZANGA 1980	36,2	33,8	30,0
KOGNERE 1956	16,1	10,7	73,2
KOGNERE 1980	37,4	5,6	57,0

Si sur les deux bassins on assiste à un recul important des jachères au profit des champs cultivés, sur le bassin de Kognere l'extension des cultures s'est également faite au détriment de la savane arborée. L'examen comparatif des photos aériennes permet également de mettre en évidence, a Kognere, une modification importante des unités cartographiques par une extension des zones très érodées (figure 3) qui sont multipliées par 20 entre les deux missions aériennes.

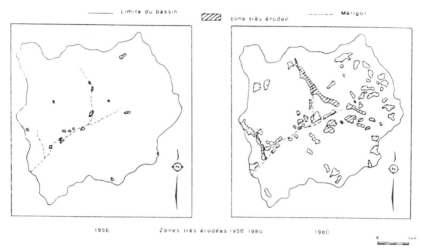

Figure 3 Evolution de la superficie des zones très érodées.

Sur le bassin de Kazanga les contours des unités cartographiques restent sensiblement les mêmes, seule leurs sous divisions en zone cultivée et en jachère est modifiée.

Dans un premier temps, pour appréhender l'effet respectif des modifications des états de surfaces et du changement climatique nous reconstituerons sur le bassin de Kognere une chronique de lames ruisselées à partir des deux cartes et de l'ensemble de la série des pluies journalières du poste voisin Boulsa.

Effets comparés de la modification du régime des pluies
et des caractères physiographiques des bassins

Sur la figure n°4 sont reportées les lames ruisselées reconstituées sur l'ensemble des observations pluviométriques pour les deux cartes d'états de surface de Kognere. On observe une homogénéité de la relation pluie-lame ruisselée journalière sur chaque reconstitution et une différence notable entre les deux reconstitutions, la carte de 1980 donnant des ruissellements plus forts. L'analyse des ruissellements sur chaque reconstitution (Tableau 3) met en évidence une diminution sensible du nombre de ruissellements moyens par an et de la hauteur de la lame ruisselée moyenne annuelle pour la periode postérieure a 1969 dans les deux reconstitutions.

Afin de reconstituer une chronique complète des lames ruisselées la plus probable, on se propose d'utiliser pour chacun des deux bassins la carte des états de surface dérivant de la première mission aérienne pour les observations pluviométriques précédent 1969 et celle dérivant de la seconde pour la période postérieure.

Les hauteurs de lames ruisselées reconstituees ont été comparées à

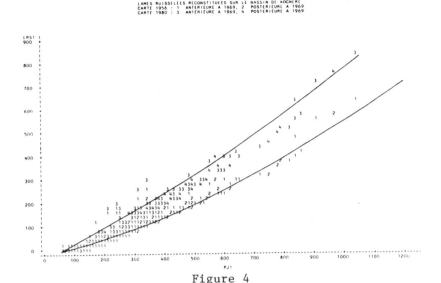

LAMES RUISSELEES RECONSTITUEES SUR LE BASSIN DE KOGNERE
CARTE 1956 : 1 ANTERIEURE A 1969, 2 POSTERIEURE A 1969
CARTE 1980 : 3 ANTERIEURE A 1969, 4 POSTERIEURE A 1969

Figure 4

Tableau 3   Lames ruisselées reconstitutées sur le bassin de Kognere

Photos	Date	Nbre moyen de lames* ruisselées par an	Hauteur ruisselée* annuelle (mm)	Valeur max de la lame ruisselée (mm)
Reconstitution d'après la carte 1956	Avant 1969	21,3	174,7	64,0
	Après 1969	15,1	116,2	37,2
Reconstitution d'après la carte 1980	Avant 1969	23,4	241,3	84,1
	Après 1969	17,2	166,7	51,9

*Seules les lames ruisselées ≥ 0,7 mm ont été prises en compte

celles observees pendant les deux periodes de fonctionnement des bassins: les coefficients de corrélation varient entre 0.75 et 0.90

Sur les figures 5 et 6 sont reportées les lames ruisselées reconstituées pour les deux périodes; on remarque:

– pour le bassin versant de Kazanga les deux nuages de points se confondent assez bien avec une légère tendance à un plus fort ruissellement pour les événements de période sèche.

– pour le bassin versant de Kognere, l'ensemble des pluies supérieures à 25 mm engendrent des lames ruisselées plus fortes en 1969–1983.

Sur les figures 7 et 8 les hauteurs de lames ruisselées rangées en classe de 0.5 mm sont reportées en fonction des fréquences expérimentales. On remarquera le peu de différence entre les distributions statistiques des échantillons de période sèche et de période humide.

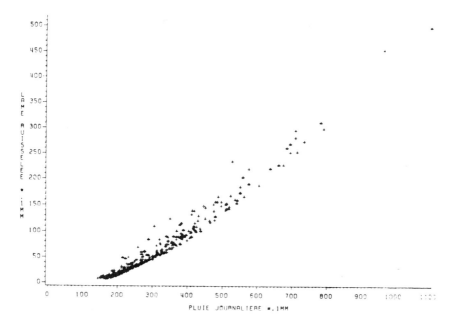

Figure 5   Lames ruisselées reconstituées sur les bassins
versants de Kazanga + période antérieure à 1969
* période postérieure à 1969.

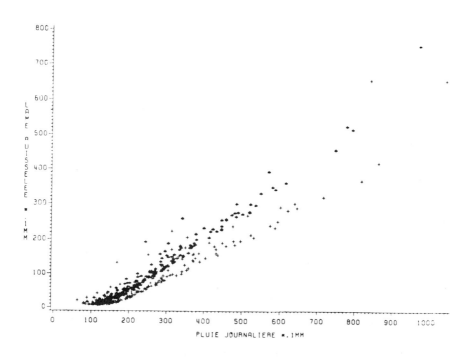

Figure 6   Lames ruisselées reconstituées sur les bassins
versant de Kognere + période antérieure à 1969
* période postérieure à 1969.

Lames ruisselées reconstituées sur le bassin de KAZANGA

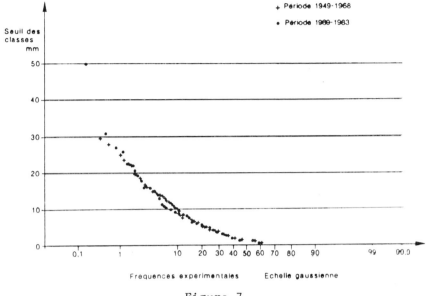

Figure 7

Lames ruisselées reconstitués sur le bassin de KOGNERE

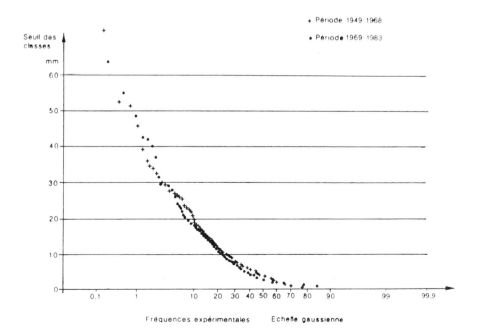

Figure 8

Discussion

L'affaiblissement du régime pluviométrique pendant la periode 1969-1983 semble être largement compensé par la modification des états de surfaces dans le fonctionnement des petits bassins versants. Ces modifications qui ont amené surtout sur les bassins au nord de l'isohyète 800 mm, des conditions de ruissellement plus favorables, sont dûes à l'action conjuguée de l'homme et des nouvelles conditions climatiques La diminution du couvert herbacé et l'extension des zones cultivées favorisent les tassements de la surface du sol et le développement de pellicules imperméables ainsi que l'extension de régions très érodées. Le bassin de Kognere présente actuellement les caractères des paysages habituels de régions plus septentrionales, vaste étendue de sol nus formant des glacis lisses ou caillouteux.

Les espèces de graminées pérennes disparaissent enfaveur des annuelles, dans la végétation arborée on remarque que les rares recrus sont toutes des epineux qui prennent la place des combrétacées.

La comparaison des distributions statistiques des lames ruisselées sur les deux périodes, tout comme les crues exceptionnelles apparues ici et là doivent mettre en garde contre une révision à la baisse des normes de sécurité pour les ouvrages en raison de la période sèche que nous vivons. L'apparente contradiction de fonctionnement entre ces systèmes hydrologiques et les plus grands rappelle l'hétérogénéité spatiale des phénomènes hydrologiques. En effet, si une forte pluie a une égale probabilité de se produire localement dans la période actuelle, elle survient temporellement dans une chronique moins pluvieuse et dans des conditions d'evaporation plus fortes. L'alimentation des nappes alluviales ainsi que celle des réserves de surface reste donc défavorisée mais ces phénomènes ne sont pas répercutés dans l'ecoulement des plus petits systèmes hydrologiques.

Références

Albergel, J., (1986) Evolution de la pluviometrie en Afrique Soudano -Sahelienne. Exemple du Burkina Faso, In: Colloque international sur la revision des normes hydrologiques suite aux incidences de la secheresse, Cieh Ouagadougo, 17.p.

Albergel, J. (Valentin, C. (1986) "Sahelisation" d'un petit bassin versant: Boulsa Kognere au centre nord du Burkina Faso; In: Colloque Nordeste Sahel, Institut des Hautes etudes d'Amerique Latine, Paris 1986.

Albergel, J. Ribstein, P., Valentin, C. (1985) Quels facteurs explicatifs de l'infiltration? Analyse sur 48 parcelles au Burkina Faso. Journees Hydrologiques de Montpellier, Coll. et sem. Orstom, 26-48.

Albergel, J. Casenave, A., Valentin C. (1985). Modelisation du ruissellement en zone soudano-sahelienne; simulation de pluies et cartographies des etats de surfaces. Journees Hydrologiques de Montpellier. coll. et sem. Orstom, 75-84.

Albergel, J. Carbonnel, J.P., Grouzis, M. (1984). Pejoration climatique au Burkina Faso. Incidences sur les ressources en eau et sur les ressources en eau et sur les productions vegetales. Cah. Orstom, ser. Hydrol. Vol XXIno. 1,3-19.

Billon, B. (1985) Le Niger a Niamey. Decrue et etiage 1985 Cah. Orstom, ser. Hydrol. Vol. XXIno. 4,3-22.

Carbonnel, J.P. Hubert, P; (1985) Sur la secheresse au Sahel d'Afrique de l'Ouest. Une rupture climatique dans les series pluviometriques du Burkina Faso (ex Haute Volta); C.R. Acad. Sc. serie Hydrologie Vol VII, tome 301 n 13, 941-944.

Leusink, A. Tyano, B. (1985) Observations du niveau de la nappe des eaux souterraines et sa composition chimique et isotopique de socle cristallin au Burkina Faso. Bull.de liaison du Cieh no. 62.

Nicholson, S.G. (1984) Rainfall fluctuations in Africa 1901 to 1973 in: Colloque OMM sur le Xeme anniversaire de l'experience Etga Dakar, decembre 1984, 103-105.

Olivry, J.C. (1983) Le point en 1982 sur la secheresse en Senegambie et aux iles du Cap Vert. Examen de quelques series de longue duree (debits et precipitations) Cah. Orstom, ser. Hydrol. Vol XX, no.1, 47-69.

Sircoulon, J. (1986) Bilan hydropluviometrique de la secheresse 1968-84 au Sahel et comparaison avec les secheresses des annees 1910 a 1916 et 1940 a 1949 in: Colloque Nordeste Sahel. Institut des Hautes etudes d'Amerique Latine, Paris - 16 au 18 janvier 1986.

Sircoulon, J. (1985) La secheresse en Afrique de l'Ouest. Comparaison des annees 1982-1984 avec les annees 1972-1973. Cah. Orstom, ser. Hydrol. Vol no. 4, 75-86.

Snidjers, T.A.B. (1983) Interstation correlations and non stationarity of Burkina Faso rainfall. Journal of climage and applied meteorology, Vol.25,524-531.

Valentin, C. (1985) Differencier les milieux selon leur aptitude au ruissellement: une cartographie adaptee aux besoins hydrologiques. Journees hydrologiques de Montpellier, coll. et sem. Orstom, 50-73.

The Influence of Climate Change and Climatic Variability on the Hydrologic
Regime and Water Resources (Proceedings of the Vancouver Symposium,
August 1987). IAHS Publ. no. 168, 1987.

The glacial and hydrological regime under climatic influence in the Urümqi River, northwest China

Yao Tandong
Lanzhou Institute of Glaciology and
Geocryology, Academia Sinica,
Lanzhou, China

ABSTRACT   The glacial and hydrologic fluctuations and
their distinct features under climate influence in the
Urümqi River are analysed.
     The glacial and hydrologic fluctuations in the Urumqi
River have the same pattern in which high discharge and
positive mass balance periods are mainly identical with
cold periods, low discharge and negative mass balance
periods identical with warm periods. A particular
temperature-wetness pattern, which could be explained by
prevailing general atmospheric circulation patterns is
responsible for the pattern.   Distinct features between
glacier and discharge under climatic influence are
distinguished. Glacial terminal fluctuates as a slow
process with low frequency while discharge fluctuates as a
rather fast process with high frequency in the basin,
which results from the time lag difference between glacier
and discharge responding to climatic fluctuation.

Variations glaciaires et hydrologiques sous
l'influence climatique dans le bassin de la
rivière d'Urümqi, dans le Nord-ouest de la
Chine

RESUME   Les variations glaciaires et hydrologiques ainsi
que leurs caractéristiques distinctes on été analysées
sous l'influence du climat dans le bassin de la rivière
d'Urümqui.
     Les variations glaciaires et hydrologiques dans ce
bassin ont la même tendance: les périodes de haut débit
et de bilan de masse positif sont principalement
identiques à la période froide, alors que les périodes de
bas débit et de bilan de masse nègatif sont identiques à
la période chaude. Ceci qui dépend principalement du
modèle particulièr température humidité, qui peut être
expliqué par la circulation atmosphérique régnant dans le
Nord. Sous l'influence du climat, les variations de la
glacière et du débit dans cette région ont des différences
évidentes: la glacière terminale fluctue avec un développement lent et une fréquence basse, tandis que le débit
fluctue avec un développement rapide et une fréquence
haute, ce qui résulte de la différence de retard de temps

entre la glacière et le débit à l'égard de la fluctuation climatique.

Introduction

Fluctuations of glacial terminal and discharge are generally related to climatic fluctuations. The mass budget process of glacier and discharge is dependent on heat balance and precipitation features which are related to climate. Some relationships between glacier, discharge and climate are discussed in the present paper by taking the Urümqi River as an example.

The Urümqi River originates from the eastern part of the Tianshan Mountain in the southern part of Urümqi. The river basin has an area of 924 km^2 above Yingxiong Bridge Hydrologic Station which is the main station measuring the outlet discharge, and an area of 46 km^2 of glaciers at the head area. The meteorologic record started in 1940 at the Urümqi Station in the lower reaches of the basin, and in 1958 at the Da Xigou Station in the upper reaches. The hydrologic record started in 1950 but it can be extended to 1940 by interpolation. The observations and records of glaciers in the basin started in 1958. The Urümqi River basin is a comprehensive one in earth science studies in northwestern China, which is suitable for a comprehensive study of the relationship between glacier, discharge and climate.

The fluctuations of glacier, discharge and their relationship to climate and general atmospheric circulation in the Urümqi river

The study of fluctuations in glacier and discharge in the basin was limited to the period since the Little Ice Age during which more data are available. The Little Ice Age started in the late sixteenth century and ended in the early twentieth century in the basin according to Chen (personal communication). In Glacier No. 1, the change in glacial area, glacial thickness, glacial length and volume, and in glacial equilibrium line fluctuation were estimated based on detailed glacial maps and radar sounding data (Table 1). Glacier No. 1 has decreased by 21% in length, 33% in area and 38% in volume from the maximum of the Little Ice Age to present. A similar feature was found in the fluctutations of all the glaciers in the basin. It has been estimated that there was a decrease of 41% in glacial volume and 30% in glacial area since the Little Ice Age in the whole basin.

Table 1 Changes in Glacier No.1 since the Little Ice Age

Period	Length (km)	Area (km^2)	Thickness (m)	Volume (10^{4}m^3)	Equilibrium line (m)
Present	2.33	1.84	49	9000	4050
the Little Ice Age maximum	2.98	2.38	60	14000	3890

There are three possibilities which could cause glacial advance in the Little Ice Age in the basin: (a) drop in temperature; (b) increase in precipitation; (c) combination of (a) and (b). There was certainly a drop in temperature on the whole hemisphere in the Little Ice Age, which must have influenced the basin. The temperature drop in the Northern Hemisphere in the Little Ice Age was about $1.0^{\circ}C$ in annual average, $0.5^{\circ}C$ in summer according to Flohn (1981), Lamb (1977) and Schuurman (1981). It was estimated by Chen (personal communication) that the temperature drop on Glacier No. 1 was about $0.6^{\circ}C$ in the Little Ice Age. The annual temperature drop was no more than $1^{\circ}C$ on Glacier No.1 if the smaller amplitude in climatic change in the mountain area in the basin and the results in other regions are considered.

According to the calculation of the relationship between precipitation, temperature and equilibrium line in Glacier No. 1 in the basin, a rise of 80 m in equilibrium line could be caused either by an increase of $1^{\circ}C$ in temperature or a decrease of 100 mm in precipitation. The equilibrium line in the Little Ice Age (3890 m) was about 160 m lower than that at present (4050 meter). It was estimated by using the above relationship that the precipitation on Glacier No. 1 during the maximum of the Little Ice Age is about 100 mm larger than that at present. Provided that the temperature is $1.5^{\circ}C$ lower during the maximum of the Little Ice Age than at present, an increase of 50 mm in precipitation was estimated. Because the estimate of temperature drop in the Little Ice Age was based on proxy data from other areas, the above values are not the exact estimation of the precipitation in the Little Ice Age. But it at least revealed a trend: not only was there a temperature drop but also a precipitation increase, characterized by cold and wet conditions, during the Little Ice Age in the basin. To support the above proposed relationship, three sets of tree ring data are analysed from the basin and nearby regions (Table 2).

Table 2 The relationship between temperature and precipitation revealed by the tree rings in the basin and nearby regions

Place	Total year in a tree (years)	Low temperature with high precipitation (years)	High temperature with low precipitation (years)	High temperature with high precipitation (years)	Low temperature with low precipitation (years)	Percentage of high T* with low P* and low T with high P (%)	Percentage of high T with high P and low T with low P (%)
Urümqi River	384	152	142	30	60	77	23
Altai	158	32	68	32	26	63	37
Hami	252	58	118	52	24	70	30
Average	265	81	109	38	37	70	30

Two points should be kept in mind when estimating the discharge in the basin during the Little Ice Age. The first is that the temperature then was lower than at present; the second point is that it is wetter then than at present, based on the above analysis. Because it is warmer and drier at present than in the Little Ice Age, 1976 (a year with the lowest temperature and highest precipitation since meteorological data were recorded) was selected to estimate the discharge in the Little Ice Age. In 1976, the temperature was 0.6°C lower and the precipitation 46.2 mm higher than the secular average at Da Xigou Station in the upper reaches of the basin; the temperature was 0.2°C lower and the precipitation 19.4 mm higher at Urümqi Station in lower reaches of the basin. The discharge at Ying Xiong Bridge Station that year was 280 670 000 m^3 which was 40 000 000 m^3 higher than the secular average. The discharge increased by 17% in the Little Ice Age than at present (or decreased by 14% at present than in the Little Ice Age) if taking the value of the discharge in 1976 as an estimation of the discharge in the Little Ice Age.

During the period with observations and records in meteorology, glaciology and hydrology, the glacial mass balance, glacial equilibrium line and discharge in the basin basically kept the same trend and experienced cycles. However, the glacial loss and discharge decrease trend is obvious during the same period. A comparison of two high discharge periods between the 1970's and the 1950's indicates that the total discharge in the 1970's was 148 million m^3 smaller than that in the 1950's, demonstrating a decreasing trend in discharge. In Glacier No. 1, the average mass loss is 155 thousand m^3 in water equivalence from 1959 to 1980 and the equilibrium line rise 10-12 m in the same period. The trends of the discharge and glacier decrease are related to climatic change in the basin. The average ten year temperature from the 1960's to the 1970's has increased by 0.3°C in Urümqi, by 0.2°C in Xiao Quzi and by 0.1 C in Da Xigou. The average ten year precipitation at Da Xigou Station which is close to glacier No. 1 has decreased by 7.2 mm from the 1960's to the 1970's. During the period in which recorded data are available, the relationship between the fluctuations of glaciers and discharge and climatic change is identical with that in the Little Ice Age. Although there are different temperature-wetness patterns (warm-dry, warm-wet, cold-dry, cold-wet), the dominant patterns were warm-dry patterns and cold-wet patterns. The analyses of the moving curve and anomalies in spring and summer temperature and spring and summer precipitation indicate that the period of high temperature-low precipitation (warm-dry pattern) and low temperature-high precipitation (cold-wet pattern) is 76% of the whole period analysed in the upper reaches and 63% of the whole period analysed in the lower reaches. The period of high temperature-high precipitation (warm-wet pattern) and low temperature-low precipitation (cold-dry pattern) is 24% and 37% respectively in the upper reaches and in the lower reaches.

It seems that the fluctuation features of glacier and discharge are dependent on the temperature-wetness feature in the basin. Nevertheless, the temperature-wetness pattern in the basin could be explained by changes of the general atmospheric circulation. The general atmospheric circulation in the Eurasia continent was classified into three types by Jiersi (1974). According to studies, precipitation would decrease and temperature rise in most regions of

middle latitude when W type prevails, there would be more opportunities for high temperature and low precipitation in most regions of
middle latitude when C type prevails, there would be low temperature
and high precipitation in most regions of the Eurasia continent when
E type prevails. In the Urümqi River, the general trend of fluctuations of climate and discharge is basically identical with the
general atmospheric circulation in the Eurasia continent (Table 3).

Table 3 Relationship between climate, discharge and general
atmospheric circulation during different periods in the
Urümqi River

Period	Circulation type	Precipitation anomaly	Temperature anomaly			Discharge fluctuation		
			Xiao Quzi	Urümqi	Da Xigou	Ying Xiongqiao	Northern of Xinjiang	Xinjiang
1941-1950	C	-7.3				Low discharge		
1951-1960	E	+12.9	-0.2	-0.4		High discharge	High discharge	High discharge
1961-1965	C	-4.5	+0.3	+0.7	+0.1	Low discharge	Low discharge	Low discharge
1966-1972	E	+6.0	-0.4	-0.1	-0.2	Low discharge	High discharge	High discharge

In the basin, E type generally corresponds to positive precipitation anomaly, negative temperature anomaly and a high discharge
period; W and C type generally correspond to negative precipitation
anomally, positive temperature anomaly and a low discharge period.

Distinct response between glacier and discharge
under climatic influence in the Urümqi River

According to what is discussed above, the glacial volume has
decreased by 40% and the discharge has decreased by 14% from the
Little Ice Age to present in the Urümqi River basin. It means that
the decrease rate of glacier is larger than that of discharge under
the same background of climatic warming, which results from the
distinctions in energy balance between glacier and discharge.
The input and output process in glacial system is expressed by the
mass balance equation

$$B = P_g - M - E_g + L \pm D \pm A \tag{1}$$

(where P_g stands for annual precipitation on glacial the surface, M
for glacial surface melting, E_g for glacial surface evaporation, L
for glacial condensation, D for snow drifting and A for avalanche).

The input and output process in the discharge system is expressed by the water balance equation

$$P_d = R + E_d \pm \Delta W + F \quad \text{or} \quad R = P_d - E_d \pm \Delta W - F \tag{2}$$

(where R stands for discharge, P_d for annual precipitation, E_d for evaporation, ΔW for underground water storage and F for seepage). To reveal the essential features of the input and output process of mass, a simplification to equation (1) and (2) is necessary.

According to heat balance curves by Budyko (1974), the evaporation process in the basin should belong to the continental climate type in middle latitudes. The feature of the process at this latitude is that evaporation is maximum in summer, dramatically changeable in spring and autumn, and negative in winter. Summer evaporation could therefore approximately substitute for annual evaporation. The rain season is during the summer in the Urümqi River basin, the intense evaporation season is also an intense condensation season in the basin. From observations on some glaciers in Central Asia (Lvovich, 1975), evaporation and condensation are almost equal in summer. Therefore, E and L in equation (1) could be omitted. Avalanche influence in the studied area could also be omitted. As a secular process, the amplitude of annual change of snow drifting is not so great and could be taken as a constant (D = W_c, for example). Equation (1) then could be simplified as

$$B = P_g - M + W_c \tag{3}$$

There are only two variables P_g and M in the equation.

As a secular process, F and in equation (2) are also relatively stable and could approximately be taken as a constant (F + W = W_c, for example). Equation (2) could then be simplified as:

$$R = P_d - E_d + W_c \tag{4}$$

There are also two variables, P_d and E_d, in equation (4).

Equations (3) and (4) demonstrate that the input of glacier and discharge are of the same form, but the output of these two systems are essentially distinct: by melting in the glacial system and by evaporation in the discharge system.

The mass output of both glacier and discharge is, in physical essence, the result of the input of heat. The main components of heat input are radiation balance (R), latent (L) and sensible heat (S). The heat balance equation in the glacial system could be expressed as:

$$Q_g = R_g + L_g + S_g \tag{5}$$

The heat balance equation in the discharge system could be expressed as:

$$Q_d = R_d + L_d + S_d \tag{6}$$

From equation (3) and (4) the heat (Q) absorbed in the glacial system is mainly consumed in glacial surface melting, therefore

$$Q_g \simeq M(Q) \tag{7}$$

The heat (Q) absorbed in the discharge system is mainly consumed in evaporation, therefore

$$Q_d \simeq E(Q) \tag{8}$$

Using latent heat of melting (80 cal.g^{-1}) and latent heat of evaporation (597 cal.g^{-1}), equations (7) and (8) could be expressed as:

$$M = Q_g/80 \tag{9}$$

$$E = Q_d/597 \tag{10}$$

Substituting equations (9) and (10) for the second item on the right side of equations (3) and (4) respectively, then

$$B = P - Q_g/80 + W_c \tag{11}$$

$$R = P - Q_d/597 + W_c \tag{12}$$

Equations (11) and (12) indicate that the response of the input and output process of mass to climate (actually to heat balance) is much more sensitive in glaciers than in discharge. Besides, evaporation in the discharge system only consumes part of the heat absorbed, the denominator in equation (12) is relatively enlarged. So, the response of water balance to climate is more sluggish than that expressed in equation (12).

The heat balance in the glacial system is, however, smaller than that in the discharge system, and the mentioned distinction has been weakened in some degree. But it is still evident according to the results from Glacier No. 1 and nearby area. Taking the average secular melting value (214 cm) at 3870 m a.s.l. on Glacier No. 1 and evaporation value (53 cm) at Da Xigou Station near Glacier No. 1, corresponding melting latent heat of 17 000 000 cal.cm^{-2}a^{-1} and evaporation latent heat of 32 000 000 cal.cm^{-2}a^{-1} were obtained. This is to say that although the heat consumed in discharge evaporation is two times that in glacial melting, mass loss in discharge is only 1/4 of that from the glacier. It can be deduced from the above discussion that the fluctuation amplitude of the glacier is larger than that of discharge in the past under the influence of climatic change in the Urümqi River basin.

There are obvious distinctions between the glacier and discharge in the basin under modern climatic influence. One of the distinctions is that a glacial terminal responds to climatic change as a slow process with low frequency while discharge responds to climatic change as a fast process with high frequency. Taking Glacier No. 1 as an example, the glacial terminal maintains its retreating trend since observations in the late 1950's. But the discharge in the basin experienced high and low discharge cycles during the same period, which is basically identical with the climatic fluctuation in short period and therefore rather sensitive to climatic change. Because temperature in the basin shows a warming trend in this century, it seems that the retreating trend of the glacial terminal is identical with the warming trend in climate and the glacial terminal is a rather stable indicator for the secular climatic trend. The reason for the distinctions could be explained by time lag differences between glacier and discharge in their response to climate.

The time lag of the glacial terminal ranges from several years (small mountain glaciers) to decades (larger mountain glaciers) (Nye, 1958, 1965) or even thousands of years (Budd and Smith, 1979). The time lag in the discharge process is much shorter, ranging from several hours to several days or 10's of days. This demonstrates that the discharge system could reflect climatic influence immediately in the lower reaches of a river, but the glacial terminal could require a rather longer period to reflect it. Many factors affect glacial time lag. The main factors are mass balance, which mainly depends on precipitation and temperature, glacial size and slope, glacial velocity and glacial temperature, if surging glaciers are neglected. There is no pattern to follow governing the importance of these factors. It is, therefore, difficult to find a model including all these factors and suitable for all types of glaciers. Glacial length is, however, one of the most important factors responsible for glacial time lag and is easy to obtain. A model could be established by using glacial length to estimate approximately glacial time lag in different sizes. Based on the statistical analysis of 35 "normal" (ie. non-surging) glaciers in the Northern Hemisphere, a simple model was established as

$$T = 13L^{0.375}$$

(where T stands for the glacial time lag, and L for glacial length). The relationship factor for the model has a significance of 95%. Because only glacial length L was introduced in the model, the time lag estimations are approximate values. It was estimated from the model that the time lag for Glacier No. 1 may be 18 ± 5 years.

The time lag of the discharge in the basin is different in different seasons, but depends mainly on the character of high water season of a year. The high water season in the Urümqi River could be classified as spring and summer. Summer high water season is characterized by precipitation discharge with a short time lag. It was calculated from the recorded data at an experimental discharge station in the Urumqi River, that the peak flood is 7-8 hours after peak precipitation from Hou Xia to Ying Xiongqiao. According to this, the time lag from the head to the lower reaches of the river is about one day. Spring high water season is characterized by snow melting discharge and is rather complicated. Generally, the amount of spring high water reflects the amount of precipitation during the last winter or even previous autumn plus the heat absorbed in the mountain area in spring. The time lag could therefore be several months. Although this is the result of interruption by other factors, it is still a feature in the discharge forming process. In addition, time lag of discharge also changes with climate at different altitudes. Usually, summer precipitation in the middle and lower parts of the mountain is in the form of rainfall and could be reflected immediately in the lower reaches, while the summer precipitation in the high mountain area is mainly snow and needs a longer time to be reflected in the lower reaches. In conclusion, the time lag of discharge in the basin ranges from several hours to several months from the lower reaches in summer to the upper reaches in spring.

If 10 years represents the order of magnitude of the time lag of the glacial terminals of most glaciers in the basin and 1-2 days represents the time lag of discharge from the head to the lower reaches of the basin, the time lag of the former is several orders of magnitude longer than that of the latter. Even taking several months as the time lag of the discharge in the basin, the time lag of the glacial terminal is still one order of magnitude longer than that of discharge. This partly explains why the glacial terminal responds to climate in low frequency while discharge responds to climate in high frequency.

The future trend of the glacier and discharge in the Urümqi River

Based on the discussion above, the glacial and discharge fluctuations in the basin are controlled by climatic change. The forecasting of the glacial and discharge fluctuation is only possible if the trend of the climatic change is forecast. But climatic change is still a subject which is beyond the forecasting ability of man. Most climatic forecasts only propose possibilities. This is also the case in the Urümqi River. Based on tree ring analysis and forecasting of the general atmospheric circulation by Jiersi (1974) and the relationship between the climatic change in the basin and general atmospheric circulation, a possibility of the future climatic trend in the basin was estimated. It was forecast by Jiersi (1974) that W+E circulation would prevail from 1979 to 1986, and W+C prevail from 1986 to 1996. A possible climatic trend accompanied with the circulation is that the precipitation from 1979 to 1986 would fluctuate near its secular average and that the precipitation from 1986 or so would decrease and last to the late 1990's. The decrease in precipitation may have started in 1985. Two conclusions could be made based on the tree ring data: (a) the present is a dryer period than the Little Ice Age; (b) this trend would continue until the 2020's-2030's. A wet period of several years may appear between the late 1990's and the early 2000's. The temperature rise in the next 10-15 years is probably $0.1-0.6°C$ if the temperature in the basin keeps the rising trend at present.

In recent years, the CO_2 greenhouse effect has been much discussed by Manabe & Stouffer (1980) and Manabe and Wetherald (1981). It was estimated by them that the average global temperature would increase by $2-3°C$ because of the increase of CO_2 and other gases in the atmosphere in the next 50 years. A temperature increase of $1.5-4.5°C$ in the next 50 years was announced at the Villach Conference in 1985. There was not much discussion about the potential temperature increase in the next 10-15 years. It may be still small before the twenty-first century and an estimation of $0.5°C$ might be suitable because there is a time lag from CO_2 content increase in the atmosphere to actual temperature rise. The time lag would be even longer if the influence of oceans is considered.

A rise of 0.5-1.0°C in temperature would be possible in the next 15 years if taking 0.5°C as the temperature increase caused by CO_2 and provided that the natural climate keeps its present warm trend. Assuming climatic warming in the next 15 years occurs with a range from 0.5 to 1.0°C, the forecast for glaciers and discharge around the year 2000 could be made (Table 4).

Table 4 Estimation of changes in glacier and discharge under the condition of temperature rise of 0.5-1.0°C in the Urümqi River around 2000

Period	Temperature increase (°C)	Precipitation decrease (mm)	Altitude of glacial equilibrium line (m)	Glacial mass balance (mm)	Discharge of the River $(10^6 m^3)$	Discharge Glacier No.1 $(10^6 m^3)$	State in glacial terminal	Glacial thinning (m)
1985-2000	0.5-1.0	10 - 30	4070 - 4100	-100- -300	22.1-23.7	1.5-2.1	retreating	2 - 8

Conclusions

The following conclusions were made from the above discussion:
(a)   The glacial and discharge fluctuations in the Urümqi River basin are related to climatic change from the Little Ice Age to present, which were demonstrated to be glacial mass balance and discharge decrease when the climate becomes warmer; glacial mass balance and discharge increase when the climate becomes colder.
(b)   The glacier and discharge fluctuations in the basin are related to general atmospheric circulation through climate. It is advantageous to glaciers and discharge in the basin when an E type of circulation prevails, but it is not advantageous to them when W+C types of circulation prevail.
(c)   Under the influence of a warming climatic trend since the Little Ice Age, glacier and discharge in the basin have both decreased, 41% in glacial volume and 14% in annual discharge. The rate of glacial decrease is larger than that of discharge decrease, which is the result of the distinction in energy balances between glacier and discharge.
(d)   Glacial and discharge fluctuations possess different climatic significance because of their distinctions in time lag and other processes. Glacial terminal fluctuation (a slow process with low frequency) could be taken as a relatively stable indicator for secular climatic change, while discharge fluctuation (a rather fast process with high frequency) could be taken as a relatively sensitive indicator for climatic fluctuation in a short period.

References

Budd, W.F. & I.N. (1979) The growth and retreat of ice sheets in response to robital radiation changes. Sea Level, Ice and Climatic

Change (Proc. Chanberra Symposium, December 1979) 369-409.

Budyko, M.I. (1974) Climate and Life, Academia Press, New York and London. QC 801 155 #18 C.1.

Flohn, H. (1981) Scenaries of cold and warm period of the past. Climatic Variations and Variability: Facts and Theories. D. Reidel Publishing Company.

Jiersi, A.A. (1972) The Secular Oscilation in General Atmospheric Circulation and the Secular Forecasting in Hydrology and Meteorology (translated into Chinese from Russian), Academia Press.

Lamb, H.H. (1977) Climate: Present, Past and Future, London.

Lvovich, M.I. (1975) A method of studying the water balance and estimating the water resources of glacial mountain area. Snow and Ice Symposium, IAHS-AISH Publ. No. 104.

Manabe, S. & R. S. Stouffer (1980) Sensitivity of a global climate model to an increase of CO2 concentration in the atmosphere. J. Geophy. R 85 (5529) C 10.

Manabe, S. & Wetherald, (1981) On the distribution of climate change resulting from an increase in CO2 content of the atmosphere. J. Atmos. Sci. 37(3) pp. 99-118.

Manabe, S. & Wetherald, R.T. & Stouffer (1981) Summer dryness due to an increase of atmospheric CO2 concentration. Climatic Change 3 347-386.

Nye, J.F. (1958) A theory of wave formation on glaciers, IASA 47.

Nye, J.F. (1965) The frequency response of glacier, J. Glaciol. 5(4) 567.

Schuurman, C.J. (1981) Climate of the last 1000 years. Climate Variations and Variability: Facts and theories, D. Reidel Publishing Company.

The Influence of Climate Change and Climatic Variability on the Hydrologic Regime and Water Resources (Proceedings of the Vancouver Symposium, August 1987). IAHS Publ. no. 168, 1987.

A primary study of the relationship between glacial mass balance and climate in the Qilian Mountain taking "July First" Glacier as an example

Xie Zichu, Liu Chaohai
Lanzhou Institute of Glaciology and Geocryology
Academia Sinica, Lanzhou, China

ABSTRACT   The average summer (from June to August) air temperature is 1° C lower and the equilibrium line 88m lower in the 18 years after 1968 than the 11 years before 1968 in the Qilian Mountain. It was indicated from calculation that an increase (decrease) of 1°C in air temperature would result in a rise (or drop) of 80 m of the equilibrium line. If taking the base station (3700 m.a.s.l.) temperatures above zero $^\circ$C and below zero $^\circ$C at the glacial terminus as a temperature index for the beginning and the end of the glacial melting period respectively, the melting period is 14 days shorter in average in the 18 years after 1968 than the 11 years before 1968 and the maximum ablation altitude has dropped by 130m in average on "July First" Glacier in the middle of the Qilian Mountain. It was found from the mass balance reconstructed according to air temperature and precipitation from 1957 to 1985 on "July First" Glacier (with an area of 3.0 km^2) that it was characterised by a negative mass balance state in the 11 years before 1968 with a cumulative negative mass balance value of $232.9 \times 10^4 m^3$ and characterized by a positive mass balance state in the 18 years after 1968 with a cumulative positive mass balance value of 773.4×10^4 m^3. In the recent 30 years, net mass balance has increased by $540.5 \times 10^4 m^3$ and average glacial thickness increased by 1.8m. Relatively larger positive mass balance appeared in 1967/68, 1975/76, 1982/83, which showed periodical cycles of 7-8 years. Fluctuation of 1 m of equilibrium line corresponds to a mass balance change of $0.68 \times 10^4 m^3$. Glaciers in the Qilian Mountain are typical glaciers accumulating in summer and which possess the following features: $c_s > c_w$, $a_s > a_w$; $b_s' > 0$, $b_w' > 0$ (accumulation area); $b_s'' < 0$, $b_w'' < 0$ (ablation area). On such kind of glaciers, which are typical continental glaciers, mass balance processes are mild and level of mass balances are low. The shortening of the ablation period and drop in summer temperature since the middle 1960's and the increase in precipitation since the 1970's have resulted in positive mass balance and a decrease of glacial retreat.

Etude sur le rapport entre le bilan de masse glaciaire
et le climat dans la montagne de Qilan

RESUME   Dans la montagne de Qilan, la température moyenne
en été (de juin à août) est de $1^{o}C$ plus basse pendant les
18 années après 1968 et la ligne d équilibre est de 88 m
plus basse que les 11 annees avant 1968. Le calcul montre
que la ligne d équilibre devrait monter (ou baisser) de
80 m, lorsque la température moyenne en été monte (ou
baisse) de $1^{o}C$. Si P'on prend à la station de base les
température moyennes journalières au-dessus et au-dessous
du zéro ^{o}C au bout des glaciers pour indices respectifs de
température au début et à la fin de la période d ablation
des glaciers, la periode d ablation est en moyenne de 14
jours plus courte pendant les 18 années après 1968 que les
11 années avant 1968, et l'altitude d ablation maximale
baisse en moyenne de 130 m dans la glacière "au 1er
juillet" au milieu de la montagne de Qilan.
    Le bilan de masse établi d après la température et les
précipitations de 1957 à 1984 dans la glacière au 1er
juillet (avec une surface de 3,00 km^{2}) montre qu on a
principalement dun état de bilan de masse négatif pendant
les 11 années avant 1968 avec une valeur cumulative du
bilan de masse négatif de $232,9x10^{4}m^{3}$ et état de bilan de
masse positif pendant les 18 années après 1968 avec une
valeur cumulative du bilan de masse positif de 773,4
x $10^{4}m^{3}$. Pendant ces 30 dernières années le bilan de masse
net a augmenté de $540,5x10^{4}m^{3}$ et 1 épaisseur moyenne des
glacières a augmenté de 1.8 m. Le bilan de masse positif
relativement large est apparu en 1967/68, 1975/76,
1982/83, avec un cycle périodique des 7-8 années. La
fluctuation de 1 m de a ligne d équilibre correspond à un
changement du bilan de masse de 0,68x10 m . Les glacières
dans la montagne de Qilian sont des glacières typiques,
accumulées en été avec les caractéristiques suivantes:
$c_{S}>c_{W}$, $a_{S}>a_{W}$; $b_{S}^{'} >0$, $b_{W}^{'} >0$ (la surface d accumulation);
$b_{S}^{''} <0$, $b_{W}^{''} <0$ (la surface d'ablation). Dans les glacières
de ce genre, qui sont des glacières continentales typiques,
le processus de la balance de masse est lent, et le niveau
de la balance de masse est bas.
    Depuis le milieu des années 60, la période d ablation
est devenue plus courte et la température en été plus
basse. Pour ailleurs, les précipitations ont augmenté
depuis les années 70, ce qui aboutit à bilan de masse
positif et à la diminution de la retraite des glacières.

Introduction

A systematic study of glacial mass balance was carried out by Lanzhou
Institute of Glaciology and Geocryology of Academia Sinica from 1975
to 1979 on 4 glaciers in the Qilian Mountain (Wang Zhongxiang, 1985).
To monitor recent glacial fluctuation, a glacial mass balance study
was made on "July First" Glacier from 1984 to 1985. Based on these

data, the relationship between glacial mass balance and climate was analysed.

Methods and results

Glacial melting was recorded using poles and snow stratigraphy was observed using snow pits. Net balance (b) at every point is the algebraic total of ice (b_i), snow (b_s) and superimposed ice (b_{si}):

$$b = b_i + b_s + b_{si} \tag{1}$$

Snow pits were used to observe glacial accumulation. Snow pits were dug in late August or early September. When the dirty layer of snow from the previous year was found, the depth of snow layer (m) and its density (d) were measured and the annual accumulation was calculated:

$$c = \Sigma dm \tag{2}$$

Glacial mass balance was calculated using the contour method. Net balance (B), net ablation and net accumulation were obtained by multiplying the area (S_{ci}, S_{ai}) between two intervals with average accumulative depth and melting depth (c_i, a_i):

$$B = \Sigma\ c_i S_{ci} + \Sigma\ a_i + S_{ai} \tag{3}$$

Based on the above observation and calculation, it was estimated that the annual mass balance in "July First" Glacier was $+67.3 \times 10^4 m^3$ in 1983/84 and $-9.1 \times 10^4 m^3$ in 1984/85. To compare, the mass balance results from 1974/75 to 1976/77 are listed in Table 1.

Table 1 The results of glacial mass balance of "July First" Glacier in the Qilian Mountain

Balance year	Altitude of zero-equilibrium line	Net accumulation			Net ablation			Net balance	
		Area (km^2)	Depth (mm)	Amount ($\times 10^4 m^3$)	Area (km^2)	Depth (mm)	Amount ($\times 10^4 m^3$)	Amount ($\times 10^4 m^3$)	Balance (mm)
1974/75	4650	2.14	367	78.6	0.90	758	68.2	+10.4	+35
1975/76	4550	2.63	502	132.0	0.41	373	15.3	+116.7	+384
1976/77	4620	2.48	532	132.0	0.56	454	25.4	+106.6	+350
1983/84	4600	2.41	364	87.7	0.57	358	20.4	+67.3	+226
1984/85	4710	1.78	288	51.3	1.20	503	60.4	-9.1	-31

The relationship between annual variation
in glacial mass balance and climate

Glacial mass balance is the direct response of climatic fluctuation
and an important indication to link air temperature(ablation
indicator) and precipitation (accumulation indicator). It therefore
can be reconstructed and forecast by temperature and precipitation.
The temperature on the glacial surface is an important parameter to
calculate ice and snow melting and glacial discharge. Recorded data
in the glacial area in the Qilian Mountain is discontinuous and the
temperature on the glacial surface was calculated. Taking "July
First" Glacier as an example, the summer temperature at the head of
East Niumaojuanzi Valley, which is 5.6 km away from the glacier, in
1976 and in 1984 was recorded and interpolated to understand the
temperature and its variation in the mountains. It was found from the
meteorologic record on "July First" Glacier that there is a relationship between the 10 days average temperature (Y) at the station and
the average temperature (X) at 650 mb in Jiuquan:

$$Y = 0.046 + 1.016X \tag{4}$$

The relationship coefficient for equation (4) is 0.978 and
standard variance is 0.3. The equation has a high significance, and
could be taken to interpolate and to extend recorded data for a short
period. During the recent 30 years, the summer (from June to August)
average temperature in the middle of Qilian Mountain is similar to
that in the Hexi Corridor. 1968 is a key point. Average temperature
from 1968 to 1985 is $1.0^{\circ}C$ lower than that from 1957 to 1967 (Figure
1). A similar temperature dropping feature was found in the high
mountain area of the eastern and western parts of the Qilian Mountain
(Ding Liangfu, 1985).

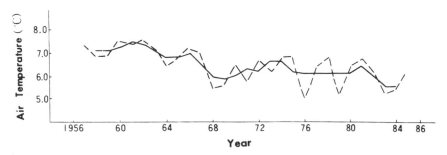

Figure 1   Curves of annual fluctuation and three year moving average
summer temperature on the "July First" Glacier.

The zero-equilibrium line is a sensitive indicator of glacier
response to climatic change, which is strictly governed by radiation
and precipitation. Glacial accumulation is equal to glacial melting
at glacial zero-equilibrium line, where glacial melting is mainly
dependant on the summer temperature. An approximate altitude of the
zero-equilibrium line could be determined by the average summer
temperature. In years with equal amounts of solid precipitation,

surplus heat would make the zero-equilibrium line rise as well as melting the solid precipitation in a year with high temperature, the zero- equilibrium line would drop in a year with a low temperature. This process reflects a dynamic equilibrium process. Therefore, the average temperature at the zero-equilibrium line not only reflects glacial melting, but also reflects glacial accumulation. It is a comprehensive indicator for glacial melting and accumulation. The freezing level over Jiuquan was selected to calculate the zero-equilibrium line on "July-First" Glacier. The analysis shows a good relationship:

$$H \quad = \quad 2968.93 \ \exp(0.0988h) \tag{5}$$

Where H stand for the glacial zero-equilibrium line (m), and h for the altitude of freezing level over Jiuquan ($\times 10^3$m). The relationship coefficient is 0.987 and standard variance 13 for equation (5). It could be used to reconstruct the zero-equilibrium line with upper air data.

The trend of the zero-equilibrium line on "July First" Glacier is basically identical with average summer temperature showing a drop of 88 m in zero-equilibrium line during the 18 years after 1968 (Figure 2). Average summer temperature fluctuation corresponds to altitude fluctuation in the zero-equilibrium line, an increase (or decrease) of 1 °C in temperature corresponds to a rise (or drop) of 80 m in zero-equilibrium line which is smaller than the temperature dependence on some glaciers and is smaller than that on Glacier No.1 in the Urümqi River (an increase of 1°C corresponds to a rise of 125m Kuhn,M. 1981). The altitude of the zero-equilibrium line on "July First" Glacier fluctuates between 4550m and 4770m with a maximum amplitude of 220m.

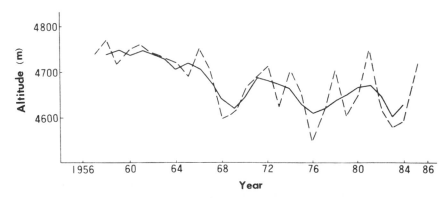

Figure 2   Curves of annual fluctuation and three year moving average of zero - equilibrium line on the "July First" Glacier.

The main source of mass for the glacier is precipitation. The precipitation in the Qilian Mountain is concentrated mainly in summer. The summer precipitation is 70-80% of annual precipitation. Because glaciers are at high altitudes, precipitation is mainly in the form of snowfall. The amount of precipitation is therefore

responsible for glacial accumulation. The precipitation in the
mountain is composed of dynamic precipitation and convective
precipitation which increase with altitude. The combination of these
two types of precipitation makes the mountains' precipitation stable
and the amplitude of annual change small. The variability factor is
generally smaller than 0.2. The stability of mountain precipitation
makes the amplitude of annual variation of total glacial accumulation
small. But net accumulation, which possesses secular water cyclic
function to glacier is mainly controlled by seasonal snow melting in
accumulative area, i.e., mainly controlled by temperature during
ablation period. Glacial coefficient (or accumulation area ratio of
glacier) which is governed by the altitude of glacial zero-
equilibrium line reflects comprehensively the influence of
precipitation and temperature on net glacial accumulation. A linear
relationship between average glacial accumulation ($\bar{c}$) and glacial
coefficient (f) was found:

$$\bar{c} = 162.12 + 58.75f \tag{6}$$

Glacial melting is a function of temperature. Krenke, A.N. (1971)
and Hoinkes, H. (1971) calculated the glacial melting and glacial
mass balance in Central Asia in the Soviet Union and in the Alps in
Switzerland respectively. Glacial melting in the Qilian Mountain
occurs mainly in the summer. It is therefore reasonable using average
summer temperature (T_S) to calculate average glacial melting
intensity ($\bar{A}$). Their relationship can be expressed as :

$$\bar{A} = 425.96 + 146.28\ T_S \tag{7}$$

The mass balance in "July First" Glacier during the past 29 years
(1957-1985) was calculated according to the above relationship
between mass balance components and the climatic factors. The trend of
mass balance, average summer temperature in the mountains and the
zero-equilibrium line altitude is basically identical, i.e., it is
characterized by negative mass balance conditions with a cumulative
negative value of $232.9 \times 10^4 m^3$ in the 11 years before 1968, while it
is characterized by positive mass balance conditions with a
cumulative positive value of $773.4 \times 10^4 m^3$ (Figure 3) in 18 years
after 1968. The glacier has thickened by 1.8m during the same period.
Relatively larger positive mass balance appeared in 1967/68 (77.2×10^4
m^3), 1975/76 ($116.7 \times 10^4 m^3$), and 1982/83 ($87.1 \times 10^4 m^3$) in "July First
"Glacier, with a cycle of 7-8 years. The drop of glacial zero-
equilibrium line (H) corresponds to an increase in mass balance (B)
with a good linear relationship:

$$B = 3\ 186.25 - 0.68\ H \tag{8}$$

A rise (or drop) of 1m in zero equilibrium line corresponds to an
increase (or decrease) of $0.68 \times 10^4 m^3$ in mass balance.

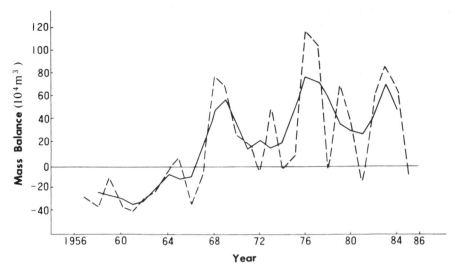

Figure 3   Curves   of   annual variation and three moving   average   of
mass balance from 1957 to 1985 on "July First" Glacier.

The relationship between variation in glacial
mass balance and climate within a year

The   seasonal   variation in glacial area in   Qilian Mountain   can   be
classified   into   cold   and   warm seasons according   to   whether   the
monthly   mean   temperature is below or above zero.   The   cold   season
(from   September   to May of the next year)   corresponds   to   the   arid
season,   in   which   the ablation and accumulation are very   small.The
warm   season   (from June to August) corresponds to   the   wet   season,
which is not only the period of glacial ablation, but also the period
of   main glacial accumulation.   Such character of summer   nourishment
makes   the   process of mass balance mild and the mass   balance   level
comparatively   low.   Furthermore,   it   also   has   influence   on   ice
temperature,   ice   formation   and glacial motive velocity as well   as
other   physical   properties.   These   features   make   the   continental
properties of the glaciers more prominent.   So the Qilian Mountain is
the   most   typical area of extra-continental glaciers in   the   world.
Taking the "July First" Glacier as an example, during the cold season
in   1974/75 and 1976/77,   there was stronger ablation on the   tongue,
i.e., the winter mass balance became negative. Observation during the
cold   season in 1984/85 also showed a negative winter   mass   balance.
Even   the abnormal phenomenon that the ablation during warm season is
smaller than that during the cold season was observed (Table2).   Thus
it came to light that for a typical glacier of warm season nourishment   the   accumulation,   ablation and mass balance have the   following
correlations (Xie Zichu, 1980):

$c_s > c_w$, $a_s > a_w$;
$b_s > 0$, $b_w > 0$ (in accumulation area);
$b_s'' < 0$, $b_w'' < 0$ (in ablation area).

Table 2   The mass balance on the ablation ? of "July First" Glacier

Interval (Year, Month)	Altitude of zero-equilibrium line (m)	Ablation area (km²)	Amount of balance (x10⁴m³)	Depth (mm)
1975,9 — 1976,5	4560	0.42	−9.22	−219
1976,6 — 8	4510	0.32	−7.65	−239
1976,9 — 1977,5	4550	0.41	−6.05	−148
1977,6 — 8	4590	0.55	−22.03	−400
1984,6 — 8	4600	0.57	−20.40	−358
1984,9 — 1985,5	4540	0.38	−5.53	−147
1985,6 — 8	4710	1.19	−56.17	−470

these characteristics of the mass balance process are just opposite from that of the cold season nourishment glaciers.

The precipitation during the cold season is no more than 20–30% of annual precipitation. The exposed glacial ice melting under solar radiation causes the negative value of mass balance in cold season. Besides, since glacial mass balance is observed on fixed dates, the ablation doesn't completely coincide with the standard summer months. This is also one of the causes for the negative value of the winter mass balance in the tongue. Thus, the air temperature at various altitudes from late May to early or middle September are especially analysed to calculate the date of the beginning and end of ablation.

According to the analysis of the observed data, the following relationship is revealed between the mean daily temperature at certain altitudes of "July First" Glacier and the corresponding daily mean temperature at the base station (3700 m.a.s.l.):

$$T_G = 0.83t_B - 0.66(H_G - H_B)/100 - 0.3 \tag{9}$$

Where T_G is the mean daily air temperature at any altitude on the glacier ($^\circ$C); t_B is the mean daily air temperature at the base station ($^\circ$C); H_G the altitude of a given point on the glacier (m); H_B is the altitude of the base station (m). It is very convenient to calculate the corresponding daily mean air temperature from equation (9) at the base station while the temperature of a given altitude of glacier is equal to $0\,^\circ$C. According to calculation, a daily mean air temperature of $0\,^\circ$C at the glacial terminus corresponds to a daily mean air temperature of $4.4\,^\circ$C at the base station, a temperature of $0\,^\circ$C at 4500 m corresponds to a temperature of $6\,^\circ$C at the base station; a temperature of $0\,^\circ$C at the glacier summit (5158 m) i.e., the whole glacier is under ablation state, corresponds to a temperature of $11.2\,^\circ$C at the base station. The corresponding daily mean temperature at the base station can be used as a temperature index indicating the beginning and end of the ablation period while daily mean air temperature at the end of the glacial tongue is above or below $0\,^\circ$C.

Summary and analysis

The results from calculation show that the ablation period in different years are distinct. The ablation period usually begins in early June, and ends in early September. The persistence of the ablation period mainly depends on the temperature conditioning in spring and autumn. The decrease of air temperature in Spring and Autumn results in the shortening of the ablation period, and vice versa. For instance, because of the higher temperature in the spring and autumn of 1963, the ablation period of the glacier was elongated to 121 days, the lower temperature in Spring and autumn in 1971 lead to the shortening of the ablation period to 71 days. With increasing altitude, the ablation period becomes shorter and shorter, up to the snow line(4650 m), the mean ablation period is 42 days. In some years, the maximum ablation altitude might get to 5150 m, which causes the lift of infiltration zone and leads to the absence of cold infiltration-recrystallization zone. Besides, during the ablation period, the weather process with strong temperature drops would also lead to great decreases of air temperature on the glacier, even making the ablation altitude lower than the glacial terminus and the meltwater would disappear temporarily.

Taking 1968 as a key point, the beginning of the ablation period on the "July First" Glacier during the 18 years after 1968 was 8 days later than that in the 11 years before 1968, but the end time was 6 days earlier i.e. the ablation period was shortened by about 14 days. The mean maximum melting altitude has dropped by 130 m or so. The shortening of the ablation period and the decrease of its temperature made the runoff of the rivers, which are mainly nourished by glaciers, decrease remarkably, meanwhile the retreat of glaciers has slowed noticeably. Glacier No. 4 in Shuiguan River of the east part of the Qilian mountain retreated 8.9 m on average from 1967 to 1984, but 16m from 1956 to 1976 each year. The "July First" Glacier in the middle part of the Qilian mountain has retreated for 1 m each year from 1975 to 1985, and has nearly got to a stable state. The tongue of Glacier No. 12 in Laohugou in west part of the Qilian Mountain becomes thicker, and the glacial area has begun to extend.

References

Wang Zhongxiang, Xie Ziuchi, Wu Guanche (1985) Mass balance of glacier in Qilian Mountains, Memoirs of Lanzhou Institute of Glacialogy and Cryopedology Chinese Academy of Sciences, No. 5.
Ding Liangfu, Kang Xingcheng (1985) The relationship between climatic variances and glacier changes, Memoirs of Lanzhou Institute of Glaciology and Cryopedology Chinese Academy Sciences, No.%.
Kuhn, M. (1981) Climate and glaciers, Sea level, ice and Climatic Changes, IAHS publ. No. 131.
Krenke, A.N. (1971) Climatic conditions of present day glaciation in Soviet Central Asia, Snow and Ice Symposium, IAHS publ. No. 104.
Hoinkes, H. and Steinacker, R. (1971) Hydrometeorological implications of the mass balance Hinterisferner, 1952-53 to 1968-69, Snow and Ice Symposium, IAHS publ. No. 104.
Xie Zichu (1980) Mass balance of glaciers and its relationship with

characteristics   of   glaciers,   *Journal   of   Glaciology   and Cryopedology*, Vol.2, No.4.

The Influence of Climate Change and Climatic Variability on the Hydrologic Regime and Water Resources (Proceedings of the Vancouver Symposium, August 1987). IAHS Publ. no. 168, 1987.

Global climatic changes and regional hydrology: impacts and responses

Peter H. Gleick
Energy and Resources Group
University of California
Berkeley , California, USA

ABSTRACT   As the atmospheric concentration of carbon dioxide and other trace gases increases, changes in global and regional climatic conditions will lead to a wide range of hydrologic impacts, including changes in the timing and magnitude of runoff and soil moisture. These hydrologic changes, in turn, will result in diverse economic, social, and political consequences.

The nature of the regional hydrologic effects depends on changes in the climatic conditions and the water-resource characteristics of the region. The research conducted to date has identified a wide range of potential problems--as well as some possible advantages--that might result from plausible changes in climate estimated by state-of-the art general circulation models.

These hydrologic changes fall into a series of distinct categories, including: changes in the timing of water availability; changes in the magnitude of water availability; changes in the hydrologic variability; and effects on water quality. Similarly, diverse societal responses to the hydrologic changes are available, including adaptation, mitigation, and prevention. Each of these responses depends on the quality of the information available on future impacts and on the perceived importance of the effects. This paper discusses the extent and character of hydrologic changes that could result from global climatic changes, together with the options available for hydrologists and water planners.

Introduction

Growing attention is being paid to climatic changes that may result from increasing atmospheric concentration of carbon dioxide and other trace gases. While the direct effects of changes in climatic conditions can be severe--as can be seen by the effects of existing climatic variability--we must also pay attention to the wide range of indirect effects, such as changes in agricultural productivity, changes in sea-level, and changes in water resources. This latter category is one of the most important and yet least well-understood consequence of future changes in climate. Hydrologic impacts may include major alterations in the timing and magnitude of surface runoff and soil-moisture availability, and changes in the quality of freshwater resources. Associated with these effects will be a wide range of economic, environmental, and societal impacts. This paper discusses the likely extent and character of important hydrologic changes that could result from global climatic changes, together with the options available to hydrologists and planners for dealing with the most severe impacts.

The limited research conducted to date has identified a wide range of potential problems—as well as possible advantages—that might result from plausible changes in climate. These hydrologic changes fall into distinct categories, including: changes in the timing and magnitude of water availability; changes in the frequency and severity of severe events, and effects on water quality. Similarly, diverse societal responses to the hydrologic changes are possible, including adaptation, mitigation, and prevention. Each of these responses depends on the quality of the information available about future impacts and on the perceived importance of the effects.

Future climatic changes

Despite the fact that hydrologists need accurate information on climatic means and variability in order to develop appropriate water-resource designs and rules of operation, details of future climatic conditions cannot yet be predicted with any high degree of confidence. The principal reasons for this inability to clearly identify future climatic changes are the complexities of the ocean-atmosphere-land interactions, the difficulties of developing satisfactory computer models to reproduce these interactions, and uncertainties about our actions that affect climatic conditions.

The problem is that, at present, while there are many ways in which climate may be affected by human actions, we are unable to see clearly either the direction of future climatic changes or nature of their societal impacts. Because we are unable to "do the experiment" directly, we must attempt to model climate and climatic changes—an imprecise alternative because of the complexity of the global climate system. Much of the effort of trying to understand the atmospheric system has focused on the development of large-scale computer models of the many intricate and intertwined phenomena that make up the climate. The most complex of these models - general circulation models (GCMs) - are detailed, time-dependent, three-dimensional numerical simulations that include atmospheric motions, heat exchanges and important land-ocean-ice interactions (see, Manabe 1969a, 1969b; Schlesinger and Gates 1980; Manabe and Stouffer 1980; Wetherald and Manabe 1981; Ramanathan 1981; Manabe et al. 1981; Hansen et al. 1983, 1984; Washington and Meehl 1983, 1984).

GCMs permit us to begin to evaluate some of the implications for global climatic patterns of increasing concentrations of radiatively-active atmospheric gases. While many uncertainties remain, a consensus is now beginning to form about the direction and magnitude of certain major impacts, such as increases in global-average temperatures and changes in the intensity and distribution of the global hydrologic cycle.

Unfortunately, state-of-the-art general circulation models are large and expensive to operate. Furthermore, while GCMs are invaluable for identifying some climatic sensitivities and changes in global climatic characteristics, they have two limitations that reduce their value to researchers interested in more detailed assessments of water resources: (1) they are unable to provide much detail on regional or local impacts, and (2) they are unable to provide much detail on small-scale surface hydrology. Until our ability to model climate improves, we must use other methods to either enhance the information available from GCMs or provide insights now unavailable from them.

Plausible future hydrologic changes

The attention focused on large-scale GCMs in recent years results in large part from their relative sophistication compared to other models. Yet this attention has also highlighted the need for new methods of hydrologic assessment. Recently there have been some serious efforts to evaluate the regional hydrologic implications of climatic changes (Schwarz 1977; Stockton and Boggess 1979, Nemec and Schaake 1982; Revelle and Waggoner 1983; Flaschka 1984; U.S. Environmental Protection Agency 1984, Cohen 1986, Gleick 1985, 1986a,b, 1987c). These works have provided the first evidence that relatively small changes in regional precipitation and evapotranspiration patterns might result in significant changes in regional water availability.

Methods for evaluating the hydrologic impacts of climatic changes include using historical data to evaluate the effects of past fluctuations in precipitation and temperature on runoff and soil moisture; determining the sensitivity of runoff and soil moisture to hypothetical changes in the magnitude and timing of precipitation and temperature; and incorporating regionally disaggregated changes in temperature and precipitation predicted by GCMs into more accurate regional hydrologic models. While none of these methods - individually - can provide much reliable information on future changes, each can provide insights into specific hydrologic vulnerabilities to climatic change.

Future hydrologic changes: what can we expect?

Changes in climate may cause changes in a variety of hydrologic variables, including the timing, location, duration, and extent of precipitation, runoff, soil moisture, and extreme events. These impacts can be categorized in a variety of different ways. One useful method, shown in Table 1, is to separate the impacts by the spatial and temporal scales involved, with additional separation for the different statistical moments of interest and the distinction between political and geophysical boundaries. In the following sections, the most plausible and worrisome changes in water availability are described. These changes are not the only hydrologic effects that will occur, and not all of these will occur at any one place or at any one time. Nevertheless, we should pay particular attention to these impacts because they are more likely to occur, they are harder to mitigate, and they may be more disruptive than other climatic effects.

Table 1 Hydrologic effects of climatic change

Hydrologic Variable of Interest

    Useful Precipitation
    Surface Runoff
    Available Soil Moisture
    Groundwater
    Temperature
    Monsoonality (Onset, Ending, Intensity, Location)
    Storm Events

Temporal Scale of Interest

    Long-Term (greater than annual)
    Annual
    Seasonal (two to six months)
    Monthly
    Daily

Spatial Scale of Interest (Political)

    Global 10^8 km^2
    Continental 10^7 km^2
    Country/Region 10^6 km^2
    Local 10^3 - 10^5 km^2

Spatial Scale of Interest (Hydrologic)

    Global 10^8 km^2
    Continental 10^7 km^2
    Regional 10^5 - 10^6 km^2
    Watershed 10^2 - 10^5 km^2

Statistical Scale of Interest

    Mean
    Variance
    Persistence
    Skew
    Higher Moments

Hydrologic Impact of Interest

    Quantity
    Quality
    Peak Events (High and Low)

Source. Gleick (1987a)

a) Precipitation

Despite the fact that all GCMs predict an intensification of the overall hydrologic cycle, particularly increases in global average annual precipitation rates, this information is only marginally useful. As the global average temperature increases, we expect an increase in the rate of evapotranspiration and precipitation. GCMs now suggest that the annual-average increase in global precipitation may be on the order of seven to fourteen percent. Far more interesting and potentially disruptive are the changes in regional precipitation patterns, which are much harder to model. At present, there is little consensus about specific regional changes.

    Two specific vulnerablities need attention: (1) changes in average precipitation rates in regions with rainfed agriculture; and (2) changes in the frequency of extreme precipitation events in areas vulnerable to flooding and storms. In the first case, an increase in precipitation in agricultural regions dependent on rainfall could have a beneficial effect, while a decrease would have the opposite effect. Similarly, floods and storms are already responsible for enormous human suffering. Such events could be exacerbated by an increase in the variability of regional precipitation.

There are a number of hydrologic effects that may be driven primarily by temperature changes, not precipitation changes. This permits the identification of certain impacts that are somewhat independent of precipitation rates. Among these impacts are changes in soil-moisture availability and changes in the timing of surface runoff. Although both of these variables depend heavily on site-specific characteristics such as soil-moisture capacity, precipitation rates, vegetation characteristics, topography, and soil depth and type, some generalizations can be made.

b) Soil-moisture availability

Soil-moisture behavior in general circulation models is very simple, and efforts to improve the representation of moisture in the soil column are now underway (Dickinson 1986; Rind 1987). For the last several years, there has been a growing interest in soil-moisture changes because of the possibility that some significant--and potentially adverse--effects on soil-moisture availability may result from increasing concentrations of carbon dioxide. In particular, some general circulation model results suggest that soil moisture in mid-continental regions in mid-latitudes may decrease during summer months, which is often the critical period for crop productivity (Manabe et al. 1981; Mitchell 1983; Manabe and Wetherald 1986; Rind 1987). Although there are disputes over the magnitude (and sometimes the direction) of these soil-moisture changes, the present active research in this area may help to resolve the uncertainties.

Recently, some detailed hydrologic models have supported the possibility of decreased summer soil-moisture availability in some regions (Gleick 1986a, 1987c). In particular, despite increases in annual and seasonal precipitation, increases in temperatures can lead both directly and indirectly to decreases in soil-moisture availability during summer months. In regions with winter snowfall and spring snowmelt (in the United States, such regions include large parts of California, the Rocky Mountains, the Pacific Northwest, and the Northeastern and Northcentral U.S.), increases in temperatures may lead to decreases in the ratio of snow to rain in winter months, increases in the speed of snowmelt in spring months, and an earlier onset of drying in early summer (Gleick 1987c). Figure 1 shows the decreases in summer soil moisture in a major California watershed that result from eight scenarios generated by three state-of-the-art GCMs. The GCMs each predict quite different precipitation, yet the regional model results using these scenarios all show decreased summer soil-moisture availability. Similar results were identified for other regions by Mather and Feddema (1986). This robust result is one example of the type of hydrologic impact that should be more carefully studied on a regional basis.

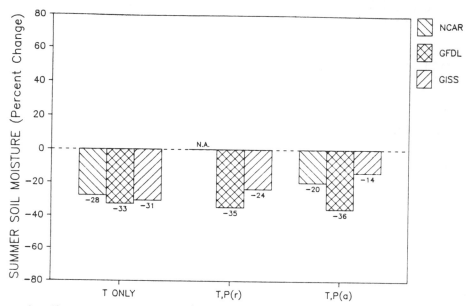

Figure 1   Change   in summer (June,   July and August) soil   moisture
           predicted   by a water-balance model of a major   California
           watershed   using precipitation and temperature   data   from
           three general   circulation models: the National Center for
           Atmospheric   Research   (NCAR),   the Geophysical   Fluid
           Dynamics Laboratory (GFDL),   and the Goddard Institute for
           Space   Studies   (GISS).   Note that all three   models   show
           decreases   in   soil   moisture.   The eight   scenarios   are:
           Temperature only,   Temperature and relative precipitation,
           and   Temperature and absolute   precipitation.   See   Gleick
           (1987a) for details of the model,   the scenarios,   and the
           uncertainties.

c) Runoff

Surface   runoff   shows   a   sensitivity   to   increases   in temperature
similar   to   the one   described above   for soil moisture.   In certain
regions,   seasonal runoff appears to be   vulnerable to changes   in the
timing of surface flows,   even when the overall annual runoff does not
change significantly.
       Figure   2   plots   historical   average-monthly runoff   in   a   major
California   watershed together with the runoff predicted by a   water-
balance   model using temperature and precipitation changes   predicted
by   a state-of-the-art GCM.   In this case,   while the   average-annual
runoff volumes do not change significantly between the two cases, the
monthly pattern has changed.   This can be seen by the large increase
in winter runoff and   the decrease in   summer   runoff.   The   physical
mechanisms   at   work   here   are similar to the   ones   described   above

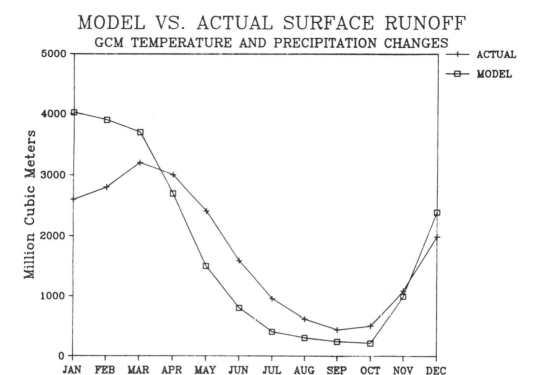

Figure 2   Average-monthly runoff: actual and model-predicted using temperature and precipitation changes developed by the Geophysical Fluid Dynamics Laboratory GCM. The annual runoff volumes for both of these runs are the same; the seasonal pattern has changed. See the text for details.

driving the soil-moisture changes—less total winter snowpack, more winter runoff, faster snowmelt in the spring, and smaller spring runoff. Figures 3 to 5 show the details of average-monthly changes in runoff using the temperature and precipitation changes from three GCMs to drive a regional water-balance model of the Sacramento Basin in Northern California - perhaps the most important watershed in California (Gleick 1987b). In all of these cases, summer runoff decreased and winter runoff increased, while average-annual runoff was only slightly changed (Gleick 1987c). These runoff changes can increase the frequency of flood events by shifting more runoff to peak runoff months, even if overall average runoff doesn't change. Similarly, regions dependent on minimum summer flows may be adversely affected.

The vulnerability of water resources to climatic conditions

The availability of freshwater for agricultural, industrial, residential, and commercial use is sensitive to existing climatic variabili-

Climate—Induced Change In Runoff
NCAR TEMPERATURE AND ABSOLUTE PRECIPITATION

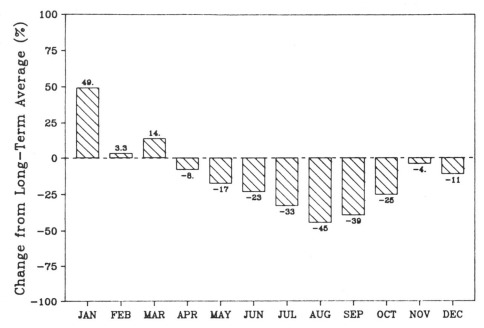

Figure 3   Percent   change   in   monthly   runoff   between   the   NCAR
temperature   and absolute precipitaion run and   the   long-
term   average   runoff.   Note the increase in winter   runoff
and the decrease in summer runoff.

ty. The sensitivity varies with supply and demand, water quality, and
the specific needs of the users. As the climate begins to change, the
most severe pressures on available water resources are likely to come
in regions where the existing water resources are already constrained
during certain times. This section discusses existing vulnerabilities
that might be either exacerbated or mitigated by climatic change.

Regions with natural deficits: Arid and semi-arid lands are, by
definition, regions with natural water deficits: the potential
evapotranspiration exceeds natural water inputs during part or all of
the year. At the same time, these lands are often thought to hold the
greatest potential for future agricultural development assuming that
water can be made available for irrigation, and that the soil quality
is high enough (Rosenberg 1981; Gleick 1987a). Improvements in the
hydrologic conditions of these regions would require increases in the
average water availability. Since evapotranspiration is likely to
increase following a doubling of atmospheric carbon dioxide, such an
increase in average availability must come through precipitation or
water transfers into the basin. At the same time, if the variability
of water resources availability were to increase, the vulnerability
of these regions to climate could remain high. For a climatic change
to be most advantageous to arid and semi-arid regions, there would
have to be an increase in mean water availability and a decrease in
the variability. It is important to note here, however, that while

Climate-Induced Change In Runoff
GFDL TEMPERATURE AND ABSOLUTE PRECIPITATION

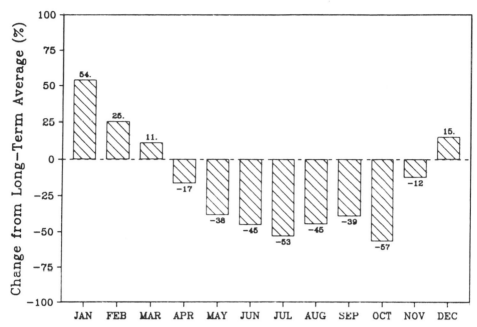

Figure 4   Percent   change   in   monthly   runoff   between   the   GFDL
temperature   and absolute precipitation run and the   long-
term   average   runoff.   Note the increase in winter   runoff
and the decrease in summer runoff.

such   changes   might be beneficial to   agricultural   productivity   or
other   human   uses,   they can lead to dramatic shifts in the   natural
character of the existing ecosystems (Gleick 1987a).

Regions with high societal demands: In many regions of the world,
the   demand for water approaches the available supply during   certain
periods.   In   these   regions,   efforts are often already underway   to
modify either the available supply or demand. Changes in climate that
exacerbate   these   demands or reduce the overall supplies   will   have
negative   consequences   for the region,   while overall   increases   in
water   availability   could   ease some problems. As   with   the   first
example,   the   most   advantageous climatic change would be   increased
mean availability and decreased variability.   An increase in variability would increase the frequency of severe events and may not result
in net benefits to the region.

Flood-prone regions: Areas prone to flooding,   such as   low-lying
floodplains,   would   benefit   from a decrease in the   variability   of
precipitation and runoff and suffer from an increase in both the mean
and the variability of water availability. In these regions, measures
are   often   taken to reduce the vulnerability of society to   floods,
such   as   the   development of flood-control   reservoirs.   While   such
facilities are often valuable,   they are also expensive to design and
build.   As   a result,   in order to properly design new   flood-control
facilities, information on the nature of future hydrology is necessa-

Climate—Induced Change In Runoff
GISS TEMPERATURE AND ABSOLUTE PRECIPITATION

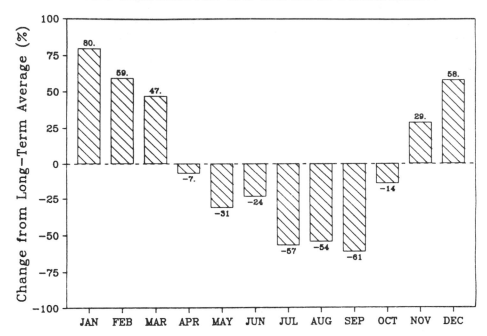

Figure 5 Percent change in monthly runoff between the GISS temperature and absolute precipitation run and the long-term average runoff. Note the increase in winter runoff and the decrease in summer runoff.

ry if a given facility is to be able to cope with future climatic conditions. A change such as the one plotted in Figure 2, which changes the seasonal patterns while not changing the annual average, would increase the risks of flooding unless changes in the flood-control system can be made.

Regions dependent on reliable seasonal supply: By far the most-often heard hydrologic truism is that the supply of water to any region is not uniformly distributed in space or time. Many regions are dependent on water supplies that arrive during particular seasons, such as the monsoons on the Asian sub-continent and winter precipitation in Mediterranean-style climates. In these regions, slight changes in the timing or magnitude of seasonality will have important consequences. Figure 2, which plots a possible change in the timing of the availability of surface runoff in a major agricultural basin, shows how the seasonality of runoff may be affected by predicted climatic changes. Unless extensive reservoir systems permit the storage and later distribution of seasonal precipitation and runoff, a change in the seasonality of water availability could stress a region. Unfortunately, as Nemec and Schaake (1982) pointed out, climatic changes could cause problems for existing reservoir systems.

Regions sensitive to lake levels: Major lake systems are sensitive to inflows and outflows, which in turn depend on both natural supply (from precipitation and runoff) and natural and artificial demand (from evapotranspiration and withdrawals). Work by Snyder and Langbein (1962) and Street-Perrott et al. (1986), among others shows this sensitivity in the context of changing climate. Although some work has been done to estimate the effects of future climatic changes on lake levels (see, Cohen 1986), no clear trends have yet been identified. Shipping, municipal and industrial water supply, recreation and natural ecosystems will be affected by both positive and negative changes in lake levels.

Regions with decreasing water quality: Deteriorating water quality due to industrial development, agricultural wastewater, and population growth will be affected by climatic changes that alter the availability of freshwater resources. Critical areas include rivers used for waste disposal that may experience decreases in minimum flows, groundwater supplies that are sensitive to pollutant inflows, and the design methods for the adequate disposal of toxic materials. In some of these cases, societal actions should be taken now to reduce the vulnerability of water quality to climatic changes. The challenge is to design such actions to be flexible enough to handle a wider range of climatic conditions than are now normally anticipated.

Perhaps the most well-known example of this is the salinity problems of the Colorado River near the U.S.-Mexican border. As more water is used for agriculture, the salinity of the river increases to the point where it becomes detrimental to further use. This problem will increase in severity in the absence of a climatic change and thus already requires mitigating actions. These mitigating actions, however, should anticipate minimum and maximum flows lower than those historically recorded because of the possibility that minimum and maximum flows could be altered by climatic changes. This added flexibility can be achieved at a lower cost now than after the actions have been designed, facilities built, and operating schemes implemented.

Regions dependent on hydroelectricity: Finally, hydroelectricity plays a major role in many regions of the world. The reliability of this energy depends on the reliability of water resources—particularly the timing and magnitude of flow rates. As the climate changes, one or another of these variables is likely to change, with the risk that alternative (and more often expensive) methods of electricity generation will be required to make up shortfalls, or that potential hydroelectricity will be lost because of incorrectly-sized and operated facilities (McGuirk 1982). At best, this suggests the need for a flexible hydroelectric system-operating style; at worst, existing facilities will have to be redesigned and new facilities evaluated.

Discussion and conclusions

It is extremely unlikely that all the hydrologic changes induced by changing climatic conditions will be beneficial. When the water-resource needs of different regions are studied, we see that different changes in the mean and variability of water resources are required in different regions and on different time scales. The probability is extremely low that the diverse changes appropriate for all regions will occur in precisely the proper location and at precisely the proper time--such as increases in means in arid regions, decreases in means in regions subject to flooding, and the appropriate changes in both annual and seasonal variability. Given this problem, attention must be focused on the vulnerability of hydrologic systems to changes in climate, so that policies to mitigate the worst effects can be implemented should negative impacts materialize.

Despite the uncertainties that surround the nature and timing of future climatic changes and their subsequent impacts, the research discussed here raises some concerns about regional water availability. In particular, certain types of impacts, such as decreases in summer soil moisture and runoff and increases in winter runoff are robust and consistent across widely-varying scenarios. This consistency suggests strongly that hydrologic vulnerabilities will make the impacts of climatic changes on water resources an issue of major concern in many regions of the world.

Some of the results described here support recent suggestions that summer soil-moisture reductions may occur in many regions of the world. The principal physical mechanisms involved--the decrease in snow as a proportion of total winter precipitation, an earlier and faster disappearance of winter snowpack due to higher average temperatures, and a more severe evapotranspiration demand during the warmer summer months--are both physically plausible and hydrologically consistent. While other, countervailing hydrometeorologic features may well exist--such as cloud cover/evapotranspiration feedbacks--the consistency of the soil moisture and runoff results described here must be considered a first warning of possible important changes in regional water availability.

Adverse hydrologic changes may, if they materialize, have serious implications for many aspects of water resources, including agricultural water supply, flood and drought probabilities, groundwater use and recharge rates, the price and quality of water, and reservoir design and operation--to mention only a few. Yet information on these changes, by itself, is unlikely to lead to major policy responses. Only by looking at the specific characteristics of water-resource problems--and their vulnerability to--the types of changes in runoff and soil moisture identified above-can details of future societal impacts be evaluated.

References

Cohen, S.J. (1986) "Impacts of CO2-Induced Climatic Change on Water Resources in the Great Lakes Basin". Climate Change 8. pp.135-154.
Dickinson, R.E. (1986) "Modeling Hydrology in a Global Climate Model: Applications to Impact of Amazon Deforestation." Presented at the American Geophysical Union Fall Meeting, San Francisco, California. December 8.
Flaschka, I.M. (1984) "Climatic Change and Water Supply in the Great Basin". Master's thesis. Department of Hydrology and Water Resources, University of Arizona.
Gleick, P.H. (1985) "Regional Hydrologic Impacts of Global Climatic Changes". in E.E. Whitehead, C.F. Hutchinson, B.N. Timmermann, and R.G. Varady, (editors). Proceedings of an International Research and Development Conference. Arid Lands: Today and Tomorrow. Office

of Arid Lands Studies, University of Arizona, Tuscon. October 20–25, 1985.

Gleick, P.H. (1986a) "Large-Scale Climatic Changes and Changes in Regional Summer Soil-Moisture Availability". Presented at the American Geophysical Union Fall Meeting, San Francisco, California. December 8.

Gleick, P.H. (1986b)"Methods for Evaluating the Regional Hydrologic Impacts of Global Climatic Changes". Journal of Hydrology 88, pp. 97–116.

Gleick, P.H. (1987a) "Global Climatic Change, Water Resources, and Food Security. Sponsored by the Indian National Academy of Science, the American Association for the Advancement of Science, and the International Rice Institute. New Delhi, India. February 6–9, 1987. 37pp.

Gleick, P.H. (1987b) "the Development and Testing of a Water-Balance Model for Climate Impact Assessment: Modeling the Sacramento Basin". In press Water Resources Research.

Gleick, P.H.(1987c) "Regional Hydrological Consequences of Increases in Atmospheric CO and Other Trace Gases". In press Climatic Change.

Hansen, J.E., Rind, D., Russell, G., Stone, P., Fung, I., Ruedy, R., and Lerner, J. (1984) "Climatic Sensitivity: Analysis of Feedback Mechanisms". In J.E. Hansen and T.Takahashi, eds. Climate Processes and Climate Sensitivity, American Geophysical Union Monograph 29. Maurice Ewing Vol. 5. American Geophysical Union, Washington, D.C.

Hansen, J., Russell, G., Rind, D., Stone, P., Lacis, A., Lebedeff, S., Rueddy, R., and Travis, L.(1983) "Efficient Three-Dimensional Global Models for Climate Studies: Models I and II".Monthly Weather Review. Vol.III, No. 4 pp. 609–662. April.

Manabe, S. (1969a) "Climate and the Ocean Circulation. I. The Atmospheric Circulation and the Hydrology of the Earth's Surface". Monthly Weather Review. Vol. 97, No.11 pp. 739–774. November.

Manabe, S. (1969b) "Climate and the Ocean Circulation. II. The Atmospheric Circulation and the Effect of Heat Transfer by Ocean Currents".Monthly Weather Review. Vol.97, No. 11. pp.775–805. Nov.

Manabe, S. and Stouffer, R.J.(1980) "Sensitivity of a Global Climate Model to an Increase of $CO2$ Concentration in the Atmosphere". J. Geo. Res.. Vol 85, No. C10, pp. 5529–5554. October.

Manabe, S. Wetherald, R.T.(1986) "Reduction in Summer Soil Wetness Induced by an Increase in Atmospheric Carbon Dioxide". Science. Vol. 232, pp.626–628.

Manabe, S., Wetherald, R.T., and Stouffer, R.J.(1981) "Summer Dryness Due to an Increase of Atmospheric $CO2$ Concentration". Climatic Change 3, pp. 347–386.

Mather, J.R. and J. Feddema. (1986) "Hydrologic Consequences of Increases in Trace Gases and $CO2$ in the Atmosphere". in J. Titus (editor). Effects of Changes in Stratosphere Ozone and Global Climate. Volume 3: Climate Change. United States Environmental Protection Agency. pp. 251–271.

McGuirk, J.P.(1982) "A Century of Precipitation Variability Along the Pacific Coast of North America and Its Impact". Climatic Change 4. pp. 41–56.

Mitchell, J.M., Jr. (1983) "An Empirical Modelling Assessment of

Volcanic and Carbon Dioxide Effects on Global Scale Temperature". American Meteorological Society, Second Conference on Climate Variations , New Orleans, Louisiana. (January 10-14.)

Nemec, J. and Schaake, J. (1982) "Sensitivity of Water Resource Systems to Climate Variation". Hydrological Sciences 27. No. 3, pp. 327-343.

Ramanathan, V.(1981)"The Role of Ocean-Atmosphere Interactions in the CO2 Climate Problem." J.Atmos.Sci Vol. 38, pp. 918-930.

Revelle, R.R. and Waggoner, P.E.(1983) "Effects of a Carbon Dioxide-Induced Climatic Change on Water Supplies in The Western United States". In Changing Climate. National Academy of Sciences. National Academy Press, Washington D.C.

Rind, D.(1987) The Doubled CO2 Climate and Future Water Availability in the United States. Goddard Space Flight Center Institute for Space Studies. Submitted to J. Geophysical Research.

Rosenberg, N.J.(1981) "Technologies and Strategies in Weatherproofing Crop Production". In L.E. Slater and S.K. Levin (editors). Climate's Impact on Food Supplies: Strategies and Technologies for Climate-Defensive Food Production. AAAs Selected Symposium 62. (Westview Press, Boulder, Colorado). pp. 157-180.

Schlesinger, M.E., and Gates, W.L. (1980) "The January and July Performance of the OSU Two-level Atmospheric General Circulation Model." J. Atmos. Sci., Vol. 37, pp. 1914-1943. September.

Schwarz, H.E.(1977) "Climatic Change and Water Supply: How Sensitive is the Northeast?". in Climate, Climatic Change, and Water Supply. National Academy of Sciences. Washington, D.C.

Snyder, C.T. and Langbein, W.B.(1962) "The Pleistocene Lake in Spring Valley Nevada".J.Geophys.Res. Vol. 67, No.6. pp. 2385-2394. June.

Stockton, C.W. and Boggess, W.R.(1979) "Geohydrological Implications of Climate Change on Water Resource Development". U.S. Army Coastal Engineering Research Center, Fort Belvoir, Virginia. May.

Street-Perrott, F.A., M.A.J. Guzkowska. I.M. Mason, and C.G. Rapley (1986) "Response of Lake Levels to Climatic Change- Past, Present and Future." in J.Titus (editor). Effects of Changes in Stratospheric Ozone and Global Climate. Volume 3: Climate Change. United States Environmental Protection Agency. pp.211-216.

U.S. Environmental Protection Agency (1984) "Potential Climatic Impacts of Increasing Atmospheric CO2 with Emphasis on Water Availability and Hydrology in the United States". Strategic Studies Staff, Office of Policy Analysis, Office of Policy, Planning and Evaluation. April.

Washington, W.M. and Meehl, G.A.(1983) "General Circulation Model Experiments on the Climatic Effects Due to a Doubling and Quadrupling of Carbon Dioxide Concentration". J. Geophys. Res. Vol.88, No. CII, pp. 6600-6610. August.

Washington, W.M. and Meehl, G.A.(1984) "Seasonal Cycle Experiment on the Climate Sensitivity Due to a Doubling of CO2 With an Atmospheric General Circulation Model Coupled to a Simple Mixed-Layer Ocean Model".J. Geophys. Res. Vol. 89, No. D6, pp. 9475-9503. Oct.

Wetherald, R.T. and Manabe, S.(1981) "Influence of Seasonal Variation Upon the Sensitivity of a Model Climate". J.Geophys. Res. Vol. 86, No. C2, pp. 1194-1204. February 20.

The Influence of Climate Change and Climatic Variability on the Hydrologic Regime and Water Resources (Proceedings of the Vancouver Symposium, August 1987). IAHS Publ. no. 168, 1987.

The impacts of CO$_2$-induced climate change on hydro-electric generation potential in the James Bay Territory of Quebec

Bhawan Singh
Professor agrege
Universite de Montreal, Canada

Introduction

CO$_2$ - induced climate change can significantly affect the economy of water - resource - based industries. Other studies in Canada, for the Great Lakes-St. Lawrence basin (Cohen, 1987, 1986; Howe et al; 1986, Sanderson et al, 1985; Bruce, 1984) have shown that changes in lake and river levels, resulting from changes in Net Basin Supply (NBS) can have far-reaching economic consequences on hydro-electric power production, lake shipping and shoreline erosion.

In this paper, we intend to focus on the impact of CO$_2$ - induced climatic change on the hydro-electricity generating capacity of the three basins, within the James Bay Territory, that are presently exploited for hydro-electric generation, namely the La Grande, the Caniapiscau and the Opinaca-Eastmain basins (Figure 1a and Figure 1b).

The hydro-generating capacity of the three basins are to be derived from actual (normals:1951-1980) and projected (2 x CO$_2$) changes in NBS using the GFDL (scenario A) and GISS (scenario B) scenarios data provided by Environment Canada (Hengeveld and Street, 1986). These data sets provide normals (1 x CO$_2$) and projected (2 x CO$_2$) temperature and precipitation conditions for several grid points within the James Bay Territory (Figure 1a and Figure 1b).

These projected changes in NBS will then be used to gauge the potential economic costs or benefits of CO$_2$ - induced climate change, based on both climate change scenarios (GFDL and GISS).

Methods, data sources and derivations

For the three drainage basins selected, namely, the La Grande, the Caniapiscau and the Opinaca-Eastmain, the NBS was calculated for both the normals(1951-80) and the projected data for the two scenarios (GFDL) and (GISS) in question.

The general form of the formula for calculating NBS for drainage basins consisting of both land and open lake and reservoir surfaces is written as:

$$NBS = \begin{array}{l} \text{Land} \\ \text{Discharge} \end{array} \times \begin{array}{l} \text{Land} \\ \text{Area} \end{array} + (P_{lake} - E_{lake}) \times \begin{array}{l} \text{Lake} \\ \text{Area} \end{array} - \text{Consumption} \qquad (1)$$

NBS is expressed in m^{3}s^{-1}, P$_{lake}$ represents the total precipitation falling on the lake or reservoir surface and E$_{lake}$ represents the total evaporation from the lakes or reservoirs.

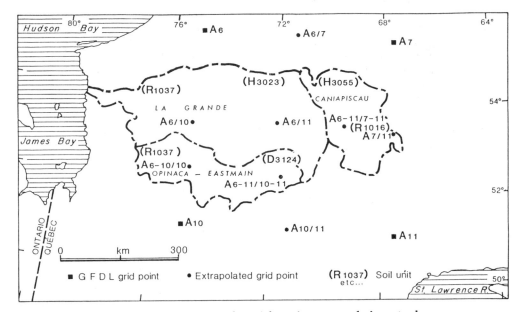

Figure 1a   Drainage basins and grid points used in study
(Scenario A).

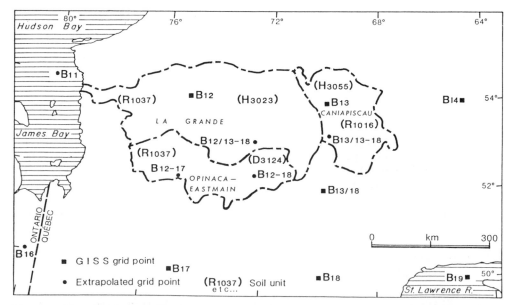

Figure 1b   Drainage basins and grid points used in study
(Scenario B).

In the analyses, land discharge was calculated as the surface runoff or water surplus (mm) using the Thornthwaite monthly water budget method as written by Environment Canada (Johnstone and Louie, 1984). The Thornthwaite water budget formula, stated simply is:

$$R_0 = P - AE \pm \Delta St \qquad (2)$$

R_0 is the monthly surface runoff (mm), P is the monthly precipitation (mm), AE is the monthly evapotranspiration depending on air tempera-ture and soil moisture and, ΔSt is the monthly change in soil mois-ture. Based on our knowledge of the average soil depth of the region (Singh and Taillefer, 1984) and on values cited elsewhere for similar regions (Cohen, 1986, 1987; Howe et al, 1986), a value of Water Holding Capacity (WHC) equal to 100 mm was used.

Because of the remoteness of the area and the very sparse popula-tion, water consumption is negligible and, because very little water is consumed in hydro-electric power generation, this small total amount of water consumed is neglected in the analysis.

Since the hydrology of the region has been modified recently by the creation of dams for hydro-electric power generation, and artifi-cial reservoirs have expanded natural lake surfaces, it was necessary to slightly modify equation (1) to reflect these changes. Total basin area, land area and reservoir area for each of the drainage basins are supplied in Table 1. In addition, Table 1 provides mean hydrolo-gic characteristics of each basin based on the period 1949-1980. These conditions essentially represent the pre-damming period since the dams were closed in November 1978 for La Grande basin, July 1980 for Eastman-Opinaca basin and, October 1981 for Caniapiscau basin (M.T. Tran-Van SEBJ, personal communication).

Table 1   Physiographic and hydrologic parameters of drainage basins
(Source: SEBJ 1986)

Drainage Basin	Total Basin Area (Km^2)	Natural Lakes Area (Km^2)	Reservoirs Area (Km^2)	Land Area without reservoirs (Km^2)	Land Area with reservoirs (Km^2)	Mean Dis- [1] charge and (Std. Dev) m^3/S	Mean dis-[2] charge/ Prec. Ratio (%)
La Grande	98,031.5	19,606.3	6,021.8	78,425.2	72,403.4	1762 (255)	77
Caniapiscau	36,881.6	7,745.1	4,273.5	29,136.5	24,863.5	788 (122)	77
Eastmain-Opinaca	40,274.5	7,652.2	1,041.2	36,622.3	31,581.1	851 (104)	84

1 Mean discharge and (Standard Deviation) for period 1949-80:
   Measured data: La Grande; 1960-79, Caniapiscau: 1962-79, Eastmain-Opinaca: 1960-80.
   Other years data is estimated (SEBJ).

2 Mean discharge Precipitation Ratio for period 1949-80.

The hydrologic data set in Table 1 (1949-1980) therefore corres-ponds to the normals (1951-1980) data and was thus used for calibrating the normal discharge data.

For the normals (1951-1980) data set, NBS was then calculated using equation (1). However for the projected scenarios data, NBS was calculated both with equation (1) which ignores the reservoirs and the following equation (3) which takes the surface area of the artificially created reservoirs into account.

$$NBS = \quad \text{Land} \quad x \quad \text{Land} \quad + \quad (P_{lake_R} - E_{lake_R}) \quad x \quad \text{Lake}$$
$$\quad \text{Discharge} \quad \text{Area}_R \qquad\qquad\qquad \text{Area}_{RX} \qquad (3)$$

Land Area$_R$ represents the reduced land area resulting from the creation of reservoirs, P_{lake_R} and E_{lake_R} represent the total of monthly lake plus reservoir precipitation and the monthly evaporation respectively, and   Lake Area$_R$ represents the sum total of lake and reservoir areas.

In equations (1) and (3), since land discharge and $P_{lake} - E_{lake}$ are in mm/year and land area and lake area are in sq. km each half of the equations, neglecting consumption, was multiplied by the conversion factor of 31.71×10^{-6} so that the final units of NBS are in $m^3 s^{-1}$. Monthly values of discharge, precipitation and evaporation were summed to derive total annual values.

As mentioned previously total annual land discharge is calculated using the Thornthwaite water budget method. Lake and lake plus reservoir precipitation was assumed to be equal to the land precipitation, unlike other other studies (Cohen, 1986,1986b) since the lake and reservoir surfaces are relatively small.

Because of the small basin areas and the coarse spatial resolution of the grid points for both scenarios (GFDL and GISS), it was necessary to perform a series of extrapolations between grid points so as to derive at least two reference points for each basin. These extrapolations were done linearly along the X (latitude) and Y (longitude) coordinates and diagonally across these X-Y vectors.

For instance, in the case of the La Grande basin, the two derived grid points for scenarios A are $A_{6/10}$ and $A_{6/11}$ and for scenario B, B_{12} and $B_{12/13-18}$ (Figure 1 and Figure 2).   Grid point $A_{6/10}$ is derived by linearly extrapolating latitudinally between GFDL grid points A_6 and A_{10} (Figure 1). On the other hand, grid point $A_{6/11}$ is derived by extrapolating diagonally between GFDL grid points A_6 and A_{11}. For scenario B, GISS grid point B_{12} is already within the basin (Figure 2). However, grid point $B_{12/13-18}$ is derived by first extrapolating latitudinally between GISS grid points B_{13} and B_{18} to derive $B_{13/18}$ and then by extrapolating diagonally between grid points $B_{13/18}$ and B_{12}.

Interpolated grid points for the other two basins, as shown in Figure 1 and Figure 2, namely Caniapiscau and Eastmain-Opinaca were derived using a similar linear extrapolation procedure.

These extrapolated grid points were then averaged so as to derive the mean basin precipitation and the mean land surface discharge on a total annual basis. This averaging was weighted so as to reflect the approximate proportion of the total basin that each extrapolated grid point represented and so as to reflect the spatial variation of precipitation and temperature (used to evaluate discharge) as shown elsewhere (Singh <u>et al</u>, 1987). These results are shown in Tables 2a, 2b, 3a, 3b, and 4a, 4b.

Table 2a   Annual input parameters for deriving unadjusted net basin
supply (NBS) and correction factor from extrapolated grid
points for La Grande basin: Scenario A (GFDL); normals
data

Extrapolated Grid Points	Land Discharge (m^3/S)	Lake Precipitation (mm)	Lake Evaporation		Unadjusted Mean NBS (m^3/S)	Underestimate of Mean Discharge %	Correction Factor
			Soil Unit	(mm)			
$A_{6/10}$	345.2	747.1	D3023	708.5			
$A_{6/11}$	447.1	844.3	R1037	459.7			
	406.4*	805.4*		559.2**	1163.8	34.0	1.514

*   Weighted mean = .6 $(A_{6/11})$ + .4 $(A_{6/10})$

**   Weighted mean = .6 (R1037) + .4 (D3023)

Table 2b   Annual input parameters for deriving unadjusted net basin
supply (NBS) and correction supply from extrapolated grid
points for La Grande basin: Scenario B (GISS); normals
data

Extrapolated Grid Points	Land Discharge (m^3/S)	Lake Precipitation (mm)	Lake Evaporation		Unadjusted Mean NBS (m^3/S)	Underestimate of Mean Discharge %	Correction Factor
			Soil Unit	(mm)			
B_{12}	313.2	705.6	D3023	708.5			
$B_{12/13-18}$	382.4	782.7	A1037	459.7			
	354.8*	752.2*		559.2**	1002.3	43.1	1.757

*   Weighted mean = .6 $(B_{12/13} - 18)$ + .4 (B_{12})

**   Weighted mean = .6 (R1037) + .4 (D3023)

Lake evaporation on the other hand was calculated according to the
Priestley-Taylor (1982) equation, written as

$$E_{lake} = \{\alpha\ S/S + \gamma(Q^* - QG)\}/\ L \tag{4}$$

where: E_{lake} is the monthly lake evaporation (mm), Q^* is the net
radiation $(MJ/m^2/month)$, QG is the soil heat flux $(MJ/m^2/month)$, S is
the air temperature-dependent slope of the saturation vapor pressure
curve $(Pa/^{o}C)$, γ is the psychrometric constant $(Pa/^{o}C)$, α is a non
- dimensional surface evaporability factor and L is the latent heat
of evaporation (MJ/kg).
     The value of α used here is 1.26, as recommended by Priestley
and Taylor (1972) and Stewart and Rouse (1977) for open water

Table 3a   Annual input parameters for deriving unadjusted net basin
           supply (NBS) and correction factor from extrapolated grid
           points for Caniapiscau basin: Scenario A (GFDL); normals
           data

Extrapolated Grid Points	Land Discharge (m^3/S)	Lake Precipitation (mm)	Lake Evaporation		Unadjusted Mean NBS (m^3/S)	Underestimate of Mean Discharge %	Correction Factor
			Soil Unit	(mm)			
$A_{6/11}$	447.1	844.3	H3055	432.9			
$A_{7/11}$	476.5	866.8	R1016	449.7			
	461.8*	855.6*		441.3**	528.4	32.9	1.491

*   Weighted mean = .5 $(A_{6/11})$ + .5 $(A_{7/11})$

** Weighted mean = .5 (H3055) + .5 (R1016)

Table 3b   Annual input parameters for deriving unadjusted net basin
           supply (NBS) and correction factor from extrapolated grid
           points for Caniapiscau basin: Scenario B (GISS); normals
           data

Extrapolated Grid Points	Land Discharge (m^3/S)	Lake Precipitation (mm)	Lake Evaporation		Unadjusted Mean NBS (m^3/S)	Underestimate of Mean Discharge %	Correction Factor
			Soil Unit	(mm)			
B_{13}	386.0	762.8	H3055	432.9			
$B_{13/18}$	451.5	859.7	R1016	449.7			
	418.8*	811.3*		441.3	477.8	39.4	1.649

*   Weighted mean = .5 (B_{13}) + .5 $(B_{13/8})$

** Weighted mean = .5 (H3055) + .5 (R1016)

surfaces. However, as cautioned by Singh and Taillefer (1986), this
value of α can increase appreciably in the presence of horizontal
warm air advection.

The term (Q*–QG) was derived from the Canadian 1951–1980 normals
data of global solar radiation (K↓) as described by Stewart (1983).
Similarly the term S was calculated from the Canadian normals (1951–
1980) air temperature data. In consequence the normals values of
E_{lake} are identical for both scenarios A(GFDL) and B(GISS).

For the projected changes in E_{lake}, Stewart (1983) assumed that
(Q* – QG) remained unchanged and that only S changed as a function
of changes in air temperature as predicted by both the GFDL and GISS
scenarios.

However the Stewart (1983) calculations of E_{lake} are located in
the middle of the soil class units for Canada (Clayton **et al**, 1977).

Table 4a   Annual input parameters for deriving unadjusted net basin
            supply (NBS) and correction factor from extrapolated grid
            points for Eastmain–Opinaca basin: Scenario A (GFDL);
            normals data

Extrapolated Grid Points	Land Discharge (m^3/S)	Lake Precipitation (mm)	Lake Evaporation		Unadjusted Mean NBS (m^3/S)	Underestimate of Mean Discharge %	Correction Factor
			Soil Unit	(mm)			
$A_{6-10/10}$	384.1	801.5	D3124	476.0			
A6–11/10–11	486.0	898.8	R1037	459.7			
	435.1*	850.2*		467.9**	542.9	36.2	1.567

*   Weighted mean = .5 $(A_{6-10/10})$ + .5 $(A_{6-11/10-11})$

**   Weighted mean = .5 (D3124) + .5 (R1037)

Table 4b   Annual input parameters for deriving unadjusted net basin
            supply (NBS) and correction factor from extrapolated grid
            points for Eastmain–Opinaca basin: Scenario B (GISS);
            normals data

Extrapolated Grid Points	Land Discharge (m^3/S)	Lake Precipitation (mm)	Lake Evaporation		Unadjusted Mean NBS (m^3/S)	Underestimate of Mean Discharge %	Correction Factor
			Soil Unit	(mm)			
$B_{12/17}$	387.1	806.4	D3124	476.0			
$B_{12/18}$	415.1	831.1	R1037	459.7			
	401.1*	818.8*		467.9**	500.0	41.2	1.702

*   Weighted mean = .5 $(B_{12/18})$ + .5 $(B_{12/17})$

**   Weighted mean = .5 (D3124) + .5 (R1037)

As is evident in equation (4) however, the soil characteristics have
no influence on the values of E_{lake}. Because our grid points for
E_{lake} were tied to the soils classification, we chose the dominant
soil classes within each basin (Figure 1 and Figure 2) and calculated
a weighted mean for E_{lake} based on the relative areas of these soil
class units. Again it must be emphasized that the weightings do not
reflect soil characteristics but moreso changes in (Q*–QG) and in air
temperature.
    The normals value of E_{lake} as shown in Tables   2a,   2b, 3a, 3b and
4a, 4b however seem a bit high for this region when compared to
results found for the Great Lakes region (Cohen, 1986a;   Howe et al,
1986).
    As a verification of the Priestley – Taylor E_{lake} calculations, we
calculated E_{lake} using a mass-transfer type equation, that is based

on the temperature dependent mean monthly vapor pressure gradient
(Pa) between the lake surface and the air and the mean monthly wind
speed (km/hr) (Richards and Urbe, 1969; Quinn and Den Hartog, 1981).
These were done for stations where the necessary lake water and air
temperature and wind speed were available, namely La Grande Rivière
(53°N and 77.5°W) in the La Grande basin and Nitchequon (54.5°N and
70.8°W) which is located within the Capiapiscau basin. Unfortunately
the data set is limited by the short time-period of water temperature
data: 1978-1984 for La Grande Riviere and 1980-1984 for Nitchequon,
which for all intents and purposes relates to the normals (1951-1980)
period.

Table 5 shows the total annual and average totals of E_{lake} using
the mass - transfer method. The average totals of E_{lake}, though
somewhat influenced by lower than average values in 1983, are in
general lower than the Priestley-Taylor E_{lake} calculations shown in
Tables 2a, 2b, 3a, 3b,4a and, 4b.

Table 5 Mass-transfer evaporation calculations (mm/year)

La Grande Rivière		Nitchequon	
Year	Total Annual Evaporation (mm)	Year	Total Annual Evaporation (mm)
1978	582.9		
1979	523.5		
1980	498.1	1980	495.6
1981	489.7	1981	424.4
1982	504.4	1982	458.7
1983	291.9	1983	307.6
1984	501.9	1984	299.5
	476.9		397.2

Our calculations of NBS for the normals (1951-1980) period for
both scenarios using equations (1) and (3) however, were substantially lower than the mean measured discharge (SEBJ, 1986) for the 1949-
1980 period. These under-estimations ranged from 32.9% for the Caniapiscau basin for scenario A to 43.1% for the La Grande basin for
scenario B (Tables 2a, 2b, 3a, 3b and 4a, 4b).

We suspect that these under-estimations are due partly to the fact
that our assumed water holding capacity (WHC = 100mm) is too high in
the calculation of land discharge by the Thornthwaite method since
substantial portions of these basins are covered by bare rock surfaces and partly due to the slight over estimation of E_{lake} by the
Priestley-Taylor method. Also our linear method of extrapolating grid
points could have biased our results.

In order to correct for these aberrations, we calibrated the
values of annual NBS by introducing a correction factor, which is
simply the ratio of the mean measured discharge (1949-1980) to that of

the non-adjusted NBS values for both scenarios (Tables 2a, 2b, 3a, 3b and 4a,4b). These correction factors ranged from 1.491 for scenario A for the Caniapiscau basin to 1.757 for scenario B for the La Grande basin (Tables 2a, 2b, 3a, 3b and 4a, 4b).

These same correction factors, derived using the normals (1951-1980) data for both scenarios (A and B) were also applied to the projected annual NBS values for both scenarios (A and B) so as to derive changes in annual NBS resulting from climate changes (Tables 6a, 6b, 7a, 7b and 8a, 8b).

Table 6a Annual input parameters for deriving unadjusted and adjusted mean net basin supply (NBS) from extrapolated grid points for La Grande basin: Scenario A (GFDL) projected data

Extrapolated Grid Points	Land Discharge (m^3/S)	Lake Precipitation (mm)	Lake Evaporation		Unadjusted Mean NBS (m^3/S)	Correction Factor as per Normals	Adjusted mean NBS (m^3/S)
			Soil Unit	(mm)			
$A_{6/10}$	398.4	947.2	D3023	806.9			
$A_{6/11}$	493.9	1034.2	R1037	493.9			
	455.9*	999.4*		619.1**	1382.5	1.514	2073.7

* Weighted mean = .6 ($A_{6/11}$) + .4 (6/10)

** Weighted mean = .6 (R1037) + .4 (D3023)

Table 6b Annual input parameters for deriving unadjusted and adjusted man net basin supply (NBS) from extrapolated grid points for La Grande basin: Scenario B (GISS) projected data

Extrapolated Grid Points	Land Discharge (m^3/S)	Lake Precipitation (mm)	Lake Evaporation		Unadjusted Mean NBS (m^3/S)	Correction Factor as PER Normals	Adjusted mean NBS (m^3/S)
			Soil Unit	(mm)			
B_{12}	368.5	837.2	D3023	820.7			
$B_{12/13-18}$	442.7	921.8	R1037	507.8	-		
	413.1*	888.0*		633.0**	1185.9	1.757	2083.5

* Weighted mean = .6 ($B_{12/13-18}$) + .4 (B_{12})

** Weighted mean = .6 (R1037) + .4 (D3023)

Since the same correction factors are applied for the normals and projected values for both scenarios, cateris paribus, the comparative effects should not be affected.

Table 7a Annual input parameters for deriving unadjusted and
adjusted mean net basin supply (NBS) from extrapolated
grid points for Caniapiscau basin: Scenario A (GFDL)
projected data

Extrapolated Grid Points	Land Discharge (m^3/S)	Lake Precipitation (mm)	Lake Evaporation		Unadjusted mean NBS (m^3/S)	Correction Factor as per Normals	Adjusted mean NBS (m^3/S)
			Soil Unit	(mm)			
$A_{6/11}$	493.9	1034.2	H3055	483.3			
$A_{7/11}$	552.5	996.6	R1016	503.2			
Weighted mean	523.2*	1015.4*		493.3**	611.5	1.491	911.8

*   Weighted mean = .5 (6/11) + .5 (7/11)

** Weighted mean = .5 (H3055) + .5 '(R1016)

Table 7b Annual input parameters for deriving unadjusted and
adjusted mean net basin supply (NBS) from extrapolated
grid points for Caniapiscau basin: Scenario B (GISS)
projected data

Extrapolated Grid Points	Land Discharge (m^3/S)	Lake Precipitation (mm)	Lake Evaporation		Unadjusted mean NBS (m^3/S)	Correction Factor as per Normals	Adjusted mean NBS (m^3/S)
			Soil Unit	(mm)			
B_{13}	429.4	885.3	H3055	497.7			
$B_{13/18}$	516.9	1006.4	R1016	517.9			
Eeighted mean	473.2*	945.9*		507.8**	544.8	1.649	898.4

*   Weighted mean = .5 (B_{13}) + .5 ($B_{13/18}$)

** Weighted mean = .5 (H3055) + .5 (R1016)

Results and discussion

Ignoring the presence of reservoirs for the moment, Tables 6a, 6b,
7a, 7b, and 8a and 8b show significant increases in NBS for all three
drainage basins, for both scenarios A and B. These increases, range
from 6.8% for scenario A to 20.3% for scenario B in the Opinaca-
Eastmain basin. For the other two drainage basins, namely La Grande
and Caniapiscau, the percent increases in NBS are quite similar for
both scenarios: 17.7% (GFDL) and 18.3%(GISS) for La Grande and 15.7%
(GFDL) and 14.0% (GISS) for Caniapiscau (Tables 9a and 9b).
   By adding the additional free water surface created by the reservoirs, these results do not change significantly (Tables 9a and 9b).
'or scenario A, the increase in NBS is reduced from 17.7% to 16.5%

Table 8a   Annual input parameters for deriving unadjusted and
           adjusted mean net basin supply (NBS) from extrapolated
           grid points for Eastmain-Opinaca basin: Scenario A (GFDL)
           projected data

Extrapolated Grid Points	Land Discharge (m^3/S)	Lake Precipitation (mm)	Lake Evaporation Soil Unit	Lake Evaporation (mm)	Unadjusted mean NBS (m^3/S)	Correction Factor as per Normals	Adjusted mean NBS (m^3/S)
$A_{6-10/10}$	409.0	929.8	D3124	569.0			
$A_{6-11/10-11}$	504.6	1016.8	R1037	493.4			
Weighted mean	456.8*	973.3*		531.2**	579.8	1.567	908.5

*   Weighted mean = .5 $(A_{6-10/10})$ + .5 $(A_{6-11/10-11})$

** Weighted mean = .5 (D3124) + (R1037)

Table 8b   Annual input parameters for deriving unadjusted and
           adjusted mean net basin supply (NBS) from extrapolated
           grid points for Eastmain-Opinaca basin: Scenario B (GISS)
           projected data

Extrapolated Grid Points	Land Discharge (m^3/S)	Lake Precipitation (mm)	Lake Evaporation Soil Unit	Lake Evaporation (mm)	Unadjusted mean NBS (m^3/S)	Correction Factor as per Normals	Adjusted mean NBS (m^3/S)
$B_{12/17}$	461.7	963.8	D3124	523.5			
$B_{12/18}$	486.4	982.4	R1037	507.8			
Weighted mean	474.1	973.1		515.7	601.4	1.702	1023.6

*   Weighted mean = .5 $(B_{12/17})$ + .5 $(B_{12}/18)$

** Weighted mean = .5 (D3124) + .5 (R1037)

for La Grande basin, decreases from 6.8% to 6.7% for Eastman-Opinaca
basin and remains unchanged at 15.7% for Caniapiscau basin. On the
other hand for scenario B, the increase in NBS is reduced in all
basins: 18.3% down to 15.6% for La Grande basin, 14.8% down to 13.0%
for Caniapiscau basin and 20.3% down to 20.2% for Eastman-Opinaca
basin.
     These minor reductions in NBS increases, caused by the addition of
reservoirs, is most likely due to the fact that the decreases in land
area discharge are greater than the increase in total lake and reservoir evaporation. This is not surprising for these drainage basins
that are covered by bare rock surfaces over significant parts of
their surface area.
     The values of the change in NBS project an increase that ranges

Table 9a Normal and projected NBS and percent change in NBS for all drainage basins: Scenario A (GFDL)

Drainage Basin	Adjusted NBS without reservoirs			Adjusted NBS with reservoirs		
	(m^3/S)		Percentage Increase (%)	(m^3/S)		Percentage Increase (%)
	Normals	Projected		Normals	Projected	
La Grande	1761.9	2073.7	17.7	1761.9	2051.8	16.5
Caniapiscau	787.8	911.8	15.7	787.8	911.8	15.7
Eastmain-Opinaca	850.7	908.5	6.8	850.7	907.9	6.7

Table 9b Normal and projected NBS and percent change in NBS for all drainage basins: Scenario B (GISS)

Drainage Basin	Adjusted NBS without reservoirs			Adjusted NBS with reservoirs		
	(m^3/S)		Percentage Increase (%)	(m^3/S)		Percentage Increase (%)
	Normals	Projected		Normals	Projected	
La Grande	1761.0	2083.5	18.3	1761.0	2035.0	15.6
Caniapiscau	787.9	898.4	14.0	787.9	890.6	13.0
Eastmain-Opinaca	851.0	1023.6	20.3	851.0	1022.7	20.2

from 6.7% (Eastman-Opinaca: Scenario A) to 20.2 percent (Eastmain-Opinaca: Scenario B). On the other hand comparable studies in the Great Lakes region (Cohen, 1986, 1987; Sanderson et al, 1985; Howe et al., 1986) project a decrease in NBS of the order of 10 percent. The explanation for this difference seems to be related to the very high projected increases in precipitation that are of the order of 30 to 40 percent in July and January for both scenarios in these basins. This order of precipitation increase masks and even over-rides the projected increases in temperature and hence lake evaporation. In the Great Lakes area both precipitation and temperature increases were smaller, and in the case of precipitation there was even a decrease in some areas. Also our methods and results, though lacking in some areas seem well founded and consistent. The consistency of our results are shown in Table 10 where it is shown that there is little change in the ratio of net basin supply to total basin precipitation. This reflects the concurrent increases in discharge, resulting from a net increase in precipitation minus evaporation, and in precipitation.

Table 10   Annual Net Basin Supply/total annual precipitation ratios
(%) from measured, normals and projected data

Drainage Basin	Measured Mean Discharge/ Prec. (%)	Adjusted Net Basin Supply / Precipitation ratio (%)					
		Scenario A (GFDL)			Scenario B (GISS)		
		Normals	Projected	Change	Normals	Projected	Change
La Grande	77*	70.4	66.7	-3.7	75.4	75.5	+0.1
Caniapiscau	77**	78.8	76.8	-2.0	79.7	81.2	+1.5
Opinaca-Eastmain	84*	78.4	73.1	-5.3	81.4	82.4	+1.0

Socio-economic costs/benefits

According to McIntyre (1984) and Sanderson et al (1986) there seems
to exist a matching relationship between NBS and the average amount
of energy generated in the Great Lakes - St. Lawrence River Basin. If
these same relationships were to apply to the James Bay area, then
according to Tables 9a and 9b, the hydro-electric generating potential should increase by 15.6 percent (GISS) to 16.5 percent (GFDL)
for the La Grande basin, by 13.0 percent (GISS) to 15.7 percent
(GFDL) for the Caniapiscau basin and by 6.7 percent (GFDL) to 20.2
percent (GISS) for the Eastman-Opinaca basin.

The present potential generating capacity of the three drainage
basins, namely the La Grande, the Caniapiscau and the Opinaca-
Eastmain, and their sum total are shown in Table 11. If one were to
assume an equivalent percentage change between net basin supply NBS
and generating capacity, then according to Table 11, the hydro-
electric generating potential of the La Grande basin would increase
by 5.0 TW/h (GISS) to 5.3 TW/h (GFDL), that of the Opinica basin by
2.7 TW/h (GISS) to 3.3 TW/h (GFDL) and that of the Caniapiscau basin
by 0.6 TW/L (GFDL) to 1.8 TW/h (GISS). The total hydro-electric
production potential of all three basins together should therefore
increase by 9.2TW/h (GFDL) to 9.5TW/h (GISS).

Since warmer conditions and higher air-conditioning demands are
projected for south-eastern Canada and for the Northeast United
States, and since air conditioners generally function on electricity,
the potential increased supply of hydro-electricity from the James
Bay area will go towards satisfying this increased demand to the
south. Of course, this will be offset by a decrease in winter heating
requirements.

Also, since electricity is generally one of the cheaper forms of
energy, increased hydro-electricity generating potential as caused
by CO_2 - induced climatic change in the James Bay area would allow
for substitution of more expensive forms of energy such as coal and
oil by less expensive hydro-electricity. These different supply and
demand changes and consequent substitution effects will have to be

Table 11   Change in net basin supply (NBS) for the sum total of all
three drainage basins and change in hydro-electric
generating capacity

Drainage Basin	Present generating capacity (TW/h)*	NBS (m^3/S) Scenario A (GFDL)					NBS (m^3/S) Scenario B (GISS)				
		Normals	Projected	Change in NBS %	Projected gen. cap. (TW/h)	Change in gen. cap. (TW/h)	Normals	Projected	Change in NBS %	Projected gen. cap. (TW/h)	Change in gen. cap. (TW/h)
La Grande	32.2	1761.9	2051.8	16.5	37.6	5.3	1761.0	2035.0	15.6	37.3	5.0
Caniapiscau	20.9	787.8	911.8	15.7	24.2	3.3	787.9	890.6	13.0	23.6	2.7
Eastmain-Opinaca	9.0	850.7	907.9	6.7	9.6	0.6	851.0	1022.7	20.2	10.8	1.8
Total	62.2	3400.4	3871.5	13.9	71.4	9.2	3399.9	3948.3	16.1	71.7	9.5

*   Source: M. Tram-Van, SEBJ, personnal communication.

quantified so as to derive economic cost/benefit figures. The
comparative advantage of the James Bay area, with respect to other
supply regions of hydro-electricity such as the Great Lakes region
and to changes in demand patterns in the south will also have to be
taken into account.

Conclusions and recommendations

CO_2 - induced climate change as projected by both the GFDL (scenario
A) and GISS (scenario B) models will supposedly increase NBS by the
order of 6.7 to 20.2 percent for the three drainage basins considered
in the James Bay territory, namely the La Grande, the Caniapiscau and
Opinaca-Eastmain basins.
     This increase in net basin supply should increase the hydro-
generating potentials of these drainage basins by about 9.2TW/h
(GFDL) to 9.5 TW/h (GISS). These increased generating potentials
should provide important economic advantages for the province of
Quebec, since hydro-electricity, being a relatively cheap energy
form, would replace the other more conventional and costly forms of
energy such as coal and oil.
     Warmer conditions that are projected for south-east Canada and
north-east United States should allow for increased electricity
demand for cooling in the summer but for less electricity demand for
heating in the winter. However milder winters should favor electricity as a heating agent.
     For subsequent phases of this segment of this research, it would
however be advisable to increase the spatial resolution of the global
circulation models (GCM's) so as to be able to better assess climate
change impacts over smaller spatial distances. Also, if extrapolation

has to be resorted to, a more objective method than mere linear extrapolation should be devised.

A more accurate estimate of the water holding capacity (WHC) of the soils of this region should also be made for calculating land discharge. For the calculation of evaporation, more valid assumptions should be made with respect to the possible changes in net radiation (Q^*) resulting from global climate change.

Finally the economic variables involved such as increased electricity generating potential, higher summer-time and lower winter time energy demand together with substitution effects and regional changes in comparative advantage should be quantified and modelled so as to derive more precise economic costs or benefits.

ACKNOWLEDGEMENTS   This paper eminates from contract funds from the Quebec region of the Atmospheric Environment Service (SEA) of Environment Canada. I also wish to acknowledge M. Richard Gilbert and M. Gerald Vigeant of SEA for their prompt and generous support in matters relating to contract details and data sources. I also sincerely thank Dr. S. Cohen (Environment Canada) for providing guidance and copies of the Thornthwaite Monthly Water Budget and of the mass transfer programs; Dr. R.B. Stewart and his assistant R. Muna for providing the Priestley-Taylor evaporation calculations, M. Thack Tran-Van of SEBJ for the hydrologic and hydro-electric generation data, Mme Lynn Gregorie for data analysis, M. Guy Frumignac for the cartography and Mme Joelle Casamajou for typing the manuscript.

References

Bruce, J.P. (1984) Great Lakes levels and flows: past and future. Journal of Great Lakes Research, 10 (1), 126–134.

Clayton, J.S., W.A. Ehrlich, D.B. Cann, J.H. Day, I.B. Marshall (1977) Soils of Canada. Vols 1 and 2. Research Branch. Canada Department of Agriculture, Ottawa, 239 p.

Cohen, S.J. (1986) Impacts of CO_2-induced climatic change on water resources in the Great Lakes basin. Climatic change, 8 , 135–153.

Cohen, S.J. (1987) Climatic change, population growth, and their effects on Great Lakes water supplies Professional Geographer, 38 (4), pp. 317–323.

Hengeveld, H.G. and R.B. Street (1986) Development of CO_2 climate change scenarios for Canadian regions. Unpublished Paper.

Howe, D.A., D.S. Marchand. C. Alpagh and P.K. Stokoe (1986) Socio-economic assessment of the implications of climatic change for commercial navigation and hydro-electric power generation in Great Lakes-St Lawrence River System. Special Report for Atmospheric Environment Service, Environment Canada,122p.

Johnstone, K.J. and P.Y.T. Louie (1984) An operational water budget for climate monitoring. Unpublished Manuscript, Canadian Climate Centre Report No. 84-3, Downsview, Ontario, 52 p.

McIntyre, J. (1984) Water-existing and possible future economic value for hydro-electric generation. In proceedings, Ontario Water Resources Conference, Futures in Water, June 12-14, 1984, Toronto,

Ontario.

Priestley, C.H.B. and R.J. Taylor (1972) On the assessment of surface heat flux and evaporation using large-scale parameters. Monthly Weather Review, 100, 81-92.

Quinn, F.H. and F. den Hartog (1981) "Evaporation Synthesis" in E.J. Albert and T.L. Richards(eds) IFYGL- The International Field Year for the Great Lakes, U.S. Dept. of Commerce, Ann Arbor, 221-245.

Richards, T.L. and J.G. Irbe (1969) Estimates of monthly evaporation losses from the Great Lakes, 1950 to 1968, based on the Mass Transfer Technique. Proceedings Twelfth Conference on Great Lakes Research, International Association of Great Lakes research, 469-487.

Sanderson, M.E., T. Choi, D.A. Howe, D.S. Marchand and P.K. Stokoe, (1985) Socio-economic assessment of the implications of climatic change for future water resources in the Great Lakes/St. Lawrence river system. Hydro-electric power generation and commercial navigation. DSS Contract NO. 02 SE.KM147-4-1414. 127 p.

SEBJ. La Grande Complex-Hydrological Studies. Unpublished paper.

Singh, B., R. Taillefer and J. Poitevin (1984) Les echanges radiatifs et energetiques et le bilan thermique du sol en Jamesie. Canadian Geotechnical Journal, 21, 223-240.

Singh, B. and R. Taillefer (1986) The effect of synoptic-scale advection on the performance of the Priestley-Taylor evaporation formula. Boundary Layer Meteorolgy, 36, 267-282.

Singh, B., L. Gregorie, D. Skiadas, G. Renaud, A. Viau and B. Cairns. (1987) Prospective d'un changement climatique our les ressources naturelles du Quebec. DSS Contract No. S-102-1-5510-4210-0000-101-1100, 291p.

Stewart, R.B. and W.R. Rouse (1977) Substantiation of the Priestley and Taylor parameter $\alpha = 1.26$ for potential evaporation in high latitudes. Journal of Applied Meteorology, 16, 649-650.

Stewart, R.B. (1983) Modelling methodology for assessing crop production potential in Canada. Research Branch, Agriculture Canada,, Contribution 1983-12E, 29p.

Tran-Van, T. (1987) SEBJ, Service Hydraulique, Personal Communication.

Climatic change and water resources systems

The Influence of Climate Change and Climatic Variability on the Hydrologic Regime and Water Resources (Proceedings of the Vancouver Symposium, August 1987). IAHS Publ. no. 168, 1987.

Climate change and water resources

A.J. Askew
World Meteorological Organization
Geneva, Switzerland

ABSTRACT   The consideration of the impact of climate on water resources is placed in a wider context: the chain reaction associated with such an impact crosses the interfaces between climate, hydrology, water-resource systems and society.   Each interface demands collaboration and   exchange of information between specialists and assurance   that   current   tools   and   approaches   are appropriate for the new environment that may evolve.   The chain reaction must be studied from start to finish if the final impact on water resources is to be assessed. Various points are presented for consideration at each stage.

Impact du climat sur les ressources en eau

RESUME   La consid\u00e9ration de l'impact du climat sur les ressources en eau est plac\u00e9e dans un contexte plus large: la r\u00e9action en cha\u00eene associ\u00e9e \u00e0 un tel impact comprend les domaines communs au climat, \u00e0 l'hydrologie, aux syst\u00e8mes de ressources en eau et \u00e0 la soci\u00e9t\u00e9.   Chaque domaine   commun   n\u00e9cessite la collaboration et l'\u00e9change d'informations entre les sp\u00e9cialistes et l'assurance que les outils et approches actuels soient appropri\u00e9s au nouvel environnement susceptible d'\u00e9voluer. La r\u00e9action en cha\u00eene doit \u00eatre \u00e9tudi\u00e9e du d\u00e9but \u00e0 la fin si l'impact final   sur les ressources en eau doit \u00eatre \u00e9valu\u00e9. Diff\u00e9rents points sont pr\u00e9sent\u00e9s pour \u00eatre consid\u00e9r\u00e9s \u00e0 chaque stade.

Introduction

The last ten to twenty years have seen a major change in the general-ly accepted view of climate and of climatology.   In the past, considerations of climate concentrated on long-term steady-state conditions. Variations in space were studied and explained, but variations in time were seen more as interesting historical phenomena than as anything of particular importance for mankind in the present age.

This all changed in a remarkably short period of time when, in the 1970's, a greatly improved capability for modelling the global atmosphere combined with increasing evidence and concern for the consequences for the atmosphere of man's activities. Some preliminary findings hinted at major changes in climate under certain conditions, others led to quite different conclusions. At times the former received considerable publicity and the scientific community recognized the need for broadly-based programmes of research and evaluation on the whole subject of climate and the potential impact of its variability and change on society. Many countries have established national climate programmes for this purpose. At the international level, the World Meteorological Organization (WMO) convened the World Climate Conference in February 1979 (WMO, 1979), with the support of UNEP, FAO, Unesco and WHO. This led later in the same year to the establishment of the World Climate Programme whose overall objectives are:

(a) to apply existing climate information for the benefit of mankind,

(b) to improve the understanding of climate processes,

(c) to monitor significant climate variations and changes.

The existence of climate variability has long been recognized, including even the possibility of quite significant variations over comparatively short periods of time. However, without any real means of predicting such variations, they were not seen as being of any great relevance to regional or national planning. Our ability to predict variations is still very limited and any predictions that are made are hotly debated. It is interesting to speculate on the cause of the current interest in climate variability and change. Certainly, the fact that man himself may be the cause of such variations, even changes, has given the whole question far greater importance and urgency.

If we had perfect knowledge and foresight of the situation, we may find that variations and changes in climate over the next 200 years are neither greater nor more sudden than those experienced during the last 200 years. It is theoretically possible, therefore, that current analytical and planning techniques are adequate for generations to come and there is no real need for concern or for major changes in the way we manage affairs. However, we have neither perfect knowledge nor perfect foresight. What is more, there is a very real possibility that man's actions will have significant negative effects on the climate. These we should try to predict, plan for and, above all, prevent wherever possible. Therefore, while the reasons behind the current concern over climate change and variability may be studied by those interested in the history and philosophy of science, society has every right to expect scientists, engineers and planners to take the subject itself with the utmost seriousness.

The basic needs of mankind are commonly taken to include food and drink, clothing and shelter, and security against physical harm. The provision of food and drink requires an adequate supply of water and security demands protection from flooding as well as from other threats to safety. Some of the most important, in fact probably the most important, impacts on society of climate variability and change are introduced through the water cycle. The hydrologist and the water-resource engineer therefore have a major role to play in studying and planning for these impacts and it is very important that such work be undertaken in a co-ordinated fashion at national and international levels.

Climate variability and change

A whole range of definitions can be presented with reference to climate variability and change. The following have been found useful in developing plans for water-related projects under the WCP (WMO, 1985a):

"Weather is associated with the complete state of the atmosphere at a particular instant in time and with the evolution of this state through the generation, growth, and decay of individual disturbances."

"Climate is the synthesis of weather over the whole of a period essentially long enough to establish its statistical ensemble properties (mean values, variances, probabilities of extreme events, etc.) and is largely independent of any instantaneous state."

"Climate change defines the difference between long-term mean values of a climate parameter or statistic, where the mean is taken over a specified interval of time, usually a number of decades."

"Climate variability includes the extremes and differences of monthly, seasonal and annual values from the climatically expected value (temporal mean). The differences are usually termed anomalies."

Water-resource systems

The simple existence of a body of water does not define it as a "water resource". For it to be a "resource" it must be available, or capable of being made available, for use in sufficient quantity and quality at a location and over a period of time appropriate for an identifiable demand. There is therefore an important distinction to be drawn between hydrology and the study of water resources.

A water resource may already be used or it may represent only a potential for the future. In either case, its current and future reliability are important factors and a firm prediction of a future reduction in quantity or reliability may deny the use of the term "water resource". In this sense, therefore, a future climate change would not only affect the magnitude or reliability of existing water resources but it would also make available resources that had not previously been considered as such or result in the total loss of many existing resources.

The hydrologic cycle is an integral part of the climate system and is therefore involved in many of the interactions and feed-back loops which give rise to that system's complexity. Water-resource systems represent man's intervention in and use of the hydrologic cycle for his own benefit. Even the simplest are subject to a number of external influences each offering avenues for the impact of climate change. They contain many interactions within them and, on a local scale, can significantly modify the hydrologic cycle. Therefore, while water-resource systems may be viewed principally as the recipients of climate impacts, through the intermediary of the hydrologic cycle, they may themselves have an impact on climate, particularly where they are very large in scale or in number or where the hydrologic cycle is in a delicate state of balance.

Any investigation of the interaction between the climate system, hydrologic cycle and water-resource systems must recognize the complexity of the relationships involved. As regards water-resource systems, one approach would be to first identify the various elements involved in each system and then to study the climatologic and hydrologic factors influencing the design and efficiency of operation of each such element. A systematic presentation of the various types of water-resource system and the hydrologic factors and techniques involved in the design and operation of each type was compiled under the auspices of the International Hydrological Programme of Unesco (1982). The emphasis of WMO's Operational Hydrology Programme is on the requirements of each type of project for hydrologic and climatologic data. These were considered by Andrejanov (1975) over ten years ago and a recent report (WMO, 1987) contains extensive tabulations setting out such requirements.

At this point in time, the principal demand is not for precise estimates of the potential impact of climate change on specific water projects, but for indications of the general nature and extent of such impacts on various types of water project. Each hydrologic factor of relevance and each data requirement represent an avenue by which climate change might have an impact on the type of project in question. Therefore, reports such as those referred to above offer a good basis for identifying not only the types of project to be considered but also the climatologic elements and their characteristics which are likely to be important in a study of the impact of climate change.

Needs for information on climate change

The distinction drawn above between "variability" and "change", while clear in principle, is by no means easy to apply in practice. Variability is not so difficult to recognize and assess, but in order to study the effect of climate change we must first be able to distinguish change from variability. The progress being made in this regard by both climatologists and hydrologists is to be noted and they should be encouraged to continue with their important work.

Likewise, encouragement should be offered to those climatologists and atmospheric physicists who study the climate system and seek to explain its variability and past changes and to predict its future behaviour. Their advice is vital to those concerned with the impact on water resources for, without some indication of likely future changes in climate, any discussion of impacts will be restricted to theory and be of limited practical value. However, the water-resource engineers and planners must be prepared to state clearly what advice they require. The climatologists and atmospheric physicists have legitimate interests of their own and, in addition, many other groups make requests of them for specific information to satisfy their own particular needs. It is not enough to express dissatisfaction with the form or content of current climate predictions, the hydrologist and water-resource engineer must clearly define their own needs and make them known by appropriate means.

This list of needs is likely to encompass such parameters as:

(a) Space scale: global; hemispherical; continental; regional; national; sub-national.

(b) Time scale: hundreds of years; tens of years; annual; seasonal; monthly.

(c) Elements: radiation (at ground surface); temperature; wind speed; precipitation; humidity.

(d) Characteristics: mean; variance; skew; probability of extreme values; spatial and temporal correlation; cycles; trends; abrupt discontinuities.

At present climate predictions concentrate on changes in the mean values of a few selected elements over periods of ten to twenty years on a global or hemispherical basis. It is not at all clear at first sight what characteristics of what elements are likely to be of greatest significance with regard to the impact on water resources. Precipitation is a prime candidate but, unfortunately, it is the element about which climate predictions say the least and say it with the least confidence. Similarly, while abrupt changes or trends in mean values are important, increased or decreased dispersion and/or probabilities of extreme values are likely to be far more critical. The mean precipitation and temperature may remain unchanged even within each season, but a modest increase in their variability could throw doubt on the viability of certain rain-fed crops or run-of-the river hydropower schemes and could leave major reservoirs either empty or threatened by floods which are greater than those for which their spillways were designed.

If the types of water-resource projects are identified together with the climate characteristics that are significant for each then a list of climate prediction requirements might be drawn up. It may not be such a difficult task for an individual with a particular interest, but it will not be easy to obtain consensus on the matter among a group of experts viewing the problem from various national or regional perspectives. It is important that any such list be as short as possible and those concerned must realize that it will be many years before even a preliminary response can be made on some items. Despite these difficulties, it is important that this task be undertaken in an appropriate context at an early date. If it is not, then the climatologists and atmospheric physicists will not be able to take due account of hydrology and water resources in their investigations and the most important channel for climate impact may be poorly treated.

One last, but important, comment on this question: the identification of needs is likely to be a dynamic and iterative process. As each climate prediction is studied and as the likely impact on a particular type of water-resource system is assessed, new information about the sensitivity of the system to such changes will be acquired which will often lead to a request for more detailed predictions or predictions concerning additional parameters. The list of needs will therefore evolve and change with time, but until a first draft list is established, little progress can be made.

The impact of climate change on hydrology

The impact of climate change follows a chain reaction. As already noted, the hydrologic cycle is part of the climate system and so the first link in the chain is the impact on hydrologic processes. Here the obvious approach is to use the hydrologic model as the basic tool, adjust various parameters and inputs to simulate climate change and study the model's response both as regards its state variables and output. Even given a specific prediction scenario, it is not easy to decide what adjustments to make to what; it is even more difficult to interpret the results when the scenarios used are as speculative as they are at present. Nevertheless the work that has already been done in this regard (e.g. Nemec and Schaake, 1982) is of great importance in that it has awakened interest in the hydrologic community to the whole subject and has laid the groundwork for future studies.

The more precise one wishes to be in any investigation, the more one should question the qualities and appropriateness of the tools used. Where the tool is inadequate for the task, the results can be of little value. Klemes (1985) has set very exacting standards for hydrologic models if they are to be used in this context. It is to be hoped that model developers will apply such tests so that any results obtained by using a model may be judged against the results of his or similar tests.

Alternative approaches which do not make use of hydrologic models have also been proposed. For example, studies in comparative hydrology are seen as offering a potential source of information. As with all analyses which trade variations in space for variations in time, these should be approached with caution. They can illustrate the types of equilibrium states that have been established in the past under various external influences. The question to be asked is how much they can tell us about the dynamic response of hydrologic systems when they are subjected to changes in time in climatic factors.

This first link in the chain also involves the interface between the atmosphere and the land surface. Inadequate modelling of the water and energy balances at this interface are held to be one of the factors which currently limit the performance of general atmospheric circulation models. It is on these models that many of our hopes depend for improved climate predictions. As hydrologists seek better predictions of climate variability and change, they should therefore be prepared to make a substantial input to the work of those who model the atmosphere. The basis for such collaboration has already been established and the work commenced (Eagleson, 1982; WMO, 1985b): a most welcome sign.

Hydrologic processes and water-resource systems

The second link in the chain is that between hydrologic processes and water-resource systems. As the systems are man-made it is easy to see this as a simple matter. The anticipated variations and changes in the climatologic and hydrologic processes can be introduced in the relevant parameters and time series inputs to mathematical models of the systems. The impact on their performance may then be investigated on the basis of the outputs and general system response. However, as mentioned earlier, the interaction between water-resource systems,

climate and hydrology may be more complex than at first expected. It
may involve feed-back mechanisms and, in particular, significant
changes in certain processes may cause the system to operate in a
manner totally different from that experienced to date and may even
result in the system being unable to serve any useful purpose. For
example, a moderate increase in the probability of below-freezing
temperatures would have a negligable influence in the hydrologic
regime but could make it impossible to sustain the production of
citrus fruits. Unless an alternative crop could be cultivated, the
irrigation scheme serving the orchards would then be of little future
value. In another situation, a shift in timing of the wet season may
permit rain-fed agriculture where irrigation was previously essential.
The more obvious examples are long term trends in mean values or
changes in probabilities of extremes which could result in empty
reservoirs or increases in water logging or the threat of flooding.

The interface between water-resource systems and the natural
processes within which they are embedded is also complicated by the
fact that, while the systems are man-made and hence reasonably
clearly defined in physical terms, the manner in which they are
operated is rarely so well understood and can change or be changed
with considerable ease. The operating policy is one of the principal
characteristics of a water-resource system. For some systems it may
be possible to amend the policy so as to counteract or even take
advantage of climate variability and change. For others, however,
the freedom to change the policy may be very limited. The actual
manner in which water-resource systems are currently operated is the
result of a complex, even stochastic, balance of factors: physical,
socio-economic, political and human. There is no reason to believe
that this will not continue to be the case in the future and this
complexity and uncertainty should always be borne in mind when
evaluating theoretical optimum policies. Despite this, it is vital
that a consideration of operating policies be included as a part of
any water-resource system study.

Water-resource systems and society

The response to predictions and indications of climate change could
be structural, such as an increase in the height of levee banks, the
construction of new storage reservoirs or the installation of
additional turbines. It could also be non-structural: the implementation of hydrologic forecasting systems to allow more optimal use of
water supplies, revised operating policies for existing systems or
changes in social habits and economic activity.

The third link in the chain, that between water-resource systems
and society, is therefore of great importance and one that is
dominated by feed-back mechanisms. Both the physical characteristics
and the operating policy of a system are designed to meet the
perceived needs of society in one way or another. A change in climate
or a change in its variability could greatly affect these needs.
Domestic and agricultural demands for water would be expected to
increase if temperatures increase, but if this is accompanied by an
increase in precipitation then agricultural demand may fall,
depending on seasonal factors. If a significant change in climate is

predicted, farmers, industrial managers and the general public may respond in a manner analogous to the response seen during recent oil crises, and the net impact on society may be far less than anticipated. Conversely, a negative or ill-judged reaction could aggravate the situation and amplify the impact.

In one sense, the water-resource system, with all its imprecision, sits as a relatively well-defined entity linked on the one hand with the natural environmental systems of climate and hydrology and on the other with the socio-economic and political systems of man. This latter interface, the third in the series, is complex, dynamic, multi-faceted and, above all, difficult if not impossible to predict as regards its future characteristics and performance. Much valuable work has been done on the multi-objective planning of water-resource systems and various techniques have been developed for rationally accounting for competing demands expressed in financial terms or in various measures of public safety and welfare. In theory, these techniques should hold for the consideration of the impact of climate change, but in practice it is likely to prove vastly more difficult to express in concrete terms the desires and limitations of each sector under the predicted future conditions than under present circumstances. What will be the priorities of a society which is faced with a change in climate where this might lead to marked increases or decreases in temperature and precipitation, in food and water supplies, in health and safety risks and in the general quality of life? Where our hydrologic models need to be tested to ensure that they will yield valid results under conditions beyond those for which they were originally derived, so too will our socio-economic models. Strictly speaking, this is outside the subject of this paper, but it is certainly of relevance to the subject.

Hydrologists have long been concerned that hydrologic forecasts are adequately disseminated and correctly interpreted so as to ensure that they are of greatest value. Those who predict the response of water-resource systems to climate change should be equally concerned that their predictions could affect the validity of the predictions themselves. It is essential, therefore, that current efforts to involve the appropriate water users and decision makers in the planning and design of water-resource systems be taken much further in the study, planning and design of systems to respond to the impact of climate change. Without true dialogue, the predictions could include grave errors and the plans could prove very ineffective.

Concluding remarks

The purpose of this paper is not to review the current state-of-the-art in the study of the potential impact of climate change on water resources. Up until mid-1986 there were not so many published papers on the subject (Beran, 1986), but the field is gaining in interest and the papers presented at the current symposium should provide a good indication as to what has been achieved and what studies are planned for the future. The aim of the author has been to put the whole field of study into its wider context, to raise certain questions that need to be answered and to propose, in some instances, what approaches might be taken.

The past concentration on thirty-year normals has been replaced in climatological circles with the study of climate variability as an important and relevant topic. To this has been added the oft dramatic predictions of climate change, either man-made or resulting from natural causes. Even the more sober analyses and consensus (e.g. WMO, 1985c) indicate that the climate may well change to a degree and within a time frame which makes it important to take such a possibility into account in future planning of major projects. Of all the possible impacts of such change, the most important is likely to be that on water resources. Hence the importance of the whole subject considered in this paper.

The renewed interest in the climate system and its interaction with other natural systems is not solely related to climate change. Examples are international projects concerning the El Nino-Southern Oscillation and related phenomena which form part of the Tropical Ocean and Global Atmosphere Programme (WMO, 1985d). These promise to provide a more complete understanding of the climate system leading to an enhanced ability to predict climate variability quite apart from climate change. Such predictions have great potential value for the better management of current water-resource systems as they offer the hope of one day being able to forecast future water supplies over extended periods. It would be wiser, therefore, to think always in terms of predicting climate variability and not just change.

Nevertheless the principal topic of concern here is the impact of climate change and, to summarize the points made earlier, it would be valuable to:

(a) categorize the important types of water-resource system:

(b) identify the climatologic and hydrologic factors which affect the design and operation of such systems;

(c) identify the climatologic characteristics which are likely to be important in a study of the impact of climate change;

(d) offer all encouragement to those studying the climate system with a view to predicting future behaviour;

(e) clearly define the needs of water-resource engineers and hydrologists for climatologic information relevant to possible climate change;

(f) test the hydrologic models and other tools used to study the impact of climate change as a guide to the reliability of the results they yield;

(g) contribute to the work of atmospheric modellers in their search for an improved representation of the atmosphere/land surface interface;

(h) take account of the real nature of the water-resource systems being studied, in particular their operating policies;

(i) persue investigations as far as the impact on society so as to account for the complex and dynamic nature of the feed-back mechanisms which link water-resource systems to their socio-economic environment.

The above tasks will not be easily accomplished, but their achievement is likely to be of the greatest importance to the future well-being of mankind. Those who embark on this work should do so with a sense of purpose and caution; above all they should avoid sensational or ill-founded speculation but firmly and widely proclaim any findings and predictions which they see as being of importance for the future planning and operation of water-resource systems.

ACKNOWLEDGEMENTS   The author wishes to express his appreciation to the Secretary-General of WMO for his permission to submit this paper.

References

Andrejanov, V.G. (1975) Meteorological and hydrological data required in planning the development of water resources, Operational Hydrology Report No. 5, WMO Publ. No. 419, WMO, Geneva.

Beran, M. (1986) The Water Resource Impact of Future Climate Change and Variability - in Effects of Changes in Stratospheric Ozone and Global Climate, Volume I, J.G. Titus, ed., UNEP/EPA, Washington.

Klemes, V. (1985) Sensitivity of water resource systems to climate variations, WCP Report No. 98, WMO, Geneva.

Nemec, J. and Schaake, J. (1982)   Sensitivity of water resource systems to climate variation, Hydrological Sciences Journal, vol. 27, pp. 327-343.

Unesco (1982) Methods of hydrological computations for water projects, B.S. Eichert, J. Kindler, G.A. Schultz and A.A. Sokolov, eds., Studies and Reports in Hydrology No. 38, Unesco, Paris.

WMO (1979) Proceedings of the World Climate Conference, WMO Publ. No. 537, WMO, Geneva.

WMO (1985a) WCP-Water Project A.2 - Analyzing long time series of hydrological data with respect to climate variability: draft project description, WMO, Geneva.

WMO (1985b) Report of the First Session of the JSC Scientific Steering Group on Land Surface Processes and Climate, WCP Report No. 96, WMO, Geneva.

WMO (1985c) Report of the International Conference on the Assessment of the Role of Carbon Dioxide and of Other Greenhouse Gases in Climate Variations and Associated Impacts, WMO/UNEP/ICSU, WMO Publ. No. 661, WMO. Geneva.

WMO (1985d) Scientific Plan for the Tropical Ocean and Global Atmosphere Programme, WCRP Publ. Series No. 3, WMO/ICSU. Geneva.

WMO (1987) Hydrological Data - User Requirements and Benefit-Cost Assessment, Operational Hydrology Report - in press, WMO, Geneva.

WMO/ICSU (1982) Land surface processes in atmospheric general circulation models, P.S. Eagleson, ed . Cambridge University Press, Cambridge.

The Influence of Climate Change and Climatic Variability on the Hydrologic Regime and Water Resources (Proceedings of the Vancouver Symposium, August 1987). IAHS Publ. no. 168, 1987.

Macroscale hydrologic models in support to climate research

Alfred Becker
Head of department, Institut für
Wasserwirtschaft, Berlin, GDR
Jarmir Nemec
Director, Hydrology and Water Resources Department
WMO Secretariat, Geneva

ABSTRACT   For simulation experiments with atmospheric general circulation models (GCMs) more realistic information is required on certain land surface related hydrologic variables, in particular soil moisture and areal evapotranspiration. After reviewing in brief the present situation in large scale hydrologic land surface modelling a proposal is made for a grid related first level modelling which fulfills the requirements of GCMs and a subsequent second-level modelling which is used for validation of the first level and is river basin oriented. An important aspect of the proposed methodology is that it can be applied to any area of interest inside or outside gauged river basins, including also to the large GCM grid areas. The proposal takes into account existing experience in the application of grid techniques to river basin modelling as well as the important requirements to and recent developments relating to macroscale oriented hydrologic land surface modelling.

Des modèles hydrologiques de grande échelle
pour les besoins de la recherche climatique

RESUME   Pour l'expérimentation avec les modèles de la circulation générale atmosphérique (GCMs) il est nécessaire d'avoir une représentation plus réaliste de certains phénomènes hydrologiques manifestant sur la surface du sol et dans ce dernier, en particulier l'humidité du sol et l'évapotransporation globals. Après une revue succincte de la situation actuelle dans la modélisation à grande échelle du cycle hydrologique au niveau est un modèle hydrologique maillé qui correspond aux besoins des GCMs. Le deuxième niveau permet une vérification du modèle de premier niveau en utilisant le débit mesuré à l'échelle d'un bassin fluvial. La caractéristique la plus importante de cette proposition prend en considération l'expérience qui existe dans la modélisation hydrologique au niveau du bassin qui utilise le procédé de maille et, en même temps, les besoins impératifs de la modélisation du cycle hydrologique au niveau de sol à grande échelle.

431

Introduction

The World Climate Programme (WCP) and, in particular the World
Climate Research Programme (WCRP), as indicated in its implementation
plan (WMO 1985) will require far reaching co-operation of national
administrations, scientific institutions and individual scientists to
bring about the necessary concentration of material resources and
scientific talents to bear on the complex problem of climate and
climate changes. It is clear that contributions from the world
community of scientists not only in the field of meteorology but also
of oceanography, hydrology and glaciology will be needed to achieve
the ambitious objectives of the programme. The range of interests of
the WCRP reaches far beyond meteorology and atmospheric sciences and
the nature of phenomena require much longer observational efforts
extending over several years or even decades.

The highest priority requirement of WCRP is for consistent long
time series of global data describing the components of the climate
system. For the atmosphere there exist sources of such series, in
particular those supplied by the World Weather Watch (WWW) of the
WMO. For the oceans there exists the Integrated Global Ocean
Services System (IGOSS) of the IOC/WMO. The only global data series
on water runoff that has been assembled for the time being for these
purposes is the set of data on runoff in selected stations in the
world for water years 1978–1980. The Global Water Runoff Data
Centre has until now received data from about 1100 stations,
collected by the WMO Secretariat. The basic data consist of daily
average discharges in $m^3 s^{-1}$. The data centre is proceeding with
quality control and transfer of the data using standard computer-
compatible support. While the collection of the data continues, the
requirement of the WCRP, as formulated in the First Implementation
Plan (WMO, 1985, para. 5.8), is the support to the operation of a
Global Water Runoff Data Centre for a ten-year period (1986-1996).

However, what do these hydrologic data represent for climate
research? The discharge measured at a stream gauging station is the
total surface and sub-surface runoff from a basin (catchment) of a
river, integrated in space and time to the measurement point by the
runoff process. The fact of having an integrated measurement of the
result of three very important processes within the hydrologic cycle:
precipitation, evaporation/evapotranspiration and runoff, is not
often sufficiently appreciated by those who try to model these three
processes not only in a vertical column (as do the majority of
general atmospheric circulation models (GCMs)), but also in a three-
dimensional space. Often the observational data of the integrated
process are by far the most accurate; for monthly (and larger time
step) averages the relative accuracy can be about five per cent for
the catchment as a whole.

Unfortunately these runoff data cannot be used directly in GCMs
(Nemec, 1986), because they include the time lag caused by the
process of flow of water through different storages in the basin.
Approximate techniques exist for separating discharges, represented
as a time sequence in the form of a hydrograph, into their main
components, thus singling out the flow through individual storages.
However, it is not possible, in particular in larger and more complex
river basins to redistribute the flow measured at a gauging station

(outflow from the basin) to different points or to hydrologically uniform unit areas of the basin from where they originated. Thus it is not possible to solve the differential water balance equations for these points or unit areas using a computational time DT and from these equations to derive and use directly in GCMs the most important hydrologic variables, namely change of soil moisture and real evapotranspiration over the time interval DT.

The GCMs of an earlier generation assumed these differential equations to be for a hydrologic balance of nearly or completely uniform land surface and soil conditions, which was not realistic and did not permit advances in general circulation modelling.

It is therefore necessary, and the climate modellers have now fully realized this need, to use more complex approaches to simulate and parameterize land surface and soil moisture processes with the help of physically based macroscale hydrologic models. These did not exist, however, until relatively recently.

Why do physically based hydrologic models satisfy climate research requirements?

The requirements on hydrologic models for simulation of land surface processes in relation to the atmospheric boundary layer have been investigated by several authors. From the hydrologic researchers it was in particular Klemes (1985) who examined the suitability of hydrologic models for investigating the sensitivity of water resources to climate processes. He formulated the following general requirements for these models:
(a) They must be geographically transferable and this has to be validated in the real world;
(b) Their structure must have a sound physical foundation and each of the structural components must permit its separate validation;
(c) The accounting of evapotranspiration must stand on its own and should not be a by-product of the runoff accounting. Precipitation and potential evapotranspiration usually form the independent input variables.
It may be said that these requirements are inherent to physically based hydrologic models, as they represent the system's components as they appear in nature. While physically based models are satisfactory as regards their structure, their use presents several problems, the first relating to the different scales of hydrological processes.

Scale in hydrology

Depending on the space and time scale of a hydrologic investigation, different models and modelling approaches may be applied. A general overview of scale ranges of meteorologic and hydrologic processes is given in Figure 1. If the fundamental differential equations of the continuum hydro- and thermo-dynamics are applied to the modelling of the hydrologic land surface processes, they can only conserve a real-world validity on a microscale, where the conditions of continuity, internal homogeneity, etc. are sufficiently fulfilled, i.e. in a "topic" dimension (point, elementary plot, hydrotope, or homogeneous

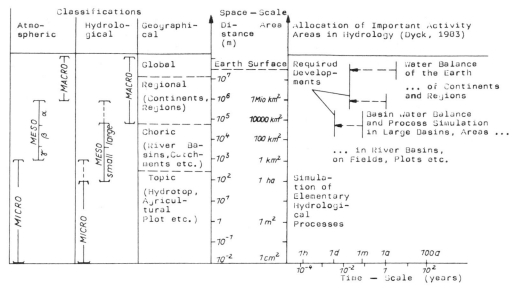

Figure 1   Classification of scales and allocation of important
           activities areas in hydrology.

hydrologic unit   area of generally less than 1 km^2 (see Figure 1,
(Becker, 1986)).   Unlike   land   surface areas these conditions   are
fulfilled in the atmosphere for much larger scales.   Thus in the GCM
scale the areas involved are 10^4 to 10^5 km^2 which is a macroscale for
hydrologic processes.
     The problem of the hydrologic macroscale is extensively discussed
by the GCM modellers themselves.   Dickinson (1985) indicated that
most GCMs used, if they introduced the land surface hydrologic
processes, Budyko's "bucket model." It represents a large scale
lumped soil reservoir with one single layer, that can be filled to
some maximum theoretical "field water capacity" (e.g. 150 mm) from
which soil water evaporates at a rate proportional to the remaining
water content. This macroscale "hydrology" conserves the mass (water)
and   net energy balance at the land atmosphere interface, but it is
oversimplified in areal integration to the extent that the results of
the   modelling   become questionable and are in many respects not in
accord with reality. Efforts were therefore directed to use physically and biologically better based models. One interesting attempt was
that of Warrilow (1986) who replaced the bucket model in a GCM by an
improved hydrologic model. However, this model could not be validated
against observed discharges within the framework of a river basin
model.
     Others attempted to take care of the complex processes in the
soil—plant continuum, including the stomatal processes (Sellers et
al., 1985). Their use and problems were characterized by McNaughton
and Jarvis (1983):
     "Plant communities are complex in ways that the atmosphere is not.
     If we had the wit and the computing power then we might solve the

equations of atmospheric motion and thermodynamics to a very close approximation, since the equations and properties of the system are quite well known. The biosphere, on the other hand, is replete with particular adaptions and special cases to such an extent that some biologists and ecologists deny that any general quantitative predictions can be made. Such people seem unduly pessimistic, of course, but they do have a point. There are no general laws of plant response to environment. Plant form and function are both highly variable. Our basic knowledge consists of a collection of empirical examples, all different but with some common trends which can be interpreted using some unifying conservation principles. Models, if they are to claim generality, must necessarily be approximate."

In consideration of Sellers' canopy model which contains 49 tunable constants, plus a leaf angle distribution function, it is further argued:

"There is a great scope for imagination here and, although it cannot be doubted that a careful selection of 49 values will give a good fit to almost any data available, the relationship between the model and the world can become tenuous. Let us, however, suppose that the model is accurate for a given canopy and soil, and that values for the constants have been selected. A GCM grid square of about 10^5 km^2 contains many such units, so we must address the problem of averaging the model over the whole range of vegetation and soil types... Then we could calculate the behaviour of each of the different canopies separately and so form an average by weighing their individual contributions according to their fractional areas. But this is of course absurd.... ...If these arguments be accepted, then many parts of a reasonably complete model, such as Sellers', would have conceptual value only, signifying real processes but not describing them in any interpretable fashion. But what, in the context of a GCM, is the conceptual significance of a root resistance or a stomatal response to saturation deficit? Surely something simpler is indicated."

On the same lines Dooge (1985) discusses the problem of scales in hydrology, but in a more general way. He argues that the large scale usually means that "the finer scale processes may be either ignored or may be represented by their statistical effects on the large scale description." He states that "it has been found in practice that the models based on the continuum mechanics are too complex to allow for the spatially variable nature of hydrologic systems to be taken into account and they have been simplified to such an extent that they became in effect simple conceptual models."

As indicated above, modelling in hydrologic macroscale has begun only recently. It has been examined by Nemec (1970), Dooge (1981), Eagleson (1982), Fiering (1982) and Dyck (1983). The statement of the latter applies best to the problem of hydrology in the GCM: "One of the assumptions frequently made is that our understanding of the microscale elements and processes (in the hydrologic cycle) can, with minor modifications, be extrapolated in principle to the understanding of the macroscale environment, thus enabling reliable predictions to be made by linking the solutions to form a causal chain. Unfortunately, it seldom happens that way. Sooner or later, at some scale or characteristic dimension, mechanistic explanation breaks down and is necessarily replaced by unverified causal hypotheses or statistical representations of the processes."

The extension of differential equations of hydro- and thermo-dynamically forced processes of moisture movement in a vertical column of the soil covered by vegetation to a basin or eventually to the large grid surface of GCMs is an excellent example of passing from a hydrologic microscale to a hydrologic macroscale. It is of course understandable to examine the microscale hydrologic processes in order to justify the prediction which will eventually have to be made on meso or macroscale. The microscale is, however, in our opinion unable to express the feedbacks, areal variabilities and other spatial integrational features needed to be included in a macroscale hydrologic land surface process model.

From the above and from our own experience it can be concluded that models based on continuum mechanics and/or on existing knowledge of transpiration control of vegetation canopy will hardly supply better results than the simplest models, such as the Budyko "bucket." The latter is oversimplified, while for the former, in the words of McNaughton "something simpler is needed." Here the knowledge of hydrologists is to be put to use and their existing conceptual physically based models could be a starting point for the "something."

The path towards an improved macroscale hydrologic land surface process modelling

The first large scale computations in hydrology were estimations of areal evapotranspiration for larger river basins, and finally for the whole land surface part of the earth, on the basis of the water balance equation in which long-term annual averages of precipitation and runoff were inserted. This is indicated in Figure 1 in the right upper part (Dyck, 1983). Later, in particular during the last decade, efforts were initiated to adopt shorter time scales as is indicated by the arrows in Figure 1.

A methodologically important attempt in this direction was the "statistical dynamic approach" of Eagleson (1978). It unifies in a coherent system a number of complex interrelated phenomena and represents them in an appropriate and efficient manner. The use of a modified infiltration equation of Philip results in an average annual local water balance equation by which the water balance is defined in terms of only four independent system parameters (three for soil, one for vegetation). Eagleson's approach is primarily oriented towards the incorporation of the vital short-term soil moisture dynamics into long-term average water balances. Thus it solves the time integration but does not deal satisfactorily with the integration of space variability.

In conceptual hydrologic models the areal non-homogeneities are in general averaged and expressed by lumped large scale parameters (including system parameters as well as inputs, state variables, etc). Such average values are then used in microscale dynamics (processes) to obtain a "lumped" representation of the macroscale system.

A very early attempt to improve this approach was the introduction of statistical distribution functions for most important parameters, e.g. a linear distribution of the soil capacity for infiltration and evaporation within a river basin (so-called "source-area" method used by Crawford and Linsley (1966)). This method makes it possible to assess "statistically" the areal variability within a river basin of important system parameters.

Later the same principle was applied to the tension water storage capacity WSM of the rooted soil zone as is shown in Figure 2 (Becker, 1975). As a result, an improved macroscale soil moisture accounting model with three parameters (WSMAX, WSC, WSGR according to Figure 2) was obtained as a separate component for river basin models. This accounting procedure appears to be advantageous in three respects:

(a) It takes into consideration, at least to a first approximation, the areal variability of the storage capacity for tension water of the rooted soil layer (in contrast to other models, such as the "Sacramento Model," where this is considered a lumped parameter (Burnash, Ferral, 1980));

(b) It relates this storage capacity to soil and vegetation parameters which are observable in the basin, in particular field capacity, wilting point and root depth;

(c) It is capable of working with different computation time steps (Δt) without loosing its physical relevance (in contrast to most infiltration excess models).

Thus, the model can be considered as physically based. In comparison with the simple Budyko bucket model, it introduces in the surface hydrologic cycle an areal distribution function and is applicable without the need for local calibration from observed rainfall and runoff series (as is usually necessary in conceptual catchment modelling). The final model equations derived by Becker (1985) may be used as a component of any large scale model (for a large grided surface or river basin area).

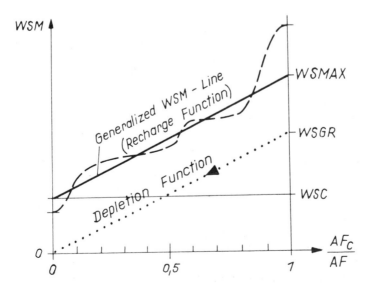

Figure 2   Real distribution (dashed line) and generalized distribution (solid line) of the soil storage capacity for tension
water (WSM) within a river basin subarea AF (permeable
soils, deep groundwater). AF_c - graction of AF.

That the model fulfills the expectations has been demonstrated by
different operational applications of it as a sub-component within
the framework of the river basin model system EGMO (Becker, 1985;
Becker and Pfützner, 1986, 1987).   It should be noted that these
applications were for those subareas of the modelled river basins
where the groundwater table is so far below the surface that plant
roots cannot reach it (deep groundwater block in Figure 3).   For the
other parts of the basins (shallow groundwater areas, impervious
areas, etc., see Table 1 and Figure 3), other specific submodels were
applied.   That for the shallow groundwater areas AN is described in
the above-mentioned paper by Becker and Pfützner (1986).

Table 1   Subareas in a river basin with significantly
          different hydrological regime

SUBAREA TYPE	SYMBOL	EVAPOTRANSPIRATION	DIRECT RUNOFF GENERATION
Open water surface	AW	potential	100% of precip.
Impervious area	AIMP	nearly zero	nearly 100%
Shallow groundwater	AN	potential	from saturated parts
Deep groundwater	AF	soil moisture dependent	nearly zero
Slopes with shallow permeable soils	AH	soil moisture dependent	interflow and infiltration

Two-level modelling approach

The approach of EGMO can be used as a basis for a two-level concept
of hydrologic modelling.   Furthermore, it is an intermediate between
the coarse large scale approach of lumping all parameters over the
river basin or other larger areas and the fine square grid based
approaches as indicated below.   It assumes the following modelling
steps:
    (a) Subdivision of the basin into hydrologically nearly homogeneo-
us uniform subareas (with regard to evapotranspiration and runoff
formation) in accordance with Table 1 and Figure 3;
    (b) Modelling of moisture variations, evapotranspiration and
"runoff production" in each subareas type by separate specific
submodels (as mentioned in the last section) which take into account
in a generalized form (statistically) the areal variability of
important land surface characteristics within the subarea, but not
the geographical position of each contributing elementary unit;
    (c) Establishing of models for each of the existing flow systems
in the river basin (surface flow and subsurface flow, possibly in
different levels) by taking into account the water divides and the
interlinks (inflows, outflows, extractions, feedbacks, etc.) between
the different subsystems.

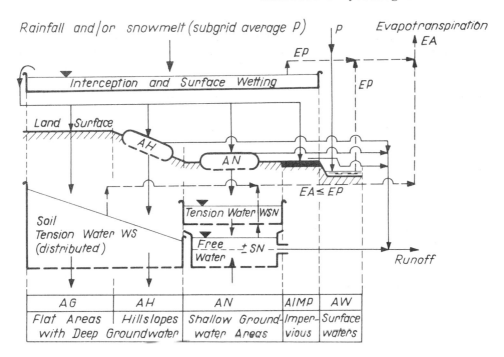

Figure 3   Schematical representation of water storage and fluxes in a subgrid area (hydrologic first-level model) according to the EGMO-system conception.

It may be noted that this approach implies on one hand a large scale oriented approximate solution of the areal integration problem, and on the other hand a step towards "two-level modelling." The first-level model is principally related to the vertical processes as indicated under (b):
   - the moisture exchange with the atmosphere,
   - the runoff production in each flow level of interest,
   - the local or subarea type related water balance.
Accordingly the first-level model can provide all information which is required for coupling hydrologic land surface models and climate models.
   The second-level model is concerned with the flow processes: surface flow, interflow and groundwater flow. Whereas the first-level model can be related to any land surface unit (river basins, small or larger uniform subareas, or grid areas), the second-level model has to take into account existing water divides and is therefore more closely related to river basins, aquifer systems, etc.
   This two-level modelling approach is particularly suitable for the coupling procedure because it makes it possible to construct that part of the hydrologic model which will be coupled (the first-level model) in such a manner that it is already adapted to the require-ments of climate models (e.g. grid structure). The other part (the second-level model) can then be related to river basins, aquifer systems, etc., as is usually done in hydrology. It is important here

to clearly define the interface between these two modelling levels, and in particular the water exchanges between them (outputs in form of infiltration excess, soil percolation, exfiltrations, water extractions, feedbacks between different subsystems, grids, etc.). The first-level model must be able to supply the required inputs for climate models on one hand and river basin or aquifer system models on the other, independently of the space scale applied in the modelling approach.

The use of flexible grids

The atmospheric models assume a computation in grid points. The upper range of GCM grids is 250 to 500 km. For some purposes, however, finer grids are used (subgrids or different mesoscales according to Figure 1) with a lower limit of 10 km.

Hydrologic models start with much smaller grids (1 km or less). Except for large scale homogeneous land surfaces (for instance the Sahara) a 10 km grid is considered as a reasonable upper limit in any hydrologic modelling. Therefore such a grid could be a meeting point acceptable for hydrologic and climatic models. To make a step forward a flexible grid technique is however required so as to satisfy smaller or larger scales.

The present state of grid technique application in larger scale river basin modelling is represented by the work initiated by Girard (1970) and Solomon et al. (1968). The rationale behind the models of Girard et al. and Solomon et al. is different but the actual applications are similar. In the work of Solomon et al. the "study area" is covered with a square grid and then considered as a matrix of squares. The system is thus adapted for computer processing and is used first to store, process and retrieve information on each grid square and then to estimate, by multiple regression, precipitation, temperature and runoff distribution on a larger scale. It presumes a hydrologic regionalization of the larger "study area" so that the squares are hydrologically homogeneous. The method has been further developed and lately used as a model for information transfer and network design in the Amazon basin in Brazil (UNDP/WMO, 1983) and for similar purposes in Canada (Ambler, 1986). A similar recent application of grid techniques was recently undertaken in Sweden (Johansson, Jutman, 1986).

The approach of Girard (1970) was later developed by Girard et al. (1972) and in his work at INRS-CEQUEAU in Quebec transformed into the so-called CEQUEAU model by his co-authors. It was finally modified by Girard et al. (1981) into the so-called "modèle couplé." The model uses a conceptualization of the moisture regime and runoff production process in the soil in each grid square. A simple routing model and a distributed groundwater model based on the Darcy equation are used for the surface and for the groundwater flow respectively. Due to the use of the grid, the system is readily usable for input from remote sensing devices, in particular satellite pictures. As is obvious, this feature makes the model particularly suitable for linking with GCMs, because the remote sensed information will, in some cases, be the only directly observed data to be used in the land surface processes model. This was clearly demonstrated in practice during the HAPEX-1 experiment in France.

Proposed approach for macroscale modelling

It is felt that the required progress in macroscale hydrologic land surface modelling can most efficiently be achieved by further development and application of the two-level modelling approach described earlier, in combination with square-griding for the first-level model. General principles of the first-level model and its application could be the following:

   (a) The land surface area of interest, e.g. a grid area of a GCM or a large river basin, is subdivided into square subgrid units as shown in Figure 4. The size of these subgrid units is dependent on the areal variability of the hydrologic and climate conditions. In relatively homogeneous conditions the subgrid units can be 50 X 50 km, in some cases, 100 X 100 km or even greater (e.g. in the Sahara, in large flatlands, prairies, pampas, tundra, etc.). In other areas, e.g. in mountainous regions and in the case of river basin modelling for testing or other purposes, it will be necessary to consider smaller subgrid units, e.g. 10 X 10 km.

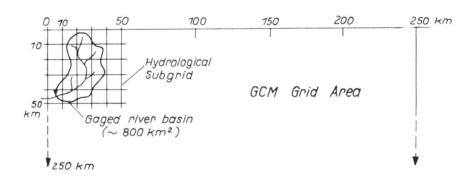

Figure 4 Schematical representation of a river basin and of a
          hydrologic subgrid within a GCM grid area.

   (b) The subgrid units are modelled separately without taking account of water divides between river basins (gauged or ungauged).
   (c) The heat and moisture fluxes at the land surface, in particular precipitation and evapotranspiration, are estimated for each subgrid unit. Precipitation and all meteorologic mesoscale parameters which control potential evapotranspiration (air temperature, humidity, etc.) are considered as lumped inputs (or as uniformly distributed within the subgrid area, e.g. according to the long-term average areal distribution).
   (d) Subareas of a significantly different hydrologic regime, in particular for evapotranspiration and runoff formation as indicated in Table 1, may be determined and separately treated within each subgrid unit (Figure 3).
   (e) Physically based large scale oriented submodels are then used. Their parameters can be estimated for each subarea in all

subgrid units from "real world" characteristics (soil, land use, vegetation, geology, geomorphology, etc.) by making use of generally available information such as maps, air photographs, satellite pictures, etc.

(f) Continuous computations of soil moisture variations, evapotranspiration and runoff formation are performed for each subarea in all subgrid units (for simplicity, subareas for which the meteorologic inputs and the model parameters are within a certain range can be grouped together in these computations).

(g) The computed subarea outputs are superimposed in each subgrid unit, to give lumped outputs (actual evapotranspiration, runoff formation from the subgrid in its main components: overland flow, interflow, percolation, see Figure 3).

(h) These outputs are then taken as inputs to GCMs and to second-level models related to river basins.

The second-level model is developed for selected test river basins where a comparison of simulated with observed river discharges is intended. The model simulates the dynamics of the most important flow systems:

- land surface and channel system;
- interflow system
- base flow system (short period and delayed component).

Submodels for the different systems may be either conceptual or distributed (hydraulics-based), grid related or not, according to the available information. The inputs into the model are outputs of the first-level model, which can be aggregated, in the presence of data as a macroscale distributed model input and in the absence of data as macroscale "statistically" lumped model input. The output of the second level model permits a validation of the whole modelling system using measured outflows from corresponding river basins, provided such measurements are at hand. In this way the global runoff data set can be used for the validation of the GCMs and their hydrologic components.

One additional interlink between both levels is the extraction of water from the above-mentioned flow systems by evapotranspiration (from open water surfaces and shallow groundwater areas, see Figure 3). A simplified solution of this problem on a larger scale has been presented in a recent publication by Becker and Pfutzner (1986).

Conclusions

After reviewing in brief the present situation in large scale hydrologic land surface modelling a proposal is made for a grid related first-level modelling which fulfills the requirements of climate models (GCMs) and a subsequent second-level modelling which is used for validation of the first-level and is river basin oriented. An important aspect of the proposed methodology is that it can be applied to any area of interest inside or outside of gauged river basins, i.e. also to the large GCM grid areas.

The critical problem in the practical application of the methodology is the requirement that the first-level models be physically based and their parameters be derived from generally available characteristics (soil, geological, geomorphological, land

use and others).   Performance tests of the grid related first-level
land surface models are possible by routing their outputs through the
flow subsystems of existing gauged river basins in the investigated
region.   When a sufficiently good fit of computed and measured river
basin discharges is at hand, this can be considered as an indication
of the reliability and correctness of the results of the first-level
model.   This permits their extrapolation to other areas in the region
(GCM grid areas, etc.) where no directly measured discharge data
exist.   It is considered that this is the only way of using measured
discharge records for the benefit of climate models.

Indeed, it would be a waste to leave aside this important data
source, which the GCMs cannot afford to do without at the present
state of ground truth availability.   The proposed model is readily
usable for remote sensed data input, as demonstrated by the
referenced literature, which however does not take into consideration
the use of the measured discharge data for the validation of the
parameters.

ACKNOWLEDGEMENT   The authors acknowledge the kind permission of the
Secretary-General of WMO and the Director of the Institut fur
Wasserwirtschaft of GDR to submit this article for publication.

References

Ambler, D.C. (1986)   Square Grid Techniques: Opportunities for Use
    in Network Evaluation and Planning.   Co-ordination Meeting for
    Implementation of WCP-Water Projects, Working Paper No. 7, WMO,
    Geneva.
Becker, A. (1975)   The Integrated Hydrological Catchment Model EGMO.
    Hydrol. Sci. Bulletin, 21 (1975) No. 1.
Becker, A. (1985)   Analysis and Simulation of Runoff from River
    Basins (in German), Stud. material "Oberflachenwasserbewirtscha-
    ftung," Ing.-Schule fur Wasserwirtschaft Magdeburg, H.7.
Becker, A. (1986)   New Requirements and Developments in Macroscale
    Hydrological Modelling (in German). Wasserwirtschaft-Wassertechnik,
    Berlin, 36, H.7.
Becker, A., Pfützner, B. (1986)   Identification and modelling of
    river flow reductions caused by Evapotranspiration Losses from
    Shallow Groundwater Areas.   In "Conjunctive Water Use" Proc. of
    the Budapest Symposium S2, July 1986, IAHS-Publication No. 156.
Becker, A., Pfützner, B. (1987)   EGMO-System Concept and Subroutines
    for River Basin Modelling.   Acta Hydrophysica, Berlin, Bd. 31,
    H.3/4.
Burnash, R.J.C., Ferral, R.L. (1980)   Conceptualization of the Sac
    ramento Model.   RA II-RA V Roving Seminar on Math. Models Used for
    Hydrological Forecasting.   WMO H/S-46/Doc. 7.
Crawford, N.H., Linsley, R.K. (1966) Digital Simulation in hydrology:
    Stanford watershed model IV. Technical report No.39, Dep. of Civil
    Eng., Stanford University.
Dickinson, R.E. (1985)   Status of the Formulation of the Land-
    Surface Hydrological Processes in GCMs.   WCP-Publ. Ser. No. 96,
    WMO/TD-No. 43, Geneva.

Dooge, J.C. (1981)   Parameterization of hydrologic processes.   In "Land surface processes in atmospheric general circulation models," Cambridge University Press, Cambridge, UK.

Dooge, J.C. (1985)   Hydrological Modelling and the Parametric Formulation of Hydrological Processes on a Large Scale, WCP-Publ. Ser. No. 96, WMO/TD-No. 43, Geneva.

Dyck, S. (1983)   Overview on the present status of the concepts of water balance models.   In "New Approaches in Water Balance Computations" (Proc. of the Hamburg Workshop, August 1983), IAHS -- Publication No. 148.

Eagleson, P.S. (1978)   Climate, soil and vegetation, Parts 1-7, Water Resources Research, 14, 5, 705-776, Washington, USA.

Eagleson, P.S. (1982)   Dynamic hydro-thermal balances at macroscale. In "Land surface processes in atmospheric general circulation models." Cambridge University Press.

Fiering, M. (1982)   Overview and recommendations.   In "Scientific basis of water resources management," National Academy Press, Washington, D.C.

Girard, G. (1970)   Essai pour un modèle hydropluviomètrique conceptuel et son utilisation au Quebec.   Cahiers de l'ORSTOM, sér. Hydrol., Vol. IV, No. 1.

Girard, G., Charbonneau, R., Morin, G. (1972)   Model hydrophysiographique.   Symp. Int. sur les techniques mathématiques appliquées aux systèmes des ressources en eau, vol. 1, pp. 190-205, Environment Canada, Ottawa.

Girard, G., Ledoux, E., Villeneuve, J.P. (1981)   Le modèle couple simulation conjointe des écoulements de surface et des écoulements souterrains sur un système hydrologique.   Cahiers de l'ORSTOM, sér, Hydrol., vol. XVIII, No. 4.

Girard, G. (1984)   Modélisation hydrologique de la Maille M100 du Sud-ouest.   Ecole Nat. des Mines de Paris, Centre d'Informatique Geol., Fontainebleau.

Johansson, B., Jutman, T. (1986)   Hydrological Maps -- Development of a System for Calculation and Presentation.   Co-ordination Meeting for Implem. of WCP-Water Projects.   Working Paper No. 8, WMO, Geneva.

Klemes, V. (1985)   Sensitivity of water resource systems to climate variations, WCP Report No. 98, WMO, Geneva.

McNaughton, K.G., Jarvis, P.G. (1983)   Predicting Effects of Vegetation Changes on Transpiration and Evaporation.   In "Water Deficits and Plant Growth," Vol. VII, 1-47, Academic Press, New York.

Nemec, J. (1970)   Scaling problem in coupling of hydrologic and general circulation models, Paper for the JPC of the GARP, ICSU-WMO, Geneva.

Nemec, J. (1986)   Global Runoff Data Sets and Use of Geographic Information Systems. Proc. ISLSCP Conference, Rome, Dec. 1985, ESA SP-248 (May 1986).

Solomon, S.I., Denouvilliez, J.P., Chart, E.J., Woolley, J.A., Cadou, C. (1968)   "The Use of a Square Grid System for Computer Estimation of Precipitation, Temperature and Runoff."   Water Resources Research, Vol. 4, pp. 919-929, No. 5, October 1968, Washington, D.C.

Sellers, P.J., Mintz, Y., Sud, Y.C. (1986)   A Brief Description of

the Simple Biosphere Model (SiB).   Proc. ISLSCP Conference, Rome,
   Dec. 1985, ESA SP-248 (May 1986).
Warrilow, D.A. (1986)   The Sensitivity of the UK Meteorological
   Office Atmospheric General Circulation Model to Recent Changes to
   the Parametrization of Hydrology.   Proc. ISLSCP Conference, Rome
   Dec. 1985, ESA SP-248 (May 1986).
UNEP/WMO (1983)   A Square Grid Hydrological Study of the Amazon
   River Basin.   Progress Report No. 8 (1022-02-1-83) on Project
   BRA/72/010, Shawinigan Eng. Comp. Ltd.
WMO (1985)   First Implementation Plan for the World Climate Research
   Programme. WCRP -- Publ. Ser. No. 5, WMO/TD No. 80, Geneva.

The Influence of Climate Change and Climatic Variability on the Hydrologic Regime and Water Resources (Proceedings of the Vancouver Symposium, August 1987). IAHS Publ. no. 168, 1987.

Variabilité spatiale et temporelle des bilans hydriques de quelques bassins versants d'Afrique de l'ouest en liaison avec les changements climatiques

Bernard Pouyaud
Directeur de Recherches
Orstom, Laboratoire d'Hydrologie
Miniparc Bât. 2, Rue de la Croix Verte
34100 Montpellier, France

RESUME   La sécheresse en Afrique de l'Ouest des 15 dernières années a un effet apparemment contradictoire sur l'hydraulicité des petits bassins versants inférieurs à 500 km^2 et sur celle des très grands fleuves, Senegal et Niger par exemple. Le présent article s'attache à définir les termes du bilan hydrique de 8 bassins versants, appartenant à un échantillon de superficies croissantes, 1 000 à 100 000 km^2 soumis à des pluviométries échelonnées de 400 à 1 000 mm. Il s'efforce particulièrement de caractériser la variabilité spatiale et temporelle de ces bilans hydriques et de les relier aux changements climatiques intervenus. Il propose enfin une approche susceptible d'expliquer comment ces bassins de surfaces intermédiaires remplacent finalement sous l'influence de la sécheresse les excédents d'écoulement de leurs parties amont par des baisses significatives et continues de l'hydraulicité de leurs exutoires qui constituent les grands bassins fluviaux. A cette échelle de superficie de bassin versant correspond   donc   la   discontinuité   structurelle   ou conceptuelle majeure de la liaison èntre la précipitation et l'écoulement.

Mots clefs:

Afrique de l'Ouest, Sahel, Sécheresse, Bassin Versant, Bilan hydrique, Ecoulement, Pluviométrie, Echelle spatiale

ABSTRACT   The drought of the last fifteen years in western Africa has a seemingly contradictory influence over the water yield of watersheds smaller than 500 km^2 and that of large river basins such as the Senegal and Niger rivers. This paper is devoted to studying the water balance components over eight watersheds, whose area ranges from 1 000 to 100 000 km^2, and rainfall from 400 to 1 000 mm. It emphasizes the spatial and temporal variability of the water balance and attempts to link it to the recent climatic changes.

     A new approach to this problem is proposed, that should help explain how the increase in runoff of intermediate watersheds gives place to the deficit observed at the outlet of the great water basins. At this scale the major conceptual discontinuity in the rainfall runoff relationship takes place.

Key words:

Western Africa, Sahel, Drought, Watershed, Water balance, Runoff, Rainfall, Areal scale.

Introduction

En Afrique de l'Ouest, sahélienne et soudano-sahelienne, au nord de l'isohyète 1 000 mm, la sécheresse des premières années de la décennie 80 est au moins aussi sévère que celle du début des années 70, sans que les années intermédiaires n'aient jamais vraiment permis un retour à la normale. L'analyse des pluviométries, à l'échelle annuelle ou journalière, en des postes spécifiques ou au plan régional, a été abordée par plusieurs auteurs (Albergel, 1984-1985-1986; Carbonnel, 1985; Olivry, 1983; Snidjers, 1983). Un consensus semble se dégager au moins sur la non-stationnarité des hauteurs pluviométriques annuelles en une station donnée, voire sur une discordance des séries pluviométriques vers les années 1969 ou 1970. Certains voient dans cette discordance une véritable "rupture" climatique, tandis que d'autres réservent leur opinion ou ne veulent y reconnaître qu'une persistance, certes singulière, mais aléatoire, d'années déficitaires. L'hydraulicité des grands systèmes fluviaux a elle aussi fait l'objet d'études documentées, essentiellement en ce qui concerne les fleuves Niger et Senegal (Sircoulon, 1985 et 1986; Olivry, 1983). L'effondrement des modules de ces grands fleuves, qui aboutit à l'assèchement historique du Niger a Niamey en 1985 (Billon, 1986) est évident. Les écoulements des petits bassins versants (moins de 500 km^2 de superficie) ont aussi fait l'objet d'études multiples (Albergel, 1985; Ribstein, 1986), toutes convergentes, qui concluent au maintien de leurs apports moyens annuels au moins, à leur niveau d'avant sécheresse. Cela s'explique, malgré la péjoration pluviométrique, par l'augmentation généralisée des taux de ruissellement, dûe autant à la raréfaction de la végétation annuelle ou pérenne, qu'à l'évolution des états de surface des sols, induite pour une bonne part par l'action anthropique.

Les études concernant des bassins versants de dimension intermediaire (1 000 à 100 000 km^2) sont beaucoup plus rares, d'abord parce qu'il est déjà difficile d'identifier dans ces zones climatiques des bassins de cette superficie qui ne soient trop hétérogènes ou touchés par la dégradation hydrographique, ensuite parce qu'il est encore peu fréquent de disposer de données fiables, hydrométriques et pluviométriques, couvrant une période d'observation suffisamment longue. Pourtant cette classe de superficies est bien la plus intéressante pour expliquer comment une augmentation avérée des écoulements sur les petits bassins versants aboutit à une diminution, évidente et considérable des apports des grands fleuves qui en sont les aboutissements.

Constitution de l'échantillon de bassins versants de référence

Le choix des bassins versants à retenir doit respecter certaines contraintes scientifiques ou techniques évidentes:
   (a) couvrir une large gamme de superficies (1 000 a 100 000 km^2).
   (b) être représentatif des pluviométries observées dans cette zone climatique au nord de l'isohyète 1000 mm.
Pratiquement l'échantillonnage se heurte à deux limitations:
   (1) Au-dessous de 400 mm de pluie la notion même de bassin versant perd tout sens physique car la dégradation hydrographique devient le caractère dominant.
   (2) Sous ces latitudes sahéliennes, quelle signification même accorder à la notion de pluviométrie moyenne annuelle, quand l'extension spatiale des événements pluvieux, qui forment le total pluviométrique annuel, reste notoirement inférieure à la maille des postes pluviométriques disponibles?
   Au plan pratique, mieux vaut donc dénommer pluie moyenne annuelle Pm, une estimation par moyenne arithmétique des pluviométries ponctuelles observées aux postes disponibles, estimation qui sera considérée comme un "indice de pluviométrie".
   C'est sur ces bases, aux prétentions réduites, qu'a été effectué le choix des bassins versants, choix résultant donc plus des données existantes et disponibles que d'une sélection réfléchie, basée sur des critères scientifiques de représentativité.
   Les 8 bassins versants retenus peuvent être classés en deux groupes, selon leur complexité:
   . des bassins versants de taille relativement réduite, relativement homogènes et représentatifs d'une zone climatique detérminée, la zone sahélienne: il s'agit des bassins du Gorouol et du Dargol et du bassin du lac de Bam.
   . des bassins versants de taille supérieure, donc plus complexes, concernés tour à tour par des conditons climatiques sahéliennes et soudano-sahéliennes: il s'agit des grands bassins de la Volta Noire, de la Volta Blanche et de la Sirba.
   En figure n°1 est rappelée la situation géographique de ces huit bassins versants qui s'étalent entre les 12° et 16° de longitude et appartiennent au bassin des Volta ou au bassin du Niger. La quasi totalité des 8 bassins est dominée par l'omniprésence du socle précambien constitué essentiellement de granites, mais aussi de schistes et de grés. Des recouvrements de roches vertes métamorphiques birrimiennes donnent localement quelque vigueur au paysage. La dégradation hydrographique est générale· c'est le cas du bassin de Beli, affluent fossile du Gorouol, ou le bassin du Sourou qui sert de deffluent à la Volta Noire, ou encore des semi-endoréïsmes comme celui du lac de Bam appartenant au bassin de la Volta Blanche.

Les résultats obtenus

Dans le but de caractériser l'hydraulicité annuelle de chacun des 8 bassins sélectionnés dans l'échantillon de référence, on a rassemblé, pour chaque bassin, les valeurs moyennes annuelles des principaux paramètres hydrologiques. Cela est fait pour la plus longue période

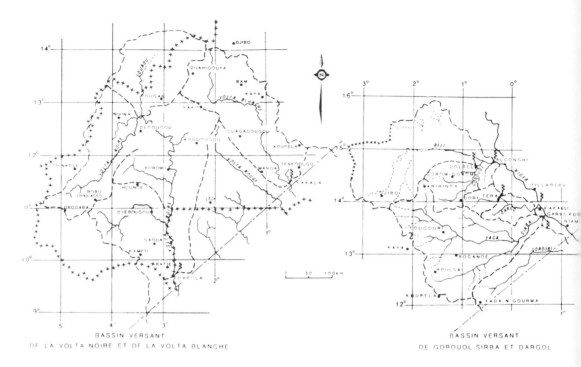

BASSIN VERSANT
DE LA VOLTA NOIRE ET DE LA VOLTA BLANCHE

BASSIN VERSANT
DE GOROUOL SIRBA ET DARGOL

Figure 1

d'observation disponible (cette période fut d'ailleurs l'un des
critères de sélection). Il est alors possible de calculer les
moyennes interannuelles de ces paramètres hydrologiques, pour toute
la période d'observation, ensuite pour la période "humide" allant
jusqu'à 1969 inclus, enfin pour la période "sèche" qui débute en
1970.
     Parmi les 8 bassins versants retenus, seule la Volta Noire à
Dapola, et dans une autre mesure le lac de Bam, sont pérennes avec un
étiage situé entre mars et mai. Néanmoins, une présentation en
années calendaires, plus simple, a été préférée à celle en années
hydrologiques qu'aurait justifiée pourtant l'utilisation d'un
paramètre tel que le déficit d'écoulement.
     Les tableaux 1 à 8 présentent les valeurs moyennes, annuelles et
interannuelles, des principaux paramètres hydrologiques:
     Pm   :pluviométrie moyenne sur le bassin, en mm
     Qe   :débit écoulé moyen, ou module, à l'exutoire, en $m^3 s^{-1}$
     Re   :lame écoulée moyenne rapportée au bassin, en mm
     Ke   :coefficient d'écoulement = Re/Pm x 100,en %
     De   :déficit d'écoulement = Pm−Re, en mm

Interprétation des résultats et tentative d'explication

Malgré les imprécisions des valeurs contenues dans les tableaux 1 à 8
et les simplifications qui ont permis leur estimation, ces résultats

Tableau 1

GOROUOL à DOLBEL

	Pm mm	Qm m³ S⁻¹	Re mm	Ke %	De mm
1961	619	13,6	57	9,2	562
1962	503	6,17	26	5,2	477
1963	637	8,63	36	5,7	601
1964	564	8,85	37	6,6	527
1965	652	6,77	28	4,3	624
1966	633	11,2	47	7,4	586
1967	470	8,98	38	8,1	432
1968	404	4,52	19	4,7	385
1969	538	10,8	45	8,4	493
Moyenne 1961-69	558	8,83	37	6,62	521
1970	390	7,26	30	7,7	360
1971	333	6,10	26	7,8	307
1972	373	5,70	24	6,4	349
1973	374	8,22	35	9,4	339
1974	435	7,70	32	7,4	403
1975	399	9,27	39	9,8	360
1976	495	7,12	30	6,1	465
1977	366	10,2	43	11,7	323
1978	499	11,4	48	9,6	451
Moyenne 1970-78	407	8,11	34	8,43	373
Moyenne 1961-78	483	8,47	35,6	7,52	447

Tableau 2

GOROUOL à ALCONGUI

	Pm mm	Qm m³ S⁻¹	Re mm	Ke %	De mm
1961	655	17,2	12,1	1,85	643
1962	483	7,22	5,1	1,06	478
1963	643	9,54	6,7	1,04	636
1964	631	16,3	11,5	1,82	620
1965	659	8,39	5,9	0,89	653
1966	587	16,0	11,2	1,91	576
1967	480	12,7	8,9	1,85	471
1968	418	5,47	3,8	0,91	414
1969	535	14,4	10,1	1,89	525
Moyenne 1961-69	566	1,9	8,4	1,46	558
1970	402	8,63	6,1	1,52	396
1971	316	7,05	4,9	1,55	311
1972	371	4,89	3,4	0,92	368
1973	390	9,98	7,0	1,79	383
1974	451	9,85	6,9	1,53	444
1975	400	8,49	6,0	1,50	394
1976	472	6,90	4,9	1,04	467
1977	375	11,5	8,1	2,16	367
Moyenne 1970-77	397	8,4	5,4	1,50	391
Moyenne 1961-77	486	10,1	7,0	1,48	481

Tableau 3

DARGOL à TERA

	Pm mm	Qm m³ s⁻¹	Re mm	Ke %	De mm
1961	727	5,46	63	8,7	664
1962	479	2,06	24	5,0	455
1963	601	3,31	38	6,3	563
1964	198	6,42	74	10,6	624
1965	545	1,92	22	4,0	543
1966	568	1,43	16	2,8	552
1967	518	3,37	39	7,5	479
1968	386	-----	----	---	---
1969	534	3,63	42	7,9	492
Moyenne 1961-69	**584**	**3,45**	**39,8**	**6,7**	**544**
1970	372	2,68	31	8,3	341
1971	371	2,46	28	7,5	343
1972	406	1,85	21	5,2	385
1973	412	2,75	32	7,8	380
1974	443	3,54	41	9,3	402
1975	361	2,58	30	8,3	331
1976	491	4,11	47	9,6	444
1977	411	4,16	48	11,7	363
1978	480	1,71	20	4,2	460
Moyenne 1970-78	**416**	**2,87**	**33,1**	**8,0**	**383**
Moyenne 1961-78	**495**	**3,14**	**36,2**	**7,3**	**459**

Tableau 4

DARGOL à KAKASSI

	Pm mm	Qm m³ s⁻¹	Re mm	Ke %	De mm
1963	542	4,60	20,9	3,8	521
1964	669	12,6	57,2	8,6	660
1965	533	5,41	24,6	4,6	528
1966	499	1,93	8,8	1,8	497
1967	537	8,23	37,4	7,0	530
1968	373	1,00	4,5	1,2	372
1969	490	3,89	17,7	3,6	486
Moyenne 1963-69	**520**	**5,38**	**24,4**	**4,6**	**515**
1970	345	----	---	---	---
1971	331	----	---	---	---
1972	393	----	---	---	---
1973	387	----	---	---	---
1974	432	5,85	26,6	6,2	426
1975	332	4,69	21,3	6,4	326
1976	505	6,81	30,9	6,1	499
1977	408	8,41	38,2	9,4	399
1978	491	4,36	19,8	4,0	487
Moyenne 1970-78	**434**	**6,00**	**27,4**	**6,4**	**428**
Moyenne 1963-78	**484**	**6,65**	**25,6**	**5,3**	**479**

Tableau 5 — **LAC DE BAM (BURKINA)**

	Pm mm	Vm Mm³	Re mm	Ke %	De mm
1966	560	13,6	5,2	0,9	555
1967	543	14,7	5,7	1,0	537
1968	588	5,9	2,3	0,4	586
1969	598	15,6	6,0	1,0	592
Moyenne 1966-69	572	12,4	4,8	0,82	567
1970	451	8,3	3,2	0,7	448
1971	579	30,0	11,5	2,0	567
1972	498	15,0	5,8	1,2	492
1973	602	31,0	11,9	2,0	590
1974	698	100,0	34,5	4,9	663
1975	630	65,1	25,0	4,0	605
1976	586	25,1	9,7	1,7	576
Moyenne 1970-76	578	39,2	14,5	2,35	563
Moyenne 1966-76	576	29,5	11,0	1,80	565

Tableau 6 — **SIRBA à GARBE-KOUROU**

	Pm mm	Qm m³ s⁻¹	Re mm	Ke %	De mm
1962	741	31,4	25,6	3,45	715
1963	720	6,7	5,5	0,76	714
1964	845	35,0	28,5	3,37	816
1965	805	41,0	33,4	4,15	772
1966	665	8,0	6,5	0,98	658
1967	666	44,7	36,4	5,47	630
1968	656	3,27	2,7	0,41	653
1969	654	19,1	15,5	2,37	638
Moyenne 1962-69	719	23,6	19,3	2,62	700
1970	502	11,5	8,4	1,67	494
1971	541	14,7	12,0	2,22	529
1972	549	6,03	4,9	0,89	544
1973	584	22,5	18,3	3,13	566
1974	673	32,4	26,4	3,92	647
1975	632	28,4	23,1	3,66	609
1976	703	12,8	10,4	1,48	693
1977	584	15,1	12,3	2,11	572
Moyenne 1970-77	596	17,9	14,5	2,38	581
Moyenne 1962-77	658	20,8	16,9	2,50	641

sont riches d'enseignements, dont le tableau 9 essaye de présenter une synthèse en faisant apparaître, aux cotes des moyennes interannuelles, sur toute la période d'observation, de Pm, Le, Ke et De, les rapports Pm_S/Pm_H, Qe_S/Qe_H et Ke_S/Ke_H entre les deux moyennes de ces trois paramètres principaux pour la période sèche ($_S$) d'après 1970, et pour la période humide ($_H$) d'avant 1969.

Les figures 2 à 5 présentent dans deux cas extrêmes (la Volta Noire a Dapola et le Gorouol à Dargol) les variations de la lame

Tableau 7 — **VOLTA NOIRE à DAPOLA (BURKINA)**

Année	Pm mm	Qm m³ s⁻¹	Re mm	Ke %	De mm
1951	1127	216,0	71	6,4	1056
1952	1093	201,0	66	6,2	1027
1953	935	114,0	38	4,1	897
1954	999	149,0	49	5,0	950
1956	1001	161,0	53	5,4	948
1957	909	105,0	35	3,9	874
1958	981	124,0	41	4,2	940
1959	987	92,2	30	3,1	950
1960	960	77,0	25	3,3	765
1961	881	117,0	39	4,1	921
1962	979	96,8	32	3,7	849
1963	1014	129,0	43	4,4	936
1964	1007	196,0	65	6,5	949
1965	968	147,0	49	4,9	958
1966	916	139,0	46	4,8	922
1967	845	88,5	29	3,2	887
1968	1031	84,6	28	3,4	817
1969	927	108,0	36	3,5	995
		154	51	5,6	876
Moyenne 1961-69	966	132	44	451	922
1970	876	149,0	49	5,7	827
1971	799	120,0	40	5,0	759
1972	758	52,0	17	2,3	741
1973	684	47,0	16	2,3	668
1974	903	127,0	42	4,7	861
1975	802	74,8	25	3,1	777
1976	878	62,7	21	2,4	857
1977	761	74,8	25	3,3	736
1978	842	60,0	20	2,4	822
1979	924	95,4	31	3,5	893
1980	841	80,3	26	3,2	815
1981	800	65,7	22	2,8	778
1982	747	48,8	16	2,2	731
1983	650	31,7	10	1,6	640
Moyenne 1970-83	805	77,8	26	3,18	779
Moyenne 1961-83	898	109	36	3,94	862

écoulée Re et du coefficient d'écoulement Ke en fonction de la pluie moyenne Pm.

L'interprétation de ces résultats peut se faire à plusieurs niveaux, en considérant d'abord les résultats moyens obtenus sur toute la période d'observation, puis en comparant ceux obtenus pour chacune des deux périodes successives: humide puis sèche.

Analyse de la variabilité des paramètres hydrologiques qui concernent la totalité de la période d'observation Pm, Re, Ke et De

Il s'agit des moyennes interannuelles pour toute la période d'observation, rappelées dans le tableau n°9, de paramètres hydrologiques se rapportant à chacun des 8 bassins ou sous bassins.

Tableau 8   VOLTA BLANCHE à YAKALA (BURKINA)

	Pm mm	Qm m³ s⁻¹	Re mm	Ke %	De mm
1956	824	51,3	51	6,1	773
1957	756	30,1	30	3,9	726
1958	794	48,2	48	6,0	746
1959	741	47,4	47	6,3	694
1960	749	20,8	20	2,7	729
1961	853	53,2	52	6,1	801
1962	848	49,4	49	5,7	799
1963	729	21,2	21	2,9	708
1964	799	45,7	45	5,6	754
1965	801	23,7	23	2,9	778
1966	727	13,2	13	1,8	714
1967	722	41,4	41	5,7	681
1968	781	12,5	12	1,6	769
1969	766	33,6	33	4,3	733
Moyenne 1956-69	778	35,1	35	4,40	743
1970	610	30,5	30	4,9	,580
1971	679	28,2	28	4,1	651
1972	687	17,5	17	2,5	670
1973	653	29,5	29	4,5	624
1974	795	51,9	51	6,4	744
1975	690	32,9	32	4,7	658
1976	751	15,1	15	2,0	736
1977	646	22,7	22	3,5	620
1978	740	11,6	11	1,5	729
1979	702	9,43	9	1,3	693
1980	708	28,8	28	4,0	680
1981	733	31,3	31	4,2	702
1982	620	14,3	14	2,3	606
1983	608	18,0	18	2,9	590
Moyenne 1970-83	687	24,4	24	3,49	663
Moyenne 1956-83	733	29,8	29,5	3,94	703

L'examen des pluviométries moyennes interannuelles Pm vérifie bien le classement des divers bassins selon leur position géographique et leur extension spatiale: les bassins de la Volta Noire, puis de la Volta Blanche, de la Sirba et du lac de Bam sont naturellement les plus arrosés, suivis du Dargol et du Gorouol plus septentrionaux. Mais l'étude détaillée des pluviométries annuelles dans les tableaux 1 à 8 montre bien une variabilité moindre à l'échelle d'un bassin versant que celle observée ponctuellement en une seule station; cela est d'autant plus évident que le bassin est de plus grande dimension et s'étend, du Nord au Sud, sur un champ plus vaste d'isohyètes.

L'Analyse des lames écoulées Re (ou des modules Qe) met en évidence les rôles simultanés et contradictoires de la superficie des bassins et de leur pluviométrie moyenne: à pluviométrie comparable la lame écoulée Re est plus importante sur le Dargol à Tera qu'à Kakassi, ou le bassin a doublé. La lame écoulée par la Sirba est moitié de celle de la Volta Blanche, les rapports de superficie et de pluviométrie jouant dans le même sens. Il faut néanmoins noter que cette analyse est rendue difficile, voire partiellement occultée, par les différences notables entre les dégradations hydrographiques des divers bassins.

456 B. Pouyaud

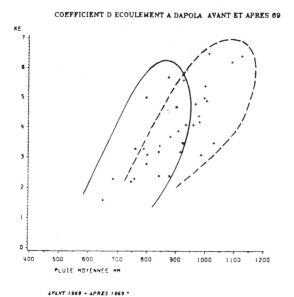

COEFFICIENT D ECOULEMENT A DAPOLA AVANT ET APRES 69

AVANT 1969 + APRES 1969 *

Figure 2

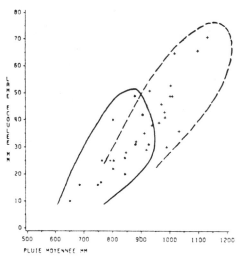

LAMES ECOULEES A DAPOLA COMPARAISON AVANT ET APRES 69

AVANT 1969 + APRES 1969 *

Figure 3

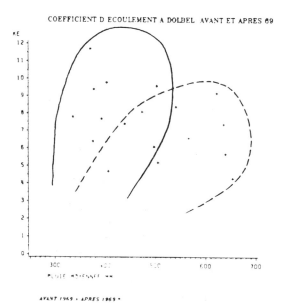

COEFFICIENT D ECOULEMENT A DOLBEL AVANT ET APRES 69

AVANT 1969 + APRES 1969 *

Figure 4

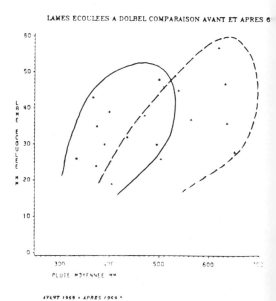

LAMES ECOULEES A DOLBEL COMPARAISON AVANT ET APRES 6

AVANT 1969 + APRES 1969 *

Figure 5

L'examen des déficits d'écoulement De est peu significatif dans ce contexte climatique, car au lieu d'être comme plus au sud représenta-tif de l'évapotranspiration réelle ou même potentielle, ce paramètre traduit ici en fait la part consommée de l'eau consommable par l'évaporation ou la transpiration végétale et est donc à ce titre davantage lié à la pluviometrie, que caractéristique du bassin en lui-même.

Tableau 9 Synthèse des paramètres hydrologiques

Bassin	Station	Période d'obs.	Superf. km2	Pm mm	Re mm	Ke %	De mm	Pm$_S$/ Pm$_H$/ %	Re$_S$/ Re$_H$ %	Ke$_S$/ Ke$_H$ %
Gorouol	Dolbel	1961-78	7500	483	35,6	7,52	447	73	92	127
Gorouol	Alcongui	1961-77	44850	486	7,0	1,48	481	70	71	102
Dargol	Tera	1961-78	2750	495	36,2	7,3	459	71	83	119
Dargol	Kakassi	1963-78	6940	484	25,6	5,3	479	83	112	139
Lac	de Bam	1966-76	2600	576	11,0	1,8	565	101	316	287
Sirba	Garbe-Kourou	1962-77	38750	658	16,9	2,50	641	83	76	91
Volta Noire	Dapola	1951-83	94000	898	36	3,94	862	83	59	71
Volta blanche	Yakala	1956-83	32000	733	29,5	3,94	703	88	70	79

Analyse de la variabilité des paramètres hydrologiques qui permettent la comparaison entre la période sèche et la période humide: Pm_S/Pm_H, Re_S/Re_H et Ke_S/Ke_H

L'analyse de ces paramètres permet véritablement une analyse plus fine: en quoi la période sèche s'est-elle traduite par une modification de la variabilité des paramètres hydrologiques traditionnellement représentatifs de l'hydraulicité des bassins? Les trois coefficients réduits Pm_S/Pm_H, Re_S/Re_H et Ke_S/Ke_H permettent une approche progressive de cette question:

Le rapport Pm_S/Pm_H est en lui-même déjà significatif: il montre que la période de sécheresse est en valeur relative plus sévère au nord de la zone sahélienne, pour des pluviométries plus faibles, qu'au sud à pluviométrie plus abondante. Pm_S/Pm_H passe de 70-75% pour le Gorouol et le Dargol à 80-90% pour les Volta Blanche et Noire et la Sirba. Il faut noter que les 100% du lac de Bam résultent du faible échantillonnage où une seule série exceptionnellement pluvieuse en août 1974 suffit à déséquilibrer l'échantillon des années disponibles.

L'analyse du rapport Re_S/Re_H est nettement inférieure a 100% pour la Volta Noire (59%), Blanche (70%), le Gorouol a Algongui (71%) et la Sirba (76%). Dans les cas du Gorouol à Dolbel (92%) et du Dargol à Téra (83%), on se rapproche déjà des 100%, dépassés nettement par le Dargol à Kakassi (112%) et le lac de Bam. L'influence de la taille du bassin versant est bien sûr ici dominante et il apparaît déjà un seuil de superficie au-dessous duquel la sécheresse s'est traduite nettement par une augmentation des écoulements absolus.

L'interprétation des variations des coefficients d'écoulement est encore plus évidente, puisque le rapport Ke_S/Ke_H n'est inférieur à 100% que pour les seuls très grands bassins: Volta Noire (71%) et Blanche (79%), Sirba (91%). Il devient supérieur à 100% dans tous les autres cas, y compris le Gorouol à Algongui (102%), encore que dans ce cas la superficie active réelle ou la surestimation de la pluie moyenne sur le bassin pourraient conduire à réviser ce chiffre à la baisse. Mais il est clair que les rapports Ke_S/Ke_H, de tous les bassins de l'échantillon, de superficie inférieure à 10 000 km^2, sont systématiquement et parfois très largement supérieurs à 100%: 119 et 139% pour le Dargol à Tera et Kakassi, 127% pour le Gorouol à Dolbel, 287% pour le bassin du lac de Bam.

Ainsi est-il clairement mis en évidence que la sécheresse a certes réduit les écoulements et encore plus les coefficients d'écoulements des bassins de superficie supérieure à 10 ou 20 000 km^2, mais agit de façon rigoureusement inverse pour les bassins de superficies inférieures. Cette gamme d'échelles de superficies de bassins versants, entre 1 000 et 100 000 km^2, semble donc bien être celle où se situe un seuil de superficie, variable avec la nature des sols, la pluviométrie et la dégradation hydrographique, seuil qui différencie les comportements classiques et contradictoires des petits et des grands bassins versants, confrontés à une période de sécheresse.

Compléments statistiques dans quelques situations particulières

Une étude statistique comparative des modules des périodes humide et sèche successives ne peut être entreprise, par suite de la brièveté des échantillons, sauf dans les cas des Volta Blanche et Noire (Pouyaud, 1976), il est intéressant d'en rappeler ici les résultats:

Dans le cas de la Volta Noire à Dapola, le module médian passe de 124 m^{3}s^{-1} pour la période 51-71 à 65 m^{3}s^{-1} (la moitié) pour la période 72-85. Cette tendance est accentuée si on passe à des occurences plus rares puisque les modules décennaux secs estimés sur les mêmes périodes passent de 89 m^{3}s^{-1} pour 1951-71 à 35 m^{3}s^{-1} (le tiers) pour 1972-85.

Il n'en va plus de même pour la Volta Blanche à Yakala, où le module médian de la période 1956-71 de 33 m^{3}s^{-1} ne chute qu'à 21 m^{3}s^{-1} (les 2/3) pour la période 1972-85. Cette proportion est conservée pour les modules décennaux secs qui passent de 17 m^{3}s^{-1} à 11 m^{3}s^{-1} pour ces deux mêmes périodes.

Il se vérifie donc ainsi statistiquement que l'impact de la sécheresse sur l'hydraulicité des bassins décroît dans le même sens que les superficies, avant de s'inverser pour les petits bassins versants.

A l'échelle des petits bassins versants, le cas du lac de Bam, pour lequel existe à la fois une chronique des assèchements et des déversements et des relevés pluviométriques depuis - 1927, est très instructif; on sait en effet (Pouyaud, 1976) que les déversements succèdent systématiquement aux périodes sèches, conclues par des assèchements. Ainsi est bien vérifiée, sur une très longue période, l'influence positive sur l'accroissement des coefficients de ruissellement des périodes sèches accompagnees de désertification.

Essai de synthèse explicative

Les constatations convergentes des paragraphes précédents permettent une synthèse explicative, fondée sur l'originalité de la géomorphologie et surtout du contexte climatique des bassins versants présentés:

Les coefficients de ruissellement élevés des têtes de bassin, avant dégradation hydrographique, s'expliquent aisément par leur géomorphologie et la nature de leurs états de surface. Sur ces bassins, de quelques dizaines à quelques centaines de km^2 , l'essentiel de l'écoulement provient du ruissellement pur ou légèrement différé.   L'absence généralisée de nappes significatives, autant parce que la géomorphologie ne s'y prête guère que parce que les crues passent trop vite pour permettre une infiltration notable, ne permet presque jamais à un écoulement de base de soutenir ces ruissellements dominants.   Dans ces conditons l'existence d'un paroxysme sec prolonge exacerbé les capacités de ruissellement, sans qu'une diminution des apports souterrains, de toutes facons insignifiants, n'entrave cette augmentation des écoulements globaux.

Puis avec l'augmentation des superficies et l'appaietion de la dégradation hydrographique, les apports provenant des têtes de bassin atteignent les vastes zones d'épandage. En années humides, les pertes par évaporation s'y exercent certes largement et longuement, mais sont équilibrées par les infiltrations, d'autant plus importantes que les écoulements sont plus prolongés, sans être nécessairement plus intenses. Dans l'écoulement global, la part des débits de base devient appréciable, voire prépondérante. En période sèche par contre, l'évaporation continue certes à s'exercer sur les épandages, mais ceux-ci ne durent pas assez longtemps pour qu'une infiltration soutenue ait le temps de s'installer et de réalimenter les nappes, qui, au fil des années déficitaires, s'épuisent. Les écoulements ne sont plus alors constitués que de la seule part du ruissellement pur, peu à peu consommée par l'évaporation et l'évapotranspiration, sans être relayée par celle des apports de drainage des nappes. Ce comportement singulier peut s'illustrer par le comportement de la Volta Noire en étiage (Albergel, 1984) qui en 1958-59 passait d'un module de 308 m^3s^{-1} en septembre à un module de 34.8 m^3s^{-1} en février; en 1980-81 elle passait au contraire de 435 m^3s^{-1} en septembre à seulement 3.05 m^3s^{-1} en février.

Conclusion et prospective

L'échantillon des 8 bassins versants traités semble donc montrer, en s'appuyant sur une analyse simple de la variabilité, avant et pendant la période de sécheresse, des principaux paramètres hydrologiques élémentaires, l'influence contrastée de cette évolution climatique sur le comportement hydrologique des bassins versants, selon leur extension spatiale. Cette intervention de l'échelle spatiale est évidemment concomitante de celle de l'échelle temporelle, puisque c'est bien la durée, liée à la superficie, des écoulements de surface et de leurs épandages, soumis à l'évaporation ou alimentant les nappes, qui détermine en definitive l'influence positive ou négative de la sécheresse et de ses corollaires sur le bilan global de l'écoulement. Ainsi s'entrouvre de fait une autre direction de recherche, qui peut constituer un lien élégant entre les deux échelles spatiale et temporelle et voit converger les approches des hydrologues, des climatologues et même des dynamiciens de l'atmosphère: elle concerne l'analyse du bilan global hydrologique de bassins versants de superficies croissantes et particulièrement la part atmosphérique de ce bilan. En effet à l'échelle annuelle ou interannuelle, si l'on estime que s'équilibrent bon an, mal an les pertes par infiltration et le drainage des nappes, on peut toujours écrire que les volumes d'eau qui s'écoulent a l'exutoire du bassin égalent les variations du stock d'eau atmosphérique qui transitent au-dessus du bassin versant. Autrement formulé, cela signifie que le coefficient d'écoulement (rapport du volume écoulé à l'exutoire au volume précipité) égale le rapport entre le volume de vapeur d'eau incidente et le volume précipité, ou encore est le complément a 1 du taux d'autorecyclage des pluies sur le bassin.

Ainsi qu'avancé précédemment, la signification de cette représen- tation évolue avec l'échelle spatiale du bassin versant et l'échelle temporelle des écoulements qui s'y produisent. Faute d'approfondis- sements théoriques et méthodologiques suffisants et compte tenu du petit nombre de résultats de ce type disponibles, cette direction d'étude reste pour l'instant embryonnaire; pourtant une connaissance plus précise des bilans hydrologiques de bassins de surfaces intermédiaires sur de longues périodes serait une etape indispensable dans la compréhension des mécanismes qui relient l'écoulement des petits bassins versants à celui des grands et des très grands bassins versants qu'ils contribuent à former.

Références

Albergel, J., Carbonnel, J.P., Grouzis, M. (1984) Péjoration climatique au Burkina Faso. Incidence sur les ressources en eau et sur les productions végétales. Cah. Orstom, Sér. Hydrol. Vol. XXI n°1, 3-19.

Billon, B. (1985) Le Niger à Niamey. Décrue et étiage 1985 Cah. Orstom, sér. Hydrol. Vol. XXI n°4, 3-22.

Carbonnel, J.P., Hubert, P. (1985) Sur la sécheresse au Sahel d'Afrique de l'Ouest. Une rupture climatique dans les séries pluviométriques du Burkina Faso (ex Haute Volta); C.R. Acad. Sc. série Hydrologie Vol VII, tome 301 n°13, 941-944.

Olivry, J.C. (1983) Le point en 1982 sur la sécheresse en Sénégambie et aux îles du Cap Vert. Examen de quelques séries de longue durée (débits et précipitations) Cah. Orstom, sér. Hydrol. Vol XX, n°1, 47-69.

Pouyaud, B. (1986) Contribution à l'évaporation de nappes d'eau libre en climat tropical sec, collection Etude et Thèse, Orstom.

Pouyaud, B. (1985) L'évaporation, composante majeure du cycle hydrologique. Séminaire Climat et Développement, 15-16 Oct. 85. Collection colloque et séminaire Orstom, 130-139.

Pouyaud, B. (1986) Estimation des apports annuels et des étiages, avant et après la récente phase de sécheresse de la Volta Noire et de la Volta Blanche. Colloque de Ouagadougou sur les normes hydrologiques. Cieh-Ouagadougou.

Sircoulon, J. (1986) Bilan hydropluviométrique de la sécheresse 1968-84 au Sahel et comparaison avec les sécheresses des années 1910 a 1916 et 1940 à 1949 in: Colloque Nordeste Sahel. Institut des Hautes études d'Amérique Latine, Paris. 16 au 18 janvier 1986.

Sircoulon, J. (1985) La sécheresse en Afrique de l'Ouest. Comparaison des années 1982-1984 avec les années 1972-1973. Cah. Orstom, sér. Hydrol. Vol n°4, 75-86.

Snidjers, T.A.B. (1983) Interstation correlations and non stationarity of Burkina Faso rainfall. Journal of climate and applied meteorology, Vol.25,524-531.

The Influence of Climate Change and Climatic Variability on the Hydrologic Regime and Water Resources (Proceedings of the Vancouver Symposium, August 1987). IAHS Publ. no. 168, 1987.

Historical variations in African water resources

J.V. Sutcliffe
Hydrological Consultant, Goring on Thames
(formerly Institute of Hydrology, Wallingford)
D.G. Knott
Associate, Sir Alexander Gibb & Partners,
Reading, UK

ABSTRACT   The variability of African water resources is studied through historic flow records. Spatial variability is linked to rainfall distribution and areas of high runoff are limited in extent. Annual runoff is sensitive to rainfall variations and seven year moving averages on a number of rivers are presented and compared. The implications for water resources planning are discussed.

Variations historiques des ressources en eau en Afrique

RESUME   La variabilité des ressources en eau en Afrique est étudiée à partir d'un historique des rélevés hydrométriques. La variabilité spatiale dépend de la distribution de la pluviométrie et les sources d'écoulements importants sont limitées en étendue. L'écoulement annuel est sensible aux variations de la pluviométrie, et les moyennes mobiles sur sept ans des débits moyens sont présentées et comparées pour des fleuves divers. Les implications sur la gestion des ressources en eau sont discutées.'

Introduction

A number of studies have dealt with recent and historical droughts in Africa (Farmer and Wigley, 1985), their meteorologic background and predictions of the future. Most of these studies have been based on rainfall records (Bunting et al, 1976), but use has also been made of river flows and lake records (Faure and Gac, 1981; Nicholson, 1980; Sircoulon, 1985; Beran and Rodier, 1985).
    The aim of this paper is to concentrate on certain aspects of African water resources as a contribution to the wider study of hydrologic variations; it is based on historic river flow records.
    A series of graphs linking average annual runoff with rainfall in different regions is compared with others showing interannual variations of runoff with rainfall. Runoff records provide a sensitive indicator of variations in rainfall. Historical variations in river flow show similar patterns over wide areas of Africa, when expressed as seven year moving averages. Even in terms of these averages, which could be related to the yield of over-year reservoirs, variations in river flow have been very wide and this should be taken

463

into account when the water resources of an area are considered.

The medium term fluctuations of available water can only be smoothed out by large reservoirs and suitable sites may not always be available on the scale required. There is a case for accepting moving targets for irrigation or hydro-electric power production in these circumstances.

Spatial variability

Although the spatial variability of African water resources is well known, it is normally illustrated by means of annual isohyetal maps. These illustrate broadly the total available water and show, for example, rainfall decreasing from 2000 mm near the coast at Abidjan to 200 mm near Tombouctou some 1300 km to the north (Figure 1(a)).

If one considers water resources as equivalent to river flows which are available for storage and use in irrigation or hydroelectric power production, then the variability of average rainfall is only a first indication of the uneven distribution of water resources. Because runoff is the residual process in the hydrologic cycle, i.e. the difference between rainfall and evaporation, it is much more variable than rainfall itself, especially in Africa where potential evaporation is high.

This sensitivity is illustrated by the series of graphs in Figure 2, in which mean annual runoff is compared with rainfall for gauging stations on tributaries within a number of basins in different parts of Africa. These comparisons, taken from recent water resources studies, confirm that spatial variations in rainfall are amplified

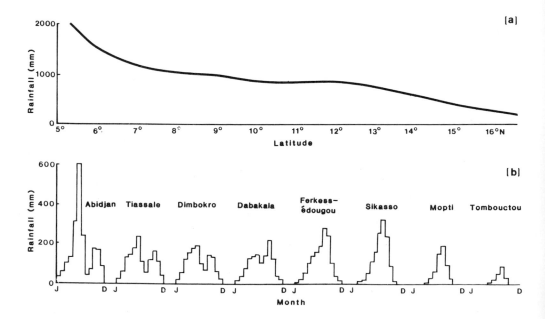

Figure 1   (a) Mean annual rainfall and (b) seasonal rainfall along 4°W longitude.

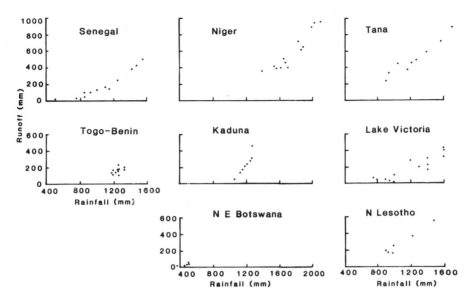

Figure 2 Mean rainfall and runoff of tributaries in different basins (each point represents the period mean rainfall and runoff of an individual tributary).

when runoff is expressed as depth over the basin.

The comparison of annual rainfall with runoff depth oversimplifies the relationship, which depends also on seasonal distribution of rainfall. Where the excess of rainfall over potential transpiration is concentrated into a single season, this excess first fills up the depleted soil moisture storage and then gives rise to runoff. Where the rainfall is spread over two seasons, transpiration losses occur over longer periods and there are two occasions for soil moisture replenishment. Therefore the runoff for a bimodal rainfall distribution is lower than the equivalent for the same gross rainfall occurring in a single season.

Figure 1(b) illustrates such a change in the seasonal rainfall distribution associated with the migration of the ITCZ, with the single rainfall season in the north and the bimodal distribution near the coast. Potential transpiration may be subtracted from rainfall to give maps of net rainfall, which have been used in Guinea and in Togo/Benin to estimate runoff at ungauged sites (Sutcliffe and Piper 1986). However, maps of net rainfall are available only for small areas of Africa, and it is therefore necessary to use maps of gross rainfall, accepting that local relations between mean rainfall and runoff can only be empirical because of the effect of seasonal distribution.

The comparisons of Figure 2 from many different parts of Africa confirm that areas with rainfall less than 1000mm provide low runoff below 100mm, while disproportionately higher runoff occurs from areas with rainfall above 1500mm or 2000mm where runoff generally exceeds

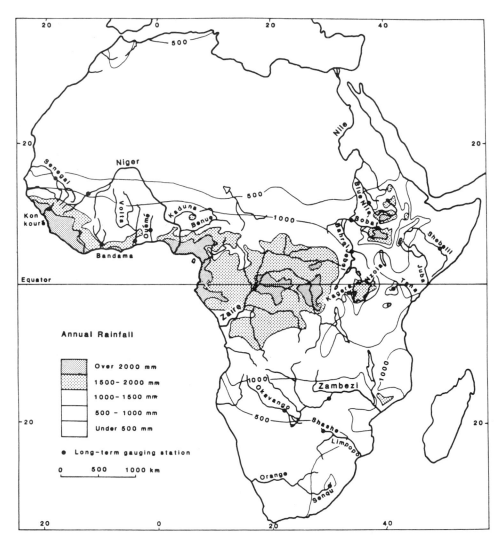

Figure 3   Mean rainfall and drainage in Africa
           (isohyets after Leroux, 1983).

400mm or 1000mm. The areas of central Africa with average rainfall above these amounts are illustrated in Figure 3.

   It follows that the areas of high runoff on the African continent are extremely limited in geographical extent, if one excepts the Zaire and adjacent basins, where large water resources exist but development has not been commensurate. In the Atlas mountains of the Maghreb, winter rainfall of the Mediterranean type gives rise to locally important water resources, but these are not discussed further in this paper.

   Many of the high rainfall areas naturally coincide with mountai-nous areas, which in turn provide the headwaters of a number of major

rivers.   The sources of a number of West African rivers,   like the
Senegal, Gambia, Niger, Konkouré and Bandama, coincide with the high
ground running east from the Fouta Djalon in Guinea which experiences
a single rainfall season.   The areas of high runoff in Togo coincide
with the high ground between Palimé and Lama-Kara,   the river Kaduna
draws much of its flow from the Jos Plateau, and the Benue depends on
the Cameroon mountains.

In eastern Africa,   the Blue Nile, Sobat, Juba, Shebelli and other
rivers have their sources in the Ethiopian highlands,   while the Tana
derives from Mount Kenya and the Aberdares.   The Lake Victoria
tributaries   and   thus   the Nile outflow depend on the   mountains   of
Rwanda   and   Burundi   drained by the Kagera,   and   the   area   to   the
northeast of the lake below the Mau escarpment.

In southern Africa,   the Okavango and Zambezi,   with other rivers,
have   their   farthest sources in the mountains of Angola,   while   the
Lesotho highlands give rise to the Orange river.

Thus the variability of a number of rivers,   whose contribution to
African water resources is important,   depends on rainfall in limited
areas of the continent.

Annual variability of river flows

Just as the sensitivity of runoff to variations in rainfall,
illustrated by Figure 2, is responsible for the concentration of
runoff generation in limited areas of high rainfall, this same
sensitivity explains the marked annual variations of river flow in
response to annual variations in rainfall and especially to the
drought of recent years.   Where rainfall is highly seasonal,   annual
runoff is the residual surplus after soil moisture replenishment and
evaporation during the wet season have been satisfied.   In the same
way   as   mean runoff varies widely according to mean rainfall from
basin   to   basin,   so the runoff from a specific basin varies widely
from year to year according to annual rainfall.   This sensitivity may
be   deduced   from flow records or by modelling (Němec and Schaake,
1982).

In   an area where seasonal rainfall distribution and   physiography
are reasonably homogeneous the relationship between mean rainfall and
mean runoff (Figure 2) provides an initial prediction of the response
of   river   flow to variations in annual rainfall.   However,   annual
rainfall   and runoff series are compared for a selection of basins in
Figure 4.   In some cases,   like the Shashe and Ouémé,   the spread of
rainfall   and   runoff in different years is greater than   the   spread
between the averages of different basins.   Nevertheless the similarities   with the relationships of Figure 2 are clear,   and interannual
variations in river flow can be said to mirror interbasin   variations
in an area.

Thus   long-term records provide a measure of water resources which
is also a sensitive indicator of fluctuations in rainfall.   As   they
represent   net   rainfall over a basin rather than at   a   point,   they
smooth   the   effect   of   random variations from point to point and
provide a valuable indication of climatic fluctuations.

The   variability   of   annual   flows   for   the   largest   rivers   is
illustrated in Figure 5 where   the available records are presented in

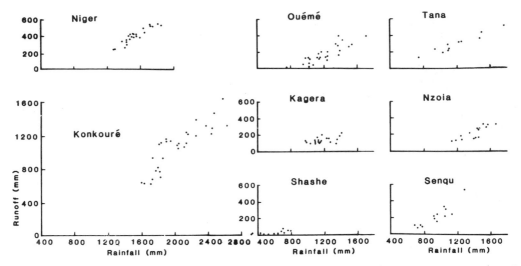

Figure 4   Annual rainfall and runoff from different basins (each point represents annual rainfall and runoff totals for individual years).

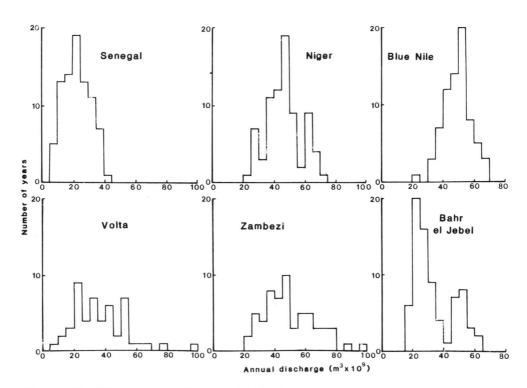

Figure 5   Histograms of annual discharges.

histogram form.   The range of annual flows is very wide even on these large rivers;   the ratio between highest and lowest years varies from 3.4:1 to 10:1.

Annual   flow   series   for several large rivers   are   presented   in Figure 6,   where the scatter tends to obscure the pattern of high and low   years which seems to be present in some   cases.   Moreover,   the flow sequence is as important as the overall spread.   Hurst (1965) had demonstrated   in   his study of Nile flows that sequences of high   and low   years   increase the size of the reservoir required   to   equalize flows beyond that required for an uncorrelated series.

The flow series for a number of African rivers (listed in Table 1) are presented in Figure 7 in the form of seven year moving   averages, covering   as   far as possible the areas which have been noted as   the main sources of runoff.

There   are   certain   striking   similarities but also   a   number   of contrasts among these series.   The flows of the Senegal and Niger are very   similar,   which would be expected as they derive from   adjacent areas   of surplus rainfall.   The range of seven year mean   flows   is itself very wide,   particularly for the Senegal where the recent mean is only a third of earlier peaks and the recent drought has been much more severe than previous droughts in 1910-1915 and 1940-1945.

The pattern of flows in other west African rivers like the   Volta, the Konkouré in Guinea,   the Bandama in Côte d'Ivoire,   and the Ouémé

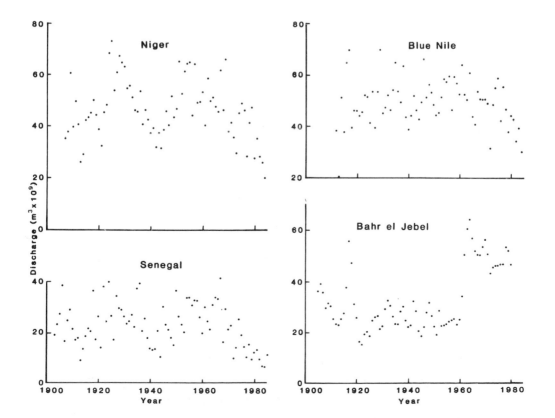

Figure 6   Annual discharges of selected rivers.

Table 1 Long-term river flow records

River	Station	Lat.	Long.	Area km²x 10³	Period	Mean Runoff m³x 10⁹mm	
Senegal	Bakel	14°54'N	12°27'W	218	1903–85	22.9	105
Niger	Koulikoro	12°51'N	7°33'W	120	1907–84	46.4	386
Konkouré	Teliméle	10°30'N	12°54'W	10.2	1951–80	10.8	1058
Bandama	Tiassale	5°54'N	4°49'W	95.5	1928–85+	9.58	100
Volta	Senchi	6°12'N	0°06'E	394	1936–85	37.2	94
Ouémé	Savé	8°00'N	2°25'E	23.6	1932–83+	3.77	160
Blue Nile	Roseires	11°52'N	34°23'E	210	1912–84	49.3	235
Shebelli	Belet Uen	4°44'N	45°12'E	212	1951–85	2.21	10
Sobat	Doleib	9°20'N	31°38'E	22.5	1905–80	13.6	60
B. el Jebel	Mongalla	5°12'N	31°46'E	450	1905–80	33.0	73
Kagera	Nyakanyasi	1°12'S	31°15'E	55.8	1940–79	6.38	114
Tana	Kamburu	0°49'S	37°41'E	9.30	1908–77+	2.99	321
Okavango	Mohembo	18°13'S	21°49'E	180	1932–76	10.6	59
Zambezi	Kariba	16°31'S	28°50'E	664	1924–85	50.3	76
Shashe	Shashe dam	21°22'S	27°26'E	3.65	1922–85+	0.074	20
Senqu	Koma Koma	29°36'S	28°41'E	7.95	1921–83+	1.88	237

+ Extended using rainfall records

in Benin, have been similar to the Niger and Senegal. It is more surprising that the flows of the Blue Nile, though less variable than the Niger, have also reflected droughts in the early 1940's and more recently.

The flows of the Bahr el Jebel or White Nile at Mongalla present a completely different pattern from the Blue Nile and reflect the rise in Lake Victoria in 1961-64. It is interesting to note that other rivers in East Africa, like the Kagera and Tana, present a similar pattern without the effect of lake storage. The flows of the Sobat which are attenuated by spill to the Machar marshes, show some similarity but the Shebelli does not do so.

The flows of the Zambezi present a quite different pattern, with increased runoff between 1950 and 1980. However, there seems to be little similarity with the other flow records from southern Africa, which are the inflows to the Okavango Swamp and the Shashe in Botswana and the Senqu in Lesotho.

Proxy series

Long term homogeneous river flow series are not available in all parts of Africa. In general, the francophone countries appear to be better provided with long-term river flow records, while the anglophone countries have shorter flow records but a number of long-term rainfall records.

The Nile basin is an exception to this generalization, as the evident importance of river flow records led the Physical Department of the Egyptian Government to establish river gauges throughout the basin at an early date, so that key records are available from about 1905.

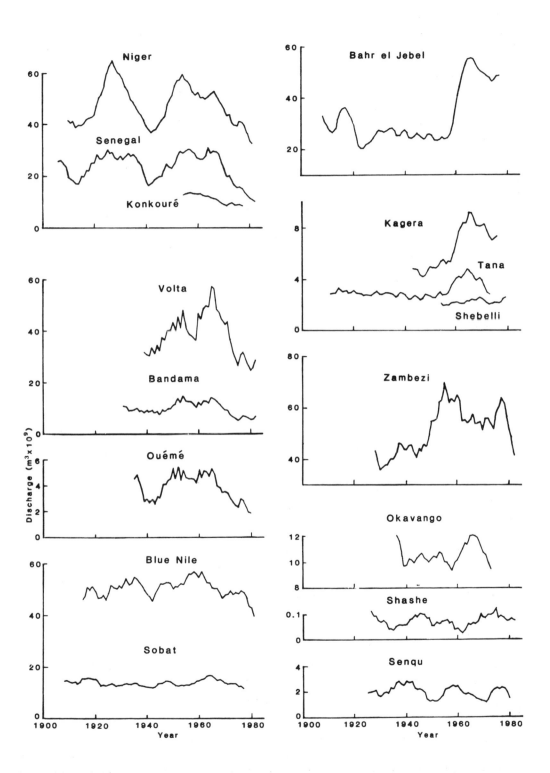

Figure 7   Seven year moving average discharges at long-term   gauging
          stations.

However, long-term rainfall series can be used to estimate basin rainfall series which can be converted to runoff series either by statistical methods or conceptual models calibrated by recent flow records. Some of these extended series have been included in the runoff series of Figure 6.

Where a river basin includes a lake for which the outflow is related to level by a fixed control, as at Lake Victoria and the Ripon Falls, or where there is no outflow and evaporation over the lake area provides the control, then a long-term series of lake levels can be used as a proxy record for a river flow series.

Such lake series are useful as historical levels are often available before the start of regular readings and these can be used to infer hydrologic conditions before river flows are generally available. For example, early accounts of Lake Victoria indicate a period of high levels and high Nile outflows for the period 1875–1895, which is supported by other evidence (Piper et al., 1986).

The rise in Lake Victoria levels in 1961–1964, which was reflected in the flows of the Kagera and Tana rivers as well as the Bahr el Jebel, has been shown to be similar to a contemporary rise in Lake Tanganyika (Kite, 1981). Zaire river levels at Kinshasa were also exceptionally high in 1961–1963 (Hirst and Hastenrath, 1983). Rises also occurred at this time in Lake Turkana (Rudolf) where previous high levels before 1895 have been deduced by reconstructing a lake level series by comparison of early maps and other evidence (Butzer 1971). Similar rises have occurred in Lakes Abaya/Chamo/Chew Bahir in the Ethiopian rift valley (Kingham, 1975).

Similar historical evidence, including such records as the Nile level series from 600 A.D. has been used to extend our knowledge of wet and dry periods over a much longer time scale (Nicholson, 1980).

Water resources planning

These characteristics of the climatology and hydrology have important implications for water resources planning.

(a) They lead to greater emphasis on overyear storage as a means of optimizing the use of water resources. The seven year moving average provides a useful target for the yields that might be provided by overyear storage. It is relevant to water resources planning that these moving averages fluctuate widely in Africa. In general the available resource indicated by the seven year moving average has varied by a ratio of 1:2 over this century. The importance attached to overyear storage capable of providing adequate regulation is reflected in the capacity of such projects as the Aswan High Dam and the Kariba hydro-electric project.

(b) The comparative concentration of the sources of runoff imposes a high degree of interdependence between those countries which control the resource and those which depend on it for their supply of water. The flows of the Nile, which are the principal water resource of Egypt and the Sudan, originate almost entirely from outside the frontiers of those countries. Similarly the flow of the River Senegal is largely derived from Guinea, while it is mainly used for irrigation in Mali, Senegal and Mauritania. This situation has led to the creation of a number of international bodies whose function is to regulate the development and use of individual river basins.   River basins in the continent of Africa which are planned by supranational authorities include the Nile, the Senegal (OMVS), the Niger and the Gambia (OMVG). The OMVS has direct control of the implementation of the Manantali dam on a major tributary of the River Senegal in Mali and the Diama barrage on the border between Senegal and Mauritania.

(c) It may be necessary to review the conventional criteria that are used in water resource planning and adopt a more flexible approach. While this may not be realistic in economic terms for urban water supply, where reservoirs may have to be supported by conjunctive use of groundwater supplies, such an approach could be useful in irrigation or hydro-electric planning.

Traditionally irrigation schemes seek to supply water with an assured reliability in, say, 4 years out of 5. If, however, there is a persistence of low flows followed by a series of high flows, it may be preferable to plan systems on the basis of substantial reductions in cropping intensities for long periods, to be increased when hydrologic conditions permit. For example, irrigation development in the Senegal Basin envisages a total area of 375,000 ha, but for certain periods there will only be sufficient water for full intensities on about 250,000 ha.

The same principle holds for hydro-electric planning. It would be economically attractive to take advantage of above average hydrologic conditions by varying the assumptions on firm energy for short periods. This could lead to decreases in thermal generation and deferment of additional capacity. An example of this approach is elaborated in the following section.

Planning of Kariba hydro-electric scheme

Planning studies of the Kariba hydro-electric scheme provide an interesting illustration of a possible response to changes in the hydrologic regime. The initial installed capacity of the scheme in 1962 was 666 MW and this capacity was increased by a further 600 MW in 1977. In the earlier years of operation a firm energy of 8500 GWh per year was adopted, based on reservoir studies for a 44 year period from 1926 to 1970. It was noted, however, that there had been a marked increase in river flows since 1950. When planning studies were carried out in 1975 for the installation of a further 600 MW at Kariba the question arose as to whether it would be appropriate to adopt a higher firm energy for planning purposes based on the more recent period of record, which some argued was representative of changed hydrologic conditions.

This would have important economic advantages as it would permit a reduction in thermal generation and some deferment in new capacity. It was concluded that for short term planning a figure of 10000 GWh per year could be adopted, although it was recommended that for longer term planning the original figure of 8500 GWh should be maintained. Any risk associated with such a policy would be mitigated by the large element of overyear storage in the Kariba reservoir.

It is perhaps fortunate that it did not prove necessary for the system to rely too heavily on the additional firm output, as the three consecutive years 1981-1983 were all extremely low. Firm output was only maintained by drawing down the reservoir to very low levels. This outcome vindicates the view that the evidence of a long period of record is to be preferred to a shorter period.

Nevertheless the principle of reacting to changes in hydrologic regime is still valid. Should there be a return to the conditions of 1950 to 1980 it might be permissible and economically advantageous to adopt a higher firm energy for short periods and optimize the planning of the system accordingly.

Conclusions

This study has concentrated on establishing a number of facts about African water resources. The variability of river flows in both space and time is reasonably well established, but the scientific and practical implications need considerable thought. It is hoped that the facts presented on the scale of variations during this century will provide meteorologists with a useful supplement to the evidence from rainfall records. The key to further advance must be in the domain of global meteorologic study. Any advice on the possible future pattern or, less ambitiously, the scale of periodic fluctuations would provide considerable benefit to the water resources planner.

ACKNOWLEDGMENTS The basic data used in this study were kindly provided by national hydrometric and international organizations, by ORSTOM, or by consultants, in the course of a number of water resources investigations or reviews in which the authors have taken part.

References

Beran, M.A. & Rodier, J.A. (rapporteurs) (1985) Hydrological aspects of drought, Unesco/WMO Studies and reports in hydrology, No. 39.
Bunting, A.H., Dennett, M.D., Elston, J. & Milford, J.R. (1976) Rainfall trends in the West African Sahel, Quart. J. Roy. Met. Soc., 102 , 59-64.
Butzer, K.W. (1971) Recent history of an Ethiopian delta, Univ. Chicago Geogr. Dept. Res. Pap. 136.
Farmer, G. & Wigley, T.M.L. (1985) Climatic trends for tropical Africa, Climatic Research Unit, Norwich.
Faure, H. & Gac, J.-Y. (1981) Will the Sahelian drought end in 1985? Nature, Lond., 291, 475-478.
Hirst, A.C. & Hastenrath, S. (1983) Diagnostics of hydrometeorological anomalies in the Zaire (Congo) basin, Quart. J. Roy. Met. Soc., 109, 881-892.
Hurst, H.E., Black, R.P. and Simaika, Y.M. (1965) Long-term storage. Constable, London, 145 pp.
Kingham, T.J. (1975) Discussion of Grove, A.T., Street, F.A. & Goudie, A.S., Former lake levels and climatic change in the rift valley of Southern Ethiopia, Geog. J., 141, 197-199.

Kite, G.W. (1981) Recent changes in level of Lake Victoria, Hydrol. Sci. Bull., 26, 233-243.

Leroux, M. (1983) Le climat de l'Afrique Tropicale, 2 vols., Champion, Paris.

Němec, J. & Schaake, J. (1982) Sensitivity of water resource systems to climate variation, Hydrol. Sci. J., 27, 327-343.

Nicholson, S.E. (1980) Saharan climates in historic times, The Sahara and the Nile, M.A.J. Williams & H. Faure (eds), A.A. Balkema, 173-200.

Piper, B.S., Plinston, D.T. & Sutcliffe, J.V. (1986) The water balance of Lake Victoria, Hydrol. Sci. J., 31, 25-37.

Sircoulon, J. (1985) La sécheresse en Afrique de l'Ouest: comparaison des années 1982-1984 avec les années 1972-1973, Cah.ORSTOM, 21, 75-86.

Sutcliffe, J.V. & Piper, B.S. (1986) Bilan hydrologique en Guinée et Togo-Bénin, Hydrologie Continentale, 1 (1), 51-61.

The Influence of Climate Change and Climatic Variability on the Hydrologic Regime and Water Resources (Proceedings of the Vancouver Symposium, August 1987). IAHS Publ. no. 168, 1987.

Climatic change and Great Lakes water levels

M. Sanderson & L. Wong
Great Lakes Institute
University of Windsor
Windsor, Ontario
Canada   N9B 3P4

ABSTRACT   World climatologists believe that increasing concentrations of CO_2 in the earth's atmosphere will result in significant changes in the world's climate during the next century. In the present paper, the predictions of monthly temperature and precipitation for the Great Lakes basin derived from a current General Circulation (climatic change) Model, are incorporated into a hydrologic model of the Great Lakes to predict future levels and flows under 2 x CO_2 conditions.

Changement de climat et niveaux d'eau dans les Grands Lacs

RESUME   Les climatologues internationaux suggèrent que les concentrations de CO_2 qui ne cessent d'augmenter dans l'atmosphère terrestre vont aboutir à des changements importants dans le climat mondial durant le siècle prochain. Dans cet article, les prévisions des temperatures mensuelles et la précipitation dans le bassin des Grands Lacs, dérivées d'un Modèle actuel de Circulation Generale (changement de climat), sont integrées à un modèle hydrologique des Grands Lacs afin de prévoir les futurs niveaux et écoulements dans les conditions de concentration double de gaz carbonique.

Introduction

The Laurentian Great Lakes contain 20% of the world's fresh water and the Great Lakes basin is home to 40 million people on both sides of the Canada–United States border. Lake levels and flows in the system have been measured for some 100 years, with extreme high and extreme low levels differing by as much as 2 metres. Low levels are of great concern for navigation and hydro-electric power companies, while high levels cause flooding and erosion problems for shore owners. However, the mean level and fluctuations in levels that have occurred in the past are the ones the users of the lakes have become accustomed to, and learned to live with. Scientists, as well as the general public, usually assume that the future of a large hydrologic system like the Great Lakes will resemble the past. However, this may not be the case if climatic change occurs.
  A consensus is emerging among world climatologists that increasing

concentrations of CO_2 in the earth's atmosphere will result in significant changes in the world's climate during the next century (Bruce 1984). For all users of the Great Lakes, an immediate question arises as to what effect this climatic change will have on Great Lakes water levels.

Throughout the earth's history, high concentrations of CO_2 have been associated with warm conditions, low CO_2 with cold, and changes in CO_2 appear to precede changes in climate (Budyko, 1977). The level of atmospheric CO_2 is thought to have been approximately 200 ppm during the Pleistocene ice age and to have risen to its pre-industrial value of 275 ppm approximately 10,000 years ago. However, man's activities are now known to be affecting CO_2 concentrations. The monitoring of atmospheric CO_2, which began at the Mauna Loa Observatory in Hawaii in 1958, has shown an increase from 312 ppm in 1958 to 345 in 1984. Similar increases of about 0.4% per annum have been recorded at Barrow Alaska, the South Pole and Alert in Canada's Northwest Territories. Concentrations of 390 to 580 ppm (depending on future energy demand) are predicted for the year 2050 (WMO 1986). In addition, at least 20 other gases, including chlorofluorocarbons, methane, nitrous oxide and ozone, have been identified as possible contributors to atmospheric warming.

Whether the effect of increasing concentrations of greenhouse gases is visible in world temperature records is a very difficult question to answer because of the great natural variability of climate from place to place and from year to year. Using the most reliable world climate records, which rarely exceed 100 years duration, climatologists have constructed the graph in Figure 1 (Hansen, et al., 1983). During the period 1880-1980, measured surface temperature as a global mean increased approximately 0.5 °C. Climatologists, using global scale general circulation models (GCMs) of the earth's atmosphere now predict significant changes in world temperature under higher CO_2 concentrations.

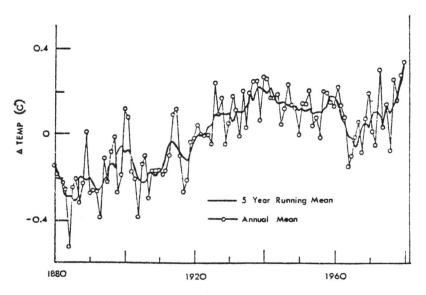

Figure 1

2 x CO$_2$ and Great Lakes climates

The Canadian Climate Centre of Environment Canada has obtained and modified the output of the Goddard Institute of Space Studies' GCM in an attempt to predict climatic conditions in Canada with an atmospheric CO$_2$ concentration twice that of pre-industrial times. There are many problems in any such large scale models and it must be noted that the outputs of all models are very tentative estimates of future climate for any specific area (Cohen 1986). The GISS model used in the present study predicts a significant increase (approximately 4.4^{o}C) in mean annual temperature in the Great Lakes area, and an average increase of 6.5% in mean annual precipitation.

   The output data from the GISS model for ten grid points in the Great Lakes basin were provided to the authors by the Canadian Climate Centre (Figure 2). Although the average temperature change is 4.4^{o}C, monthly temperatures are predicted to vary considerably – from 5°C in the winter months to 3°C during the summer in the Great Lakes basin. According to the model annual precipitation will increase by 6-17% in the northern basin but decrease by 2 – 4% in the south eastern basin. Table 1 presents the model output for temperature and precipitation for stations 1 and 10. Although acknowledging the

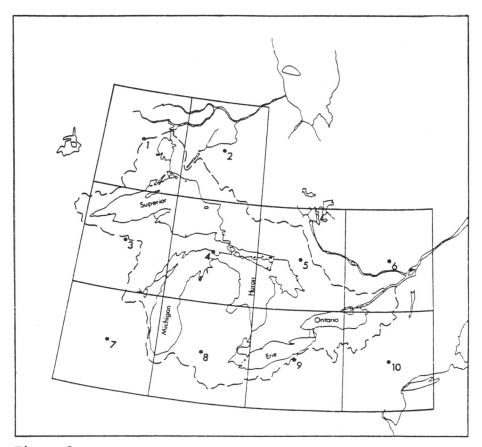

Figure 2

Table 1 Present and projected future mean monthly temperature and precipitation in the Great Lakes region (grid points 1 and 10)

	(1) Temperature °C			(1) Precipitation (cm)			(10) Temperature °C			(10) Precipitation (cm)		
	NORM	CH	2xCO2	NORM	EFFECT	2xCO2	NORM	CH	2xCO2	NORM	EFFECT	2xCO2
Jan	-20.2	5.6	-14.6	1.2	117.6	1.4	-4.5	5.6	1.1	2.4	104.2	2.5
Feb	-17.5	5.6	-11.9	1.1	116.7	1.2	-5.5	5.1	-0.4	2.7	104.2	2.8
Mar	-9.5	5.3	-4.2	1.2	109.7	1.3	-2.5	4.8	2.3	3.2	100.0	3.2
Apr	0.2	4.5	4.7	1.4	107.7	1.5	4.5	4.5	9.0	3.5	100.0	3.5
May	7.8	3.6	11.4	2.1	103.2	2.2	13.0	4.0	17.0	3.4	100.0	3.4
Jun	14.5	3.0	17.5	3.0	110.0	3.3	18.0	3.8	21.8	3.3	103.3	3.4
Jly	17.0	2.9	19.9	3.0	111.1	3.3	21.0	4.0	25.0	4.0	103.8	4.2
Aug	14.8	3.1	17.9	2.9	113.0	3.3	18.5	4.3	22.8	3.7	95.2	3.5
Sep	9.5	3.7	13.2	2.9	110.0	3.2	15.0	4.4	19.4	3.3	77.8	2.6
Oct	4.0	4.6	8.6	2.1	116.7	2.5	10.0	4.5	14.5	3.2	70.0	2.3
Nov	-6.0	5.2	-0.8	1.8	123.5	2.3	1.5	5.0	6.5	3.3	78.3	2.5
Dec	-15.5	5.5	-10.0	1.3	123.5	1.6	-4.0	5.6	1.6	2.4	88.0	2.1
MEAN		4.4			113.5			4.6			93.7	

NORM = 1951-80 normals

CH = magnitude of change between 1951-80 normals and 2xCO2 normal

2xCO2 = projected normals under 2xCO2 conditions

uncertainty of the numbers predicted, approximate analogies to present climates may be useful. For example, northwest of Lake Superior, where the mean annual temperature is now 0°C, the model predicts a temperature of 4°C, similar to the climate of Prince Edward Island. Similarly, in the warmest part of the basin, south of Lake Erie, which presently has a mean annual temperature of 10°C the model predicts a future temperature of approximately 14°C, similar to the present climate of Southern Kentucky.

The data suggest that north of Lake Superior, months with average temperatures below freezing will be reduced from five to three and areas to the south of Lake Erie that now have three months with below freezing average temperatures, will have no month with this condition.

Predicted future precipitation for the two stations in the Great Lakes basin (Table 1) shows the region north of Sault Ste Marie to have a precipitation increase of 17%, with the largest increases in fall and early winter. On the other hand, a decrease in annual precipitation of 6% is predicted for the area east of Lake Ontario.

Coupling the climate change model with a Great Lakes hydrologic model

Hydrologists who have studied the Great Lakes system for many years have stated that if the measured data on levels and flows are used, the contribution of each lake to the whole system, the net basin supply (NBS), can be determined as follows:

NBS = I - O $\pm$ ΔS

where I is the inflow from the upper lake, O the outflow to the downstream lake and ΔS the change in storage. Although past net basin supplies can be determined in this manner, projections of future NBS rely on quantifying the individual constituents of this parameter:

NBS = P - E + R

where P is precipitation on the lake surface, E is evaporation from the lake and R is the runoff from the surrounding watershed. The task of quantifying these constituents for such large bodies of water is not an easy one.

Much has been written on the problems of estimating the precipitation that falls on the lake surfaces (Bolsenga, 1977; Sanderson, 1966). The usual practice is to use some lake/land precipitation ratio to estimate over-lake precipitation. The present study made use of the following equation for estimating over-water precipitation (Southam and Dumont 1985)

$P_w - P_1$ x (P_w/P_1)

where P_w is over water precipitation, P_1 is land precipitation and P_w/P_1 is the ratio for the period 1951 - 1976. These ratios were found to be the following: Superior, 1.00; Michigan-Huron 0.99: Erie, 0.99; and Ontario, 0.926. These values were applied to recorded monthly overland precipitation values for 1951-1976 to develop a set of overlake data.

Estimates of present and future lake evaporation were taken from Cohen's work for the Canadian Climate Centre using the mass transfer technique (Cohen, 1986). Under $2xCO_2$ conditions, mean annual lake evaporation for 1951-1980 is projected to increase by approximately 35% on Lakes Superior, Michigan, Huron and Ontario and 25% on Lake Erie. Mean annual increases are as follows: Superior from 507 to 684 cm; Michigan from 576 to 784 cm; Huron from 681 to 905 cm; Erie from 784 to 981 cm; and Ontario from 559 to 738 cm. To estimate the runoff to each lake under current and predicted future climate, an adaptation of Witherspoon's (1970) and Morton's (1982) models was used.

The mean monthly net basin supply (NBS) was calculated for the years 1951 to 1976 using the changes in climate predicted by the GISS model for $2xCO_2$ conditions (Southam and Dumont 1985). These values were compared to recorded NBS values for 1951-76 and the differences calculated. These values were then applied to NBS values for the entire period of reliable data, 1900-1976.

The analysis resulted in the production of three scenarios of Great Lakes levels and flows (Table 2). The first, the basis of comparison (BOC), gives the levels and flows which would have occurred during 1900-1976 under current diversions, regulation practices (Plan 77 for Superior and Plan 1958-D for Ontario) and physical conditions of the lakes and connecting channels (eg. capacity of St. Clair and Detroit Rivers). The diversions include 142 m^3s^{-1} into Lake Superior via Long Lac and the Ogoki River, 91 m^3s^{-1}

Table 2 Great Lakes levels and flows: three scenarios

	BASIS OF COMPARISON Scenario 1		CLIMATIC CHANGE Scenario 2		CLIMATIC CHANGE + CONS. USE Scenario 3	
	LEVEL (m)	OUTFLOW (CMS)	LEVEL (m)	OUTFLOW (CMS)	LEVEL (m)	OUTFLOW (CMS)
LAKE						
Superior						
Mean	183.01	2204	182.80	2209	182.71	2198
Monthly Max	183.47	3483	183.46	3483	183.45	3455
Monthly Min	182.48	1557	182.27	1557	182.16	1557
Range	0.99	1926	1.19	1926	1.29	1898
Michigan-Huron						
Mean	176.26	5236	175.67	4599	175.43	4357
Monthly Max	177.41	6570	176.53	5720	176.28	5635
Monthly Min	175.40	3171	174.74	3143	174.46	2917
Range	1.74	3398	1.79	2577	1.82	2718
Erie						
Mean	173.97	5867	173.53	5086	173.29	4642
Monthly Max	174.83	7646	174.39	6739	174.14	6258
Monthly Min	173.16	4304	172.63	3540	172.36	3115
Range	1.67	3342	1.76	3199	1.78	3143
Ontario						
Mean	74.57	6856	-	-	-	-
Monthly Max	75.78	8778	-	-	-	-
Monthly Min	73.64	5324	-	-	-	-
Range	2.14	3454	-	-	-	-

out of Lake Michigan at Chicago and 198 m^3s^{-1} from Lake Erie to Lake Ontario through the Welland Canal. These data (scenario 1) thus reflect the influence of climate alone on levels and flows in the Great Lakes.

The average ranges in lake levels are seen to vary from 1 m for Lake Superior to 2.14 m for Ontario. The average outflows from each lake range from 2204 m^3s^{-1} for the St. Mary's River (Lake Superior) to 6856 m^3s^{-1} for the St. Lawrence River (Lake Onterio).

Using the same restrictions as for the BOC condition (i.e. the same diversions, regulation practices and physical conditions of the system), the data for Scenario 2 indicate that under $2xCO_2$ conditions, mean lake levels will drop slightly for Lakes Superior, Michigan, Huron and Erie. The range between maximum and minimum levels would increase to 1.2 m for Lake Superior and 1.76 m for Lake Erie. Mean flows would be decreased 12% in the St. Clair River (5236 to 4599 m^3s^{-1}) and 13% in the Niagara River. It was found that if the current regulation plan for Lake Ontario was strictly applied with the $2xCO_2$ climate scenario, levels would drop dramatically during certain periods (e.g. the 1930's and 1960's). Realistically, of course, such changes in level would not be allowed to occur and the regulation practices would be altered to allow for such periods of low water supply. For this reason, the output for Lake Ontario was not utilized in this study.

Scenario 3 (Table 2) includes the impact of future increased consumptive use in addition to climatic change. The International Joint Commission most likely projections of consumptive use in 2035 for each of the Great Lakes were applied throughout the 77-year period (IGLSB, 1981). Although the consumptive use projections have recently been revised downward (IJC, 1985), these data were not available prior to completion of this study. However, the projected climatic change will result in increases in consumptive use (Bruce, 1984; Cohen 1985a,1985b, 1986), so our Scenario 3 may be a reasonable forecast. In this scenario, a small change is predicted in the level and outflow of Lake Superior, the mean level of Michigan-Huron is seen to decrease by 80 cm and Erie by 70 cm from BOC conditions. The flow in the Detroit River is seen to decrease by 20% and in the Niagara River by 6%. Again, Plan 77 was strictly applied to all cases.

While Table 2 presented average conditions for the three scenarios, Figures 3 and 4 show the annual levels that would have occurred in the Great Lakes during the 77 year period under the three scenarios. To put the changes in levels and flows into perspective, Table 3 shows the changes in the predicted frequency of occurrence of low levels and flows for Lakes Superior, Michigan-Huron and Erie. Because

Table 3   Frequency of 1963-65 mean levels and outflows

	MEAN LEVEL (m)	% OF YEARS WITH LEVEL AT OR BELOW 1963-65 MEAN	MEAN OUTFLOW (CMS)	% OF YEARS WITH OUTFLOW AT OR BELOW 1963-65 MEAN
LAKE SCENARIO				
SUPERIOR				
1 BOC	183.01	10	2204	52
2 CLIM. CH.	182.80	61	2209	55
3 CLIM. CH. + CU	182.71	79	2198	55
MICHIGAN-HURON				
1 BOC	176.26	8	5236	3
2 CLIM. CH.	175.67	57	4599	43
3 CLIM. CH. + CU	175.43	77	4357	65
ERIE				
1 BOC	173.97	5	5867	5
2 CLIM. CH.	173.53	38	5086	38
3 CLIM. CH. + CU	173.29	77	4642	71

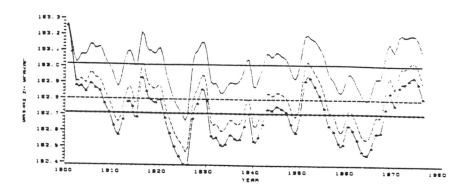

YEARLY MEAN LEVELS - LAKE SUPERIOR

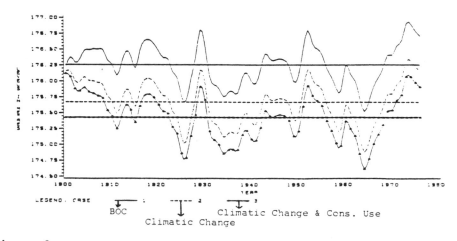

LAKE HURON—MICHIGAN

Figure 3

of the problems with the regulation plan, Lake Ontario data were not included. For example, for Lake Erie during the BOC period 1900–1976, levels as low as those of 1963–65 occurred 5% of the time. Under the predicted 2 x CO_2 climate they would occur 38% of the time, and with the addition of consumptive use forecasts, the model suggests that such low levels will occur 77% of the time.

Further modeling of Great Lakes water levels

The approach described above represents a first attempt to model future Great Lakes Water levels under 2 x CO_2 climatic conditions. Other factors such as the interbasin transfer of water and increased consumptive use may also affect lake levels in the future. A more comprehensive model which can be used to evaluate the combined

YEARLY MEAN LEVELS - LAKE ERIE

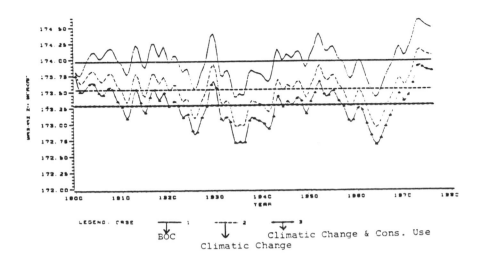

Figure 4

effects of all natural and man-made factors on lake levels is now being developed at the Great Lakes Institute. The model consists of three submodels as shown in Figure 5. The model is being applied using current conditions to produce a set of BOC lake levels. The factors will then be modified based on a range of possible future conditions to produce new scenarios of future lake levels. An example of the output from the Basin Runoff model for Lake Superior (Figure 6) shows that under the GISS predicted climate change, basin runoff will not only be decreased from BOC conditions, but maximum runoff will occur a month earlier. Further outputs of the model, using

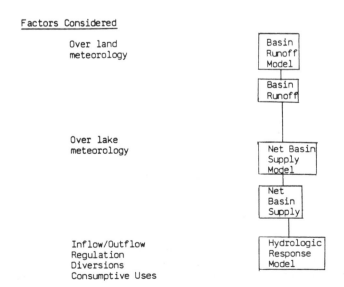

Figure 5

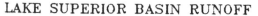

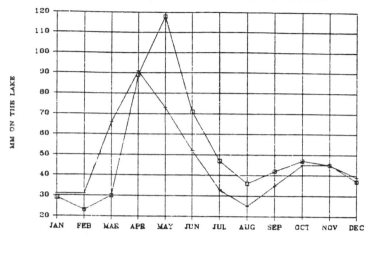

Figure 6

various scenarios of future basin conditions, will be presented at the conference in Vancouver.

ACKNOWLEDGEMENTS   The research reported here was supported by the Atmospheric Environment Services, Environment Canada and the Donner Canadian Foundation.

References

Bolsenga, S. (1977)   Lake-land precipitation relationships using Northern Lake Michigan data. J. Appl. Meteor. 16(11), 1158-1164.
Bruce, J.P. (1984)   The Climate Connection. Proceedings, Ontario Water Resources Conference: Futures in Water, June 12-14, 1984. Toronto, Ontario. Ontario Ministry of Natural   Resources.
Budyko, M.I. (1977) Climatic Changes. American Geophysical Union.
Cohen, S.J. (1985a)   Effects of climatic variations on water withdrawals in Metropolitan Toronto. Canadian Geographer. 29, 113-122.
Cohen, S.J. (1985b)   Projected Increased in Municipal Water Use in the Great Lakes due to CO_2-Induced Climatic Change. Unpublished paper, Canadian Climate Cebtre, Atmospheric Environment Service, Environment Canada.
Cohen, S.J. (1986)   Impacts of CO_2-induced climatic change on water resources in the Great Lakes Basin. Climatic Change, 8, 135-153.
Hansen, J. et al. (1983) Climatic effects of atmospheric CO_2 Science, 223, 873-875.
International Joint Commission   (1985)   Great Lakes Diversions and

Consumptive Uses. A Report to the Governments of Canada and the United States under the 1977 Reference.

International Lake Erie Regulation Study Board. (1981). Lake Erie Water Level Study. International Join Commission.

Morton, F.I. (1982) Integrated basin response – a problem of synthesis a problem of analysis? _Proc._ _Can._ _Hydrol._ _Symp._ _82._ Fredericton, N.B. June 21–22.

Sanderson, M.E. (1966) The 1958–63 water balance of the Lake Erie Basin. _Proc._ _Ninth_ _Conf._ _on_ _Great_ _Lakes_ _Research._ Great Lakes Research Division, University of Michigan, 274–282.

Southam, C. and S. Dumont (1985) Impacts of Climate change on Great Lakes Levels and Outflows. Canada Centre for Inland Waters, Environment Canada, Internal Report.

Witherspoon, D.F. (1970) A hydrological model of the Lake Ontario Local drainage basin. Tech. Bull. No. 31. Inland Waters Branch. Environment Canada.

World Meteorological Organization (1986) Report of the International Conference on the Assessment of the Role of Carbon Dioxide and of Other Greenhouse Gases in _Climate_ _Variations_ _and_ _Associated_ _Impacts._ WMO Publication No. 661.

The Influence of Climate Change and Climatic Variability on the Hydrologic Regime and Water Resources (Proceedings of the Vancouver Symposium, August 1987). IAHS Publ. no. 168, 1987.

Sensitivity of water resources in the Great Lakes region to changes in temperature, precipitation, humidity, and wind speed

Stewart J.Cohen
Canadian Climate Centre
Atmospheric Environment Service
Downsview, Ont.
M3H 5T4

ABSTRACT   Scenarios of global climatic warming ("Greenhouse Effect"), based on atmospheric models, historical analogues, and hypothetical cases, were used to determine the sensitivity of Great Lakes water resources to changes in several climatic elements. Results tend to confirm earlier studies that climatic warming would lead to decreases in net basin supply and soil moisture. However, when original temperature and precipitation scenarios were modified by changes in humidity and wind speed, results changed significantly, indicating the importance of lake evaporation to net basin supply in this region. More research is needed in deriving regional wind and humidity scenarios within the context of global climate warming.

Sensibilité des ressources en eau de la région des Grands Lacs aux changement de temperature, de precipitation, d'humidité, et de vitesse du vent

RESUME Les scénarios des changements climatiques mondiaux (l'effet de serre),dérivés des modèles atmosphèriques, d'analogies historiques, et de cas hypothètiques, ont été utilisés pour déterminer la sensibilité des ressources en eau des Grands Lacs aux changements de quelques elements climatiques. Les résultats confirment les études précédentes à savoir que le réchauffement climatique résulterait des réductions de la quecentilé d'eau du bassin et de l'humidité de la terre. Mais quand les températures et les précipitation originales sont modifiées par les changements d'humidité et de vitesse du vent, les résultats ont changé de manière significative et ont indiqué l'importance de l'évaporation des lacs par rapport à la quantite d'eau du bassin dans cette région. De plus amples recherches sont necéssaires pour décrire des scenarios régionaux d'humidité et de vitesse du vent dans le context d'un réchauffement climatique mondial.

Introduction

The Great Lakes basin in North America contains the world's largest
system of fresh water lakes. This is a region with a broad mixture of
urban and rural land uses, and a drainage area of approximately
774,000 sq.km, of which 246,000 sq.km (32%) is lake surface.

Climatic fluctuations have had a significant impact on mean annual
Net Basin Supply (NBS) and soil moisture levels in this region.
Annual NBS, which is equivalent to mean annual discharge at the mouth
of the basin, has varied widely in recent years, from 5530 m^3s^{-1} in
1965 to 8730 m^3s^{-1} in 1973. The latter value may be exceeded in 1986,
a year in which record high lake levels have caused flooding, erosion,
and damage to shoreline properties. However, previous research has
shown that two scenarios of projected climatic warming, caused by
increased concentrations of atmospheric carbon dioxide and other
trace gases ("Greenhouse Effect"), would reduce annual NBS to 1965
levels by the mid-21st century (Cohen, 1986). If this were to occur,
significant losses would be felt by several water-related sectors in
the regional economy, including hydro-electric power production and
commercial shipping (Allsopp and Cohen, 1986).

Regarding soil moisture, Cohen (1986) noted that soil moisture
deficit (SMD) would increase by more than 100% in the two scenarios
mentioned above. This would mean reduced soil moisture levels,
particularly in summer when the SMD is generated. Schlesinger and
Mitchell (1985) reviewed soil moisture estimates of several studies
which projected widespread decreases in the mid-latitudes. Similar
results were also obtained by Manabe and Wetherald (1986). Such
decreases would have significant implications for agriculture and
water consumption.

The purpose of this discussion is to determine the sensitivity of
annual NBS and soil moisture levels in the Great Lakes basin to
changes in certain climate parameters, over a wide range of climatic
warming scenarios. It is assumed that

$$NBS = P(lake) - E_O + R \tag{1}$$

where P(lake) is overlake precipitation (mm), E_O is open water
evaporation (mm), and R is runoff from the land surface (mm). In this
study, all elements are converted to units of discharge (m^3s^{-1}).
Existing diversions and groundwater are assumed to be of negligible
importance at this scale. Consumptive use is significant and will
have a major impact on future NBS (Cohen 1986), but is not considered
here.

We are particularly interested in temperature (T), P, atmospheric
vapor pressure (VP), and wind speed (u). The decreases in annual NBS
and soil moisture, described above, were calculated despite projected
increases in annual P. Since there is uncertainty regarding the
magnitude of the Greenhouse Effect warming, and considerable
uncertainty about changes in other climatic elements, a sensitivity
analysis appears to be a very useful exercise to undertake. For
example, how much additional P is needed to overcome a warming of
2°C? Would changes in humidity and u make a reduction in NBS more or
less likely in a warmer climate?

In the next section, the various climatic warming scenarios are described. This is followed by a discussion of methods, and the resulting estimates of annual NBS and SMD.

Scenarios of climatic change

Scenarios derived from atmospheric models

Fourteen scenarios were derived from changes in T and P projected by two global-scale general circulation models: the Goddard Institute for Space Studies (GISS) model and Geophysical Fluid Dynamics Laboratory (GFDL) model. A detailed review of these and other atmospheric models can be found in Schlesinger and Mitchell (1985).
  There are a number of uncertainties in the modelling and parameterisation of terms, and neither model, as of this time, perfectly simulates the present climate, nor generates detailed regional estimates of climatic changes. As a result, it was necessary to adapt the original models' outputs for use in this regional study. This involved adding model projected anomalies at the models' grid points to normals estimated by the Canadian Climate Centre at those same points. These normals were determined from isoline maps based on actual station data (Cohen, 1986).
  In the Great Lakes region, both models project warmer T for all months and higher annual P (Table 1). Certain months, particularly late summer and autumn, show reductions in P in the central region. A projection of monthly u, available from the GFDL output, indicates decreases of 15-20% during the autumn season, increases of 6-15% during winter, and a mixture of increases and decreases during spring and summer.

Table 1 Changes in T, P, and u projected for each month by GISS and GFDL in the Great Lakes Basin (1984 models)

	GISS		GFDL		
	T(°C)	P(%)	T(°C)	P(%)	u(%)
J	5.6-5.9	4 to 18	1.1-5.0	-15 to 10	7
F	5.1-5.8	4 to 17	3.6-6.0	13 to 67	15
M	4.8-5.4	0 to 17	2.4-6.4	-19 to 7	8
A	4.1-4.6	0 to 15	0.9-3.6	-6 to 34	6
M	3.2-4.0	0 to 12	1.8-4.0	-5 to 13	-7
J	2.8-3.8	3 to 12	1.7-4.7	-25 to 11	-11
J	2.9-4.0	3 to 16	2.4-3.3	-27 to 5	4
A	3.1-4.3	-5 to 27	2.5-5.3	-32 to 6	40
S	3.7-5.0	-22 to 42	2.0-3.3	-16 to 29	-20
O	4.4-5.3	-30 to 39	1.7-4.4	-6 to 53	-15
N	5.0-5.5	-22 to 25	4.6-6.7	-4 to 28	-16
D	5.4-5.9	-12 to 24	1.9-3.6	-20 to 30	-19

  For each of the two atmospheric model scenarios, as well as the other scenarios described in the following sections, seven scenarios developed, using different combinations of changes in the u and VP (Table 2). This was done so as to provide a more complete picture of the possible changes in annual NBS. These two elements are important in this exercise because of the large size of the lakes' open water surface. Earlier work has shown that E_O is of similar magnitude to R and P(lake) in this basin (Cohen, 1986).

Table 2 Scenarios of changes in wind and vapour pressure

u/VP	Comment
N/N	base case
-20%/N	u changed all months
GFDL/N	see Table 1 for u
N/+10%	VP changed all months
N/-10%	VP changed all months
GFDL/+10%	as above
GFDL/-10%	as above

Present normals of u and VP were derived from data obtained from the Canadian Climate Archive and processed by the Marine Statistics data system (MAST) of the Canadian Climate Centre.

These were hypothetically adjusted for each month in the various scenarios, as indicated in the table. The exceptions were the three scenarios that utilized GFDL winds.

Historical analogues

Eight historical analogues were used to generate a total of 56 scenarios (seven for each analogue, as in the previous section). The eight analogue cases (Brown and Walsh, 1986) represent either blocks of years, or several individual years, in which the T difference between the warmest and coldest years were assumed to represent scenarios of future warming(Table 3). It is important to note that in certain cases,T decreases are indicated in individual seasons (Table 4). All eight analogues show increased annual T, but these are not as high as the warming projected by GISS or GFDL.

Table 3 Data used to derive historical analogues

HIST-	YEARS	COMMENTS
A	20-year blocks	hemispheric, 1901-1980 data
B	10-year blocks	Arctic, post-1925 data
C	5 individual years	mid-latitude, 1901-1980 data
D	7-year blocks	mid-latitude, post-1946 data
E1-E4	5-year blocks	mid-latitude, 1901-1980 data

Source: Brown and Walsh (1986).

Having defined which years would be used, seasonal P anomalies were determined. Table 4 shows that many of these anomalies are negative. Thus, unlike the general circulation model scenarios, most of the historical analogues show decreased annual P. When comparing the

Table 4 Seasonal changes in T and P for historical analogues

HIST-	DJF	MAM	JJA	SON
		T(^{o}C)		
A	0.9-1.2	0.0-0.3	0.6-1.0	0.2-0.5
B	0.1-1.2	0.4-0.5	0.8-1.4	0.1-0.3
C	3.0-3.4	1.9-2.5	1.3-1.6	1.0-1.4
D	1.0-1.7	-0.2 to 0.3	0.5-0.6	0.1-0.3
E1	0.6-2.3	0.9-1.1	0.1-0.5	1.2-1.5
E2	-0.8 to -0.4	0.0-0.4	0.2-0.6	0.3-0.4
E3	0.0-0.8	0.5-1.0	0.3-0.5	0.5-0.7
E4	0.3-1.2	0.5-0.8	0.0-0.4	1.0-1.3
		P(mm)		
A	-3 to 6	-4 to 7	-3 to 5	0-5
B	-16 to -5	0-10	-16 to -5	-8 to -2
C	-11 to 8	-2 to 5	-9 to 21	-6 to 5
D	-5 to 9	-7 to 11	-10 to -3	-26 to -12
E1	-4 to 1	-7 to 11	0-3	-7 to 2
E2	-12 to -4	-4 to 2	-23 to -10	-21 to -7
E3	-22 to -8	-5 to 20	-18 to -9	-17 to -2
E4	0-6	-11 to -2	-7 to 8	-12 to -2

analogues to GISS and GFDL the question is: will the resulting NBS estimates be lower for the slightly warmer and drier historical analogues, or the much warmer and wetter GISS and GFDL scenarios?

Hypothetical scenarios

Ten hypothetical warming cases were used to generate a total of 70 scenarios. The hypothetical cases are similar to those used by Gleick (1986) in his study of a river basin in the western United States. Every month of the year is assumed to experience the same warming, 2°C or 4°C and the same change in precipitation, -20%, -10%, 0%, +10%, or +20%.

As a tool for a sensitivity analysis, hypothetical scenarios represent a useful baseline reference when a number of atmospheric models and analogues are being compared. For example, it may be possible to say that in terms of impacts on annual NBS in the Great Lakes, the GISS scenario is equivalent to a hypothetical scenario of +2°C, -5% P, with no change in u or VP.

Methodology

Climatic water balance

Estimates of R and SMD for the Great Lakes basin as a whole were obtained by utilising a modified version of the monthly Thornthwaite climatic water balance model, available at the Canadian Climate Centre (Johnstone & Louie, 1984). The model estimates monthly R and

SMD as residuals after accounting for precipitation and evapotranspiration, the latter being dependent on T and daylength. Previous work has demonstrated its applicability to climatic impacts problems of this scale (Cohen, 1986; Mather and Feddema, 1986). Despite the empirical nature of the Thornthwaite approach, it has been found to provide reasonably reliable estimates of water balance components in most climates (Mather, 1978)

A detailed description of the analysis method is available in Cohen (1986). The assumption made in that particular study, including a soil moisture storage capacity of 100mm and a minimum monthly temperature for snowmelt of $0.0^{\circ}C$, are maintained herein. The model does not consider wind speed, or possible changes in plant transpiration rates that may result from higher CO_2 concentrations (Idso and Brazel, 1984).

Lake evaporation

A mass transfer approach was used to obtain estimates of E_0 for each of the five Great Lakes. In this model, the vertical movement of water vapour to or from the lake surface is dependent on u and the magnitude of the atmosphere-lake VP gradient. This approach has been used in earlier studies in the basin (Richards and Irbe, 1969; Quinn and den Hartog, 1981: Cohen, 1986).

Estimation of the vertical VP gradient required data on lake surface temperature (T(lake)) and the dew point (T(dew)), as well as T over the lake surface. In all scenarios, it was assumed that T(lake) and T(dew) would increase by the same amount as the projected increases in T, except during months when the normal T(lake) was near $0.0^{\circ}C$, in which case, the increase in T(lake) was assumed to be less than the change in T.

Overlake precipitation

There are no direct measurements of P(lake), although data are available for several islands in the lakes. A 1972 field study of Lake Ontario concluded that P(lake) was 8% less than over the surrounding land surface (Wilson and Pollock, 1981). Cohen (1986) had assumed that P(lake) for all the lakes was 8% less than for land areas. The same assumption is used here.

Results

Net basin supply

Comparison of scenarios

Results of calculations of annual NBS for all 140 scenarios are shown in Table 5. The first column, marked "N/N", represents the 20 basic scenarios without any changes in u and VP. The other six columns include the effects of these changes, to be discussed later.

Table 5 Projected changes in NBS (%). Normal=7477 m^3s^{-1}.
        Consumptive use not included

SCENARIOS	N/N	−20%/N	GFDL/N	N/+10%	N/−10%	GFDL/+10%	GFDL/−10%
GISS	−23.6	−7.0	−20.0	3.9	−51.1	6.7	−47.5
GFDL	−16.9	−2.7	−14.0	8.5	−42.3	11.3	−39.2
HISTA	6.2	19.2	8.3	24.7	−14.9	29.3	−12.6
HISTB	−16.1	−3.3	−14.1	5.2	−37.4	7.0	−35.2
HISTC	1.5	14.2	3.5	24.5	−21.4	23.7	−19.2
HISTD	−16.4	−3.8	−14.4	4.5	−37.3	6.4	−35.1
HISTE1	−8.9	4.8	−6.1	13.1	−30.3	15.3	−27.5
HISTE2	−26.5	−14.0	−24.4	−5.9	−47.1	−4.0	−44.8
HISTE3	−24.0	−11.2	−21.8	−2.9	−45.1	−0.9	−42.6
HISTE4	−8.8	4.2	−6.4	12.4	−30.0	14.5	−27.3
T2/−20P	−58.2	−44.5	−55.6	−34.9	−81.4	−32.6	−78.6
T2/−10P	−36.3	−22.6	−33.7	−13.0	−59.5	−10.7	−56.7
T2/N	−12.5	1.1	−9.9	10.7	−35.8	13.1	−32.9
T2/+10P	12.1	25.8	14.7	35.4	−11.1	37.7	−8.3
T2/+20P	37.7	51.3	40.2	60.9	14.4	63.3	17.2
T4/−20P	−70.9	−55.9	−68.0	−44.4	−97.5	−41.7	−94.3
T4/−10P	−50.4	−35.4	−47.5	−23.9	−77.0	−21.2	−73.8
T4/N	−28.4	−13.3	−25.5	−1.8	−54.9	0.8	−51.7
T4/+10P	−4.7	10.4	−1.8	21.8	−31.3	24.5	−28.1
T4/+20P	19.7	34.8	22.6	46.3	−6.8	48.9	−3.9

The "N/N" column reveals that the two atmospheric models project significant decrease in NBS, −23.6% and −16.9% respectively. To put this in perspective, NBS in 1965 represented a decrease of about −25% (including consumptive use).

These values are slightly different from those in Cohen (1986), which listed −20.8% and −18.4% respectively. A higher "normal" NBS is used (7477 m^3s^{-1} instead of 7412 m^3s^{-1}), reflecting the higher values recorded in 1983 and 1984, now included in the normal. Also, R has been recalculated for both scenarios using the same " normal" R. In Cohen (1986), different "normals" had been estimated for the two scenarios because their grid points were in different locations[*]. The new estimates use the percentage change in results from Table 1 of Cohen (1986) and the normal R listed in Table 4 of Cohen (1986).

Of the eight analogues, decreases are calculated for six of them. HISTE2 and HISTE3 register the largest decreases, −26.5% and −24% respectively, because of relatively low R and P(lake).Their estimates of E_0 are actually somewhat lower than in the other analogues, as well as the atmospheric models. Recall that HISTE2 includes a decrease in T during winter (Table 4). Thus, we have two cases where significant reductions in annual NBS have been caused primarily by large decreases in P. Two other analogues , HISTB and HISTD, also showed large decreases in R and P(lake), so losses here have occurred because of relatively high T, which increases E . In terms of how the NBS loss has been generated, these two analogues come closest to
* see Figure 1 Cohen(1986)

simulating the losses projected by GISS and GFDL.

HISTA and HISTC are the only analogues that show increased NBS. In both cases, higher P is the main cause. However, the increase in HISTC is not as large because it experiences a higher increase in T. This results in lower R, which has increased the evapotranspiration loss. It is interesting to note that HISTC is actually the warmest of the eight analogues, but it also includes the highest P. Also, during autumn, the peak season for E_O, the increase in T is less than 1.5^OC, compared to an increase greater than 3.0^OC for GISS and GFDL.

The 10 hypothetical cases show, not surprisingly, that higher T and lower P result in greater losses in NBS. More importantly, we can see that for NBS to remain unchanged, P would have to increase by about 5% and 12% to overcome 2^OC and 4^OC warming scenarios, respectively. These are equivalent to annual P increases of about 40mm and 100m respectively.

Using the hypothetical cases for reference, the NBS impacts projected by GISS are most similar to a 2^OC warming with -5% change in P, or a 4^OC warming with P+2%. For GFDL, the P changes would be -2% and +5% respectively. The analogue cases have wide ranging results, from HISTE2 with the greatest loss to HISTA with a small gain. Compared with the 2^OC and 4^OC hypothetical warming scenarios, HISTE2 would be similar to P changes of -5% and +1% respectively, while for HISTA, it would be +7% and +13% respectively.

Effects of wind and humidity changes

The remaining six columns in Table 5 show the results from the various scenarios modified by changes in u and VP. The column "-20%/N" represents a 20% decrease in mean monthly u. This reduces E_O, thereby leading to smaller losses, or in some cases, larger increases in NBS. Results for the GFDL wind scenario ("GFDL/N") are similar in direction, though the magnitude of the change is only about 2-3% different from "N/N". If VP is increased by 10%, E_O is reduced significantly, so that both atmospheric models and six of the eight analogues would show increases in NBS. On the other hand, a 10% decrease in VP would generate NBS losses in every scenario except T+2^OC, +20%P.

The final two columns combine the GFDL wind scenario with changes in VP. Overall, the 10% change in VP has a much greater influence on results, so these columns bear a strong resemblance to the "N/+10%" and "N/-10%" columns.

Prevailing theory suggests that the global temperature increase will be accompanied by increased atmospheric VP and relative humidity because of higher global scale evaporation (Schlesinger and Mitchell, 1985). Table 5 indicates that a global increase of VP superimposed over the Great Lakes would reduce the loss in NBS because of reduced E from the lakes themselves. Many scenarios, including GISS and GFDL, could project NBS gains under such conditions, depending on the magnitude of the increase in VP and relative humidity.

The combined effects of higher VP and higher T on R could not be determined from the Thornthwaite model. If we assume for the moment that R does not change significantly from the "N/N" scenario calculations, it would appear that NBS losses would not be as great as indicated in the "N/N" column. Will changes in NBS be similar to the

"N/+10%" or "GFDL/+10%" projections? The 10% change may be too large. There may be a feedback effect of this higher humidity on local cloud cover, lake effect precipitation, and lake surface temperatures, which might also influence the NBS estimates. A great deal more work needs to be done in determining the impacts of CO_2-induced climatic warming on NBS at this scale.

Soil moisture deficit

Another aspect of the climatic water balance that may experience significant impacts is the annual soil moisture deficit. In the Great Lakes basin, evapotranspiration generally exceeds P from May to September, and as soil moisture is withdrawn by the vegetation, a small deficit occurs. In some years, there have been dry spells up to six weeks duration, which have led to localized water supply problems and damaged crops(Brown and Wyllie, 1984).

On the global scale, both atmospheric models have projected significant decreases in summer soil moisture in North America. However, GFDL projects a continent-wide decrease, while GISS shows increased soil moisture in certain areas, including the northwest portion of the Great Lakes basin (Schlesinger and Mitchell, 1985).

The Thornthwaite climatic water balance approach was used to calculate summer soil moisture deficit (SMD) for all the "N/N" cases. Results are shown in Table 6. For the results reported herein, all scenarios are based on the same "normal" computed using the GISS grid points.

Table 6 Projected changes in SMD (%). Normal = 477 m^3s^{-1}

SCENARIOS	%change	SCENARIOS	%change
GISS	116.6	T2/-20P	508.5
GFDL	166.4	T2/-10P	364.5
HISTA	13.7	T2/N	252.9
HISTB	111.0	T2/+10P	163.3
HISTC	31.8	T2/+20P	88.3
HISTD	70.2	T4/-20P	777.0
HISTE1	-6.9	T4/-10P	655.3
HISTE2	109.2	T4/N	470.9
HISTE3	102.3	T4/+10P	353.0
HISTE4	-27.1	T4/+20P	258.1

Results show that 18 of 20 cases project increases in summer soil moisture deficit,implying decreased soil moisture, in this case during May to September. Only HISTE1 and HISTE4 show reduced deficits, due to small increases in summer warming of only 0.1°C in most areas in the basin. The GFDL deficit increase is greater than for GISS. HISTB, HISTE2, and HISTE3 also show similar deficits, equivalent to about 1000 m^3s^{-1}.

Nine of the ten hypothetical cases show larger deficits than any of the other scenarios. Only the T+2°C/+20% P case projects a deficit below 1000 m^3s^{-1}. This appears to indicate primarily the influences of greater increases in summer T. Although GISS an GFDL show greater warming in some months, the effect is not as great as a consistent warming of $2^{\circ}C$ every month, let alone $4^{\circ}C$. Note also that this occurs despite the increased P in some of the hypothetical cases.

Conclusion

This study has presented estimates of Great Lakes net basin supply for 140 climate warming scenarios, derived from 20 cases composed of two atmospheric models, eight historical analogues, and ten hypothetical cases. The majority of these project decreases in net basin supply. Increases in summer soil moisture deficit are projected for 18 of the 20 original cases. The above results often occur despite increases in annual precipitation.

When the scenarios include hypothetical or model-based changes in wind speed and humidity, net basin supply estimates can change significantly. This reflects the importance of lake evaporation to net basin supplies in this region. More research is needed in deriving regional wind and humidity scenarios within the context of the Greenhouse Effect. For the Great Lakes, this necessitates the derivation of scenarios of lake surface temperatures as well. Air temperature and precipitation data alone are not enough to completely answer all the questions about the impacts of future climate warming. Although the general indications of decreased water supplies and summer soil moisture levels appear to be confirmed by this study, more precise answers will require additional effort from researchers interested in the climate/water resources interface.

ACKNOWLEDGEMENTS My thanks to R. Brown and D. Etkin for their assistance in obtaining data for the historical analogue scenarios.

References

Allsopp, T.R. & Cohen, S.J. (1986) CO_2- induced climate change and its potential impact on the province of Ontario. In: Preprints of the Conf. on Climate and Water Management- A Critical Era and Conf. on the Human Consequences of 1985's Climate(Asheville, August 1986), 285-290. American Meteorological Society.

Brown, D.M. & Wyllie, W.D. (1984) Growing season dry spells in southern Ontario. Climatol. Bull. 18(1), 15-30.

Brown, R. & Walsh, K. (1986) Canadian seasonal climate scenarios for a CO_2-induced global warming. Canadian Climate Centre Report No. 86-14, Atmospheric Environment Service, Downsview, Ontario, Canada.

Cohen, S.J. (1986) Impacts of CO_2-induced climatic change on water resources in the Great Lakes Basin. Climatic Change, 8, 135-153.

Gleick, P.H. (1986) Regional water resources and global climatic change. International Conf. on Health and Environmental Effects of Ozone Modification and Climatic Change, (Washington, D.C., June 1986), U.N. Environment Programme and U.S. Environmental Protection Agency.

Idso, S.B. & Brazel, A.J. (1984) Rising atmospheric carbon dioxide concentrations may increase streamflow. Nature, 312, 51–53.

Johnstone, K.J. & Louie, P.Y.T. (1984) An operational water budget for climate monitoring. Canadian Climate Centre Report No. 84-3, Atmospheric Environment Service, Downsview, Ontario Canada.

Manabe, S. & Wetherald, R.T. (1986) Reduction in summer soil wetness by an increase in atmospheric carbon dioxide. Science, 232, 626–628.

Mather, J.R. (1978) The Climatic Water Balance in Environmental Analysis. Lexington Books, Lexington, MA, USA.

Mather, J.R. & Feddama, J. (1986) Hydrologic consequences of increases in trace gases and CO_2 in the atmosphere. International Conf. on Health and Environmental Effects of Ozone Modification and Climatic Change, (Washington, D.C., June 1986), U.N. Environment Programme and U.S. Environmental Protection Agency.

Quinn, F.H. and den Hartog, G. (1981) Evaporation synthesis. In: IFYGL- The International Field Year for the Great Lakes (ed. by E.J. Aubert & T.L. Richards), 221–245. U.S. Dept. of Commerce, Ann Arbor, Michigan, USA.

Richards, T.L. & Irbe, J.G. (1969) Estimates of monthly evaporation losses from Great Lakes 1950 to 1968 based on the mass transfer technique. Proc. Twelfth Conf. On Great Lakes Res., 469–487. Inter. Assoc. of Great Lakes Res.

Schlesinger, M.E. & Mitchell, J.F.B. (1985) Model projections of the equilibrium climatic response to increased carbon dioxide. In: Projecting the Climatic Effects of Increasing Carbon Dioxide (ed. by M.C. MacCracken & F.M. Luther), 81–148. U.S. Dept. of Energy, Washington, D.C., USA.

Wilson, J.W. & Pollock, D.M. (1981) Precipitation. In: IFYGL-- The International Field Year for the Great Lakes (ed. by E.J. Aubert & T.L. Richards), 51–78. U.S. Dept. of Commerce, Ann Arbor, Michigan, USA.

The Influence of Climate Change and Climatic Variability on the Hydrologic Regime and Water Resources (Proceedings of the Vancouver Symposium, August 1987). IAHS Publ. no. 168, 1987.

Les conséquences durables de la sécheresse actuelle sur l'écoulement du fleuve Sénégal et l'hypersalinisation de la Basse-Casamance

J.C. Olivry
Orstom 213 rue La Fayette 75010 Paris, France

RESUME   Dans les régions soudano-sahéliennes, l'actuelle sécheresse se distingue des événements déficitaires antérieurs par ses conséquences durables sur les régimes hydrologiquws de basses-eaux liées à l'épuisement des aquifères. Ainsi, les hydrogrammes du fleuve Sénégal montrent après la crue annuelle une période de tarissement exponentiel dont le coefficient très stable pendant 70 ans ($a = 0.0186$ j^{-1}), est devenu de plus en plus fort au cours des dernières années ($a = 0.038$ j^{-1} en 1984).   De même, l'intrusion marine en Basse-Casamance a considérablement augmenté en saison sèche depuis 1973. Les marigots ont un gradient de salinité positif vers l'amont leur évaporation n'étant plus compensée par l'alimentation des nappes souterraines aujourd'hui taries (sursalure jusqu'à 150‰ contre 35‰ auparavant). Les précipitations plus abondantes 1984-85 n'ont pas renversé la tendance. L'éventuel retour de conditons climatiques favorables n'entrainera pas le rétablissement immediat des régimes antérieurs. Une réversibilité totale suppose la reconstitution des aquifères qui exigera une longue période humide non interrompue.

Mots-clefs: Sénégal, sécheresse, réserves souterraines, tarissement, intrusion saline.

Persistent consequences of the present drought
on the flows of the Senegal River and the
hypersalinisation of the Lower Casamance

ABSTRACT In Sahelian and Sudanese regions, the persistent consequences of the present drought differentiate it from other dry spells of the 20th century. Two investigations carried out in Senegal showed the part of a major depletion of groundwater storage. After the annual flood, the hydrographs of the Senegal River showed discharges falling off exponentially according to a depletion coefficient without large change during seventy years (mean value: a = 0.0186 day^{-1}).   This trend has been altered over the present drought showing an increasingly rapid decline (a = 0.038 day^{-1} in 1984). Marine intrusion and tidal influences characterize the Lower Casamance. In dry season, until 1973, salinity decreased going upstream (maximum < 35‰). Actually, the salinisation gradient has become positive going upstream (maximum > 150‰) because evaporation is not compensated by groundwater reserves which are almost fully depleted. The more abundant rainfalls in 1984-85 did not reverse this trend. These examples demonstrate that even if climatologic conditions improve, the return of former hydrologic regimes will not be immediate. It requires the reconstitution of groundwater reserves (several wet years without further dry spells).

Keywords: Senegal, drought, groundwater reserves, depletion, salt intrustion.

Introduction

Le caractère exceptionnel du phénomène actuel de la sécheresse dans les régions soudano-sahéliennes n'est plus à démontrer. L'étude des déficits pluviométriques de la zone intertropicale de l'Afrique a montré depuis une quinzaine d'années un phenomene global de péjoration climatique. Ces déficits se sont largement répercutés et généralement amplifiés, dans l'écoulement des bassins fluviaux. Par l'intégration spatiale du régime des précipitations qu'elle suppose sur l'ensemble d'un bassin versant, la variabilité de l'écoulement annuel constitue un paramètre de choix dans l'étude des fluctuations climatiques. Dans ce cadre de recherches, le Sénégal offre à Bakel la plus longue série chronogogique de modules en Afrique de l'Ouest. Diverses études ont montré que l'événement actuel se différenciait nettement des autres épisodes déficitaires du XXème siècle par son intensité et sa durée (Sircoulon, 1985 - Olivry, 1983). Malgré des fluctuations de large amplitude qui ont pu faire penser à des variations pseudocycliques, on a mis en évidence sur le fleuve Sénégal une tendance générale à la baisse depuis le milieu du XIXème siècle (Olivry et al. 1986).

Sans préjuger d'une poursuite de c ette tendance ou au contraire d'un retour à une période humide, il apparaît nettement que la sécheresse actuelle aura des conséquences durables sur certains paramètres hydrologiques. Après la relative amélioration des conditions pluviométriques observée en 1985 et 1986, un "effet mémoire de la sécheresse" est nettement apparu dans l'écoulement des grands bassins fluviaux. Mais il est encore trop tôt pour y voir le poids de modifications pérennes du milieu physique (états de surface et végétation) sur l'ensemble des composantes des régimes hydrologiques. Au niveau des paramètres principaux de l'écoulement (modules et maximum annuels), on assiste à une péjoration des valeurs médianes sans grande incidence au niveau des lois de distribution et s'apparentant davantage à une dérive de l'échantillon d'observation qu'à un bouleversement fondamental du régime hydrologique (Olivry, 1986a).

Un retour à des conditions très humides entraînerait probablement un retour à des conditions d'hydraulicité comparables à celles déjà observées dans le passé. Par contre, au niveau des basses-eaux et des phases de tarissement, "l'effet mémoire" de l'actuelle période déficitaire est particulièrement déterminant: l'éventuel retour à des conditions climatologiques favorables n'entraînera pas le retablissement immédiat du régime hydrique antérieur. L'appauvrissement considérable des aquifères a entraîné une modification du régime des basses-eaux qui se traduit sur le fleuve Sénégal dans l'écoulement qui suit l'hydrogramme de crue et en Basse-Casamance par l'hypersalinisation des marigots en saison sèche (cf. carte de situation figure 1).

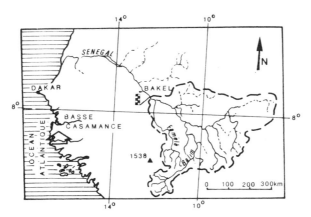

Figure 1 Carte de situation.

Le tarissement du fleuve Sénégal

Les débits du fleuve Sénégal sont observés depuis 1903 à la station de Bakel (A: 218 000 km^2). Diverses corrélations entre les données historiques et les observations hydrologiques ont permis d'évaluer le module interannuel de la seconde moitié du XIXème siècle à environ 900 m^3 s^{-1} (op. cit.). Pour les 84 années d'observations, ce module interannuel n'est plus que de 702 m^{3}s^{-1} ; avec l'actuelle sécheresse le module moyen des quinze dernières années tombe à 390 m^{3}s^{-1} ; il ne dépasse pas 284 m^{3}s^{-1} pour les 5 dernières années, la maximum minimorum ayant été atteint en 1984 (205 m^{3}s^{-1}). En 1985 et 1986, les précipitations reçues par le bassin se sont rapprochées de la normale mais avec une répartition telle que l'écoulement est resté très déficitaire.

Ce sont d'ailleurs ces effets de la répartition des pluies au cours de la saison humide qui expliquent que l'on a rarement sur un grand bassin fluvial de très bonnes corrélations hydropluviométriques. La chronique des modules du Sénégal à Bakel est presentée dans la figure 2 (cf. aussi tableau 1).

Dans l'étude du fleuve Sénégal, C. Rochette (1974) a montré qu'après une rapide phase de décrue en septembre-octobre on abordait une période de tarissement où la décroissance des débits suit les lois exponentielles classiques ($Q_i = Q_0 e^{-at}$). Cette première phase de tarissement couvre généralement les mois d'octobre à janvier pour une gamme de débits de 500 à 50 m^{3}s^{-1} et le coefficient de tarissement moyen sur 60 ans est de 0.018 j^{-1}. C. Rochette décrit une seconde phase de tarissement beaucoup plus rapide en dessous de 50 m^3 s^{-1}. Appelée phase d'épuisement, cette période ne concernerait plus que

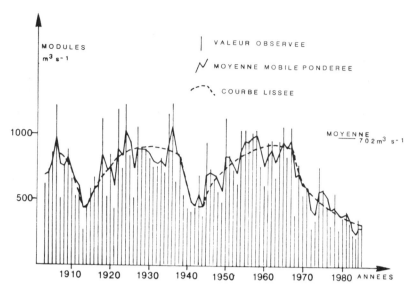

Figure 2   Modules du Sénégal à Bakel.

les réserves de la vallée alluviale coupées des ressources des aquifères amont. Le coefficient de tarissement est de 0.045 j^{-1}. C'est cette phase qui a conduit ces dernières années à une interruption de l'écoulement du fleuve de quelques jours à quelques semaines.

La première phase de tarissement, par les volumes qu'elle implique et sa répresentativité de l'ensemble des aquifères du bassin constitue une caractéristique importante du régime hydrologique du fleuve Sénégal. L'étude des coefficients de tarissement (tableau 1) montre jusqu'aux années 1972-73 une rémarquablé regularite des valeurs. Sur 70 ans, la moyenne est de 0.0186 j^{-1} et l'écart type de 0.0014 j^{-1} (Cv = 0.075). Les minimums (0.0.16 j^{-1}) correspondent aux périodes humides (années 1925-28, 1935-37, 1955-60 et 1964-68) avec une richesse et une extension maximale des aquifères. Les maximums (0.022 j^{-1}) correspondent aux periodes sèches (1913, début des années 70) mais aussi à des années abondantes comme 1922-23 ou 1950-53; les déficits des années 40 ne sont pas marqués au niveau du tarissement. Sur cette période, les variations contiennent donc une part aléatoire qui ressort probablement de l'hétérogénéité spatiale du comportement des aquifères d'une année sur l'autre; l'incidence cumulée d'années successives doit déjà être prise en compte.

Les choses changent du tout au tout au cours des dernières années. Avec 14 valeurs supplémentaires, la moyenne du coefficient de tarissement (84 ans) passe à 0.0202 j^{-1} et l'écart-type à 0.0042 j^{-1} (multiplié par 3) (Cv = 0.208). L'échantillon de la période 1973-87 a pour valeur moyenne 0.0178 j^{-1} avec un ecart-type de 0.0053 j^{-1} (Cv = 0.191). Il s'agit donc d'un véritable bond des valeurs du coefficient de tarissement, avec un maximum de 0.0380 j^{-1} en 1984, bien proche de la phase d'épuisement évoquée plus haut.

La figure 3 illustre ces variations et le changement majeur du

Tableau 1 Fleuve Sénégal à Bakel: module débit maximum et
coefficient de tarissement principal

Année hydro	Module (m3/s)	Q max (m3/s)	Coef.taris 10^{-2} j^{-1}	Année hydro	Module (m3/s)	Q max (m3/s)	Coef.taris 10^{-2} j^{-1}
1903-04	631	3 560	1.73	1945-46	945	6 480	1.89
04-05	737	4 790	1.89	46-47	745	4 460	1.95
05-06	874	3 840	1.82	47-48	666	4 360	1.77
06-07	1 233	9 340	1.77	48-49	572	3 590	1.84
07-08	521	2 850	1.89	49-50	467	3 760	1.77
08-09	767	4 200	1.89	50-51	1 152	7 630	2.11
09-10	902	5 490	1.84	51-52	842	5 340	2.05
10-11	670	3 840	1.87	52-53	718	5 060	2.17
11-12	537	3 330	1.95	53-54	631	4 180	1.95
12-13	564	3 290	1.87	54-55	1 068	6 610	1.92
13-14	270	1 040	2.33	55-56	1 049	5 260	1.77
14-15	444	1 885	1.80	56-57	952	6 050	1.66
15-16	592	3 140	1.90	57-58	1 029	5 660	1.89
16-17	691	4 200	1.87	58-59	1 037	8 170	1.71
17-18	647	4 960	1.92	59-60	788	5 460	1.76
18-19	1 144	7 300	1.67	60-61	621	3 550	1.82
19-20	530	3 560	1.84	61-62	944	7 030	1.74
20-21	834	3 630	2.04	62-63	969	4 410	1.93
21-22	431	2 850	1.87	63-64	666	3 760	1.98
22-23	1 219	9 070	2.07	64-65	970	7 180	1.68
23-24	754	4 670	1.93	65-66	1 048	7 000	1.74
24-25	1 247	6 350	1.87	66-67	841	5 450	1.72
25-26	841	4 610	1.57	67-68	1 037	5 820	1.62
26-27	521	2 290	1.89	68-69	397	2 880	1.89
27-28	1 075	6 460	1.72	69-70	764	3 770	2.11
28-29	904	5 490	1.92	70-71	542	3 440	1.92
29-30	899	5 490	1.89	71-72	598	4 320	2.15
30-31	839	4 610	1.74	72-73	263	1 430	2.02
31-32	739	4 300	1.80	73-74	361	2 550	2.09
32-33	770	4 850	1.78	74-75	760	5 780	2.62
33-34	838	5 490	1.93	75-76	602	5 000	2.56
34-35	700	5 340	1.78	76-77	470	2 500	2.40
35-36	1 164	6 680	1.81	77-78	324	2 700	2.17
36-37	1 234	7 600	1.57	78-79	523	3 250	2.28
37-38	644	3 590	1.90	79-80	301	1 760	2.84
38-39	807	5 630	1.80	80-81	402	3 640	2.84
39-40	559	3 400	1.86	81-82	423	2 840	2.78
40-41	430	2 760	1.82	82-83	303	2 280	2.56
41-42	417	2 890	1.87	83-84	220	1 240	3.28
42-43	437	3 590	1.92	84-85	205	917	3.80
43-44	666	3 480	1.87	85-86	356	2 370	3.43
44-45	330	1 740	1.78	86-87	(337)	2 681	3.28

comportement hydrogéologique du bassin versant depuis la période sèche récente. Les moyennes mobiles pondérées des figures 2 et 3 sont obtenues par l'expression:

$$\ddot{x}_o = \frac{1}{2} \sum_{t=0}^{n} x_i e^{-0.7(t_o-t_i)}$$

Les aquifères sont considérablement affaiblis en ressources (baisse du niveau piézometrique parfois supérieur à 15 m) et en extension spatiale (a dépend par la loi de Darcy des caractéristiques géométriques des aquifères).

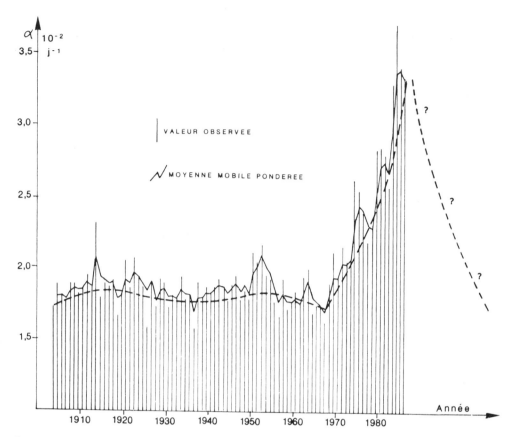

Figure 3   Evolution des coefficients de tarissement du fleuve
           Sénégal à Bakel.

Actuellement, le déficit des ressources souterraines disponibles
pendant la phase principale du tarissement est de l'ordre de $10^9 \, m^3$ ce
qui correspond à un débit moyen sur la période de 90 à 100 $m^3 s^{-1}$ et un
déficit de 25 $m^3 s^{-1}$ sur le module qui aurait pu être observé dans des
conditons hydrogéologiques normales.   Pour relativement minime qu'il
soit, ce déficit survient en saison sèche et ne peut être modifié
dans l'évaluation de la ressource disponible a cette époque.
   Les années 1985 et 1986 n'ont pas apporté d'amélioration sensible
des conditons hydrogéologiques et l'on peut craindre que celles-ci ne
se rétablissent pas avant de longues années en supposant que l'on
retrouve rapidement des années de fortes précipitations.   Si la
tendance sèche s'inverse aujourd'hui, un retour à un coefficient de
tarissement de l'ordre de 0.019 à 0.020 j^{-1} suppose au minimum une
durée égale à celle du phénomène de dégradation soit près de 15 ans.
   Il faudra probablement davantage car on ne peut comparer l'appauvrissement des nappes phréatiques par écoulement à leur recharge par
cheminement intersticiel de la lame d'eau filtrée à travers une
tranche aérée du matériau aquifère (dont l'épaisseur a considérablement augmenté par suite de la baisse du niveau piézométrique.

L'hypersalinisation de la Basse-Casamance

La Basse-Casamance, située au sud du Sénégal et de la Gambie, est une region caractérisée par le développement des milieux paraliques consécutifs à la profonde pénétration des eaux marines dans des rias aux nombreuses circonvolutions. Ces rias et les vallées qui les prolongent, creusées lors de la régression préholocène (15000-30000 B.P.) dans le plateau du continental terminal sont plus ou moins comblées depuis la transgression nouakchottienne (5000 B.P.) (P. Michel, 1960, 1971). Elles forment une constante du paysage avec des bas-fonds larges de quelques kilomètres à quelques hectomètres, où la mangrove a pu se développer aves ses sols caractéristiques et ses zones de "tannes" sursalés, et en bordure desquels s'est installée une riziculture traditionnelle qui a fait la richesse économique de cette région jusqu'à la crise climatique actuelle. Des interfluves ne dépassant pas une altitude de 30 à 40 m, couverts d'une forêt sèche plus ou moins claire complètent ce paysage essentiellement marqué par la présence de l'eau et l'absence de reliefs.

En complément d'études générales sur le bief principal de la Casamance (Y. Brunet, Moret 1970), des recherches de détail ont porté sur ses tributaires nord: le Marigot de Baila (A: 1700 km^2) et le Marigot de Biguona (A: 800 km^2 à Elora).

Depuis ces dernières années, on observe une modification fondamentale des mécanismes réglant les variations spatiales et saisonnières de la salinité en Basse-Casamance; les phénomènes peuvent être décrits suivant les deux périodes d'observation:

(1) Mécanismes de la salinisation de la Basse-Casamance en période "normale" de précipitations moyennes à fortes:
Les études antérieures à l'actuelle période déficitaire (Olivry et al., 1981) ont mis en évidence, pendant la saison sèche, un fonctionnement classique de progressive salinisation du réseau hydrographique soumis à l'influence des marées; cette progressive salinisation, dans le temps et de l'aval vers l'amont, est due à une compensation des pertes par évaporation par un excédent des volumes du flot (venu de l'aval) par rapport aux volumes du jusant (ces volumes sont induits par l'onde de marée).

En saison des pluies, precipitations directes et apports des écoulements de surface suffisent à un retrait marqué sinon total de la salure; le "rincage" des marigots est d'autant plus fort que l'on est près de la source des apports.

Les concentrations en sels suivent, en toute saison, et de l'aval vers l'amont, un gradient négatif, les maximums dépassant rarement 35 g/l.

Les bilans hydrologiques ont montré de faibles coefficients d'écoulement de surface pour la partie continentale des bassins (6% en année moyenne pour 1200 à 1400 mm de précipitation et beaucoup moins en années sèches) et un terme non négligeable au niveau de l'infiltration. Ce terme comprend la recharge proprement dite des nappes et la lame d'eau reprise par évapotranspiration pendant les mois de saison sèche. Mais la nappe phréatique est mal connue (niveau piézométrique à 10 ou 15 m sous la surface topographique au coeur des interfluves, amplitude annuelle de 1 à 2 m) et l'importance de ses apports potentiels n'a pu être évaluée; toujours est-il que ces apports souterrains compensent partiellement l'évaporation sur les surfaces d'eau libre du marigot en limitant ainsi la progression du sel vers l'amont.

On verra dans la figure 5(a) (années 70-71) que les concentrations restent généralement inférieures a 35g/l sur le marigot de Baila en saison sèche sauf éventuelle sursalure dans l'estuaire et rejoignent celles d'eaux douces au cours de la saison des pluies.

(2) Hypersalinisation de la Basse-Casamance en période "sèche" et ses conséquences durables:

La succession d'années sèches observées depuis 1972, et surtout depuis 1979 a modifié complètement les conditions écologiques de la Basse-Casamance; la mangrove meurt, les rizières sursalées sont abandonnées; la forêt claire s'étoile, les nappes baissent (de plus de 10 m parfois).

La chronique des précipitations annuelles relevées à Bignona pour les bassins étudiés est donnée dans le tableau 2.

Tableau 2

Année	Total en mm	Année	Total en mm	Année	Total en mm
1954	1806	1968	827	1980	619
1955	1633	1969	1465	1981	1030
1956	1585	1970	1163	1982	933
1957	1391	1971	905	1983	656
1958	2189	1972	452	1984	915
1959	1176	1973	1054	1985	1125
1962	1274	1974	1084	198	1000
1963	1100	1975	1352	Moy.	1196
1964	1118	1976	1209	E.T.	388
1965	1766	1977	848	Cv	0,324
1966	1247	1978	1499		
1967	1795	1979	873		

Depuis 1979, un déficit chronique apparaît malgré une légère reprise en 1985 et 1986. Il est responsable d'une évolution tout à fait differente des concentrations en sel. Pour une même station les valeurs sont beaucoup plus fortes qu'en période normale: c'est l'hypersalinisation qui apparaît et reste présente pratiquement en toute saison. La figure 4 montre l'extraordinaire progression de la salinité à partir de 1980, même si 1982 amène un quasi-adoucissement du marigot à Djibidione. Les maximums absolus dépassent 100 g/l depuis 1981 et atteignent même 154 g/l en 1984 à la station de Djibidione située à 158 km de l'Océan.

Les variations saisonnières sont illustrées dans la figure 5 (b). On retrouve bien le gradient négatif de salinité de l'aval vers

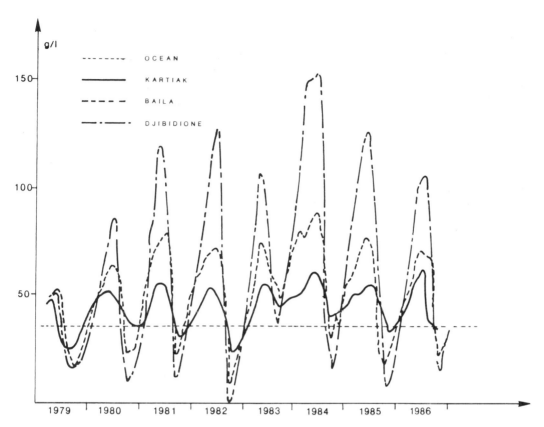

Figure 4   Variations de la salinité sur trois stations du marigot
de Baila.

l'amont observé en période normale pour la saison des pluies (sauf en
1983 vraiment très déficitaire); les phases d'adoucissement sont
directement liées à l'importance des apports de saison des pluies.
On observe surtout en saison sèche une renverse du gradient qui
devient largement positif; par suite les amplitudes de variation de
concentration sont d'autant plus grandes que l'on remonte vers
l'amont.

Gallaire (1981) a relevé en 1979 des valeurs relativement stables
d'une station à l'autre; entre 90 et 158 km de la mer les maximums
tournaient autour de 50 g/l. Cette année 1979 paraît être le tournant
pour le marigot de Baila de l'inversion du gradient de salinité.

La figure 6 montre les différentes courbes annuelles des maximums
de salinité relevés aux différentes stations de marigot de Baila
(Olivry, Dacosta 1984). Le gradient devenu positif depuis 1980 ne
cesse d'augmenter jusqu'en 1984. Il n'y a pas de liaison simple
entre celui-ci et la pluviométrie de l'année précédente (plutôt avec
un cumul des pluies des années antérieures).

Le mécanisme d'hypersalinisation du marigot correspond au
fonctionnement d'un marais salant.

Le rôle plus ou moins compensateur des apports des nappes
phréatiques pendant la saison sèche ayant pratiquement disparu,

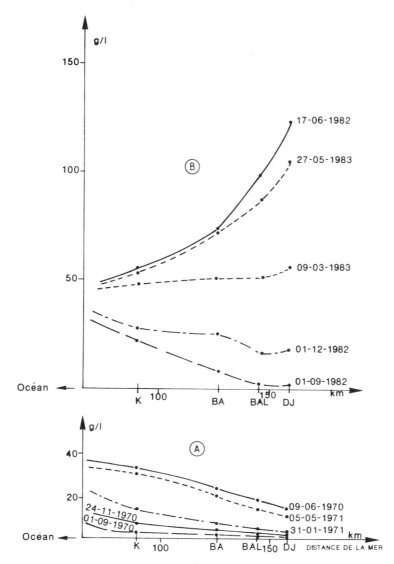

Figure 5   Evolution saisonnière des salinités sur le marigot de
Baila: (a) en période normale et humide année 1970-1971;
(b) en période déficitaire année 1982-1983.

l'évaporation régulière du marigot appelle de l'aval des volumes
chargés en NaCl toujours plus importants par rapport au volume du
bief concerné.   Ceci explique que la progression de la sursalure
augmente en remontant vers l'amont.

La détérioration des conditons hydrogéologiques, résultat de
quinze années de déficits pluviométriques, constitue de toute
évidence un phénomène durable.   Les années 1984, 85 et 86 ont été
plus abondantes sans toutefois atteindre la normale; elles n'ont
cependant pas apporté d'amélioration vraiment notable au niveau des

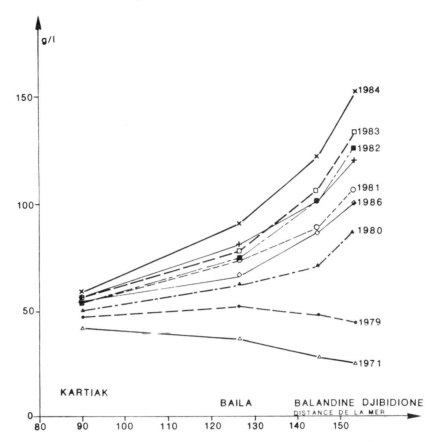

Figure 6 Evolution des concentrations maximales observées en fin
de saison sèche sur le marigot de Baila.

concentrations en sel de saison sèche les apports occultés de la
nappe aux marigots sont probablement toujours aussi faibles, et la
recharge des nappes demandera de nombreuses annees.

Dans des conditons climatiques optimales la Basse-Casamance peut
rapidement retrouver un adoucissement de ses eaux en saison humide;
mais elle restera marquée, dès la saison sèche, pendant encore de
longues années par une hypersalinisation des marigots.

Références

Brunet-Moret, Y. (1970) Etudes hydrologiques en Casamance. Cah.
Orstom, sér. hydrol. Paris vol XII-2.
Gallaire, R. (1980) Etude hydrologique du marigot de Baila, Ortom
Dakar.
Michel, P. (1960) Recherches géomorphologiques en Casamance et en
Gambie meridionale. BRGM Dakar.
Michel, P. (1971) Les bassins des fleuves Sénégal et Gambie. Etude

géomorphologique. Thèse Doctorat d'Etat. Fac Lettres Strasbourg.

Olivry, J.C., Chouret, A. (1981) Etude hydrologique du Marigot de Bignona – campagne 1970-71. Orstom-Dakar.

Olivry, J.C. (1983) Le point en 1982 sur la sécheresse en Sénégambie et aux Iles du Cap Vert. Cah. Orstom. Sér. hydrol. Vol. XX n°1.

Olivry, J.C., Dacosta, H. (1984) Le Marigot de Baila. Bilan des apports hydriques et évolution de la salinité. Orstom Dakar.

Olivry, J.C., Chastanet (1986) Evolution de l'hydraulicité du Fleuve Sénégal et des précipitations dans son cours inférieur depuis le milieu du XIXeme siècle. Colloque Nordeste-Sahel. IHEAL Paris, Janv. 86.

Olivry, J.C. (1986) Possibilités d'allegement des réseaux hydrométriques dans les pays en voie de développement après réalisation de synthèses hydrologiques régionales. Proceedings of the Budapest Symposium July 86. IAHS Publ. n°158.

Rochette, C. (1974) Le bassin du Fleuve Sénégal. Monographies hydrologiques ORSTOM Paris.

Sircoulon, D. (1985) La sécheresse en Afrique de l'Ouest. Comparaison des années 1982-84 avec les années 72-73. Cah. ORSTOM. Sér. Hydrol. Vol. N°4 85-85.

The Influence of Climate Change and Climatic Variability on the Hydrologic Regime and Water Resources (Proceedings of the Vancouver Symposium, August 1987). IAHS Publ. no. 168, 1987.

Effect of climate changes on the precipitation patterns and isotopic composition of water in a climate transition zone: Case of the Eastern Mediterranean Sea area

Joel R. Gat & Israel Carmi
Isotope Department
Weizmann Institute of Science, Rehovot, Israel

ABSTRACT   In a climatic transition zone such as the Eastern Mediterranean Sea area, in which storm tracks and moisture origin is quite varied, the most important cause for a shift in the isotopic composition of meteoric waters appears to be a modification in the weighting factor in the annual average for each contributing synoptic pathway (each with its distinct isotopic signature).   Changes   in local meteorologic parameters affect the isotopic composition of rain only to a limited extent, with rain intensity as the most effective parameter.   The effects of global temperature   and   humidity   changes   express   themselves primarily by means of the "d"-excess parameters.   All-in-all, the reduced role of the Mediterranean Sea as a dominant vapor source for the region's precipitation in the past, appears to be the dominant factor in the change of isotopic composition between present-day precipitation and the paleowaters of the Levant.

Effet des changements climatiques sur les modes de   précipitation   et   sur   la   composition isotopique de l'eau dans une zone climatique transitaire - Cas de la région de la Mer Méditerranée Orientale

Resume   Dans une zone climatique transitaire, tel que la Mer Méditerranée   Orientale dans lequel le cours des tempêtes et l'humidité sont d'origines variées, la raison principale d'une variation de la composition isotopique des eaux météoriques est le changement du facteur-poids de la moyenne annuelle des pluies pour chaque trajet meteorologique (ayant   chacun sa propre   empreinte   isotopique). Les changements des paramètres meteorologiques locaux n'affectent que dans une certaine mesure la composition isotopique des pluies, l'intensité des pluies étant le paramètre le plus influençant.   Les effets de la température du globe et des changements d'humidité sont exprimés par le parametre "d-excèss". En fait, la Mer Méditerranée, dont le role de principale source de vapeur dans la région était mineur dans le passé, devient le facteur dominant du changement de la composition isotopique entre les précipitations actuelles et les eaux anciennes dans le Levant.

Introduction

Present-day precipitation at a given location is characterized by a rather well defined stable isotope composition, which reflects its geographic situation, especially in relation to the source of moisture (Dansgaard, 1964; Gat, 1980). This mean isotope composition is correlated with temperature (Yurtsever & Gat, 1981). Paleowaters are encountered in many locations with distinctly different isotopic composition than the local present-day precipitation (Munich & Vogel, 1962; Fontes, 1981).

In this paper we will address the question of the isotopic change to be expected with changes in climate, and the related (inverse) question, namely how to translate a measured isotopic change into a quantitative climate description. The discussion will focus on the situation in the Eastern Mediterranean Sea area and on the Mediterranean climate, generally. This area is a zone of sharp climatic transitions, in which fluctuations from arid to pluvial occurred during the late Pleistocene and the Holocene (Gat & Magaritz, 1980).

The pattern of precipitation in the Eastern Mediterranean Sea and its isotopic signature

The Mediterranean climate is characterized by a dry summer and a rainy winter season; it is the expression of the dominance during summer of a high pressure belt which steers storm tracks away from the area, and of subsiding (warm) air masses which stabilize the air column and prevent the buildup of rain clouds in spite of the proximity of a perspiring sea. The latter vapor source is utilized only during winter when the sub-tropical high-pressure belt moves southwards, exposing the Mediterranean air space to invasion by cold, continental air; the meeting of cold and dry air with the relatively warm seawater induces enhanced evaporation instability and cyclogenesis (Trewartha, 1961). The most favoured storm path, which accounts at present for about half of the precipitation events in the Levant (Gagin & Neumann, 1974), is through the Balkans which interacts air masses over the Eastern Mediterranean or Aegean seas under conditions of a pronounced moisture deficit; this process imprints an isotopic signature with a large deuterium excess (*), typically $d > 20°/\infty$ (Gat & Carmi, 1970), (Figure 1).

The air-sea interaction over the Western Mediterranean (activated by storm tracks over France) is somewhat less intense and the deuterium-excess values which are engendered are about $d = 15°/\infty$. This moisture contributes to precipitation over the Levant in the case of two extreme situations, illustrated in Figure 2:

(a) under "high index" circulation, when the precipitating air masses are advected along the axis of the Mediterrannean Sea; this gives an opportunity for further exchange of moisture across the air-sea boundary layer, resulting in an air mass characterized by a low

---

(*)   Deuterium excess, d, is defined as $d = \delta^2H - 8.\delta^{18}O$,   (also referred to as the 'd' – parameter); it relates the isotopic composition to the appropriate "Meteoric Water Line", i.e. a line of slope of 8 on the δ^2H vs. $\delta^{18}O$ diagram.

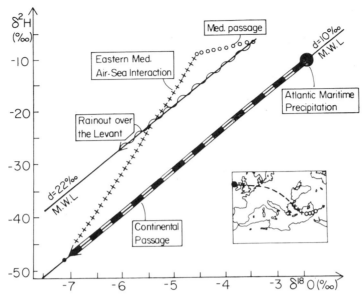

Figure 1   Evolution   of   the   isotopic composition over   the   Eastern
Mediterranean   under winter conditions (storm track   shown
in   inset)   giving   rise   to   precipitation   with   a   large
deuterium   excess.   The   diagram   shows the   precipitation
formed,   or   potentially   formed   from   the   air   masses
concerned, i.e. the values $(\delta_a + \varepsilon^*)$.

value of the 'd'-parameter and rather enriched δ- values.

   (b) under conditions of a "low index" circulation,   the precipitating air masses swing over the North-African continent.   By the time
they approach the Israeli coast from the south or SW,   the air masses
are   depleted   in   the heavy isotopic species (Leguy   et   al.,   1983)
indicating a considerable degree of rainout en-route.   Not unexpectedly, rain with this synoptic history is more prevalent in the southern
part of the country.

   A less common "rain pattern" is that of the advection of monsoonal
air masses from the SE, carrying moisture originating from the Indian
Ocean.   This pathway at times is responsible for strong showers over
the   Negev   and   Sinai deserts,   for   example   the   precipitation   of
February 1975,   which resulted in large scale flooding throughout the
region.   These rains can be distinguished by very low tritium levels
and a pristine maritime stable-isotope composition.

   The   source of moisture for the Eastern Mediterranean winter rains
is thus quite varied.   However another potential source,   namely the
Atlantic   air   masses   with   an   exclusive   continental   trajectory
(typified by precipitation at Ankara with $\delta^{18}O=-7.8°$ /∞ ;   d = 10°/∞
(IAEA,   1981))   do not seem to contribute to a significant degree   to
the present-day precipitation in the Levant.

   A   summary of the different   storm   patterns   and trajectories   was
given   by   Rindsberger & Gat (1985).   One notes that each   of   these

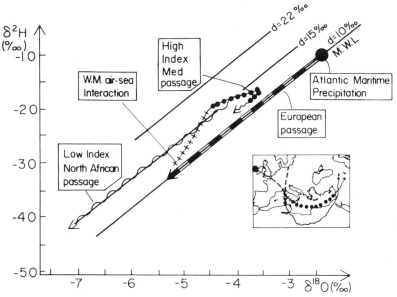

Figure 2 Evolution of the isotopic composition of rain originating
          from air–sea interaction over the Western Mediterranean,
          for two trajectories (shown in the inset) of the 'high'
          and 'low' index circulation, respectively.

pathways is associated with a relatively well defined range of
isotopic compositions, which are quite distinct from one another.
     The averaged isotopic composition for a period of rain (which
constitutes the input into the hydrologic systems) obviously will be
the amount–weighted average of the individual storm events.   The
temporal variability of this average for any given location, be it
seasonal or annual, is the result of the scatter of δ–values within
any synoptic grouping; more important in most instances, however, it
reflects a shift in the relative weight of the different rain
patterns which make up the "climatic" average. Similar arguments
apply when one compares the mean isotopic composition of precipitation from place to place.   Some of the effects which were described
by Dansguard (1964), namely the effects of altitude, latitude, etc.,
come about through the modification of an air mass by further rainout,
by orographic mixing or by effects related to rain intensity and
cloud height, such as evaporation from the falling raindrops (Gat,
1980).   There are, however, also effects due to the different weight
to be assigned to precipitation from air masses with different
synoptic histories at the different locations. As an example, compare
the synoptic analysis for precipitation in the Negev (Leguy et al.,
1983) with that at Gaaton in Northern Israel (Rindsberger et al.,
1983).

The effect of changes in meteorologic variables and
climate on the isotopic composition of rainfall

These effects must be considered on three levels:
   (a) For a given synoptic scale setting (i.e., when the air masses
which reach the region are unaltered), what are the effects of local
changes in temperature, surface conditions, etc., be they natural or
man made?
   (b) For a worldwide heating or cooling trend and humidity changes
(but assuming for the sake of simplicity that the general circulation
pattern is unchanged), what will be the effects on the precipitation
originating from the different source areas?
   (c) For changes in the general circulation, what could be the
effects on the local climate and, in particular, its rainfall pattern?

Local changes in environmental parameters

Due to the rapid isotopic exchange between the falling raindrops and
the ambient moisture, one finds the isotopic composition of rain to
be determined, to a first approximation, by the isotopic composition
of the atmospheric moisture; the latter represents the sum-total of
its synoptic history (Rindsberger & Gat, in prep.). Local climate
parameters and details of the cloud structure do not come into play
prominently, with the possible exception of the rain intensity; some
"amount effect" on the isotopic values can often be identified
(Dansgaard, 1964).   Both the evaporative enrichment of very light
showers (say less than 2mm/day) due to the evaporation from the
falling droplets beneath the cloud base, or on the other end of the
scale, the depleted isotopic values often associated with the most
intense downpours (Levin et al., 1980), contribute to the effect.
Further, the special case of snow and hail has to be considered,
since water in the solid phase does not take part in the isotopic
exchange with the ambient moisture and retains the outstanding
depleted isotopic values which are typical of the in-cloud situation.
In many other cases, however, the basis for the "amount" effect is
not to be sought in the local condition but in the correlation
between the synoptic history of the precipitating system and its rain
intensity.
   A local temperature change per-se will not have a far reaching
effect on the isotopic composition; the temperature effect on the
isotopic fractionation factor for phase transition, α , is quite
negligible in the overall context. Presumably the most important
temperature effect would manifest itself through a change in the
prevalence of snow.
   A local change in surface roughness or albedo has the potential of
affecting the isotopic composition by mixing of surface air with
higher layers, thus possibly affecting isotopic properties of the air
mass which enters the cloud. A coastal city, for example, can produce
a semi-orographic effect, similar to one documented for natural
conditions in the northern coastal plain of Israel (Gat & Dansgaard,
1972), where the mixing of surface air into the cloud layer resulted
in the lowering of the 'd'-excess parameter in the resultant precipitation.

An interesting special case concerns the effects of cloud seeding, a man-made climate change which is being practised in Israel for rain enhancement.   The isotope effect produced will depend critically on the change in precipitation pattern.   If the effect is the extension of the rained-on area without a significant increase in rain intensity (as seems to be the case for the Israeli cloud-seeding experiment (A. Gagin, pers. comm.)), then the expected isotope effect is minor. A dramatic local increase in rain intensity, on the other hand, would shift the isotopic composition along the "Meteoric Water Line" towards more depleted isotopic values.   Conversely, if the effect is one of hail suppression (hail always showing more depleted isotopic values than the accompanying rain) then evidently the shift in isotopic composition would be in the opposite sense.   Figure 3 shows some of these expected changes.

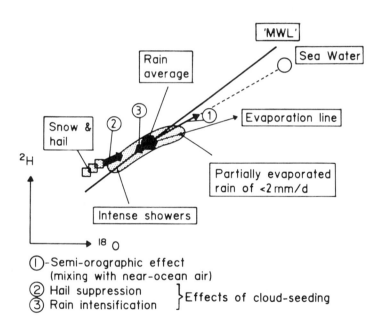

Figure 3   The expected effects on the isotopic composition of precipitation due to local anthropogenic changes and cloud seeding. The normal range of isotopic composition of rain is shown by a shaded area on the Meteoric Water Line (MWL).

Worldwide temperature and humidity changes

The foremost effect of a temperature change of the world ocean is on the vapor gradient across the liquid-air boundary layer, across which evaporation takes place. This is immediately being translated into a change of the 'd' - excess which is imposed on the atmospheric moisture at its source (Craig & Gordon, 1965; Gat, 1980).   As pointed out by Merlivat & Jouzel (1979) the present-day saturation deficit over the Atlantic Ocean of about 20% (which engenders the $10^o/oo$ MWL), was reduced during the Pleistocene to only 10%, based on an estimate of d = 5°/∞ for the then Meteoric Water Line, as shown by data for the paleowaters of North Africa (Sonntag et al., 1978) which were presumably derived from the same source. In the Mediterranean, using a similar argument, we would expect a decrease in the value of the 'd' - parameter at times when the temperature of the sea was lower (only partially offset by a freshening of the surface waters). However, the immediate result of a decrease in the Mediterranean Sea's temperature would be the dampening of its cyclo-genetic effect, so that this source of moisture would be severely curtailed and rather ineffective (to be discussed).

One must realize that it is not only the temperature of the ocean water which fixes the 'd' -value (by means of the saturated vapor pressure value at the air-sea boundary), but that also the humidity in the "free atmosphere" comes into play. Indeed, even in the present atmospheric water cycle, the dry European air which interacts with the Mediterranean waters in winter produces the outstanding high values of the duterium excess (namely d>20°/∞) although the tempera-ture of the winter Mediterranean is comparable to summer conditions in the Northern Atlantic, where moisture of d = 10-15°/∞ is being formed. The humid conditions over the tropical seas, of course, show the opposite effect, with 'd' - values as low as those which are usually associated with a cold sea surface.

Lower temperatures over land affect the rainout rate from the air masses and obviously will result in more depleted (negative) δ - values. One cannot, however, simply apply the temperature-δ correla-tions which pertain to the present-day isotopic data (Yurtsever & Gat, 1981), since the pertinent parameter to use in the Rayleigh equation (which describes the change of δ with rainout, (Dansgaard, 1964), is the ratio of the moisture content in the airmass to its initial state; the latter quantity being determined by the conditions in the source region of moisture and is in itself a function of the temperature.

Finally, would there be an effect on the isotope composition resulting from the rain-enhancement program suggested by Assaf (1982)? This is based on the idea of increasing rainfall by increasing the heat content of the Eastern Mediterranean by mixing of its surface waters.   Since this scheme does not envisage a large change of the temperature of the surface waters one would not, a priori, expect a different isotopic value in the rain than for moisture picked up normally by interaction with the sea.   However as this scheme purports to increase the total amount of rainfall originating in the Eastern Mediterranean in the yearly total, the envisaged effect of such a program would be one of the shifting of the seasonal average isotopic composition towards that which is characteristic of the Eastern Mediterranean moisture source.

Changes of the large scale circulation pattern

During 1971-74 there occurred a notable shift in the annual isotope composition in the Eastern Mediterranean Sea area relative to the mean isotope composition of previous years (Figure 4, Carmi & Gat, 1978). The averaged annual values in our network of rain sampling are based on the data from periodic sampling of 10 days duration

throughout the rainy season.   A comparison of the histograms of
distribution of the δ-values for the 10 day samples from Bet Dagan,
Israel, during the pre-1971 and the 1971-74 periods, respectively,
(Figure 5) is very revealing: there is no widening of the total range
of isotopic data in these sets of samples, but there is a shift in
the distribution within this range.   This is interpreted as a shift
in the relative weight of contribution of the different synoptic
pathways to the annual average.   During these particular years the
overall effect was one of strengthening the Mediterranean character
of the isotope composition, notably an increase in the value of the
'd' parameter.
    During this period, which can be taken as a model for a worldwide
change of circulation, there was no major change of either temperature or humidity and the dominant pattern of change was one of
reshuffling of familiar synoptic elements.   During periods in the
past when the world-wide circulation pattern was changed, one must
reckon with different synoptic pathways, as well as a modification of
the familiar ones, as described above.
    It is believed that the storm tracks which nowadays inundate the
European continent were displaced southwards, during the Pleistocene,
bringing Atlantic moisture into Northern Africa and possibly also to

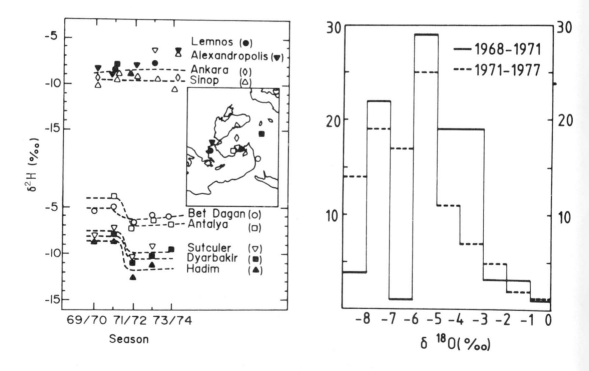

Figure 4   The change of the annual averaged isotope composition of
(LEFT)   precipitation in the Eastern Mediterranean Sea   region
         during the years 1971-1974.

Figure 5   Histograms of δ18O values for precipitations in Bet Dagan,
(RIGHT)   Israel, between 1969-1971 and between 1972-1977.

the Middle East.   In support of this thesis it is cited   that   the
inland isotopic gradient of the Saharan paleowaters parallels that of
modern European precipitation (Sonntag et al., 1978).   As discussed
before, the lower oceanic temperatures at that time are expressed by
a lower value of the 'd' excess in these paleowaters.

What would be the fate of an air mass passing over the   cold,   and
possibly also less saline, Mediterranean Sea?   Presumably not much
would happen: the air mass could .pick up some moisture to make up for
losses by previous rainout, resulting in an isotopic composition with
a low 'd' value and not too depleted δ-values;-alas not too   distinctive an isotopic composition. Indeed we would expect a rather similar
isotopic composition for precipitation which originated in the Indian
Ocean (on the assumption of rainout having occurred from these
monsoonal air masses over the desert regions; judging from the rather
negative δ values reported by Tongiorgi (private comm.) for Arabian
paleowaters, this was indeed the case).   All these proposed changes
are   schematised in Figure 6.

Obviously there is no justification to discuss the isotopic
composition of paleowaters in the Levant in terms of a slight   shift
or perturbation on the present-day patterns.   Rain at that time must
rather be viewed as resulting from storm patterns which are   quite
different from the present ones, some of which (such as Atlantic
disturbances) are not represented in the modern   rainfall of our
region.

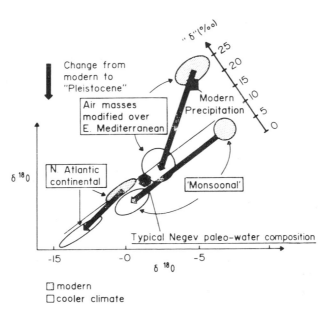

Figure 6   Postulated   change in the isotopic composition of precipitation   for air masses originating in the Atlantic, the
Eastern Mediterranean or the Indian Ocean, respectively.
For the present day climate: full areas, and for the
Pleistocene-open circles. The average isotope composition
for rain in Israel's coastal plain and a typical value for
paleowaters of the Negev, are given for comparison.

Interpretation of paleo-isotope data

Most of the paleo-isotopic evidence is derived from minerals, mainly carbonates, which only contain the Oxygen isotopic record and are therefore of limited use for a full paleo-hydrologic interpretation. The discovery of paleowater sources in the Negev (Issar, 1985) with their distinctive isotopic composition (Gat, 1971; Gat & Galai, 1982) enables one to relate these waters to the meteoric water within this region.

Unlike the situation in many parts of the world where the isotopic composition of paleowaters is only slightly shifted from that of modern waters (in low latitude locations, no shift or even a reverse one has been reported), in the Middle East the change is more dramatic. Paleowaters are not only somewhat more depleted in $\delta^{18}O$, but are characterised by a large shift in the value of the 'd' parameter, from d>20°/∞ to an average of just d = 7°/∞. In other words, the distinguishing Mediterranean character is not present in these paleowaters.

As was discussed, it was to be expected that the rain forming role of the Mediterranean would be severely curtailed, as it evidently was. On the other hand it is clear from the isotopic data that rains in the Negev are not simply an inland projection of the Saharan paleowaters produced by Atlantic-based storms, as then we should have seen a more depleted isotopic composition such as are typical of precipitation in Eastern Europe. Horovitz and Gat (1984), based on the similarity of the data to the European summer precipitation and on palynologic evidence, interpret these waters as derived from summer rains (originating in the Atlantic). As can be seen in Figure 6, however, the isotopic composition of the paleowaters are also consistent with other storm trajectories, e.g. with monsoonal air masses or Atlantic-based disturbances modified over the Mediterranean Sea. Probably the isotopic composition of the paleowaters represent a weighted average of a number of different water types, just as is the case nowadays.

Evidently one cannot do more in such a climate transition zone than document a change in the meteorologic condition and exclude some suggested scenarios on the basis of incompatibility of the isotopic evidence. It is only when other evidence or models suggest the preponderance of a particular synoptic scenario that one can hazard a more quantitative reconstruction of the climatic conditions at the source area for the moisture or in the region discussed.

References

Assaf, G. (1982) Artificial sea mixing. In Charney workshop on Weather Modification, publ. by Ormat Turbines Ltd. Yavne.
Carmi, I. & J.R. Gat (1978) Changes in the isotope composition of precipitation of the eastern Mediterranean Sea area:- a monitor for climate change? In: Israel Meteorological Research Papers Vol II Israel Meteorological Service: 124-135.
Craig, H., & R.L.I. Gordon (1965) Deuterium and Oxygen-18 variations in the ocean and marine atmosphere. In: table Isotopes in Oceanic Studies and Paleotemperatures, Lab. Geol. Nucleare, Pisa, p.9-130.

Dansgaard, W. (1964) Stable Isotopes in Precipitation. Tellus 16: 436-468.

Fontes, C. (1981) Paleowaters. In: Stable Isotope Hydrology (Gat & Gonfiantini, eds) IAEA Technical Series no 210, IAEA, Vienna pp. 273-253.

Gagin, A. & Y. Neumann (1974) Rain stimulation and cloud physics in Israel. In: Weather & climate modification, W.N. Hess (edit), Wiley, N.Y.: pp. 454-494.

Gat, J.R. (1971) Comments on the stable isotope method in regional groundwater investigation. Wat. Resour. Res. 7: 980-993.

Gat, J.R. (1980) The relationship between surface and subsurface waters: water quality aspects in areas of low precipitation. Hydrol. Sci. 25: 257-267.

Gat, J.R. & A. Galai (1982) Groundwater of the Arava valley: an isotopic study of their origin and interrelationships. Israel J. Earth Sci. 31:25-38.

Gat, J.R. & W. Dansgaard (1972) Stable isotope survey of the fresh water occurrences in Israel and the northern Jordan rift-valley J. Hydrol. 16: 177-211.

Horowitz, A. & J. R. Gat (1984) Floral and isotopic indications for possible summer rains in Israel during wetter climates. Pollen et Spores 26: 61-68.

IAEA (1981) Statistical treatment of environmental isotope data in precipitation. Technical reports series no 206, IAEA, Vienna, pp. 255.

Issar, A. (1985) Fresh water under the Sinai-Negev peninsula. Scientific American 253:82-88.

Kukla, G.J. & H.J. Kukla (1974) Increased surface albedo in the northern hemisphere. Science 183: 709-714.

Leguy, C., M. Rindsberger, A. Zangivil, A. Issar & J.R. Gat (1983) The relation between the ^{18}O and deuterium contents of rain water in the Negev desert and air-mass trajectories. Geologia Applicate e Idrogeologia XVIII; 257-266.

Levin, M.J., J.R. Gat & A. Issar (1980) Precipitation flood- and ground-water of the Negev highlands: an isotopic study of desert hydrology In: Arid-Zone Hydrology: Investigations with Isotope Techniques, IAEA, Vienna, 3-22.

Merlivat, L.S. & J. Jouzel (1979) Global climatic interpretation of the deuterium-oxygen 18 relationship for precipitation. J. Geophys. Res. 84: 5029-5033.

Munich, K.O. & J.C. Vogel (1962) Untersuchungen an Pluvialen Wassern der Ost Sahara. Geol. Rundschau 52: 611-624.

Rindsberger, M., M. Magaritz, I. Carmi & D. Galed (1983) The relation between air mass trajectories and the water isotope composition of rain in the Mediterranean Sea area. Geophys. Res. Lett. 10: 43-46.

Sonntag, Ch., E. Klitsch, E.M. Shazli, Ch. Calluke & K.O. Munich (1978) Paleo klimatische information im isotopengehalt ^{14}C-datierter Saharawasser;-Kontinental Effekt in D und ^{18}O. Geol. Rundschau 67: 413-424.

Trewartha, G.T. (1961) The Earth's Problem Climates The Univ. of Wisconsin Press, Madison.

The Influence of Climate Change and Climatic Variability on the Hydrologic Regime and Water Resources (Proceedings of the Vancouver Symposium, August 1987). IAHS Publ. no. 168, 1987.

Climatic variability and regolith groundwater Regime in Southwestern Nigeria

E. O. Omorinbola
Department of Geography, University of Ife
Ile-Ife, Nigeria

ABSTRACT This paper illustrates how the regolith ground-water regime in the Basement Complex of southwestern Nigeria responds spontaneously to the variability of the prevailing humid tropical climate. The rainfall and the groundwater regimes have a remarkably similar degree of variability. The mean monthly rainfall and the mean monthly depth to the water table have a highly significant inverse correlation with a coefficient of about −0.80 and a relationship aptly described by simple linear regression equations. The monthly variation pattern of depth to the regolith groundwater table is optimally described by polynomial regression equations. The applicability of the study results is specified in respect of predicting mean depths to the water table at given points in time, determining groundwater level fluctuation rates by differential calculus, estimating the proportion of the available rainfall used in groundwater replenishment, and the correct timing of well-sinking operations so as to increase their success ratios and ensure adequate and perennial well yields.

Introduction

The Basement Complex of southwestern Nigeria (Figure 1) is located entirely within the humid tropics as defined by Fosberg et al. (1961). Although the mean annual rainfall in this geological region is quite considerable, there exists quite distinct and alternating wet and dry seasons of unequal duration in the year. Generally the wet season prevails between March and October when most of the rainfall occurs while the dry season prevails from November to February when the precipitation is nil or negligibly small (Udo, 1970; Oyebande and Oguntoyinbo, 1970). This rainfall regime attributable largely to the meridional oscillation of the Intertropical Discontinuity (Adefolalu and Oguntoyinbo, 1985) constitutes the most spectacular phenomenon of the regional climate and significantly affects the surface and sub-surface water resources. In fact rainfall variability is considered as the principal component of the climatic variability in the region. Deep chemical weathering of crystalline Basement rocks in this humid tropical region has produced an overburden of generally thick regoliths (De Swardt, 1955; Thomas, 1966; Faniran, 1974) in which water table aquifers of generally considerable lateral extent occur (Faniran and Omorinbola, 1980a; Omorinbola, 1982). The response of the regolith groundwater to the prevailing rainfall regime in the

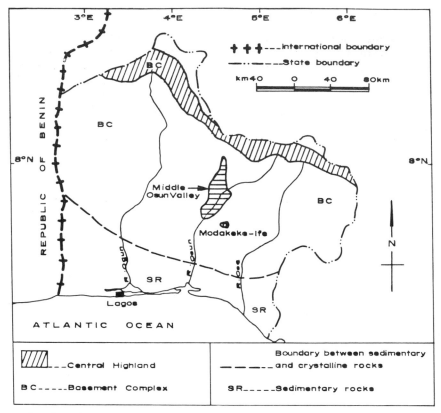

Figure 1   Southwestern   Nigeria   study areas.

region is the subject matter of this paper. A proper study of the nature and implications of such hydrologic response to climatic variability is undisputably required for a better understanding of humid tropical geohydrology and is quite vital to the development of the regolith groundwater resources for specific socio-economic purposes. Consequently, by reference to two study areas, attempts have been made in this paper to achieve the following specific objectives:

(a) To understand a quantitative comparison of the degree of variability of the rainfall regime and that of the regolith groundwater regime.

(b) To determine the nature and strength of the correlation between the mean monthly rainfall values and the mean monthly depths to the water table in the regolith.

(c) To derive empirical equations that optimally describe the temporal variation pattern of depth to the regolith groundwater table as induced by the variability of the contemporary climate.

(d) To identify the major applications of the study results from both theoretical and practical standpoints relating to the effects of climatic variability on the geohydrology of humid tropical regolith.

The study areas and sources of data

The two areas in the Basement Complex of southwestern Nigeria from which data were obtained for this study are the Middle Osun Valley (796 sq.km) and Modakeke-Ife which spatially is an integral part of the ancient and University town of Ile-Ife. The locations of the two study areas are shown in Figure 1. As the two areas are in many environmental respects typical of the Basement Complex of southwestern Nigeria (Omorinbola, 1986a), the findings from this study could logically be taken as typical of the nature or pattern of response of the regolith groundwater regime to the climatic variability in the entire geological region.

The regoliths in both areas are considerably thick (Faniran and Omorinbola, 1980b; Omorinbola, 1982) and from them groundwater is tapped mostly by means of dug wells for mainly domestic purposes. The spatial density of dug wells in the two areas is quite high (Loehnert, 1981; Omorinbola, 1986a). The monthly fluctuation of groundwater level in the regolith of the Middle Osun Valley was monitored from January, 1974 to May, 1976 at the sites of 30 observation wells widely distributed over the area (Figure 2). In Modakeke-Ife 21 observation wells were used for the same purpose during the period of January, 1981 to July, 1983. The actual geohydrologic variable on which data were collected in both study areas is depth from the ground surface to the regolith groundwater table. The range of the diameters of the observation wells (0.91 to 1.22 m) made them quite suitable for monitoring groundwater level changes in the regoliths (Ward, 1967). The data for each study area covered approximately three consecutive dry seasons and three consecutive wet seasons.

The monthly rainfall data for the corresponding periods of observation in the two study areas were obtained from two properly maintained meteorologic stations. For the Middle Osun Valley the data from the meteorologic station operated by the Nigerian Meteorological Services Department at Osogbo were used while the climate station maintained by the Geography Department of the University of Ife within the University Campus for teaching and research purposes is the source of the data recorded in Table 1 for Modakeke-Ife. The mean annual rainfall in the Middle Osun Valley is about 1480 mm. The double maximum rainfall typical of the region is reflected in the data recorded here for Modakeke-Ife.

Table 1 Mean monthly rainfall in Modakeke-Ife from 1981 to 1983

Month	Mean rainfall(mm)	Month	Mean rainfall(mm)
January	0.0	July	278.0
February	20.0	August	216.0
March	57.9	September	231.2
April	152.9	October	102.6
May	181.6	November	21.8
June	221.0	December	13.6

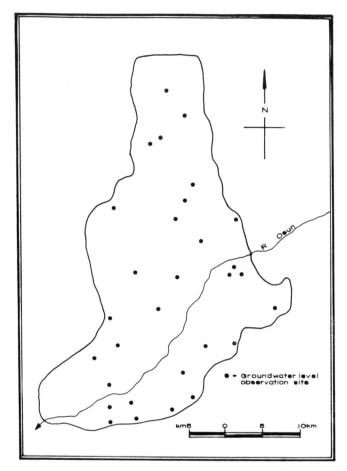

Figure 2   Location of groundwater level monitoring
           sites in the Middle Osun Valley.

Data analysis and results

Suitable descriptive and inferential statistical methods were applied
to   analyze   the data obtained on rainfall and depth to the   regolith
groundwater   table in the two areas studied.   In each study area   the
pattern   of the groundwater level fluctuation was remarkably   similar
at   all the observation well sites.   Consequently the mean   value   of
depth   to   water table for each month of the period of the study   was
reliably   obtained   in   each study area by dividing the   sum   of   the
sample data in each month from all the observation sites by the total
number of sites. The data thus obtained are presented in Table 2 from
which   the   pattern   of the temporal variation in   the   magnitude   of
depth   to the regolith groundwater table in the two areas during   the
periods of study could be deciphered.
    As   they   are   analogous to mean monthly areal rainfall the data   in

Table 2 Mean monthly values of depth (meters) to regolith
groundwater table in the study areas

Month	Middle Osun Valley			Modakeke-Ife		
	1974	1975	1976	1981	1982	1983
January	5.21	5.18	5.20	4.62	4.60	4.63
February	5.55	5.56	5.54	4.94	4.91	4.98
March	5.33	5.32	5.34	4.75	4.70	4.74
April	4.94	4.85	4.90	4.33	4.36	4.35
May	4.32	4.31	4.34	3.64	3.59	3.61
June	3.67	3.62		2.85	2.82	2.80
July	3.02	3.00		2.21	2.20	2.24
August	3.33	3.36		2.52	2.54	
September	3.66	3.71		2.87	2.82	
October	3.95	3.98		3.17	3.11	
November	4.32	4.34		3.56	3.48	
December	4.71	4.75		4.04	4.12	

the table thus became a strong platform in this study for (i)
comparing the degree of variability of the rainfall regime and that
of the regolith groundwater regime, (ii) correlating rainfall and
depth to water table, (iii) undertaking a graphical illustration of
the relationship of the rainfall and groundwater regimes, (iv)
performing a time-series analysis of depth to the regolith
groundwater table using a polynomial regression technique, and (v)
estimating the proportion of the mean annual rainfall used for
regolith groundwater accretion.

An analysis of variance and the t-test of significance (Davis,
1973; Walpole, 1968) revealed that in each study area the annual
sample data sets on mean monthly rainfall did not differ significantly among the study years. A similar result was also obtained in the
analysis of the sample data on depth to the regolith groundwater
table in each of the two areas. Using the Lorenz-curve technique as
demonstrated by Hammond and McCullagh (1977), it was also found that
rather than being evenly distributed the monthly rainfall in each
area was markedly concentrated within a number of months and thus
revealing that a markedly wet season and a markedly dry season
prevailed annually in the two study areas. By this technique the
index of rainfall concentration, having zero and unity as its minimum
possible and maximum possible values respectively, was computed as
0.81 for the Middle Osun Valley and 0.79 for Modakeke-Ife. With
respect to depth to water table the values of its index of temporal
concentration are 0.66 for the Middle Osun Valley and 0.62 for
Modakeke-Ife. These results indicate quite clearly that, barring any
major climatic change, the seasonal variability of rainfall
associated with the currently prevailing climatic regime in the study
areas is largely consistent from year to year and that the regolith
groundwater regime responds spontaneously to this climatic
variability.

A bivariate least-squares correlation analysis further revealed
the remarkable responsiveness of the regolith groundwater regime to
the prevailing rainfall regime in the two areas studied. The values

of the product—moment coefficient of correlation between mean monthly
areal precipitation (P in mm) and depth to the water table (H in mm)
are −0.83 for the Middle Osun Valley and −0.80 for Modakeke-Ife.
Both of these coefficients are highly significant. The corresponding
simple regression equations are

$$H = 5.2644 - 0.0072P \tag{1}$$

for the Middle Osun Valley, and

$$H = 4.5905 - 0.0069P \tag{2}$$

for Modakeke-Ife. The inverse relationship indicated by these
equations is perfectly in order as, under natural conditions, depth
to water table tends to decrease (i.e. the groundwater level rises)
as more rain falls and more water is made available for recharging
the regolith aquifers (Omorinbola, 1986b). A graphical illustration
of this relationship as shown in Figure 3 for the Middle Osun Valley
provides a clear visual impression of this geohydrologic phenomenon.
    A time-series analysis of the sample data on depth to the regolith
groundwater table was undertaken using the technique of polynomial
regression (Davis, 1973) and following the guidelines prescribed by
Cooley and Lohnes (1971). For the purpose of this analysis, the
monthly depth to the water table (H) was the response variable
regressed on the independent or predictor variable t, (t − time in

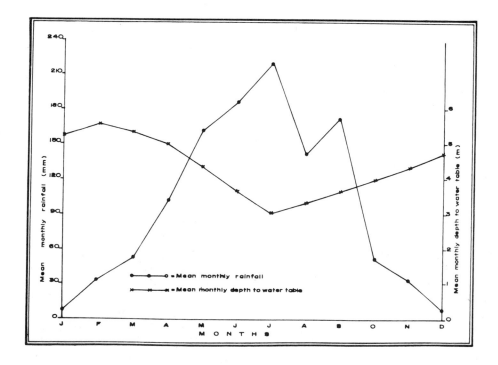

Figure 3   Relationship between mean monthly rainfall and mean
           monthly depth to water table in the Middle Osun Valley.

months), using the data in Table 2. In both study areas the maximum values of the dependent geohydrologic variable occurred in February while the minimum values occurred in July. The predictor variable was assigned values of consecutive positive integers from 1 to n (i.e. $1 \leqslant t \leqslant n$) with 1 representing the February of the first year of the observation period and n, a positive integer, corresponding to the February of the last year of the observation period. Thus for the Middle Osun Valley the t values range consecutively from 1 for February of 1974 to 25 for February of 1976 while for Modakeke-Ife the values range from 1 for February of 1981 to 25 for February of 1985.

The equations listed in Table 3 were computed using a polynomial regression computational algorithm available at the Ife University Computer Centre. For each of the study areas the fourth-degree equation is the optimal polynomial regression equation for the sample data set on the mean monthly depth to the regolith groundwater table. Both equations are statistically highly significant as indicated by the values of their goodness of fit (R^2) and the corresponding F-test statistic. In both study areas the computed quartic regression equations amply describe the temporal variation pattern of depth to the regolith groundwater table which is perfectly in sympathy with the prevailing rainfall regime. Figure 4 illustrates this temporal variation pattern of depth to the water table in Modakeke-Ife.

Table 3 Polynomial regression equations of the temporal pattern of mean monthly depth to water table during the study periods

Study Area	Polynomial regression equation (H_t = depth to water table in metres at time t)	Goodness of fit (R^2)	F-test statistic
Middle Osun Valley	$H_t = 7.71711 - 1.87516t + 0.28939t^2 - 0.01669t^3 + 0.00032t^4$	0.83664	18.76029*
Modakeke-Ife	$H_t = 7.41665 - 2.10833t + 0.32667t^2 - 0.01890t^3 + 0.00036t^4$	0.85022	19.29451*

* Significance level: $P < 0.01$.

Applications of study results

The results obtained from this study show clearly that the regolith groundwater regime in the study areas responds spontaneously to the variability of the prevailing climatic regime of which the most spectacular parameter is the seasonally incident rainfall. Although the time periods covered by the study may be considered relatively short for monitoring rainfall and groundwater level variations, the results are nevertheless reliable as the study periods can each be regarded in statistical sense as a random and independent sample of years from the population of years with characteristically similar variations in climatic and geohydrologic regimes. The quantitative methodology adopted in this study provides an objective basis for

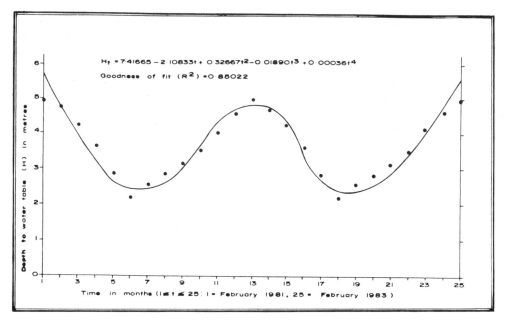

Figure 4   Temporal   variation pattern of mean monthly depth to water
table in Modakeke-Ife.

identifying   the   remarkable   similarity in the   variability   of   the
climatic   regime and the regolith groundwater regime,   the nature and
strength   of   the relationship between rainfall and   depth   to   water
table,   and   fluctuation   associated   with   the   variability   of   the
comtemporary climate.   The spontaneous response of the   geohydrologic
environment to the prevailing rainfall regime provides a good example
of how a dynamic equilibrium is maintained within physical hydrologic
systems   (Ward,   1975).   It also has,   particularly for inland   water
resources   management,   several important practical   implications   of
which the most readily outstanding are briefly examined below.

This   study   throws   considerable   light   on   the   possibility   of
accomplishing   a   reliable   quantitative   evaluation   of   groundwater
recharge   in humid tropical regoliths,   especially by the   method   of
°groundwater   evaluation from first principles' (Omorinbola,   1984).
For example,   in the Middle Osun Valley where the climatic variability,
as   indicated by the prevailing rainfall regime,   has produced in the
regolith   an   annual   groundwater recharge zone of   2.40m   on   average
(Omorinbola,   1986b);   the quantity of drainable water content in the
zone can be directly computed if reliable data are available on   such
other relevant hydrogeologic parameters like the specific yield,   the
areal extent and volume of the saturated regolith.

As   a   strong   inverse   correlation   exists   between   mean   monthly
rainfall   and mean monthly depth to the regolith groundwater table   a
good   opportunity   is   offered   under   this   condition   for   reliably
undertaking a rainfall-recharge correlation in deeply weathered humid
tropical   terrains.   Also   the   proportion   of   the   available   annual
rainfall   that   contributes   to   groundwater   accretion   in   regolith

overburden can be estimated. For the Middle Osun Valley the proportion has been calculated as approximately 36%. This information can provide an objective basis for upholding or refuting largely subjective views of some workers (e.g. Asseez, 1972) on groundwater recharge in humid tropical regoliths. It is also vital in providing appropriate guidelines for regulating the exploitation of the regolith aquifers with a view to preventing over-abstraction and its numerous attendent severe consequences (Todd, 1980).

Since the temporal variation pattern of depth to the regolith groundwater table is optimally described by empirical equations indicates that the mean depth to the water table in an area can be reliably predicted at a given point in time and the rate of groundwater level fluctuation within the weathered mantle estimated by means of differential calculus. This is vital information required in deeply weathered tropical terrains for correctly timing the sinking of water wells so as to increase their success ratios and ensure adequate and perennial yields (Buchan, 1963; Watt and Wood, 1977).

References

Adefolalu, D. O. & Oguntoyinbo, J. S. (1985) On rainfall distribution and agricultural planning in Nigeria. Malaysian J. Trop. Geogr. 11(1), 1-11.

Asseez, L. O. (1972) Rural water supply in the Basement Complex of western State, Nigeria. Hydrol. Sci. Bull. 17(1), 97-110.

Buchan, S. (1963) Geology in relation to groundwater. J. Instn. Water Engrs. 17(3), 153-164.

Cooley, W. W. & Lohnes, P. R. (1971) Multivariate Data Analysis, pp.77-78. John Wiley & Sons, New York.

Davis, J. C. (1973) Statistics and Data Analysis in Geology, pp.106-110. John Wiley & Sons, New York.

De Swardt, A. M. J. (1955) The geology of the country around Ilesha. Bull. Nigeria Geol. Surv., No.23.

Faniran, A. (1974) The extent, profile, and significance of deep weathering in Nigeria. J. Trop. Geogr. 38(1), 19-30.

Faniran, A. & Omorinbola, E. O. (1980a) Evaluating the shallow groundwater reserves in Basement Complex areas: a case study of southwestern Nigeria. J. Mining & Geol. 17 (1), 65-79.

Faniran, A. & Omorinbola, E. O. (1980b) Trend surface analysis and practical implications of weathering depths in Basement Complex rocks of Nigeria. Nigerian Geogr. J.23 (1 and 2), 113-136.

Fosberg, F. R., Garnier, B. J. & Kuchler, A. W. (1961) Delimitation of humid tropics. Geogr. Rev. 51, 333-347.

Hammond, P. E. & McCullagh, M. J. (1977) An introduction to Quantitative Techniques in Geography, pp.145-146. Ardold, London.

Loehnert, E. P. (1981) Groundwater quality aspects of dug wells in southern Nigeria. In: Studies in Environmental Science (ed. W. vanDuijvenbooden, P. Glasbergen & H. van Lelyveld), pp.147-153. Elsevier, Netherlands.

Omorinbola, E. O. (1982) Verification of some geohydrological implications of deep weathering in the basement Complex of Nigeria. J. Hydrol. 56 (2),347-368.

Omorinbola, E. O. (1984) Groundwater resources in tropical African regoliths. In: Challenges in African Hydrology and Water Resources (Proc. Harare Symp. July, 1984), 15-23. IAHS Publ. no.144.

Omorinbola, E. O. (1986a) Pattern and persistence of groundwater level fluctuation in humid tropical regoliths.Geo Journal 12, 423-431.

Omorinbola, E. O. (1986b) Empirical equations of groundwater recharge patterns. Hydrol. Sci. J. 31 (1), 1-11.

Todd, D. K. (1980) Groundwater Hydrology (2nd. edn.), pp.363-364. Wiley & Sons, New York.

Thomas, M. F. (1966) Some geomorphological implications of deep weathering patterns in crystalline rocks in Nigeria. Trans. Inst. Brit. Geogr. 40, 173-193.

Udo, R. K. (1970) Geographical Regions of Nigeria, pp.2-3, Heinemann, London.

Walpole, R. E. (1968) Introduction to Statistics, pp.228-229. Collier-Macmillan, London.

Ward, R. C. (1967) Principles of Hydrology (1st. edn.), pp.270. McGraw-Hill, London.

Ward, R. C. (1975) Principles of Hydrology (2nd. edn), pp.8-13. McGraw-Hill, London.

Watt, S. B. & Wood, W. E. (1977) Hand Dug Wells and Their Construction, pp.18-20. Intermediate Technology Publications Ltd., London.

The Influence of Climate Change and Climatic Variability on the Hydrologic Regime and Water Resources (Proceedings of the Vancouver Symposium, August 1987). IAHS Publ. no. 168, 1987.

Long-term changes in water quality parameters of a shallow eutrophic lake and their relations to meteorologic and hydrologic elements

H. Behrendt & R. Stellmacher
Academy of Sciences of the GDR
Institute of Geography and Geoecology
Department of Hydrology
Muggelseedamm 260, Berlin, 1162, GDR
M.Olberg
Humboldt-University Berlin
Section of Physics
Department of Meteorology and Geophysics
Muggelseedamm 256, Berlin, 1162, GDR

ABSTRACT Changes of climatologic conditions influence not only hydrophysical, but also hydrochemical and hydro-biological parameters of a water body. These parameters of water quality are connected with meteorologic and hydrologic factors directly or indirectly.
    Time series were collected from lake Muggelsee, the main reservoir for the drinking water supply of Berlin, G.D.R. Long-term changes of the following water quality parameters of this shallow eutrophic lake and meteorologic and hydrologic elements were analysed by statistical methods: total seston content, chemical oxygen demand, chloride concentration, oxygen content and water temperature of lake Muggelsee, discharge of the river Spree at Beeskow, global radiation in Potsdam and sum of negative daily mean temperature in Berlin.
    One-channel and multi-channel autoregressive spectral analyses are used to check the spectral behaviour of these parameters and to test for the existence of relations between water quality and meteorologic or hydrologic parameters respectively. For the time period 1946-1984 the following periodicities were detected: short periods between 2 and 3.5 years, middle periods at 5 years and long periods between 12 and 20 years.

RESUME Un changement de conditions climatiques influence directement ou indirectement non seulement les paramètres hydrophysiques, mais aussi les processus hydrochimiques et hydrobiologiques des eaux. Par la suite on étudie les liasons entre ceux-ci en utilisant des méthodes statistiques.
    Citons l'example du lake Müggelsee, situé près de Berlin R.D.A. Les séries chronologiques de séston, de D.C.O, de concentration en oxygène et en chlorures, de rayonnement à Potsdam, de sommes des températures moyennes journalières négatives à Berlin, de température de l'eau

et   du débit à Beeskow couvrant la période de 1946 à   1984
servent   de matériel de base.   On a effectué l'analyse   de
séries   chronologiques en utilisant les méthodes d'analyse
spectrale autorégressive uni- et multidimensionelle.

Les   résultats les plus importants,   valables   pour   la
periode   1946-1984,   sont les suivants: Trois des séries-
celles   du   séston,   du   rayonnement   global   et   de   la
concentration   en chlorures - possèdent une   tendance.   En
général,   les   allures   des   fonctions   spectrales   se
ressemblent   plus ou moins.   On y trouve des   périodicités
entre 2 et 3.5 ans (toutes séries considérées),   d'environ
5 ans (séston,   concentration en oxygène,   température   de
l'eau,   rayonnnement global, débit),   de 8 ans (température
de l'eau et rigueur de l'hiver),   de 12 ans (D.C.O.)   et
d'entre 15 et 20 ans (séston,   concentration en oxygène et
en chlorures, débit).

L'analyse   de   cohérence   caractérisant   les   relations
entre   différentes séries dans le domaine de la   fréquence
met en évidence une corrélation entre certaines bandes   de
fréquence et differentes combinaisons de facteurs.

Introduction

At   present   the majority of investigations in the field   of   aquatic
ecosystems concentrate on the improved understanding of processes.   A
deeper   knowledge   of   processes controlling the   balances   of   water
quality   state variables and their modelling is a necessary   supposi-
tion for ecosystem management, although not sufficient.

The complexity of processes,   interactions and feedbacks is neither
completely   explicable   nor describable   on the basis of   our   current
knowledge.   It must be pointed out that the analysis of an ecosystem,
including   not   only   anthropogenic   reactions   but   also   natural
variations,   is not possible if based solely on the study of   short-
term processes.

The   analysis of long-term fluctuations of the ecosystem with   the
support of long time series of state variables ought to be considered
if   the development of water quality is to be estimated for the   past
and   for   the future.   But the reality is quite different.   The   main
reason   for this contradiction is the absence of long time series   of
water quality parameters in most cases.   Only a few time series   have
been published with length greater than twenty years (Thomas,   1971;
Lund,   1972; Lund 1978; Ahl, 1979; Makarewicz & Baybutt, 1981; Schanz
& Thomas,   1980;   Sutcliffe et al.,   1982; Zimbalewskaya, 1983; Brooks
et al., 1984).

Long   term variations of ecosystem state variables might be caused
by human,   meteorologic and hydrologic factors.   Separation of   the
effects of these factors seems to be possible only if time series are
available   for   the   water   quality parameters,   and   for   the human,
meteorologic   and hydrologic factors too.   In most cases such   series
for   all   factors do not exist and authors postulate that   long-term
trends   are   caused by human activities and short-term variations   by
meteorologic or hydrologic factors.

An approximate comprehensive analysis of long-term behaviour of water quality parameters exists only in Ahl (1979) and Zimbalewskaya (1983). We could not find similar investigations in the literature for shallow polymictic lakes even though strong dependences of ecosystem parameters on meteorologic and hydrologic factors are to be expected especially for lakes of this type. This paper concentrates on the analysis of these dependences in a shallow lake using statistical methods.

Material and methods

The object of our investigations is Lake Müggelsee, a shallow eutrophic lake in Berlin, G.D.R., with a mean depth of 4.85 m and surface area of 720 hectares. The mean residence time is about 40 days. Lake Muggelsee is the main reservoir for the drinking water supply of Berlin. A detailed description of the lake is given by Schellenberger (1981).

Time series for the following biological and chemical parameters of lake water were available:
- total content of particulate material (total seston content)
- chemical oxygen demand (C.O.D.)
- chloride concentration and oxygen content.

The time series consisting of daily measurements span the 39 years from 1946 to 1984 . Chemical analyses were carried out according to Legler, 1969. Annual means of water quality parameters were used in the statistical analyses. The time series of total seston content and chloride concentration exhibit trends, and therefore the residuals from the trend functions were analysed. This is preferable, because our interest is in the relations between water quality parameters and meteorologic or hydrologic factors. On the other hand the length of our time series is too short to discover periodicities greater than 20 years.

Figure 1 shows the time series of four water quality parameters. Data of C.O.D. and oxygen content were standardized by the maximum of the record. Annual means of global radiation in Potsdam, water temperature of lake Müggelsee, discharge of the river Spree at Beeskow and the sum of negative daily mean temperature in Berlin represented meteorologic or hydrologic factors. The time series of global radiation in Potsdam was published and discussed by Olberg (1979) who found, among other things, a long period with a length of about 78 years. In the time span considered here the negative half wave of this period appears as a trend (see Figure 2), therefore this trend was eliminated.

The series of the sum of negative daily mean temperature in Berlin was taken from Knoch (1947) and Tiesel (per.comm.). All time series of meteorologic and hydrologic elements are shown in Figure 2.

Univariate autoregressive spectral analysis and multichannel autoregressive coherence spectral analysis were applied as statistic methods for the investigations. These methods were described in detail by Olberg & Rakoczi (1984), Olberg & v.Schoenermark (1983) and Olberg (1982).

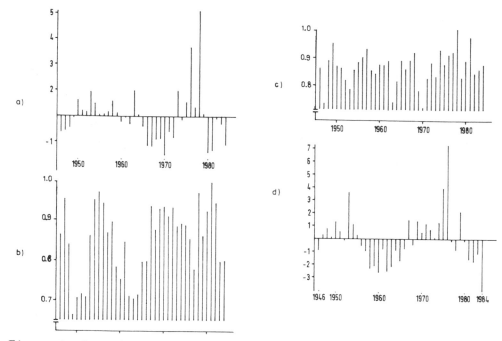

Figure 1   Annual   values   of water quality parameters for   the   time
period   1946-1984.   a) Deviations of the   seston   content
from   a   harmonic   trend;   b) Relative   values   of   C.O.D.
(standardized on the maximum; c) Relative values of oxygen
content (standardized on the maximum; and d) Deviations of
chloride concentration from a linear trend.

Results

Figures   3 and 4 represent the result of the one channel   autoregressive spectral analysis of water quality,   meteorologic and hydrologic
parameters.   Several   spectral   peaks   appear in the spectra   in   all
cases.   The   95   per   cent   significance level is marked   to   aid   in
distinguishing   significant   and   not   significant   peaks   and   the
periodicities of the different time series are listed in Table 1.
Altogether   three   groups of periodicities were detected:
(a) short periods with a spectral energy peak at 2.0 - 3.4 years,
(b) middle periodicities between 4.7 and 5.7 years and
(c) long periods with a length between 12 and 20 years.
The spectral energy peak of 8 years for the water temperature and the
sum of negative daily mean temperature seems to occupy an exceptional
position.
Some spectral features are shared by the water quality   parameters
although   it   is evident from Figure 3c that the spectral   energy   of
oxygen   content is concentrated more on shorter periodicities and   of
the   others on the long periodicity.
It appears from a visual comparison of Figure 3 and Figure 4   that
strong   connections exist between the discharge,   the seston   content
and the chloride concentration.   Multichannel autoregressive spectral
analysis was used to test this hypothsis. Up to present the coherence

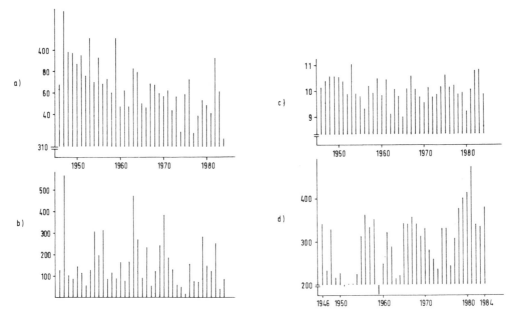

Figure 2 Annual values of meteorologic and hydrologic elements of 1946-1984. a) Global radiation in Potsdam; b) Sum of negative daily mean temperature in Berlin; c) Water temperature of Lake Müggelsee; and d) Discharge of the river Spree at Beeskow.

Table 1 Structure of different time series for the period 1946-84

Parameter	Periodicities		
Total seston content, trend	2.4 yr+,	5.0yr,	15 yr
Oxygen content	3.2 yr,	5.7yr,	16 yr
Chloride concentration, trend	3.4 yr,		20 yr
C.O.D	3.2 yr,		12 yr
Water temperature	2.0 yr,	4.7yr,	8 yr
Discharge in Beeskow	2.3 yr,+	5.0yr+,	17 yr
Global radiation in, trend, Potsdam	2.0 yr,	3.2yr,	5 yr+
Sum of negative daily mean temperature in Berlin.	3.2 yr,		8 yr

+ periodicities are not significant at the significance level of 95%

analysis has been carried out for seston content on the one side and the meteorologic factors with the exception of global radiation on the other side. The coherence connection between the seston content and the global radiation is to be reported in Behrendt et al. (1987).

Figure 5 shows the hypothesised coherence between total seston content and discharge to be significant for periods of about 5 and 15 years (Figure 5c). Figure 5b shows significant spectral coherences

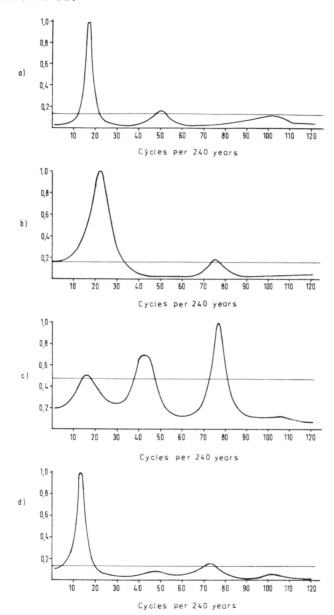

Figure 3   Standardized   autoregressive   spectral functions   of   time
series   of a) Deviations of seston content   from   harmonic
trend;   b) C.O.D.;   c) Oxygen content; and d) Deviations of
chloride concentration from a linear trend.

between   the   seston content and the sum of the negative   daily   mean
temperature for periods near to three years and for the long spectral
energy peak too. Figure 5a shows seston content and water temperature
to   be   significantly coherent only for   the   five-year   periodicity
although long periodicity coherence approaches significance.

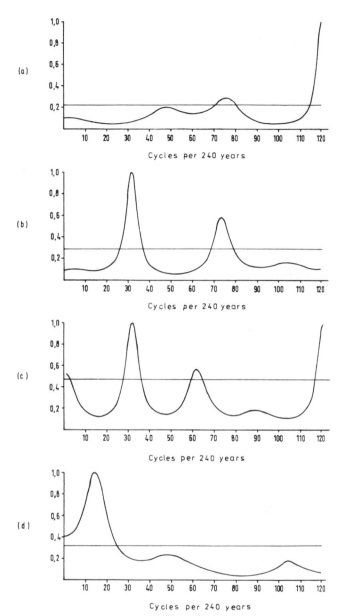

Figure 4   Standardized autoregressive spectral functions of time series of meteorologic amd hydrologic elements: a) Deviations of global radiation from alinear trend;   b) Sum of negative daily mean temperature;   c) Water temperature; and d) Discharge.

It must be considered that the coherence between seston content and meteorologic elements were achieved only on the basis of phase lags of different magnitudes. These phase lags amount to 5.2 and 6.5 years for the coherences between seston content and discharge or sum

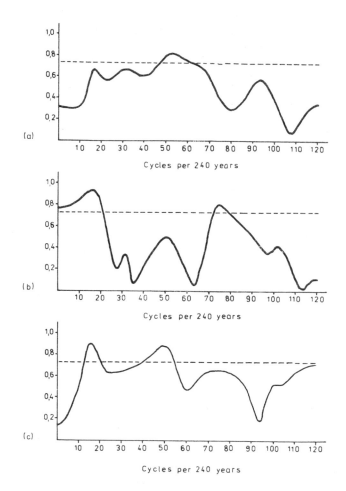

(a)

Cycles per 240 years

(b)

Cycles per 240 years

(c)

Cycles per 240 years

Figure 5   Coherence relations between the deviations of seston content from a harnomic trend and a) water temperature, b) sum of negative daily mean temperature, and c) discharge.

of negative daily mean temperature respectively in the range of a long period. Significant coherence between seston content and discharge in the range of the period of about five years is induced by a phase lag of 2.3 years. The other visible coherences (see Figures 5a, b, c) exist without phase lags.

Discussion

The following restrictions must be considered for a proper evaluation of the results of spectral energy and coherence analysis:
   (a) The spectral behaviour of the time series depends on the time period considered and on the form of the trend function.
   (b) Because the length of the time series is only 39 years, the identified long periodicities are insecure in comparison to shorter spectral energy peaks.
   Shifts of spectral behaviour depending on the period of record may

be obtained from the global radiation data. Olberg (1979), for example, estimated for the time series of global radiation in Potsdam from 1893 to 1976, spectral energy peaks of 78, 22, 4.7 and 2.2 years. Our results (Table 1) differ from these considerably. Investigations of the sum of the negative daily mean temperature in Berlin showed for different time periods that a significant periodicity occurs with a length greater than 8 years if the time series is extended from 1946 to 1936.

Our analyses confirm that strong connections seem to exist between long-term behaviour of water quality parameters of a shallow lake and meteorologic or hydrologic elements. This general statement may be deduced alone from the results of one channel autoregressive spectral analysis (Figures 3 and 4 and Table 1).

In connection with the estimation of human influences on water quality, it must be considered that all water quality parameters show long spectral energy peaks, too. On the basis of this result it is questionable whether anthropogenic changes in water quality may be deduced from trend investigations. Such a procedure seems to be dangerous if the time series is less than about five years.

Several authors have reported similar results for different water bodies. Zimbalewskaya (1983) concluded on the basis of analysis of zooplankton time series, that the short-term periodicities of ecosystem parameters seem to be caused by meteorologic or regional climatic elements. Similar statements were published by Lund (1972) and Lund (1979). Ahl (1979) pointed out that the 6-7 year periodicity of several water quality parameters corresponds to a spectral peak in the water level of Lake Malaren suggesting strong relations between climatic and water quality parameters.

This statement is confirmed by our coherence analysis between seston content and meteorologic and hydrologic elements. The middle and long periodicities of seston content show sufficient coherences to the discharge and the water temperature. Coherences with global radiation appears in the short period range (Behrendt et al. 1987).

If phase lags are taken into consideration, then it is possible to distinguish between direct and indirect effects of meteorologic factors. A direct connection between seston content and discharge with a phase lag of more than two years would be explicable only if the travel time of water between Beeskow gauging station and Lake Muggelsse was of such a magnitude, however measurements indicate a travel time of a few days only.

On the other hand a direct influence of water temperature on the seston content may be assumed present because the coherence analysis shows no phase lags. The same assumption is possible for the coherence between seston content and the sum of the negative daily mean temperature in the short period range. Negative daily mean temperature is an indicator of the ice period on the lake. Both factors, ice period and water temperature, influence the seston content in the same direction. Cold winters (high sum of negative daily mean temperature or ice period and low annual mean water temperature) indicate a low seston content. This result supports the assumption, that the seston content is determined by meteorologic conditions of winter and spring (see Behrendt et al.,1987).

Further investigations must be carried out on the influence of different time periods of water temperature on the seston content.

AKNOWLEDGEMENT   We are indebted to colleagues of the Berlin Freidrichshagen water works for collecting and providing the data in water quality parameters and water temperature.

References

Ahl, T. (1979) Natural and human effects on trophic evolution. Arch. Hydrobiol.Ergebn.Limnol. 13,259-277.

Behrendt,H., Olberg,M. & Stellmacher,R. (1987) Investigations on long-term changes of the seston content in a shallow eutrophic lake and their relations to meteorological conditions. Acta hydrophys. 31,(in press).

Brooks, A.S., Warren,G.J., Boraas,M.e., Scale,D.B. & Edginton,D.N. (1984) Long term phytoplankton changes in lake Michigan: Cultural eutrophication or biotic shifts. Verh. Internat. Verein. Limnol. 22, 452-45 9.

Knoch,K. (1947) Uber die Stenge der Winter in Norddeutschland nach der Berliner Beobachtungsreihe 1766-1947. Metorol.Rundschau 1, 137-140.

Legler, C. (ed.) (1969) Selected methods of Water Analysis, Vol. 1. Gustav-Fischer-Verlag, Jena (in German).

Lund, J.W.G. (1972) Eutropication. Proc. R.Soc. Lond.B. 180, 371-382.

Lund, J.W.G. (1978) Changes in the phytoplankton of an English lake, 1945-1977. Hydrobiolical Journal 14, 6-21.

Makarewicz, J.C. & Baybutt, R.I. (1981) Long-term (1927-1978) changes in the phytoplankton community of lake Michigan at Chicago. Bull. Torrey Bot.Club 108, 240-254.

Olberg, M. (1979) Langperiodische Schwankungen der Strahlunsbilanz, des Niederschlages und des Trockenheitsindex in Potsdam. Z.Meteor. 29, 50-55.

Olberg, M. (1982) Statistische Analyse meteorologischerklimatogischer Zeitreihen. Abn.meteorol. Dienst d.DDR 17 (128), 129-141.

Olberg, M. & v. Schoenermark, M. (1983) Numerische Experimentemit Koharenz- und gleitenden Maximum-Entropie-Spektren. Z.Meteor. 3, 111-115.

Olberg, M. & Rákóczi, F. (1984) Informationstheorie in Meteorologie und Geophysik. Akademie Verlag, Berlin.

Schanz, F. & Thomas, E.A. (1980) Die Durchsichtigkeit des Zuricher-Seewasserss von 1897-1980. Vierteljahresschr. Naturforch. Gesell. Zurich 125(3), 239-248.

Schellenberger, G. (1981) Hydrologie und Okologie des Muggelsees. Geographische Berichte 99 (2), 115-122.

Sutcliffe, D.W., Carrick, T.R., Heron,J.,Rigg,E.,Talling,J.F., Woof,C. & Lund, J.W.G.(1982) Long-term and seasonal changes in the chemical composition of precipitation and surface waters of lakes and tarns in the English Lake District. Freshwater Biology 12, 451-506.

Thomas, E.A. (1971) Oligotrophierung des Zurichsees. Vierteljahresschr. Naturforsch.Gessell.Zurich 116, 165-179.

Zimbalewskaya, L.N. (1983) Successions, monitoring and predictions of water ecosystems. Hydrobiological Journal 21, 3-9. (in Russ.)

The Influence of Climate Change and Climatic Variability on the Hydrologic Regime and Water Resources (Proceedings of the Vancouver Symposium, August 1987). IAHS Publ. no. 168, 1987

Hydrologic design criteria and climate variability

Robert A. Clark,
Adjunct Professor
Department of Hydrology and Water Resources
University of Arizona
Tucson, Arizona, USA

ABSTRACT   The effects of climate change on hydrologic de-sign criteria for flood events are discussed. In particu-lar, change could impact the magnitude of Probable Maximum Precipitation estimates easily by 10 to 20 per cent. Such increases would in turn, severely affect the design of large reservoirs. Climatic change effects of moderate frequency floods are also discussed.

La variabilité climatique et les critères des projets hydrauliques

RESUME   Les conséquence des changements climatiques sur les critères normales des projets concernant les problèmes de débordement sont discutées. En particulier, de tels changements peuvent facilement exiger un impact de 10 à 20 pourcent sur l'estimation de la précipitation maximale probable. De telles augmentations peuvent, à leur tour, affecter très sérieusement les projets de grands réservoirs. Une discussion sur les débits d'inondation de fréquence modérée est aussi présentée.

Introduction

We will limit our discussion to hydrologic design criteria based on flood values and not to long term design of water resource systems. These criteria are of importance in spillway design, determination of reservoir flood control capacity, reservoir surcharge and the design of flood control units.
    Landsberg (1958) states "one may define climate as the collective state of the earth's atmosphere for a given place within a specified interval of time." He further points out that collective states are in contrast to individual states which we commonly call "weather." For hydrologic design we normally consider the climatic element-rainfall. Depending on the criteria for design one may be concerned with means or possibly, extremes. Further, the definition relates to a specific locality and to a specified period of time-which incorporates the concept of "place" and the possibility of climatic variability. When one considers extreme events, current technology that employs a hydrometeorolgic approach involves maximization of a

545

combination of all of the appropriate physical parameters involved in the development of the extreme values. This approach normally utilizes the concept of developing a Probable Maximum Flood (PMF) based on Probable Maximum Precipitation (PMP). Design of low hazard projects utilize conditions based on events with moderate (100 years or less) return period.

The problems attendant to these design concepts in the light of possible climatic variability and/or change will be discussed. Since extreme events typically involve probabilities of occurrence of 10^{-3} to 10^{-6} , the return periods involved are longer than recorded history and may relate to geologic time – certainly since the Pleistocene Epoch which includes at least four ice ages in the last 450,000 years. Events of moderate frequency present a much different problem, viz., that associated with an inadequate period of hydro-meteorologic sampling.

Chinese scientists have assembled data on dryness/wetness during the last 5,000 years in northern and northeastern China. Their studies have indicated periods of average temperature 2°C higher than at present and January temperatures 3°C to 5°C higher than today (Daniel, 1980). The current or Halocene Epoch represents an inter-glacial period that has lasted 10,000 years and may last another 10,000 years before the beginning of another ice age. Thus, it appears that it is climatic fluctuations on a time scale of centuries rather than millenia that are particularly relevant to current activities of man.

Extreme events

The importance of designing for rare events has been emphasised by the failure of a rather large number of dams over the last century. Table 1 lists the causes of dam incidents from 1900 through 1979- over-topping representing 26 percent of reported failures and 13 percent of all accidents plus failures.

Table 1   Causes of Dam Incidents (Natural Research Council, 1983)

Cause	Concrete F	Concrete A	Embankment F	Embankment A	Other F	Other A	Totals F	Totals A	F+A
Over-topping	6	3	18	7	3		27	10	37
Flow Erosion	3		14	17			17	17	34
Slope Protection Drainage				13				13	13
Foundation Leakage	5	6	11	43	1		17	49	66
Embankment Leakage			23	14			23	14	37
Other Causes	5	10	11		3				
Total	19	19	77	163	7		103	182	285

F = Failure   A = Accident – an accident in which failure was averted by remedial action

In most instances over-topping was the result of inadequate spillways. However, a few over-topping failures were due to spillway blockage and structural settlement.

Probable Maximum Flood (PMF) estimates for design were developed during that same period from (1) design judgement based on historical facts and engineering judgement, to (2) a period of using regional discharge measurements to develop essentially an envelope curve approach and/or statistical frequency analysis, then (3) the use of storm transpiration techniques that included enveloping rainfall rather than discharges, to (4) the practice of using Probable Maximum Precipitation (PMP) to develop a PMF.

The PMP concept represents a judgement by experts on the largest rainfall that "could happen" on a particular basin (Myers, 1986). It is a hypothetical but plausible storm that could occur over the basin (NRC, 1985), but no estimates of risk or probability are possible, or even plausible. Because of the method of derivation, PMP application to the derivation of PMF makes the risk issue even more intractable.

The National Dam Inspection Program (PL 92-367) developed an inventory of about 68,000 dams in the United States that were classified according to their potential loss of life and property damage (U.S. Army Corps of Engineers, 1982). Almost 8,800 of these dams were considered "high hazard", many of which needed remedial actions. In a large number of cases the problem lay with inadequate spillway capacity. Since about one-half of the dams inspected were built prior to 1960, a large portion were built to hydrologic design criteria somewhat lower than current PMP standards.

The PMP concept and climatic variability

Myers (1967) prepared an excellent summary of the evolution of spillway design flood determination from what he calls the "early period" (based primarily on judgement) to the PMP period. He presented two definitions for PMP:

(1) The theoretical greatest depth of precipitation for a given duration that is physically possible over a particular drainage area at a certain time of the year. In practice this is derived over flat terrain by storm transpiration and moisture adjustment to observed storms (AMS, 1959).

(2) PMP is that magnitude of rainfall over a particular basin which will yield the flood flow of which there is virtually no risk of being exceeded.

Both definitions emphasize the concept of a physical upper limit in nature with essentially zero probability of occurrence. He discussed physical limitations in meteorlogic terms on rate of precipitation over a basin as follows:

(1) A limit on the humidity concentration above the basin or maximum moisture.

(2) A limit on the inflow wind velocities.

(3) A limit on the efficiency of the precipitation process.
Let's examine each of these in terms of climatic variability.

Maximum Moisture Atmospheric moisture is generally measured in terms of the liquid equivalent of the total water vapor in a verticle column – or precipitable water. Since most of the atmospheric moisture originates over the ocean, Table 2, slight changes in the sea-surface temperatures can have a large impact on atmospheric moisture.

Table 2   Global Water Movements in the Hydrologic Cycle (Nace, 1976)

Location	Evaporation km^3 / year	Percent of Total	M / year
Oceans	449,000	86	1.23
Land	72,000	14	0.48
Total	521,000	100	1.71
	Precipitation		
Oceans	416,000	80	1.14
Land	105,000	20	0.72
Total	521,000	100	1.86

A temperature rise of $1^{o}C$ increases the saturation vapor pressure over the sea about 10 percent. Since the instantaneous global atmospheric moisture averages about 13,000 km^3 , it is obvious that atmospheric moisture must be replaced approximately 40 times during the year, or once each 9 days, to account for the total volume of precipitation or 521,000 km^3 . Thus, a slight increase in evaporation could significantly increase the total atmospheric moisture and total precipitation. Further, since the process for derivation of PMP assumes moisture maximization, increased sea surface temperatures could raise the PMP values significantly. A model developed by Showalter and Solot (1942) indicated an increase in precipitation by about 9 percent for a $1\ ^{o}C$ increase in the sea level dewpoint temperature.

The current technique for adjusting storms is to use the ratio of maximum moisture to storm observed moisture. Thus a slight change (1° or $2^{o}C$) in sea surface temperatures over a period of a few centuries could significantly impact PMP values – easily by more than 10 percent. That could represent 40 to 50 mm for a 24 hour storm over 25,000 km^2 in the southestern United States (approximately $1.25 \times 10^9 m^3$ or 1×10^6 acre feet).

Recent projections related to the increase of carbon dioxide (CO_2) in the atmosphere indicate that an increase of 25 percent over the next 20 years could increase the global temperatures by about $0.5^{o}C$. If allowed to continue increasing at that rate, the next century could see temperatures warmer than at any time in the past 1000 years. Such an increase in global temperature would, of course, significantly impact sea surface temperatures.

Maximum Inflow Winds   The maximum amount of moisture of precipitable water in the atmosphere rarely exceeds 85 mm. It is obvious that since 24 hour precipitation amounts in the United States in excess of 300 mm over 25,000 km^2 have been observed, that inflow wind velocities to the major storms must be significant. For smaller areas, less than 100 km^2, precipitation amounts in 24 hours have exceeded 800 mm – again the maximum atmospheric moisture is less than 85mm.

Unfortunately, wind velocities have been measured adequately over the United States for only about 45 years. Certainly, not long enough to give a good indication of maximum velocities that might occur during a storm that produces PMP. Atmospheric covergence is, of course, related to inflow wind velocities. Convergence in the surface layers is highly correlated with the production of precipitation. Convergence, however, is even more difficult to measure than winds over the generally small sized river basins requiring spillway design floods.

Meyers (1967) pointed out that a solution to these difficulties has been to use storm precipitation itself as a means of measuring or estimating convergence. Through storm transposition it is assumed that very intense storms have been observed with intensity sufficient to approach "maximum" conditions. Convergence associated with some of these storms should approach the maximum that nature can produce. Comparison of generalized estimates of PMP with maximum observed storms (Riedel and Schreiner,1980) has revealed that 56 storms east of the 105th meridian in the United States have exceeded 50 percent of PMP estimates for a duration of 24 hours and 50,000 km^2. Four storms exceeded 70 percent. For areas as small as 500 km^2 three storms exceeded 80 per cent. These storms have undoubtly approached covergence conditions approaching a maximum in nature. It is of interest that many of these large storms were associated with hurricances in either an active or decadent stage. Since an important factor in hurricane formation is the sea-surface temperature, 26° to 27° C or greater, the effect on hurricanes of increased sea-surface temperatures due to climatic variation is difficult to postulate.

Precipitation efficiency   The mechanism by which water vapor is converted into liquid water and precipitation is well understood but difficult to evaluate in terms of efficiency. Studies by Gilman and Peterson (1958) of east coast storms has indicated that as much as 80 percent of inflow moisture is converted into rainfall during major storm situations. Again it has been postulated that a number of major storms have been observed that approach natures maximum efficiency. Thus, transposition of storms from one location to another with similar climatologic characteristics should account for the problem of designing for maximum efficiency. The question that arises: How are transposition limits determined –particularly during a period of climate change? Current technology is obviously very subjective with regards to both limits and change.

It is of interest that while evaporation may be almost of the same magnitude in both Continental Polar (CP) and Maritime Tropical (MT) air masses, it has been estimated that only 10 percent of Mississippi Valley precipitation comes from CP air masses and 90 percent from MT air masses (Gilman, 1964). The mechanism for conversion of precipitation is obviously somewhat more efficient in MT air. An inadequate storm sample from which to base maximum convergence and efficiency values and our inability to recognize transportation limits may create more serious errors in PMP estimation than small climatic variations over periods of several centuries.

Moderate frequency events

An interesting discussion of the sensitivity of water resource systems to climate variation has been prepared by Klemes (1985). In his paper he cited a number of studies that have examined the impact of climatic variations on their design. His analyses were primarily concerned with utilizing hydrologic modelling to simulate streamflow runoff under conditions of climatic change of such variables as air temperature, precipitation, evapotranspiration and snow cover. He points out that simulation models based on a data base of no greater than 20-years cannot be expected to simulate adequately, say a 100 or 200 year flood. Thus, for flood design, one should not expect simulation models to reveal effects associated with climate variability with any degree of confidence.

Normally, for moderate frequency events, stationary climatic conditions are assumed, i.e., physical conditions that do not change with time. The Inter-agency Advisory Committee on Water Data (1982) prepared detailed guidelines for determining flood flow frequency. They point out:

> There is much speculation about climatic changes. Available evidence indicates that major changes occur in time scales involving thousands of years. In hydrologic analysis it is conventional to assume flood flows are not affected by climatic trends or cycles..

They recommend data should be adjusted for short, incomplete and broken records, outliers, mixed populations (snow and rain floods) and zero flood years. For flood records, watershed changes due to man's activities may also prove serious. All of these factors may cause problems in the estimation of flood flow frequencies-problems of a more serious nature than variable climatic conditions.

One of the most interesting studies related to hydrometeorologic record length was that by Hurst (1951) who examined the variability of flows on the Nile River and their effect on long-term storage. His studies showed a relationship between record length and variability that indicated some long-term persistence in natural time series. Numerous studies have been made by many investigators to attempt to replicate mathematically Hurst's results or what is frequently referred to as the Hurst "phenomenon." Whether this phenomenon really exists for any hydrologic event let alone extreme precipitation events is difficult to say. Certainly the concept of using a stationary time series is of importance in defining hydrologic design for moderate frequencies. Numerous "100" year floods have been observed over some drainage basins in the past 20 years. So numerous as to raise doubts on the validity of using past historical events to determine the magnitude and frequency of these "so called" 100 year events. One only has to consider the billions of dollars of storm and cross-drainage structures constructed each year to understand the importance of attempting to standardize and stabilize hydrologic criteria for their design.

References

American Meteorological Society, <u>Glossary</u> <u>of</u> <u>Meteorology,</u> ed. by

Ralph E. Huschke, Boston, Mass., 1959.

Daniel, Howard, "Man and Climatic Variability", report no. 543, the World Climate Program, World Meteorological Organization, Geneva, 1980, 32 pp.

Gilman, C.S. and K.R. Peterson, "Northeastern Floods of 1955: Meteorology of the Floods," Journal of the Hydraulics Division, ASCE, HY3: 1661, June 1958.

Gilman, Charles S., "Rainfall," Section 9 in Handbook of Applied Hydrology, Edited by Ven Te Chow, McGraw-Hill Book Company, New York, 1964, pp. 9-1 to 9-68.

Hurst, H.E., "Long-term storage capacity of reservoirs", Trans. Am. Soc. of Civ. Eng., 116; 770-799, 1951.

Interagency Advisory Committee on Water Data, "Guidelines for Determining Flood Flow Frequency," Hydrology Subcommittee, Bulletin # 17B, Revised September 1981, Editorial Corrections, March 1982.

Klemes, V., "Sensitivity of Water Resource Systems to Climate Variations," World Climate Program, Report WCP-98, World Meteorological Organization, Geneva, Switzerland, May 1985.

Landsberg, H., Physical Climatology, Second Edition, Gray Printing Co., Inc., Dubois, Pa., 1958, 443 pp.

Myers, V.E., "Meteorological Estimation of Extreme Precipitation for Spillway Design Floods," Weather Bureau, Washington, D.C., 1967, 29 pp.

Myers, V.E., "PMP for Spillway Design-Origin and Evolution of Practice," Paper presented at the fall Meeting of American Geophysical Union, San Francisco, California, December 1986.

Natural Research Council, "safety of Dams-Evaluation and Improvement," Committee on the Safety of Existing Dams, National Academy Press, Washington, D.C., 1983, 354 pp.

National Research Council, "Safety of Dams- Flood and Earthquake Criteria," Committee on Safety Criteria for Dams, National Academy Press, Washington, D.C., 1985, 321 pp.

Riedel, John T. and Louis C. Schreiner "Comparison of Generalized Estimates of Probable Maximum Precipitation with Greatest Observed Rainfalls," NOAA Technical Report NWS 25, Office of Hydrology, National Weather Service, Washington, D.C., March 1980, 66 pp.

Showalter, A.K. and S.B. Solot, "Computation of Maximum Possible Precipitation," Trans. of the American Geophysical Union, 1942. Part 2, pp. 258-274.

**Man's influence on the
hydrologic regime**

The Influence of Climate Change and Climatic Variability on the Hydrologic Regime and Water Resources (Proceedings of the Vancouver Symposium, August 1987). IAHS Publ. no. 168, 1987.

Influences possibles activités humaines modifiant le climat sur le cycle hydrologique

André Junod
Institut suisse de météorologie
CH-8044 Zurich, Suisse

RESUME Voici quelque 25 ans, les variations climatiques ont été reconnues comme un facteur technique majeur auquel doivent faire face les gestionnaires des ressources en eau. Comment faut-il projeter et exploiter des ouvrages hydrauliques censés durer 100 ans et plus en tenant compte des modifications climatiques prévisibles ou possibles?

Quittant l'hypothèse de stationnarité du climat sur laquelle les hydrologistes se sont longtemps appuyés, on met en évidence les différentes forms des variations climatiques, aux différentes échelles de temps et d'espace: tendances, cycles à long terme, changements abrupts. Compte tenu des besoins spécifiques des hydrologistes en matière de caractérisation statistique, on introduit les éléments climatiques déterminants dans la fonction de transfert climat – processus hydrologiques – ressources en eau.

Alors que bon nombre des variations climatiques du passé sont attribuées aux processus internes du système climatique couplé – et sont encore imprévisibles – les modifications climatiques d'origine humaine retiennent toujours plus l'attention et, en tant que processus externes, sont accesibles à la modélisation numérique. On passe en revue, en signalant la fiabilité actuelle des évaluations, les causes et mécanismes des principales atteintes de l'hame au climat: effets globaux du gaz carbonique et des autres gaz à effet de serre, déboisement accéléré des forêts tropocales, modifications étendues de l'état du sol dues à l'urbanisation, à l'industrialisation et à certaines pretiques culturales. On souligne l'importance des chaînes de rétroaction, tant compensattoire (élevation de température – nébulosité) qu'amplificatrices (désertification – albedo – subsidence – sécheresse).

L'exposé se termine par des considérations stratégiques sur l'utilisation de scénarios d'évolution possible du climat et de ses incidences, sur l'estimation des risques encourus ainsi que sur les besoins et limites en matière de décisions. Un facteur-clé de la maîtrise (relative) du futur réside dans la recherche et le développement des connaissances et méthodes d'application relatives au climat, où l'on doit veiller à répartir judicieusement les ressources investies dans les grands programmes internationaux et dans les investigations nationales.

The possible impacts of climatic variations
on water resources due to human activities

ABSTRACT   About   25 years ago climatic variations were
recognized as major technical factors which had to be
considered by the water-resources managers. How are water
resources projects, which are supposed to work a century
or even longer, to be designed and operated, taking into
account the predictable or possible climatic   modifications?

The assumption of climatic stability, on which hydrologists have founded their work for a long time, must now
be abandoned and the various forms of climatic variations
for different scales in time and space be considered:
trends, long term cycles, abrupt changes. Considering the
specific needs of the hydrologists in the field of statistical description, the determining climatic elements are
introduced in the transfer function climate - hydrologic
processes - water resources.

While a certain number of climatic variations of the
past - which are still not predictable - are assigned to
internal processes of the coupled climatic system, climatic variations due to human activities increasingly
attract attention and, as external processes, are accessible for numerical modeling. One reviews and first points
out the reliability of actual evaluations, the causes and
mechanisms of the main human interferences on climate:
global effects of carbon dioxide and of other greenhouse
gases, rapid deforestation of tropical forests, important
modifications of the soil due to urbanisation, to industrialisation and to certain agricultural practices. The
importance of feedback processes either compensatory
(increase of temperature - nebulosity) or amplifying
(desertification - albedo - subsidence - drought) are also
pointed out.

Some   considerations on strategic   utilisation of
possible evolution scenarios of climate and their impacts
on the estimation of incurred risks and the needs for and
limits of decision making conclude the paper. A key factor
for a (relative) control of the future is the research and
development of the knowledge and of application methods in
relation to climate where the involved resources have to
be judiciously distributed between the major international
programmes and the national investigations.

A complete version of the paper will be presented
at the symposium.

*The Influence of Climate Change and Climatic Variability on the Hydrologic
Regime and Water Resources* (Proceedings of the Vancouver Symposium,
August 1987). IAHS Publ. no. 168, 1987.

A methodology for distinguishing between the effects of human influence and climate variability on the hydrologic cycle

Jens Chr. Refsgaard
Danish Hydraulic Institute
Agern Alle 5, DK-2970 Horsholm, Denmark

ABSTRACT   The hydrologic cycle is influenced both by the
climate and by the catchment characteristics such as
topography, geology, soil, vegetation, water development
etc. A methodology is presented for distinguishing between
the effects of man's influence and climate variability on
the   hydrologic cycle,   in particular streamflow and
groundwater. The methodology is based on the application
of deterministic hydrologic models. It is illustrated in
two case studies in which a lumped, conceptual and distributed, physically based model, respectively, are applied
to predict the effect of climate variability (sequences of
wet and dry years) and of human activity (groundwater
development) on the aquifer piezometric heads and streamflows.

Méthodologie pour différencier entre les
activité conséquences de l'humaine et la
variabilité climatique sur le cycle hydrologique

RESUME   Le cycle hydrologique est influencé autant par le
climat que par les caractéristiques du bassin (le relief,
la géologie, les sols, la végétation, les développements
des ressources en eau etc.).   Une méthodologie est
présentée pour différencier entre les effets de l'activité
humaine et la variabilité climatique sur le cycle hydrologique, en particulier les débits des eaux de surface et
souterraines. La base de cette méthodologie est l'application des modèles déterministes hydrologiques.   Les
conséquences de l'activité humaine (développement des eaux
souterraines) et la variabilité climatique (succession des
années humides et seches) sont examinées selon deux études
de cas. Ces illustrations utilisent respectivement un
modèle global conceptuel et un modèle physique distribues
pour prévoir les influence sur les hauteurs piézométriques
et sur les débits des rivières.

Introduction

With the increase in population, agriculture and industrial
production, man's influence on the hydrologic cycle is becoming more
and more important. In addition to the direct influence caused by
land use changes, river regulation and groundwater development, for

557

example, man's influence on the climate will also affect the hydrologic cycle. The term climate change is used in this paper when long-term (>100 years) climate variables are significantly changed, while climate variability denotes natural variations from year to year.

Long period climate variability, as for example, drought and wet periods, has an effect on water resources which is often larger than the direct effect of man's activity. Long historic time series can be analysed with the purpose of making inferences regarding climate variability and its effects on the hydrologic regime. In this connection it is necessary to be able to identify and to eliminate the effects of man's activity.

It is therefore important to be able to detect historic hydrologic changes as well as to predict future changes as a consequence of both climate changes and man's activity. Many well-documented examples are given in the literature regarding detection of changes, (see Unesco 1980). There are also many examples of predictions of the effects of man's activity. However, only a few have been able to test the outcome of the prediction in real life subsequently. In one such test Konikow and Person (1985) studied the accuracy of a 10 year old prediction showing a long-term increase in salinity due to man's activity (irrigation). They concluded that the original prediction was invalid, because it did not account for a short-term annual trend in the river flow caused by climate variability, which took place during the one-year calibration period. There are only a few examples of prediction of the effects of climate changes. Some of these studies are reported by Klemes (1985), who stresses the importance of appropriate model testing in connection with such non-observable predictions.

A variety of techniques can be used for detection of hydrologic changes ranging from Box-Jenkins models, (see Hipel et al (1977) and Sharma (1985)), to deterministic rainfall-runoff and groundwater models, (see Fleming (1975), Refsgaard (1980), and Hyden et al (1980)). On the other hand, the techniques which are applicable for the much more difficult predictions of future changes are basically restricted to the advanced deterministic models of the distributed, physically based type, with the exception that the effect of minor climate changes also can be predicted by the use of the lumped, conceptual models.

This paper presents an attempt to establish a methodology for distinguishing between the effects of man's influence and climate variability on the hydrologic cycle, in particular on streamflow and groundwater. The methodology is here confined to the application of deterministic hydrologic models.

Deterministic hydrologic models

The deterministic hydrologic models which can be utilized for this purpose are able to simulate (part of) the land phase of the hydrologic cycle over a period of several years. In these models the catchment characteristics, including the man-induced changes of land use and water development are represented somehow in the model parameters, while the climate variables are the main input data (for example, meteorologic time series).

Deterministic models can be classified according to whether the model gives a lumped or a distributed description of the considered area, and whether the description of the hydrologic process is empirical, conceptual, or fully physically based. As most conceptual models are also lumped and as most fully physically based models are also distributed, the following three main groups of deterministic models emerge:
- Empirical models (black box)
- Lumped, conceptual models (grey box)
- Distributed, physically based models (white box)

One group of black box models which may be used for detecting hydrologic changes are some of the statistically based time series models, for example the Box-Jenkins models, which in some instances can be utilized as deterministic transfer models to simulate streamflow series from meteorologic time series, for example.

The best known watershed model of the lumped, conceptual type is probably the Stanford Model, cf. Crawford and Linsley (1966). During the decade following the development of the Standford Model several other models of the same type were developed. Many of these models are applied on an operational basis today for a variety of tasks.

A first attempt to outline the potentials and some of the elements in a distributed, physically based model was made by Freeze and Harlan (1969). Today, several physically based models treating only single hydrologic processes have been developed and extensively applied. Almost all groundwater models for instance are of the distributed, physically based type. Examples of models covering the entire land phase of the hydrologic cycle are reported by Weeks et al (1974), Wardlaw (1978), Refsgaard and Hansen (1982), and Miljostyrelsen (1983). These models are all more or less specific purpose models while the SHE model, cf. Abbot et al (1986) and Storm (1986) is one of the few general purpose modelling systems which today are applicable under a wide variety of hydrologic and hydrogeologic conditions.

Methodology

All the three deterministic model types can be applied for the detection of historic hydrologic changes whereas only the distributed, physically based models can be used for predicting future changes due to man's activity. In the following outline of the methodology the statistically based black box models are not considered.

The lumped, conceptual models

The lumped, conceptual models can be utilized for analyzing the hydrologic effects of climate variability. In this way, they can implicitly be used for testing whether observed changes or trends in hydrologic time series can be explained entirely by the experienced climate variability. These models would typically be applied in cases where the observed changes/trends in the hydrologic time series occur simultaneously with both man activity and climate variability, and where it is not evident whether the hydrologic changes/trends are caused by either the activity or the climate variability (or both).

In such a case, the modelling approach would be as follows:
(a) Select two time periods denoted, for example the reference

period and the test period for the modelling analyses. The reference and   test   periods would be periods before and after   the   hydrologic changes and the human activity occurred, or vice versa. The reference period   should,   if   possible,   comprise   both dry   and   wet   climate segments.   The   length of the two periods should be sufficient for   a proper model calibration and validation,   i.e. at least in the order of   3-10   years depending on the particular model and the   particular catchment.

(b)   Calibrate   and test the model using historic   data   from   the reference   period.   Preferably a so-called 'differential split-sample test' should be carried out as follows, cf Klemes (1986). If the test period   is   generally more wet than the reference   period the   model should   be calibrated on a dry segment and validated on a wet segment of   the reference period,   or vice versa. If it is not   possible   to identify   such   segments with significantly   different   climate parameters   within the reference period the model should be tested in a substitute catchment in which a differential split-sample test   can be   carried out,   and an ordinary split-sample test should be carried out in the catchment of interest.   Only by such a comprehensive model testing,   it   can   be ensured that the calibrated model   is   able   to simulate   other   periods with the same degree of accuracy as   in   the calibration period,   provided no changes in catchment characteristics take place.

(c)   Make   model   simulations for the test period by use   of   the parameters   obtained from calibration on the reference period and   by use of the climate data from the test period.

(d)   Use   an   adequate statistical test   to   analyse   whether   the simulated   hydrologic time series from the test period can be assumed to be identical to the recorded time series from the same period.   If this hypothesis is accepted,   all the observed changes /trends in the hydrologic time series can be explained by climate variability. If on the other hand,   the hypothesis is rejected it can be concluded   that some   significant   changes   in   the   catchment   characteristics   have occurred.   Further,   the   effect   of these catchment changes   can   be estimated as the differences (where significant) between the recorded and the simulated time series.

(e) If it is concluded that some significant changes in   catchment characteristics   have occurred further analyses may be carried out to explain   these changes qualitatively as consistent with the   observed human activity.   Such analyses may include a model recalibration   for the   test   periods and a hydrologic interpretation of the changes   in model   parameters.   If some obvious inconsistency emerges   here,   the entire analysis (a)-(e) has to be reconsidered.

The above procedure is illustrated in Case Study 1. It should   be noted   that a lumped,   conceptual model cannot beforehand predict the hydrologic   changes of human activity,   because the model   parameters cannot be assessed directly from catchment information, but has to be estimated through calibration.

The distributed, physically based models

The distributed, physically based models can be utilised for   analyzing the hydrologic effects of both climate variability and human activity. As those models can explicitly account for the influence of human activity, they can be applied for more comprehensive analyses as compared to the lumped, conceptual models. The modelling analyses would typically be carried out in cases where changes or trends have been observed in the hydrologic time series. The modelling approach outlined below is slightly different from the approach for the lumped, conceptual models:

(f) Select two time periods, the reference and the test period, respectively, as outlined in (a) above.

(g) Calibrate and validate the model using historic data from the reference period. With regard to the test of the models' ability to predict the effects of climate changes the procedure outlined in (b) above is used.

(h) Make model simulation for the test period. The human-induced changes in catchment characteristics are accounted for directly in the model parameters. This model simulation serves as a test of the model's ability to predict the effects of changes in catchment characteristics, another test of the differential split-sample type.

(i) Analyse the effects of the man-made changes by comparing two simulations, in which the man-induced changes are incorporated in one, but not in the other simulations. The analyses can be carried out for both the reference and the test period.

(j) Analyse the effects of climate variability by comparing two simulations with the same catchment characteristics from the reference and the test period, respectively.

The above procedure is illustrated in Case Study 2. Thus, by using a distributed, physically based model a clear distinction can be made between the effects of climate variability and human influence. Furthermore, it is possible to quantify the effect of both, and it is possible, for example, to simulate the hydrologic time series which would have occurred if the human-made changes had not taken place.

Contrary to the lumped, conceptual models, analyses can be made with the distributed, physically based models in cases, where sufficient historical data do not exist from a period before the human-induced changes started. In this case, only one simulation period is possible for model calibration and validation, and this period is also applied for analysing the effects of the human-made changes, cf. (i) above. If further historical climate time series exist, the effects of climate variability can also be analysed, cf. (j) above. However, it should be noticed that a model validation based on these data alone is not sufficient here. Hence, it is necessary further to carry out the two difference split sample tests for climate change and catchment mentioned in (b) and (h) above in other catchments.

Case study 1: The Graese Å catchment

This case study illustrates the application of a lumped, conceptual rainfall-runoff model for analysing the hydrologic consequences of groundwater development. For more details about the study reference is made to Refsgaard (1980).

The Graese Å catchment is situated approximately 35 km north-west of Copenhagen and covers an area of 26 sq.km. Geologically, the area is composed of a prequaternary confined limestone aquifer, overlayered by glacial deposits of a predominantly morain clay and alluvial sand. The groundwater development, which started in 1955, takes place from the upper zone of the limestone aquifer.

The hydrologic model applied for this study was a traditional lumped, conceptual rainfall-runoff model, namely the NAM-model originally developed by Nielsen and Hansen (1973). The NAM-model has been applied to numerous catchments throughout the World under a wide range of climate conditions.

The model application can be described briefly according to the above procedures (a) to (e) as follows:

(a) Hydrologic and meteorologic data were collected for two periods, viz 1950-1955 and 1969-1974, which were characterized by no groundwater abstraction and abstraction respectively. The hydrologic data indicated a significant reduction in the low flows in the later period, whereas the meteorologic data indicated that the average annual rainfall was about 15% lower in the later period as compared to the earlier, reference period.

(b) The model was calibrated and tested on the data from the reference period, cf. the simulation shown in Figure 1, differential split-sample test as required in (b) above was carried out within this study. However, such tests showing the model's ability to predict the effects of climatic variations have been carried out on other catchments.

GRÆSE Å

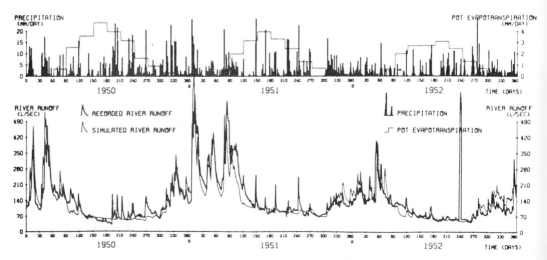

Figure 1   NAM calibration result for Graese Å before the groundwater development started (reference period) (Refsgaard, 1980).

GRÆSE Å

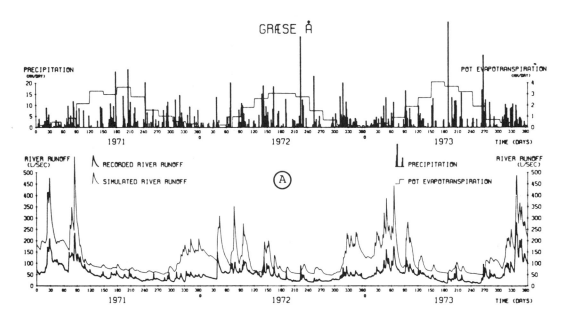

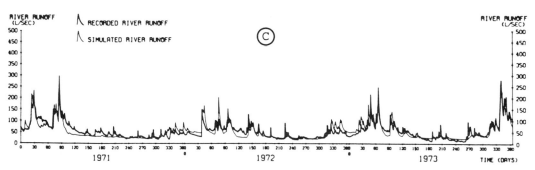

Figure 2   NAM simulations for Graese Å after the start of the
groundwater development (test period) using the same parameters as in the reference period (A), and using the
parameters calibrated from the test period (C).
Figure from Refsgaard (1980).

(c) Using the same model parameters as from the reference period
the simulation for the test period was carried out. Part of this
simulation is shown in Figure 2A.

(d) The simulated hydrograph in Figure 2A represents the river
runoff which would have been expected, if no groundwater abstraction
had taken place. The recorded runoff is seen to be significantly
reduced as an effect of the abstraction. The effect of the groundwater abstraction can be estimated as the difference between the two
hydrographs in Figure 2A.

(e) The model was calibrated on the test period with the results
shown in Figure 2C. The changes in the model parameters indicated,
not surprisingly, that the part of the catchment where groundwater

flow discharged into the river was reduced significantly. Further-more, it was concluded that the groundwater recharge was increased significantly as an effect of the groundwater abstraction and the lowering of the piezometric heads in the lower confined aquifer.

Case study 2 : The Susa catchment

This case study illustrates the application of a distributed, physically based model for analysing the hydrologic consequences of groundwater development. For more details about the study, reference is made to Refsgaard and Hansen (1982).

The modelling area including the catchments of Susa and the neighbouring catchments Køge Å, Vedskolle Å and Tryggevaelde A cover about 1000 sq.km. A regional confined aquifer is overlayered by glacial deposits, predominantly consisting of moraine clay. A groundwater development was initiated in 1964 in the north-eastern part of the area and gradually increased to 16 mill. m^3/year in the mid 1970's.

A distributed, physically based model was developed for the Susa catchment. The model is a result of an integration of an integrated finite difference groundwater model, an aquitard model, a model for the unconfined phreatic aquifers and root zone model. The model area, the topographic divides and the groundwater model polygonal mesh are shown in Figure 3.

The model application can briefly be described according to the above procedures (f) - (j) as follows :

(f) Hydrologic and meteorologic data were collected for the entire period 1950-1980 for which a continuous model simulation was carried out. For the present example data from the Køge Å catchment were selected for two periods 1961-1963 and 1974-1978 as test period (without groundwater abstraction) and reference period (with abstraction), respectively. The average annual rainfall was about 10% lower in the reference period as compared to the test period. The streamflow data showed a significant reduction of the low flows.

(g) + (h) The model tests were not carried out with Klemes (1986) requirements in mind. However, tests of comparable strengths were carried out on the very comprehensive data material available. The model was calibrated on data from the period 1974-1980 and subsequently validated on the remaining period 1950-1973. The groundwater model for the regional confined aquifer was calibrated for the entire area, while the calibration of the surface water component was only based on data from three small subcatchments of the Susa covering about 200 sq.km. The calibrated parameters were then successfully transferred to the other catchments This was possible, because the most important model parameters related to soil, vegetation, and topography were fully physically based and could be obtained directly from maps, field measurements, etc. Thus the model validation on Koge A for the period 1961-1963 is a very powerful test, because the model was partly non-calibrated and because the calibrated part (confined aquifer) was calibrated on significant changed groundwater conditions due to the start of the groundwater abstraction within the catchment in 1964. For the surface water component this model test belongs to the type which Klemes

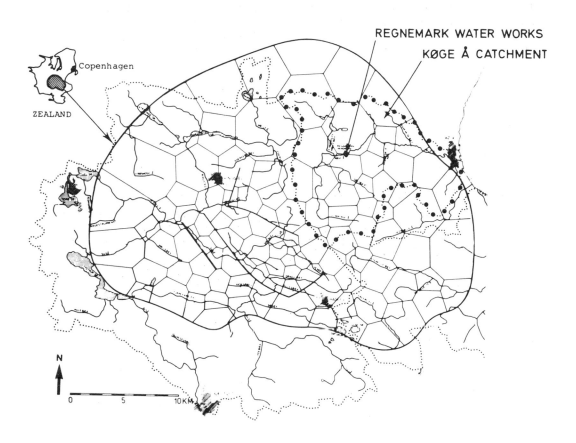

Figure 3   Model area, topographic divides, and groundwater polygonal
mesh. Figure from Refsgaard and Hansen (1982).

(1986) denotes 'proxy-basin differential split-sample test.' The
model validation was carried out on both discharge data and piezometric heads. In addition comprehensive tests were successfully carried
out on some of the individual model components by use of soil
moisture data from different vegetation types and discharge data from
tile drains. Simulation results for the two periods are shown in
Figure 4 and Figure 5 for streamflow and piezometric heads respectively. The influence of the groundwater abstraction is noticed in
terms of zero streamflow in low flow situations and a significant
decrease in heads close to the water works.

(i) The effect of the groundwater abstraction was evaluated by
comparing the two streamflow simulations shown in Figure 6, in which
Q_{sim}^{1} is the streamflow, as it would have been increased since 1980.
The difference $Q_{sim} - Q_{sim}^{1}$ is thus the streamflow depletion caused
by the increased groundwater abstraction. All streamflow values in
Figure 6 have been smoothed by a 15-day moving average filter, in
order to remove high frequencies and thus to clarify the seasonal
pattern.

(j) 1975-1976 was a drought period in Denmark with average annual
rainfall 30% below normal. Hence, there was a very intense discussion

KØGE Å

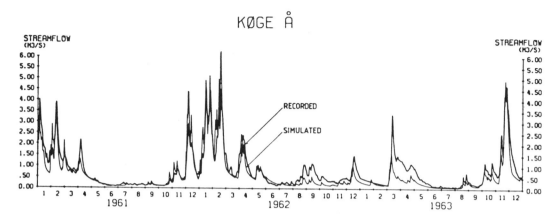

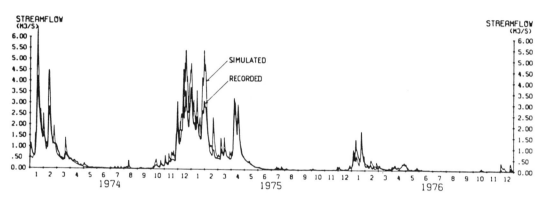

Figure 4   Comparison of simulated and recorded streamflow for the 133 sq. km Køge Å catchment. The streamflow in the period 1974-1976 is influenced by a comprehensive groundwater abstraction within the catchment, started in 1964. Figure from Refsgaard and Hansen (1982).

about whether the drying out of the rivers in the summer months, which was first really noticed by the public in 1975 and 1976, was due to the dry summers or due to the groundwater abstractions, which had then reached full capacity. The effect of the climate variability can be evaluated by comparing the Q^1_{sim} simulation of the very dry 1976 (Figure 6) with the hydrographs from the period 1961-63 (Figure 4). Q^1_{sim} shows a minimum flow in August 1976 of about 55 l/s whereas the minimum flows of 1961, 1962 and 1963 are 65 l/s, 140 l/s and 65 l/s respectively. 1962 was a wet year and 1961 and 1963 had about average rainfall. Thus, it can be concluded that extreme low flow (=0) of the river in 1976 is primarily due to the groundwater development in the area, while the climate variation from a normal year (1961, 1963) to a very dry year is only responsible for a relatively small part of the flow reduction, i.e. from 65 l/s to 55 l/s.

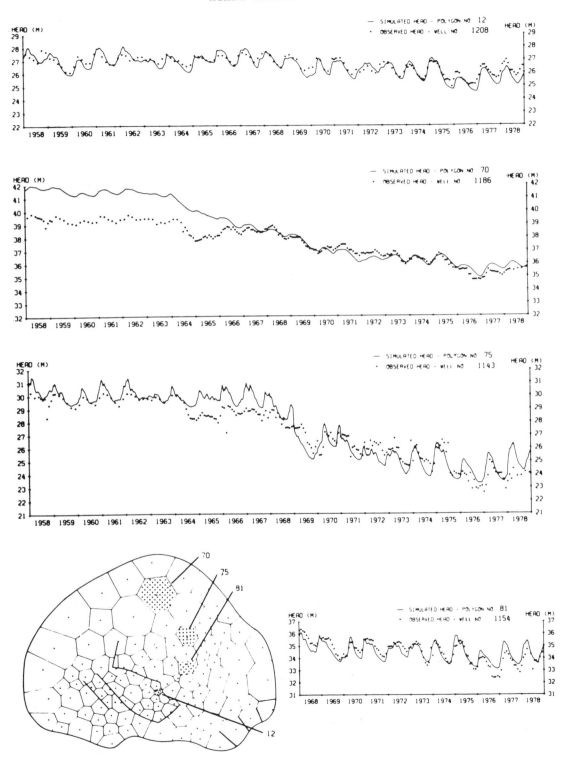

Figure 5   Comparison of simulated and observed hydraulic heads of the confined aquifer. Figure from Refsgaard and Hansen (1982).

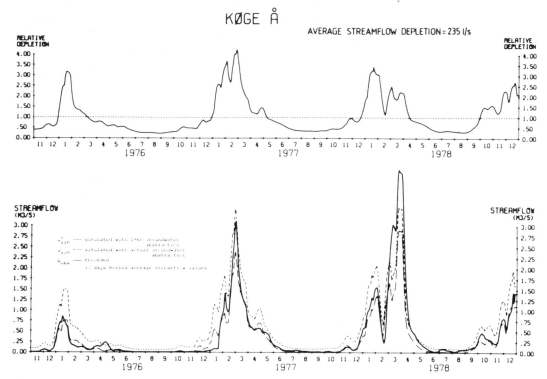

Figure 6   Comparison of 15 day moving average streamflows for Køge Å
1976-1978 (lower) and the relative streamflow depletion
caused by the groundwater abstraction (upper). Figure from
Refsgaard and Hansen (1982).

Summary and conclusions

A methodology is presented for the application of deterministic
models for distinguishing between the effects of human's influence
and climate variability in the hydrologic cycle. The lumped,
conceptual models are applicable for detecting historical hydrologic
changes and for predicting the effects of minor climate changes. The
distributed, physically based models are applicable both for
detection and for prediction of future changes due to climate change
as well as human activity.

In case of major climate changes, the lumped conceptual models are
not considered to give a sufficiently correct simulation, because
the climate change may have important effects on vegetation cover,
land use, extent of wetlands, etc. In principle, these secondary
effects can be accounted for in a realistic way only in a ꞌistributed,
physically based model.

Some of the man-induced changes in catchment characteristics occur
gradually over a long period of time. This is typical for changes in
land use, urbanization, drainage etc., while, for example, dam

construction and groundwater development schemes often introduce relatively quick changes in the hydrologic regime. If the changes occur gradually over a period of a century or more, the approach of using lumped conceptual models is not feasible, because reliable historical data of more than a century of age are usually not available. In such a case, only a thoroughly tested distributed, physically based model would be able to predict the hydrologic changes.

It is emphasized that a comprehensive model test is crucial for the reliability of the model results for these purposes. An ordinary split sample test is not sufficient. Some of the higher level tests recommended by Klemes (1986) should be carried out.

In general, the best suited model type for this purpose is the distributed, physically based models. It is realized that setting up such a model for a large catchment is a comprehensive task requiring a lot of data and work. However, in many cases, the application of such a model is the only scientifically feasible approach for quantitatively distinguishing between the effects of man's influence and climate variability.

ACKNOWLEDGEMENT   The paper is based on work carried out under IHP Project 6.3 to which the author has been appointed as the principal rapporteur. The inspiration from the co-rapporteurs and from the division of Water Sciences, UNESCO in this connection is acknowledged.

References

Abbott, M.B., Bathurst, J.C., Cunge, J.A., O'Conell, P.E. & Rasmussen, J. (1986): An introduction to the European Hydrological System-Syteme Hydrologique European- SHE- 1: History and philosophy of a physically-based distributed modelling system, and 2: Structure of a physically-based distributed modelling system. Journal of Hydrology, 87, pp 45-77.

Box, G.E.P. and Jenkins, G.M. (1970): Time series analysis, forecasting and control. Halden-Bay Inc., San Francisco.

Crawford, N.H. and Linsley, R.K. (1966): Digital simulation in hydrology, Stanford watershed model IV. Dept. of Civ. Eng., Stanford Univ., Technical Rep. 39.

Fleming, G. (1975): Computer simulation techniques in hydrology. Elsevier, New York.

Freeze, R.A. and Harlan, R.L. (1969): Blueprint for a physically-based, digitally-simulated hydrological response model. Journal of Hydrology, 9, pp 237-258.

Hipel, K.W., Mcleod, A.I., and Mcbean, E.A. (1977): Stochastic modelling of the effects of reservoir operation. Journal of Hydrology, 32, pp 97-113.

Hyden, H, Leander, B, and Voss, C. (1980): The alnarp groundwater system- A mathematical model study. Proceedings to Nordic Hydrological Conference, Vemdalen, Sweden, Part II, pp 195-219.

Konikow, L.F. and Person, M. (1985): Assessment of long-term salinity changes in an irrigated stream-aquifer system. Water Resources Research, 21, pp 1611-1624.

Klemes, V. (1985): Sensitivity of water resource systems to climate variations. World Climate Organization, WCP-98.

Klemes, V. (1986): Operational testing of hydrological simulations models. Hydrological Sciences Journal, 31, pp 13-24.

Miljøstryelsen (1983): An investigation of the hydrological consequences of irrigation in the Karup Å catchment. The Danish Environmental Protection Agency, Miljø-projekter, No. 51 (in Danish).

Nielsen, S.A. and Hansen, E. (1973): Numerical simulation of the rainfall-runoff process on a daily basis. Nordic Hydrology, 4, pp 171-190.

Refsgaard, J.C. (1980): Hydrological consequences of groundwater development. Two case studies. Proceedings of the 6th Nordic Hydrological Conference, Vemdalen, Sweden, Volume 2, pp 343-356.

Rafsgaard, J.C. and Hansen, E. (1982): A distributed groundwater /surface water model for the Susa catchment. Part I: Model description, and Part 2: Simulations of streamflow depletions due to groundwater abstractions. Nordic Hydrology, 13, pp 299-322.

Sharma, T.C. (1985): Stochastic models applied to evaluating hydrologic changes. Journal of Hydrology, 78, pp 61-81.

Storm, B. (1986): SHE. Syteme Hydrologique Europeen. A short description. Danish Hydraulic Institute.

UNESCO (1980): Casebook of methods of computation of quantitative changes in the hydrological regime of river basins due to human activities. Studies and reports in hydrology, 28.

Wardlaw, R.B. (1978): The development of a deterministic integrated surface/subsurface hydrological response model. Ph.D. Thesis, Univ. Stratchclyde, Glasgow.

Weeks, J.B. et al (1974): Simulated effects of oil-shale development on the hydrology of the Piceance basin, Colorado. US Geological Survey, Professional Paper 908.

The Influence of Climate Change and Climatic Variability on the Hydrologic Regime and Water Resources (Proceedings of the Vancouver Symposium, August 1987). IAHS Publ. no. 168 1987.

The impact of increasing atmospheric carbon dioxide concentrations upon reservoir water quality

B. Henderson-Sellers
Department of Mathematics and Computer Science
University of Salford, Salford M5 4WT, U.K.

ABSTRACT   The potential changes in a water body resulting from increasing atmospheric carbon dioxide can be related to enhanced radiative and non-radiative fluxes (especially evaporation) at the air-water interface. The initial perturbation due to enhanced CO_2 levels can be parameterised as an increase in the downwelling longwave radiative flux expressed as a function of time and CO_2 concentration (the so-called "transient CO_2 experiment"). The only experiments presented to date relating to the impact CO_2 levels have on water resources have concentrated on the oceanic environment using simple box-diffusion models- models which cannot take into account seasonal and daily changes in water temperature and the consequent non-linear effects of heat storage that occur. To investigate these non-linear effects, an eddy diffusion thermal stratification model, which includes specification of all the meteorologic variables on timescales of 1-24 hours, has been used to investigate further the impact of rising atmospheric carbon-dioxide concentrations on the aquatic environment. For a simulation period of 50 years, the net decrease in reservoir levels is found to be 2.05m. If, simultaneously, stream inflow rates increase, this is likely to result in enhanced nutrient loading and accelerated eutrophication.

Impact d'un accroisement des concentrations de dioxide de carbone sur la qualité de l'eau dans les réservoirs.

RESUME   On peut relier les changements potentiels d'une masse d'eau résultant d'un accroissement de gaz carbonique dans l'atmosphère a l'augmentation des flux radiatifs et non-radiatifs (specialement l'evaporation) à la surface de séparation eau-atmosphère. Il est possible de paramétrer la pertubation initiale due à l'augmentation de gaz carbonique par un accroissement du flux radiatif dans les ondes longues. Ceci est fonction du temps et de la concentration en gaz carbonique ("expérience du CO_2 transitoire"). Les seules expériences, qui ont été faites jusqu à présent sur l'impact des concentrations de CO_2 sur les ressources en eau, ont été concentrées sur l'environment océanique. Dans ces expériences, on a

utilisé des modèles de diffusion simple qui ne peuvent
tenir compte des changements saisoniers et journaliers de
la température de l'eau et par conséquent des effets non-
linéaires causés par ce reservoir de chaleur. Pour étudier
ces effets non-linéaires, un modèle de diffusion par
tourbillons pour la stratification thermique a été utilisé.
Ce modèle inclue toutes les variables météorologiques sur
une échelle de temps de 1 a 24 heures. Ce modèle permet
d'étudie l'impart dun accroissement des concentrations en
CO_2 sur l'environment aquatique. Pour une simulation de 50
années, on trouve que la diminution nette du niveau de
reservoir est de 2.05 metres. Si, en même temps, il y a
une augmentation du débit des eaux affluentes, il est
possible d'observer un accroissement d'accumulation des
nutriments ainsi qu'une accélération de 1 "eutrophication"
du lac.

Notation

A	cross-sectional area of lake.
C_m	heat capacity of mixed layer (ML) in box-diffusion model
c_p	specific heat at constant pressure
K_H	eddy diffusion coefficient for heat
q	penetrative short wave energy flux
t	time
T	temperature
z	depth
ΔF	diffusive energy flux into deep ocean
ΔQ	perturbed flux due to increasing atmospheric carbon dioxide
ΔT	water surface temperature increase due to the impact of increasing atmospheric carbon dioxide
λ	feedback parameter ($W m^{-2}K^{-1}$)
λ_o	value of λ for "no-feedback" case
ρ	water density

Introduction

The climatic and hydroclimatic effects of increasing atmospheric
carbon dioxide are of great concern to mankind. Since the details of
the feedbacks operating within the Earth-atmosphere system are not
well understood, it is difficult to implement a numerical model of
the total system and research has been centred on specific areas
considered likely to be of greatest influence. In this paper a
quantitative analysis is presented of the potential impact of
increasing atmospheric carbon dioxide on water levels and water
quality in inland lakes and reservoirs, use of which is frequently
made for public water supply (cf. discussion of Coutant, 1981).

Box-diffussion models

To date, the impact on water bodies of increasing atmospheric carbon
dioxide has been largely confined to studies of the global ocean
using box models. For example, two recent studies (Wigley and
Schlesinger, 1985; Hansen et al. 1985) both utilise the simple box-
diffusion model of Oeschger et al. (1975). This model is represented,
essentially, by the equation

$$C_m \, d\Delta T/dt = \Delta Q - \lambda \Delta T - \Delta F \tag{1}$$

where the perturbed surface energy balance difference term, ΔQ, is
related to the ambient carbon dioxide concentration and its temporal
evolution. For example, Wigley and Schlesinger (1985) suppose that
over the period 1850-1980 this term is given as

$$\Delta Q = 34.6534 \times 10^{-4} t \, \exp(8.686 \times 10^{-3} t) \tag{2}$$

In any perturbation experiment, this box-diffusion model approach
contains an implicit timescale of one year. It assumes a fixed mixed
layer (ML) depth throughout each year, identical for all latitudes.
However in water bodies there exists a strong seasonal (and indeed
diel) cycle. The mixed layer (ML) depth deepens a little during the
summer but is totally absent during the winter season. At this time
strong convection can occur in the water body, thus mixing heat
downwards. These non-linearities are also evident in the seasonal
hysteresis cycle (Gill and Turner, 1976), which is observed but is
not simulated by box-diffusion models.

In addition to retaining a fixed ML depth, box models also possess
a "fixed lid"; such that there is no change in water level, nor any
possibility of calculating changes in evaporation rates. That such
changes may occur as a result of increasing atmospheric carbon
dioxide, is self evident, yet has not been simulated quantitatively.
Some indications of the potential importance on global, regional and
local water resources of changes in evaporation rates is given by
Wigley and Jones(1985). Although they make no attempt to quantify
such effects directly, in their analysis of the sensitivity of runoff
to changes in precipitation and evaporation, they highlight the
potential changes for a number of perturbation scenarios. Conversely
it can be deduced from this study that a CO_2 perturbation is likely
to affect runoff and hence inflows to lakes and reservoirs. There are
thus two impacts on reservoir levels: i) enhanced evaporation and ii)
probable increased runoff (and hence inflow).

In this paper a study is undertaken of the changes in lake level
and evaporation rate due directly to increased downwelling infrared
radiation using a high temporal and spatial resolution eddy diffusion
model.

Thermal stratification model

The lake/reservoir thermal stratification model U.S.E.D. (Henderson-
Sellers, 1985) includes a specification of meteorologic variables on
timescales of 1-24 hours and calculates the thermal structure on a

daily timescale. The heat transfer equation

$$A(z) \; \partial T / \partial t = \frac{\partial}{\partial z} A(z) \; \{\alpha + K_H(z,t)\} \; \frac{\partial T}{\partial z} + \frac{\partial(Ao)/\partial z}{\rho c_p} \tag{3}$$

is solved together with a full evaluation of the surface energy budget (Henderson-Sellers, 1986). Indeed the inclusion of the surface energy budget is vital to the successful incorporation of both diffusive and convective mixing necessary for successful longterm simulations. The model U.S.E.D. utilised here is found to be stable, with cyclic climatic forcing, over periods of centuries and hence is considered suitable for climatic simulations, as described below.

Increased atmospheric CO_2

Since the model U.S.E.D. contains a full meteorologic description of surface energy exchanges, it is relatively easy to include a ΔQ-type perturbation. Equation 2 was proposed as being representative of the impact of atmospheric carbon dioxide over the last 130 years. Extrapolating this for a simulation period of 50 years into the future gives:

$$\Delta Q = 34.6534 \times 10^{-4}(t+130) \; \exp[8.68 \times 10^{-3}(t+130)] - 1.3934 \tag{4}$$

where t runs from 0 to 50 in the experiment described below. This formulation results in perturbation flux which begins as zero and rises to 2.98 W m^{-2} after a 50 year period. In the U.S.E.D. simulations, the air is allowed to change by such an amount that the perturbed downwelling (atmospheric) infrared radiation is equal to the ΔQ value given by Equation (4). Feedbacks then occur naturally since this perturbed air temperature is itself directly responsible for a change in the vapour pressure gradient across the air-water interface. Thus evaporation (and consequently sensible) heat fluxes change, in addition to enhanced longwave radiation resulting from increased water surface temperatures. It is found that this parametrisation produces a response which is approximately two to three times greater than the "no-feedback" simulation - in good agreement with the value of a feedback ratio of 2.4 proposed by Hansen et al. (1985).

Lake level simulation

The lake simulation pertains to a hypothetical lake situated at $54^{\circ}N$, with a maximum depth of 40 m and realistic bathymetry. Values of lake level changes from the control are stored and then compared to values calculated in the perturbation simulation. At the end of the 50 year simulation, the lake level depression is 2.05 m. In terms of water quantity, even such an apparently small decrease in lake and reservoir level can be a relatively large (and therefore costly) loss to the reservoir manager in terms of volumetric loss. Over the simulation period of 50 years, a level decrease of 2.05m can be evaluated in terms of volumetric loss for different sized water bodies. For example, for a water surface area of 20 km^2, the total quantity lost, over the 50 year period, is $4.1 \times 10^7 m^3$ (i.e. an average of $0.8 \times 10^6 m^3$ yr^{-1}). For a typical depth of 40 m, then this is a volumetric percentage loss over 50 years of 5.1%. Many water supply reservoirs are smaller. For a 10 m deep lake of say 2 sq.km in area, the percentage loss is 21%.

However these calculations concern only evaporative losses and do not consider the probability that another impact of increased atmospheric carbon dioxide is a change in the hydrologic cycle. Wigley and Jones (1985) evaluated possible changes in runoff for an atmosphere in which the CO_2 content has been doubled. Such changes are directly reflected in inflow rates. For a typical residence time of 50 years, the inflow to a lake is approximately 2 % of its total volume. In this case, for an x% change in inflow rates, the lake volume will change by 0.02x% per year. Over this 50 year period, it might be reasonable to anticipate, from Wigley and Jones (1985) figures for doubled CO_2, that inflow rates could increase by an average of approximately 10%. In that case, inflow changes might increase lake volumes by 10% - potentially more important than the evaporative loss. However outflow rates could also increase and an overall rise may well be limited by the lake bathymetry and the surrounding topography. Furthermore increased inflow rates are likely to alter the annual thermal stratification cycle. This could lead to a further change in evaporation resulting from changed surface temperatures due to higher inflow rates.

Secondly, changes in nutrient concentrations in the lake will be related to changes in inflow rates and lake volume. If runoff is increased, the leached nutrients (phosphorus and nitrogen especially) are likely to increase roughly at a parallel rate. This would cause ambient lake concentrations to increase, further compounded if water levels were to fall. (For example, Wigley and Jones (1985) allow for the fact that runoff may decrease due to increased atmospheric CO_2 viz x<0 in the above calculation). For example, for a ~25-50% increase in nutrient inflow the trophic state could change dramatically. The increase in nutrient level between oligotrophic and eutrophic states is typically an increase of 100% in ambient phosphorus concentrations. Calculations must, at this stage, remain as an order of magnitude analysis, since without better streamflow forecasts, which are likely to be site specific (Wigley and Jones, 1985), nutrient loading is difficult to assess with any accuracy. Suffice it to observe that the potential impact of increasing atmospheric carbon dioxide on the trophic status of freshwater bodies is unlikely to be negligible.

Conclusions

Initial quantitative assessment of evaporative loss from freshwater lakes and reservoirs, made using a highly resolved thermal stratification model, suggests a significant decrease in storage, although this deficit could be ameliorated by increased streamflow (but may possibly be exacerbated in areas where streamflow decreases) as a result of the impact of increasing atmospheric CO_2. Furthermore, if inflow rates increase, so will the nutrient loading and this is highly likely to lead to accelerated eutrophication. Further studies, using such well-resolved models, are needed, as is further information on the likely perturbations to the hydrologic cycle, especially (in this context) runoff.

References

Coutant, C.C. (1981) Foreseeable effects of CO_2-induced climatic changes: freshwater concerns, Env. Conservation, 8, 285-297.

Gill, A.E. & Turner, J.S. (1976) A comparison of seasonal thermocline models with observations, Deep Sea Res., 23, 391-401

Hansen, J., Russell, G., Lacis, A., Fung, I., Rind, D. & Stone, P. (1985) Climatic response times: dependence on climate sensitivity and ocean mixing, Science, 229, 857-859.

Henderson-Sellers, B. (1985) New formulation of eddy diffusion thermocline models, Appl. Math. Model., 9, 441-446.

Henderson-Sellers, B. (1986) Calculating the surface energy balance for lake and reservoir modeling: review, Rev. Geophys., 24, 625-649.

Oeschger, H., Siegenthaler, U., Schotterer, U. & Gugelmann, A. (1975) A box diffusion model to study the carbon dioxide exchange in nature, Tellus, 27, 168-192.

Wigley, T.M.L. & Jones, P.D. (1985) Influences of precipitation changes and direct CO_2 effects on streamflow, Nature, 314, 149-152.

Wigley, T.M.L. & Schlesinger, M.E., (1985), Analytical solution for the effect of increasing CO_2 on global mean temperature, Nature, 315, 649-652.

The Influence of Climate Change and Climatic Variability on the Hydrologic Regime and Water Resources (Proceedings of the Vancouver Symposium, August 1987). IAHS Publ. no. 168, 1987.

Who is to blame for the desertification of the Negev, Israel?

A. Issar
The Jacob Blaustein Institute for Desert
Research and Department of Geology
H. Tsoar
Department of Geography
Ben Gurion University of the Negev, Beer Sheva,
Israel

ABSTRACT   Geological and geographical evidence suggests that the Negev Desert, and most probably the Levant, enjoyed a more humid climate from ca. 100 B.C. to ca. 300 A.D. Later it became more and more arid. This is in contrast to the conventional hypothesis which puts the whole blame of its desertification on the Arab nomads who conquered this region in the 7th century A.D.

RESUME Les données géologiques et géographiques suggerènt que des ca. 100 B.C. jusqua ca. 300 A.D., le désert du Négev et probablement le Levant, jouissaient dun climat plus humide que celui de nos jours. Plus tard le climat devint de plus aride. Cela est en contradiction avec l'opinion conventionelle selon laquelle la désertification du Néguev fut causée par les Arabes qui conquirent cette région au 7 eme siecle A.D.

Introduction

In the Negev, the arid southern part of Israel, there are six once prosperous but now deserted Byzantine towns. Each is surrounded by terraced agricultural fields, which were once irrigated by a sophisticated water harvesting system (Reifenberg, 1953; Gluck, 1959; Kedar, 1967; Evenari et al., 1971). These towns were founded by the Nabateans in the 1st century B.C. and flourished until the 7th century A.D.. However, within a few decades following the area being conquered by the Moslem-Arab armies (636 A.D.), the towns deteriorated and were totally abandoned. No signs of destruction by war were detected; all evidence speaks for a slow process of desertion (Evenari et al., 1971).
    The majority of students of this region (Reifenberg, 1953; Gluck, 1959; Kedar, 1967; Evenari et al., 1971; Schechter and Galai, 1980; Evenari, 1981; Hillel, 1982) have claimed that there are no signs of a major climatic change which might have given rise to aridization and thus explained the desertion as a result of the deterioration of the Byzantine socio-economic system owing to the conquest by the Moslem-Arabs, an explanation challenged in the present article.
    The first and most important place of evidence is the invasion by

sand dunes (Issar, 1968; Issar and Bruins, 1983; Issar et al., in press). Two main groups of sand dunes are known in Israel and Northern Sinai. One comprises the inland sand dunes of northern Sinai between the Suez Canal in the west and the Northwestern Negev in the east. The other, a distinct group of active coastal sand dunes, extends parallel to the shore line from the eastern part of the Bardawil lagoon in northern Sinai up to Tel-Aviv and is also found in some isolated areas north of Tel-Aviv as far as Haifa Bay. The inland dunes are stabilised in the Negev where they underwent pedogenic development and were later covered by the coastal dunes.

The source of all the aeolian sands, which are quartziferous in composition, are beach sands derived from the Nile Delta that were transported by longshore currents (Emery and Neev, 1960; Goldsmith and Golik, 1980; Nir, 1982).

It is very clear from field observations that the coastal sand dunes are younger than those found inland. In the inland dune areas there are indications of ancient and defunct drainage channels blocked and covered by the coastal sand dunes. The color of the inland sand in Northern Sinai and the Negev becomes progressively redder with distance from the coast (Tsoar, 1976). Studies of red dune sand from other deserts show that sand becomes redder with increased age and greater distance of transport (Norris, 1969; Walkers, 1979).

Archeological finds in the inland sand dunes of the Negev indicate Upper Paleolithic and Epipaleolithic ages for the fixed reddish sands (M. Nigel-Gorrin, pers. Comm., 1984). Abundant implements and pottery yielding ages as young as late Byzantine and as old as Chalcolithic (E. Oren, pers. Comm., 1983) were found on the surface of the fixed inland dunes north of the Haluza dune field in the Negev. Today's coastal sand dunes overlay the ancient fixed inland dunes. Issar (1968) found that the coastal dunes covered red loams containing Byzantine artifacts all across the central coastal plain of Israel.

From the above findings it can be concluded that the invasion of the inland sand dunes date from between 14,000 to 7,000 B.P. It is possible that this invasion took place in several phases during this long period. It is assumed that between 5,000 and 1,300 B.P. no invasion of sand occurred and that the inland dunes became stabilized. After 600 A.D. and before the 9th to 10th century A.D., another phase of invasion started which is still continuing today.

The reddish, older sands in the Northern Sinai and Negev cover a thick deposit of loess (Issar and Bruins, 1983; Issar et al., in press). This deposition of loess characterizes the uppermost Pleistocene from about 80,000 B.P. to about 14,000 B.P. (Issar, 1968; Issar and Bruins, 1983) and is attributed to a more humid climate caused by the southward shift of the belt of cyclonic rainstorms. After passing over northern Africa, these dust-laden rainstorms reached the Negev depositing thick layers of loess (Issar and Bruins, 1983; Issar et al., in press). At the end of the Pleistocene the massive deposition of loess stopped and deposition of sands began. This change is explained by an aridization phase due to the northward shift of the ITCZ (Inter-Tropical Convergence Zone). The latter, in turn, is believed to be connected with a shift northward of the Eastern African Monsoonal Zone as evidenced by the high levels of lakes in East Africa and of the Nile of that period (Butzer, 1980; Gasse et al., 1980; Livingstone, 1980; Nicholson, 1980; Nicholson and Flohn, 1980; Williams and Adamson, 1980; Rossignol-Strick, 1985). The more humid conditions over the catchment basin of the Nile resulted in the transport of more sediments by this river, which were subsequently transported to the coast of Israel. Assuming that the mechanism which caused the sand invasions of the post Byzantine-early Moslem period is similar to that which caused the sand invasion of the post Pleistocene Early Holocene, then a climatic change of aridization must have taken place. in this region at the end of the Byzantine period. This, of course, could explain the desertion of the towns and agriculture in the Negev as a gradual process and hence the lack of any sign of destruction by war or sacking by Bedouin tribes.

The question that arises whether additional evidence can be observed in the Levant for such an aridization phase during the 7th century. Another question is whether or not the flourishing of the Negev towns and agriculture was a result of a special humid phase or even a mini-ice age, a precursor of the "Little Ice Age".

Though interpretations are contradictory, one undisputed fact emerges from evidence on the Levant presented by various authors (e.g. Gluck, 1959; Neev and Emery, 1967; Evenari et al., 1971; Lamb, 1972), namely that at about the beginning of the 7th (or end of the 6th century) A.D. a major environmental change occurred in the Levant, which in modern nomenclature falls under the term desertification. Most investigators (as did the first author in his earlier work on the sand dunes of the coastal plain (Issar, 1968) favoured an anthropogenic explanation. The same evidence, however, reexamined and viewed from a regional paleoclimatic point of view fits better into a natural climatogenic model.

Following is some of the evidence supporting this claim:

(a) The Nile and Lake Abhe in Ethiopia show maximum peaks at about the middle of the 1st millenium A.D. (Butzer, 1980; Nicholson, 1980; Nicholson and Fohn, 1980), implying a shift northward of the monsoons and maximum transport of sand by the Nile.

(b) Palynological evidence from a lake bottom core of the Sea of Galilee (Baruch, 1983; Stiller et al., 1984; and Figure 1) shows an increase in the ratio of pollen of olive and a decrease of oak starting about 300 B.C. During the first centuries A.D. a gradual diminution starts, which reaches its lowest values during the 7-8th century. The same core shows that parallel to the decrease in olive pollen there is an increase in $CaCO_3$ a slight increase in Oxygen 18 and Carbon 13 (Stiller et al., 1984; and Figure 1.) which indicates slightly warmer water temperatures and higher nutrient content, possibly the result of higher erosion rates of the soil cover in these regions. Although the team which studied this core suggest that these changes are the ultimate reflection of heavy taxation (Baruch, 1983; Stiller et al., 1984), the present authors suggest a natural cause, namely, a gradual climatic change which brought about the gradual abandonment of the olive plantations in the semi-arid eastern Galilee and western Golan. These plantations were replaced by a "maqui"-type habitat.

(c) The only dendroclimatologic evidence available from Israel from this period is from the Roman siege ramp at Massada near the Dead Sea (72 A.D.), where a scarcity of Acacia and the abundance of Tamarix was found. This points to higher humidity than that of the

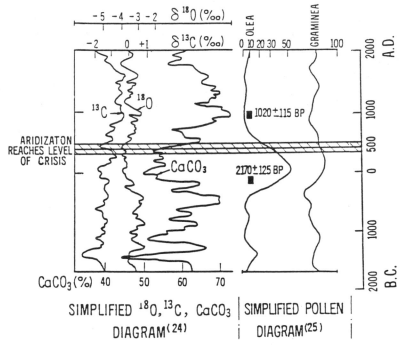

Figure 1

present (Lipschitz et al., 1981).

(d) Evidence from the Dead Sea shores is controversial. Neev and Emery (1967) report an abrupt rise in the Dead Sea level in the 7th century A.D., which they attribute to the destruction of the agriculture by the Arab-Moslem conquerors. Klein (1982), who studied the historical shorelines of the Dead Sea in detail has proof for a significant rise only between 100 B.C. and 20 A.D..

(e) In a detailed study of sediments of a river bed in central Israel, Rosen (in press) found evidence that "Roman Byzantine farmers settled the region at a time when the water table was higher than that of today and the wadi bed was about 2.5-3.5 m above the present channel. The climate may have been more moist than the present".

(f) A warmer climate phase throughout the Levant together with a northward movement of the ITCZ between the 6th and 7th centuries A.D, may have been connected with a deglaciation phase and should have then left its mark on the sea level. Indeed, observations from ancient port constructions along the coast of Israel show higher post Roman-Byzantine sea levels (Neev et al., 1973; Neev, 1974). These changes of sealevel, however, were regarded by Neev et al (1973) and Neev (1974) as evidence of post Byzantine tectonic movements along the shoreline. Vita Finzi (pers. comm. 1985) reported similar phenomena of a post Roman sea level rise along the coast of Tunisia.

(g) Lamb (1972, 1983) suggests that at the time of Christ, Palestine "enjoyed a benign period as regards the tendency of the climate.There were evidently considerable similarities to the climate of our own time, except for the continuance of a somewhat moister regime in North Africa and the Near East ". On the other hand, he suggests periods of drought which seem to have two maxima, around 300-400 A.D. and at 800 A.D.

(h) Vita - Finzi (1969), in his study on the Mediterranean valleys, observed widespread alluviation processes throughout the Near East from ca.500 to 1500 A.D., suggesting better distribution of rainfall (Vita - Finzi, 1969). Taking into account that the "Little Ice Age" started at about 1500 A.D., the question arises whether this alluviation phase is not simply evidence for accelerated erosion in the higher watershed areas due to deforestation triggered by a climatic change.

In conclusion, it is suggested that the conventional hypothesis, which puts the whole blame of the desertification of tne Negev solely on the Arab tribes that conquered this region in the 7th century A.D., be reviewed. Most evidence suggests that a climate moister than the present prevailed in the Negev from ca. 100 B.C. to ca. 300 A.D. This climate slowly became more arid and reached a critical level around 500 A.D. (Figure 1).

It is not the purpose of the present paper to aquit the nomads of the desert of all responsibility in the desertification process of the Near Eastern deserts. The difference in vegetation cover between the two sides of the Sinai and Negev border speaks for itself (Otterman, 1974, 1977). The overreadiness to put all the blame on man, however, has obscured the true objective picture, namely, that the causa causarum was nature's triggering off of a self accelerating process of desertification in which man played his role alongside all the other natural agents.

AKNOWLEDGEMENTS   Thanks are due to Prof.C. Vita-Finzi and Dr. I. Gilead for their comments, to Dr. Arlene Rosen for making available the results of her study, to Dr. Y. Bartov, Director of the Geological Survey of Israel, for the permission to publish the data on the core from the Sea of Galilee and to Mrs. B. Katz for editing the article.

References

Baruch, U.(1983) The Palynology of a Lake Holocene core from Lake Kinneret. MSc Thesis, Hebrew Univ., Jerusalem, Israel.
Butzer, K.W. (1980) Pleistocene history of the Nile valley in Egypt and lower Nubia. In: The Sahara and the Nile (ed. by M.A.J. Williams & H.J.Faure), 253-280. Balkema, Rotterdam.
Emery, K.D.& Neev, D.(1960) Mediterranean beaches of Israel. Geol. surv. Israel. Bull. No.26.
Evenari, M. (1981) Desert agriculture past and future. In : Settling the Desert (ed. by L. Berkofsky, D. Faiman & J.Gale), 3-29. Gordan and Breach Sci. Publ. Inc., New York.
Evenari, M., Shanan, L. & Tadmor, N. (1971) The Negev: The Challenge of a Desert. Harvard Univ. Press, Cambridge, Massachusetts.
Gasse, F., Rognon, P. & Street, F.A. (1980) Quarternary history of

the Afar and Ethopian Rift lakes. In: The Sahara and the Nile(ed. M.A.J. Williams and H.J. Faure), 461-400. Balkema, Rotterdam.

Gluck, N. (1959) Rivers in the Desert: A History of the Negev. W.W. Norton & Company Inc., New York.

Goldsmith, V. & Golik, A. (1980) Sediment transport model of the southeastern Mediterranean coast. Marine Geol. 37, 147-175.

Hillel, D. (1982) Negev. Praeger Publishers, New York.

Issar, A.S. (1968) Geology of the central coastal plain of Israel. Israel J. Earth Sci. 17(1), 16-29.

Issar, A.S. & Bruins, H.J. (1983) Special climatological conditions in the deserts of Sinai and the Negev during the latest Pleistocene. Paleogeo. Paleoclimat. Paleoecol. 43, 63-72.

Issar, A.S., Tsoar, H., Gilead, Y. & Zangvil, A. (1985) A paleoclimatic model to explain depositional environments during late Pleistocene, in the Negev, Israel. Desert Environments (Proc. UCLA-BIDR, Symp.) Rowman & Allanheld, Publishers (in press).

Kedar, Y. (1967) The Ancient Agriculture in the Negev Mountains. Bialik Institute, Jerusalem, Israel. (In Hebrew.)

Klien, C. (1982) Morphological evidence of lake level changes, western shore of the Dead Sea. Israel J.Earth Sci. 31, 67-94.

Lamb, H.H. (1972) Climate, Present, Past and Future. Methuen, London.

Lamb, H.H. (1982) Climate, History and the Modern World. Methuen, London.

Lipschitz, N., Lev-Yadun, S. & Weisel, Y. (1981) Denroarcheological investigations in Israel (Masada). Israel Exploration J. 31, 231-234.

Livingstone, D.A. (1980) Environmental changes in the Nile headwaters. In: The Sahara and the Nile (ed. by M.A.J. Williams & H.J. Faure), 339-359. Balkema, Rotterdam.

Neev, D. (1974) On the stability of the Mediterranean coast of Israel since Roman times: Reply. Israel J. Earth Sci. 23(4), 150-151.

Neev, D., Balker, N., Moshkovitz, S., Kaufman, A. & Magaritz, M. (1973) Recent faulting along the Mediterranean coast of Israel. Nature 245, 254-256.

Neev, D. & Emery, K.D. (1967) The Dead Sea, depositional processes and environments of evaporites. Geol. Surv. Israel Bull. No. 41.

Nicholson, S.E. (1980) Saharan climates in historic times. In: The Sahara and the Nile (ed. by M.A.J. Williams & H.J. Faure), 173-200. Balkema, Rotterdam.

Nicholson, S.E. & Flohn, H. (1980) African environmental and climatic changes and the general circulation in the Late Pleistocene and Holocene. Climatic Change 2(4), 313-348.

Nir, Y. (1982) Asia, Middle East, coastal morphology: Israel and Sinai. In: The Encyclopedia of Beaches and Coastal Environments (ed. by M.L. Schwarts, 86098. Hutchinson Ross Publishing Company, Strudsburg, Pennsylvania.

Norris, R.M. (1969) Dune reddening and time. J. Sed. Petrol. 39, 7-12.

Otterman, J. (1974) Baring high-albedo soils by overgazing: hypothesized desertification mechanism. Science 186(4163), 531-533.

Otterman, J. (1977) Anthropologic impact on the the albedo of the earth. Climatic Change 1(2), 137-156.

Riefenberg, A. (1953) Desert research. Res. Counc. Isra. Bull. Jerusalem 3, 378-391.

Rosen, A.M. Cities of the Clays. The Geoarcheology of Tells. Univ. of Chicago Press. (in press).

Rossignol-Strick, M. (1985) Mediterranean Quarternary sapropels, an immediate response of the African monsoon to variation of isolation. Paleogeo. Paleoclimat. Paleoecol. 49, 237-263.

Schechter, Y. & Glai, C. (1980) The Negev- a desert reclaimed. In: Desertification: Environmental Science and Application. Vol. 12 (ed. by M.R. Biswas & A.K. Biswas), 255-308. Pergamon Press, Oxford.

Stiller, M., Ehrlich, A., Pollinger, U., Baruch, U. & Kaufman, A. (1984) The Late Holocene sediments of Lake Kinneret (Israel)- multidisciplinary study of a 5m core. In: Geological Survey of Israel, Ministry of Energy and Infrastructure, Jerusalem, Israel.

Tsoar, H. (1976) Characterisation of sand dune environments by their grain size and surface texture. In: Geography in Israel(ed. by D.H.K. Amiram & Y. Ben-Arie), 327-343. IGU, Jerusalem, Israel.

Vita-Finzi, C. (1969) The Mediterranean Valleys, Geological Changes in Historical Times. Cambridge Univ. Press.

Walkers, T.R. (1979) Red color in dune sand. In: A Study of Global Sand Seas (ed. by E.D. Mckee), 61-81. U.S. Geol. Surv. Prof. Paper, 1052.

Williams, M.A.J. & Adamson, D.A. (1980) Late Quarternary depositional history of the Blue and White rivers in Central Sudan. In: The Sahara and the Nile (ed. by M.A.J. Williams & H.J. Faure), 339-359. Balkema, Rotterdam.

The Influence of Climate Change and Climatic Variability on the Hydrologic Regime and Water Resources (Proceedings of the Vancouver Symposium, August 1987). IAHS Publ. no. 168, 1987.

Some possible impacts of greenhouse gas induced climatic change on water resources in England and Wales

J.P. Palutikof
Climatic Research Unit,
University of East Anglia,
Norwich, NR4 7TJ,
U.K.

ABSTRACT   Two methods are employed to derive scenarios of runoff changes in ten drainage basins in England and Wales due to increasing atmospheric concentrations of the greenhouse gases. The first method compares reconstructed riverflow data for the warmest and coldest twenty-year periods this century. It suggests riverflow will decrease in southern England and Wales and increase in northern areas. The second method attempts to introduce the direct effects of CO_2 on plant transpiration into the analysis. In this scenario riverflow could be expected to increase throughout the country. The results from the two analyses are compared and discussed in the context of changing water use patterns in England and Wales.

Quelques impacts possibles de changement
climatique produit par les gaz effet de la
serre pour les ressources en eau de
l'Angleterre et du Pays de Galles

RESUME   Deux methodes sont employees pour obtenir des scénarios concernant les changements des ruissellement, à cause des concentrations augmentées des gaz a effet de serre pour dix bassins versants en Angleterre et au Pays de Galles. La premier méthode est une comparaison des données de débits réconstituées pour les périodes (de vingt ans) les plus chaudes et les plus froides de ce siecle. Cette méthode suggére que les debits des rivières vont diminuer dans les régions sud de l'Angleterre et du Pays de Galles, et vont augmentes dans les régions Nord. La deuxieme méthode éssaye d'introduire, dans l'analyse, les effets directs du CO_2 sur la transpiration des plantes. Dans ce scénario on peut s'attendre a ce que les débits vont augmenter partout dans le Pays. Les resultats des deux analyses sont comparés et discutés dans le contexte des changements de l'utilisation habituelle des ressource en eau en Angleterre et au Pays des Galles.

Introduction

Atmospheric CO_2 concentrations have risen from a pre-industrial level of between 260 and 280 ppm (Neftel et al., 1985) to around 345 ppm today due largely to anthropogenic causes which are likewise responsible for rapid increases in other radiatively-active trace gases, such as methane and carbon monoxide. All these gases, although transparent to incoming short-wave solar radiation, absorb outgoing long-wave terrestrial radiation. This is the so-called "greenhouse effect", whose net effect is expected to be to cause changes in the spatial and temporal distribution of rainfall, temperature and pressure.

This paper is particularly concerned with the effect of increasing atmospheric CO_2 on surface water resources in England and Wales. The meteorologic variables which determine the availability of surface water are precipitation, evaporation (from an open water or bare ground surface), and evapotranspiration (from a vegetated land surface).

Precipitation and evaporation are meteorologic variables, dependent upon atmospheric processes, however transpiration is determined by the physiology of the plant. In controlled laboratory experiments increased atmospheric CO_2 has been found to have direct effects on vegetation, including an increased rate of photosynthesis, suppression of photorespiration in C_3 species (those in which some of the captured solar energy is wasted in respiration), and partial closure of the stomata. The net result is generally an increase in the size and dry weight of the plant (Acock & Allen, 1985). The most important of these three effects for the discussion here is the closure of the leaf stomata which will, irrespective of the effect upon the plant, reduce the transpiration rate. Should the results of these laboratory experiments apply in the environment, the direct effects would lead to an increase in runoff in a world of increased atmospheric CO_2.

In this paper two methods are employed to investigate some possible impacts on water resources in England and Wales. The first uses reconstructed riverflow data to investigate possible riverflow changes for the early decades of a greenhouse-gas-induced warming. These reconstructed riverflows are based on precipitation data only, with a prescribed cycle for evapotranspiration. The second approach attempts to incorporate the direct effect on transpiration into estimates of percentage runoff changes in a high-CO_2 world.

The analogue scenario of climatic change

Before any investigation of riverflow changes can be made, the first requirement is an estimate of the regional climatic change for England and Wales that may be expected to occur in a high-greenhouse-gas world. There are two methods for arriving at such an estimate. The first is to use a general circulation model (GCM) to predict climate changes arising from increased atmospheric CO_2. Problems with this approach are that the horizontal resolution is generally of the order of 5° latitude by 10° longitude, and the state of the art of GCM modelling is such that they cannot be expected to generate accurate forecasts of climate changes at the regional level (Wigley, 1986). Bearing these caveats in mind, some attempts have been made to apply GCM results to forecasting regional impacts on water resources of increased levels of greenhouse gases (Rind & Lebedeff, 1984; Cohen, 1986).

The second method is to construct an analogue model, which may be based on a paleoclimatic reconstruction (Butzer 1980; Kellogg, 1977), or on instrumental data (Lough et al., 1983; Palutikof et al., 1984). Palutikof et al. (1984) used the twentieth century Northern Hemisphere temperature record of Jones et al. (1982) to generate regional analogues of climatic change in a high-CO_2 world. In their preferred scenario they compared the warmest block of twenty consecutive years, 1934-53, with the coldest twenty-year block, 1901-20. The average conditions for the cold block were then subtracted from the average conditions for the warm block in order to arrive at an estimate of early CO_2 -induced climatic changes.

Clearly, anthropogenic sources of trace gases have affected and will continue to affect atmospheric concentrations over many decades. The resulting climate changes are gradual rather than abrupt, and as such they must be accompanied by substantial adjustments in the oceans and cryosphere. By taking blocks of years Palutikof et al. are attempting, within the constraints of the data set, to incorporate these adjustments. The major disadvantage of the selection procedure is the small temperature difference that is obtained: an annual temperature change of $0.3°C$ for the whole hemisphere. Although statistically significant at the 1% level, this difference is very small compared with GCM predictions of $1.5°$ C to $6.5°$ C for a doubling of CO_2 (Manabe & Stouffer, 1979), and the results can only be taken to be indicative of changes in the early decades of a greenhouse-gas-induced warming. However, these analogue-based scenarios provide the current "best guess" for the direction and spatial distribution of regional climatic changes in a warm world.

Figure 1, taken from Lough et al., (1983), shows the precipitation changes in the British Isles which were arrived at by comparing station data for the sets of warm and cold years. In this presentation the warm-minus-cold differences are expressed in terms of their present-day variability by normalizing them with respect to the standard deviation of the long-term rainfall series. The diagrams indicate that it is the "shoulder" seasons, spring (MAM) and autumn (SON) which show the most regionally coherent trends: in spring towards drier conditions and in autumn towards greater precipitation amounts. Summer rainfall is expected to decrease over most of England and Wales and southern Ireland but increase over northern Britain and Ireland. Winter precipitation totals should generally be lower except in eastern Scotland, southern Ireland and North Wales. Thus, although the trends are not clear-cut, it appears that conditions will be drier for much of the study area except in autumn. In the next sections we investigate what implications these results might have for water resources.

Scenarios from reconstructed river flow

Jones et al. (1984) present reconstructed riverflows for ten drainage

Figure   1   Precipitation change scenarios based on the warmest and
coldest twenty-year periods this century (after Lough et
al.). Changes are warm-minus-cold differences expressed
as multiples of the long-term standard deviation.

basins in England and Wales, derived from an empirical model
developed by Wright (1978).   The input data are values of areal
catchment rainfall, and multiple regression is employed to estimate
riverflow amounts.   A set of constant long-term average monthly
values of actual evaporation are used. The justification is that the
regression model predictions were more accurate than when evaporation
was calculated separately for each individual month.   In test
catchments the variances explained by the model ranged between 84%
and 91%.

The locations of the ten drainage basins are shown in Figure 2.
Catchments 1 to 7 are located in the upland regions of England and
Wales whereas 8 to 10 are in low-lying areas. The ten catchments are
divided into three sub-groups: northern rivers (Tees, Tyne, the two
Eden catchments and the Wharfe); western rivers (Wye and Exe); and
eastern rivers (Thames, Ely-Ouse and Wensum).   Jones et al. (1984)
present reconstructed riverflows, (in m^3s^{-1}) for these ten catchments going back to as early as 1769 for the Thames. The longest
reliable reconstruction, including corrections for snowfall, extends
to 1853 (Thames) whereas the shortest goes to 1908 (Wensum).

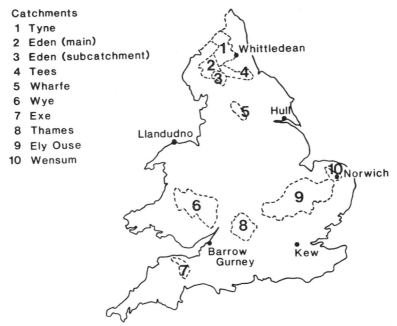

Catchments
1 Tyne
2 Eden (main)
3 Eden (subcatchment)
4 Tees
5 Wharfe
6 Wye
7 Exe
8 Thames
9 Ely Ouse
10 Wensum

Figure 2   Location of the ten drainage basins and six rainfall
stations used in this study.

The change in riverflow between the twenty coldest years (1901–
1920) and the twenty warmest years (1934–1953) has been investigated
in an attempt to determine the possible impacts of increased
atmospheric concentrations of the greenhouse gases. The monthly mean
riverflow was calculated for each twenty-year block at each station.
Then the cold-period mean was subtracted from the warm-period mean.
The difference was expressed as a percentage of the cold-period
average, which represents the baseline against which the change
occurs.   Standardization also permits inter-catchment comparison
since the mean riverflows are very different.

As a first step we examine the percentage change in the annual
mean riverflow which gives a measure of the overall change in
riverflow.   Eight of the drainage basins showed a decline in runoff
in the warm twenty-year block.   The exceptions were two northern
rivers: the Tees and the Tyne, which indeed lie in an area of annual
precipitation increase in Figure 1. The greatest negative changes are
observed in two Southern basins, the Thames (−7.1%) and the Wensum
(−8.2%).   The only other river to exhibit a difference between the
warm and cold periods of over 5% is the Eden subcatchment (−5.5%).

Monthly and seasonal variations between the cold and the warm
years may be at least as important as those at the annual level. In
Figure 3 the monthly warm-minus-cold period differences are shown as
a percentage of the cold period mean.   The catchments are grouped
into the three geographical regions: north, west and east.   In the
northern catchments the patterns are very similar and only three are

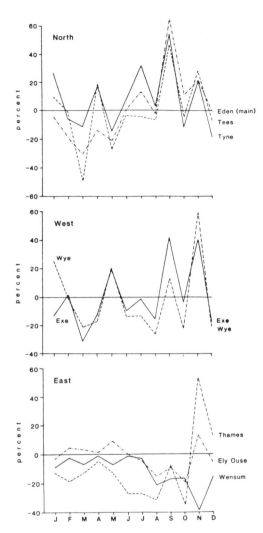

Figure 3   Monthly warm—minus—cold differences in riverflow as a
percentage of the cold period mean.

displayed here to avoid confusion.   It can be seen that there is a
definite seasonal cycle: from December to May riverflow is expected
to be lower in a high-greenhouse-gas world whereas from June to
November an increase in runoff is indicated.   In the west, the Exe
and the Wye show very variable percentage changes, with no clear
seasonal pattern emerging. Of the eastern rivers the Thames and Ely-
Ouse share similar seasonal profiles. The Thames has lower riverflow
in the warm decades in every month except November and December.
Spring and November discharges are greater for the Ely-Ouse. The
Wensum riverflow is expected to be reduced throughout the year in a
high-greenhouse-gas world, on the basis of this scenario.

It is also possible to analyze changes in the variability of the parameter under consideration. The coefficient of variation is used here as a measure of variability, given by 100 $s/\bar{x}$, where s is the standard deviation and $\bar{x}$ is the mean. Since this coefficient is already standardized a simple warm–minus–cold difference may be compared between catchments. Coefficients of variation were calculated for the Wye, Thames and Wensum in the south and Wharfe and Eden subcatchment in the north. The remaining five catchments present very similar patterns to their regional neighbours.

The variability in annual mean riverflow, as shown in Table 1, was found to be lower in the warm years for all catchments except the Wensum. The changes were small except in the Wye catchment (−7%). The monthly differences failed to show any clear trends and therefore seasonal averages were calculated, and are given in Table 1. This shows the clear predominance of negative differences in most seasons. In the two northern basins the only season when variability is expected to increase in a warm world is winter. In the three southern basins riverflow should be more variable in spring (and in the Wye valley in autumn also).

Table 1 Warm–minus–cold differences in the coefficient of
        variation of riverflow (percent)

Catchment	Spring	Summer	Autumn	Winter	Annual
Wye	11.2	−7.7	42.8	−6.3	−7.13
Thames	14.0	−33.4	−15.5	−4.8	−2.10
Wensum	17.5	−6.9	−63.6	−4.7	2.24
Wharfe	−3.6	−17.8	−10.0	18.0	−1.34
Eden	−10.2	−2.4	−19.9	0.4	−4.46

Direct effects of carbon dioxide on evapotranspiration

As already discussed in the introduction, CO_2 has direct physiological effects on plants, which in turn may have impacts for riverflow if atmospheric concentrations of the gas increase. The most relevant effect for this discussion is the partial closure of the plant stomata in a high–CO_2 atmosphere, which should lead to a reduction in the evapotranspiration rate and hence, at the basin level, to increased runoff. Thus the evapotranspiration rate in a high-CO_2 world (E_A) is given by:

$$E_A = (1 - af) \; E_O \qquad (1)$$

where E_O is the original evapotranspiration rate, a is the proportional decrease in evapotranspiration due to direct CO_2 effects, and f is the fractional area of the basin which is vegetated (Idso & Brazel, 1984).

Revelle & Waggoner (1984) studied the effects on runoff arising from CO_2 – related climate changes in twelve drainage basins in Arizona. They concluded that large reductions in riverflow could be expected to accompany increasing concentrations of the greenhouse gases. Idso & Brazel (1984) repeated this analysis taking into account the supposed direct CO_2 effect. For a scenario of a 10% decrease in precipitation they found that runoff in the five wettest basins could be expected to increase by 58% for no change in temperature, and by 41% when the temperature was increased by $2°$ C. This compares with a runoff reduction of 23% for both cases found by Revelle & Waggoner. Wigley & Jones (1985) suggest that one reason for the very large estimated impact of the direct CO_2 effect lies in the runoff ratio of 0.16 (proportion of total precipitation which becomes runoff used by Idso & Brazel). For temperate latitudes Wigley & Jones suggest a ratio of 0.4 to be more appropriate.

As part of their analysis, Wigley & Jones present graphs relating change in precipitation to change in runoff for three different runoff ratios and for reductions in the evapotranspiration rate ranging between zero and 30% (the maximum direct effect). These graphs have been applied here to the problem of estimating the direct and indirect impacts of increasing atmospheric concentrations of the greenhouse gases on riverflow in England and Wales.

In the Palutikof et al. study mean annual precipitation was found to decrease between the cold and warm blocks of years over most of England and Wales. The only substantial area of increase was over northern England, which would account for the overall increase in runoff in the Tyne and Tees catchments noted in the previous section. The Palutikof et al. (1984) study related warm-minus-cold precipitation changes to the long-term variability in the record. Figure 2 shows the six precipitation records used in that study whose data have been reworked, the precipitation differences being expressed as a percentage of the cold twenty-year mean. Again, the cold period is visualized as a base upon which greenhouse-gas-induced climate changes occur.

The percentage differences in mean annual precipitation between the warm and cold blocks of years range between -0.6% at Llandudno and +6.7% at Whittledean. Whittledean was the only site for which an increase in precipitation is suggested in a warm world. The impacts on runoff of these changes in precipitation for different direct effects of CO_2 are shown in Table 2. For the construction of this table a runoff ratio of 0.4 was assumed. The parameter b indicates the strength of the direct effect: where b = 0 there is no change in evapotranspiration and only the effect of precipitation is considered; b = 0.5 represents a reduction in evapotranspiration of 15%; b = 1 corresponds to a 30% reduction. The percentage reductions in runoff are shown to the nearest 5%, the maximum possible resolution from the Wigley & Jones (1985) graphs.

Table 2 indicates clearly that lower rainfall in a high-greenhouse-gas world may not necessarily imply a lower riverflow if accompanied by a reduction in evapotranspiration. For a maximum direct effect, even where rainfall decreased by as much as 8-9%, as at Llandudno, there is still a 20% increase in runoff. In this scenario there is no area of England and Wales that would not experience greater riverflows in a high - CO_2 world. In the North of England, a high rainfall area where further increases are predicted in a warm high CO_2 world, a greater frequency of flood events could create problems. For the intermediate scenario of a 15% reduction in evapotranspiration there is still no decline in riverflow at any of the six stations considered here. It is only when the direct effect

Table 2 Warm-minus-cold percentage changes in precipitation and their impacts on runoff. For an explanation of b, see text

Rainfall station	% change precipitation	% change in runoff for:		
		b = 0	b = 0.5	b = 1
Llandudno	-8.6	-25	0	+20
Kew	-7.2	-20	+5	+25
Norwich	-4.2	-15	+10	+30
Barrow Gurney	-3.9	-10	+15	+35
Hull	-1.9	-5	+20	+40
Whittledean	+6.7	+15	+35	+60

is ignored completely that substantial reductions in runoff result. The major shortcoming of this method is that it neglects seasonal variability, which may be at least as important as the overall change in catchment runoff in a high - CO_2 world. However, the results point very clearly to the need to consider possible direct effects.

Impacts on water resources

Figure 4 shows the geographical boundaries of the water authorities in England and Wales. Data on water use in eight of the ten authorities were extracted from the Digest of Environmental Pollution and Water Statistics (Department of the Environment, 1980, 1985) for the years 1976 to 1984. These data were used to construct the time series depicted in Figure 5, which show that in the southern authorities (Anglian, Thames and Southern) there has been a distinct and persistent increase in total water use. Elsewhere, water use has either stagnated or declined. There is no reason to suppose that these trends, which reflect economic and demographic factors, will not continue in the foreseeable future.

Clearly these water use time-series have implications for a world in which riverflow is distorted by the impacts of increased atmospheric concentrations of the greenhouse gases. In the first analysis of this paper reconstructed riverflows in the warmest and coldest twenty-year blocks this century were compared. It was found that there were only two basins, out of the ten analyzed, where riverflow increased in the warm years. These two were both located in the North of England, where water use may be expected to continue to decline in the future. Of the eight basins which experienced a decrease in riverflow, three are located in the southern water authorities where a pronounced upward trend in water use has been noted. Of these three, the Thames and the Wensum are both predicted to have lower flows in virtually every month of the year, except November and December along the Thames. The third river, the Ely Ouse, has reduced discharges in the late summer and autumn. Groundwater abstractions fulfill a substantial proportion of the water demand in the southern water authorities: 42% for the Thames region, 50% for Anglian, and 71% for Southern (Department of the Environment, 1985). However, a decrease in precipitation in what is already a dry part of the country (the mean annual rainfall at Kew is around 600mm) will be reflected not only in riverflows but also in a decline in groundwater reserves.

Thus the implications for water resources in England of increasing atmospheric concentrations of the greenhouse gases appear to be severe if only the climate changes are considered. The largest declines in riverflow will occur in the drier south where consumption is gradually rising. In the wetter northern areas the predicted increases in riverflow are of little significance since water use is actually declining.

Figure 4   Water authorities of England and Wales (after Department of the Environment, 1985).

The picture changes dramatically once the possible direct effects of increasing CO_2 are taken into account. If these lead to only a 15% reduction in evapotranspiration, the rainfall changes predicted by the Palutikof <u>et al</u>. scenario will be accompanied by an increase in riverflow rather than any decline. However, the use of instrumental scenarios can only be used to examine the possible range of climate changes, and their impacts, in the very early years of a greenhouse-gas-induced warming trend. The persistence of such a trend may well lead to changes in precipitation patterns which will overwhelm the direct effects of CO_2 and place a severe strain on water resources in southern England.

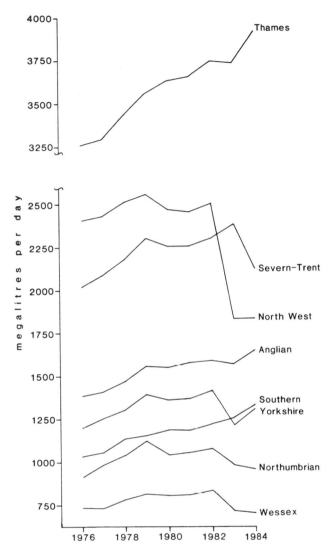

Figure 5   Water use in eight water authorities, 1976-1984
(megalitres per day).

References

Acock, B. & Allen, L.H. Jr. (1985) Crop responses to elevated carbon
   dioxide concentrations.   In: Direct Effects of Increasing Carbon
   Dioxide on Vegetation (ed. by B.R. Strain & J.D. Cure), 53-97.
   DOE/ER-0238, U.S. Dept. of Energy, Washington, D.C., USA.
Butzer, K.W. (1980)   Adaptation to global environmental change.
   Professional Geographer 32, 269-278.
Cohen, S.J. (1986) Impacts of CO_2-induced climatic change on water

596 J.P. Palutikof

resources in the Great Lakes Basin. Climatic Change 8, 135-153.

Department of the Environment (1980) Digest of Environmental Protection and Water Statistics 3, HMSO, London, UK.

Department of the Environment (1985) Digest of Environmental Protection and Water Statistics 8, HMSO, London, UK.

Idso, S.B. & Brazel, A.J. (1984) Rising carbon dioxide concentrations may increase streamflow. Nature 312, 51-53.

Jones, P.D., Wigley, T.M.L. & Kelly, P.M. (1982) Variations in surface air temperature: Part 1, Northern Hemisphere 1881-1980. Mon. Weath. Rev. 110, 59-70.

Jones, P.D., Ogilvie, A.E.J. & Wigley, T.M.L. (1984) Riverflow Data for the United Kingdom: Reconstructed Data back to 1844 and Historical Data back to 1556. CRURP8, Climatic Research Unit, University of East Anglia, Norwich, UK.

Kellogg, W.W. (1977) Effects of Human Activities on Climate. WMO Technical Note 156, WMO No. 486, World Meteorological Organization, Geneva, Switzerland.

Lough, J.M., Wigley, T.M.L. & Palutikof, J.P. (1983) Climate and climate impact scenarios for Europe in a warmer world. J. Clim. Appl. Meteor. 22, 1673-1684.

Manabe, S. & Stouffer, R.J. (1979) A CO_2-climate sensitivity study with a mathematical model of the global climate. Nature 282, 491-493.

Neftel, A., Moor, E., Oeschger, H.& Stauffer, B. (1985) Evidence from polar ice cores for the increase in atmospheric CO_2 in the past two centuries. Nature 315, 45-47.

Palutikof, J.P., Wigley, T.M.L. & Lough, J.M. (1984) Seasonal Climate Scenarios for Europe and North America in a High-CO_2Warner World. DOE/EV/10098-5, U.S. Dept. of Energy, Washington, D.C., USA.

Revelle, R.R. & Waggoner, P.E. (1983) Effects of a carbon dioxide-induced climatic change on water supplies in the western United States. In Changing Climate, 419-432, National Academy Press, Washington, D.C., USA.

Rind, D. & Lebedeff, S. (1984) Potential Climatic Impacts of Increasing Atmospheric CO_2 with Emphasis on Water Availability and Hydrology in the United States. EPA 230-04-84-006, U.S. Environmental Protection Agency, Washington, D.C., U.S.A.

Wigley, T.M.L. & Jones, P.D. (1985) Influences of precipitation changes and direct CO_2 effects on streamflow. Nature 314, 149-152.

Wigley, T.M.L. (1986) The impact of climate on resource use and availability. In: Resources and World Development (ed. by D.J. McLaren and B.J. Skinner), Springer Verlag, Berlin (in press).

Wright, C.E. (1978) Synthesis of riverflows from weather data. Tech. Note 26, Central Water Planning Unit, Reading, UK.

The Influence of Climate Change and Climatic Variability on the Hydrologic Regime and Water Resources (Proceedings of the Vancouver Symposium, August 1987). IAHS Publ. no. 168, 1987.

Detection of natural and artificial causes of groundwater fluctuations

Frans C. Van Geer
TNO-DGV, Institute of Applied Geoscience
P.O. Box 285, 2600 AG Delft, the Netherlands
Peter R. Defize
ITI-TNO, Institute of Applied Computer Science
P.O.Box 214, 2600 AG Delft, the Netherlands

ABSTRACT   Fluctuations and trends in the groundwater level are the result of a number of natural and artificial causes. Transfer/noise modelling has been applied to decompose observed groundwater level series in such a way that each cause of influence corresponds with a component of the groundwater level. In this paper two applications are presented: i) Detection of the influence of a river stage on the groundwater level. ii) Detection of an artificial trend in the groundwater level. To solve practical problems, easy to use methods are desired. Therefore transfer/noise modelling appears to be a useful addition to the existing geohydrologic modelling methods.

Détection des causes naturelles et articielles de la fluctuation de l'eau souterraine

RESUME   Fluctuations et changements permanents dans le niveau de l'eau souterraine sont le résultat dun nombre de causes naturelles et artificielles. Le modèlage transfer /bruit a été appliqué de manière que chaque cause ou influence correspond avec un constituant du niveau de l'eau souterraine.

Deux applications sont présentées dans cette communication: (i) Detection de l'influence du niveau d'une rivierè sur le niveau de l'eau souterraine. ii) Détection dun changement permanent dans le niveau de l'eau souterraine. Pour résoudre des problèmes pratiques, des méthodes simples sont désirées. C'eot pourquoi le modèlage transfer /bruit semble être une extension utile des méthodes de modèlage géohydrologiques.

Introduction

The presence and availability of groundwater is important for many human activities, such as agriculture and water supply. To obtain information about the behaviour of groundwater, observation wells have been placed at many locations. In the Netherlands, a primary groundwater monitoring network, maintained by TNO-DGV Institute of Applied Geoscience, serves as a reference network for regional groundwater management. The primary groundwater monitoring network consists of about 17,000 observation wells with 40,000 measurement screens. In 50% of the observation wells, the water level is measured 24 times a year, resulting in measurement series with time steps of approximately 15 days.

The changes and fluctuations in the groundwater level are the result of many causes. Those causes can be split into "natural" causes (e.g. the fluctuation in precipitation, water stages in rivers) and "artificial" causes (e.g. groundwater abstraction). Often it is important to detect which part of the groundwater fluctuations is due to each particular cause. Therefore, a method is needed to decompose the groundwater level into components, representing the influence of natural and artificial causes on the groundwater. Because any method gives only an approximation of the real process, an estimate of the reliability of that approximation is also required.

A well known method in systems analysis that meets these requirements is transfer modelling. The TNO-DGV Institute of Applied Geoscience in co-operation with the ITI-TNO Institute of Applied Computer Science has tested transfer modelling for its applicability to groundwater measurement series. The method proves to be quite meritorious (Defize & Rolf 1985, Van Geer, 1986).

In this paper, a brief review of the methodology is given and some specific problems encountered in applying it to groundwater are discussed. The possibilities for application of transfer modelling to groundwater are pointed out. Two case studies are presented: the detection of the influence of the stage of the river IJssel on the groundwater level and the detection whether or not the groundwater level has been lowered during the last decade in an area in the province of Noord-Brabant. Finally, some general conclusions are given.

Methodology

Transfer modelling is described in detail by Box and Jenkins (1976). Here only the basic conception of transfer modelling is given. The modelling process consists of three steps.

(a) The causes that influence the groundwater are split into presumed dominant and minor causes with respect to the groundwater level fluctuations. The dominant causes are modelled explicitly. Each dominant cause yields a component in the groundwater level. Measurements of explicitly modelled causes are called input series. The minor causes are modelled as one single cause: the noise component.

(b) All components are modelled as linear input-output models. The input of the noise model is supposed to be a zero mean white noise series, with an unknown variance. The models for the components with known input series are called transfer models, while the model with white noise as input is called the noise model. The components are the outputs of the linear models.

(c) All the components and a constant are added to obtain the groundwater level. All model coefficients, including the constant and the variance of the white noise are estimated simultaneously by means of maximum likelihood. The accuracy of the estimated coefficient is given by means of the asymptotic variance-covariance matrix. The validity of the model should be analysed by means of the residuals.

The structure of the model or transfer/noise model, described in the three steps above, is given schematically in Figure 1.

A transfer/noise is a model for discrete time series. It assumes that the input(s) and output are measured at equidistant time points. The general form of a transfer model (transfer model 1) is:

$$h_1(t) = \delta_1 h_1(t-1) + \ldots + \delta_r h_1(t-r) + \omega_o x_1(t) - \ldots - \omega_s x_1(t-s) \quad (1)$$

where: $h_1(t)$ a component of the groundwater level at time t.
       $x_1(t)$ an input at time t
       $\delta_1 \ldots \delta_r, \omega_o \ldots \omega_s$ parameters

In the model (1) the terms $\delta_i h_1(t-1)$ (i=1....r) represent the inertia of the groundwater system. Such model structures are also known in hydrology, for example, in recession curves. The terms $\omega_i x_1(t-1)$ (i=o...s) represent the driving force of the system. A hydrologic example of such a model structure is the unit hydrograph.

The noise model has a similar form as the transfer models:

$$n(t) = \phi_1 n(t-1) + \ldots + \phi_p n(t-p) + a(t) - \theta_1 a(t-1) - \ldots - \theta_q a(t-q) \quad (2)$$

where: $n(t)$ the noise component at time t
       $a(t)$ the realization of a white noise at time t
       $\phi_1 \ldots \phi_p, \theta_1 \ldots \theta_q$ parameters

In principle, a transfer/noise model assumes stationarity. However, two types of non-stationarity can be dealt with. If the time series have seasonal fluctuations the transfer/noise model can be extended with a seasonal model. An example of a seasonal model is:

$$n(t) = \phi n(t-1) + a(t) - \theta a(t-s) \quad (3)$$

where s is the seasonal period.

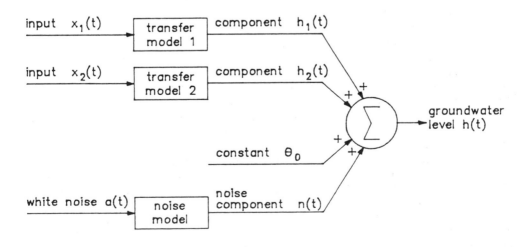

Figure 1   Transfer/noise model.

The second type is a polynomial trend that can be accounted for by differencing the original series a number of times, depending on the degree of the trend. A linear trend, for instance, can be modelled by taking the differences between two successive observations:

$$z(t) = h(t) - h(t-1) \tag{4}$$

A quadratic trend can de dealt with by:

$$z(t) = h(t-1) - 2h(t) + h(t+1) \tag{5}$$

In case of a polynomial trend the difference operation is carried out first and then the series $z(t)$ is modelled with a transfer/noise model.

It should be noted that the structure of a transfer/noise model agrees with most common mathematical models in geohydrology, where a number of linear components (precipitation, abstraction) are superimposed.

Application to groundwater

A priori data manipulation

Often the input series of the transfer models are abstractions of groundwater, precipitation excess (precipitation-evapotranspiration) or surface waterlevels. For convenience, throughout this discussion the input will be assumed to be precipitation.

The fluctuations of the precipitation may show a totally different pattern compared with the fluctuations in the groundwater level. It is our experience that transfer modelling can be carried out successfully if the input series is transformed in such a way (by moving average, for example) that the fluctuation patterns of input and output series are of the same nature or- in other words- have the same time scale. Also, the measurement frequency of the input series may be different from that of the groundwater series. For instance in the Netherlands, most groundwater levels are measured with a frequency of 24 times per year, whereas the precipitation is measured daily. Because the groundwater measurement frequency is the lowest, this frequency (or a lower one) is used in the transfer model and the precipitation series is converted to that frequency. The measurement series of groundwater level is often smooth in comparison with the measurement series of the precipitation. Distinction can be made between two cases:

(a) Shallow groundwater, where the time in which the precipitation is routed through the unsaturated zone is in the order of days.

(b) Deep groundwater where the time in which the precipitation is routed through the unsaturated zone is in the order of months.

In case of shallow groundwater, the fluctuations in groundwater level at time t are the result of the volume of precipitation measured in the days just before that time point. The number of days to be taken into account depends on the geohydrologic conditions. If the time scale of the groundwater fluctuations is larger than that of the converted precipitation a lower frequency can be taken (for example, 12 times per year instead of 24 times per year). Then, the volume of precipitation taken into account can be up to a monthly total.

In addition to the conversion of measurement frequency, in case of deep groundwater a delay occurs between the time of precipitation and the reaction of the groundwater level. Moreover, the recharge due to precipitation will be smoothed when routing through the unsaturated zone. Similar problems occur for semi-confined groundwater. The delay time is taken into account in transfer models as an explicit parameter. The smoothing can be modelled with a moving average of the precipitation, with or without weighting factors. The time interval of the moving average and the weighting factors depend on the geohydrologic conditions.

To manipulate the data in such a way that transfer models give sensible results, some experience in time series analysis as well as in geohydrologic modelling is required. There is not much guidance in the literature on this aspect.

The use of a transfer/noise model

The most obvious way of using transfer/noise models in geohydrology is decomposing the groundwater level time series into components related to measured input series. Very important components of groundwater levels are the fluctuations due to precipitation excess and drawdown caused by groundwater abstractions. With the aid of the decomposition, the fluctuations in and the drawdown level caused by each input can be quantified, as well as the reliability of the estimated fluctuations and drawdown.

Transfer/noise models can also be used to forecast groundwater levels including confidence intervals of those forecasts. In particular, forecasting is sensible if some delay time occurs between the input and the groundwaterlevel, because then the actual measurements of the input (e.g. precipitation) can be used in the forecasting. If no delay time occurs, first forecasts of the input series are needed. Generally, with forecasted input series the confidence intervals of the forecasts will be too wide to have acceptable results.

A third possible application of transfer/noise models is simulating the effects of alternative input series (e.g. for groundwater abstractions) on the groundwater level. The simulation will only give sensible results if the model structure is known in advance and will not be affected by the changes of the input.

Case studies

In the Netherlands, the TNO-DGV Institute of Applied Geoscience has applied transfer/noise modelling to several geohydrologic problems (Defize & Rolf 1985, Van Geer 1986). In this paper two examples are presented.

(a) detection of the influence of the river IJssel on groundwater fluctuation in two observation wells.

(b) detection of a possible trend in the groundwater level at a location in the province of Noord-Brabant.

The location of the observation wells near the river IJssel (33G-83 and 33H-27) and Noord-Brabant (51E-2) are located in Figure 2.

Figure 2   Locations of the observation wells.

Influence of the river IJssel on the groundwater level

The observation wells 33G-83 and 33H-27 are located at distances from the river IJssel of 0.8 and 6 km respectively. The presumed dominant causes of groundwater level fluctuations are the precipitation excess and the stage of the river IJssel. The available information for the inputs were:
(a) daily measurements of the river stage
(b) ten-day totals of the precipitation and Penman evaporation
The groundwater level is measured 24 times per year (according with a time step of about 15 days). Both inputs have been converted to input series with a frequency of 24 times per year. The precipitation excess is calculated by: Precipitation - 0.8* Penman Evaporation. As input series for the precipitation excess the total volume during one step of 15 days is taken (Figure 3).   The input series of the river stage is obtained by taking the average stage of three days, preceding the groundwater observation date (Figure 4).
The results of the decomposition of the groundwater level in observation wells 23G-83 and 33H-27 are presented respectively in Figures 5 and 6. As can be seen from Figure 5, the groundwater level fluctuations in point 33G-83 are dominated by the river stage. The influence of the precipitation excess is only of minor importance. In well 33H-27 the opposite is the case (Figure 6). Here, the groundwater level is influenced little by the river stage, while the component due to the precipitation excess is a substantial part of the groundwater level.

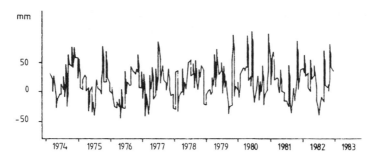

Figure 3 Input series for precipitation excess.

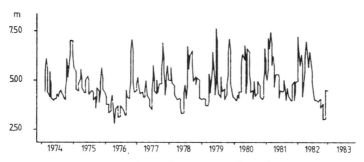

Figure 4 Input series for the water level.

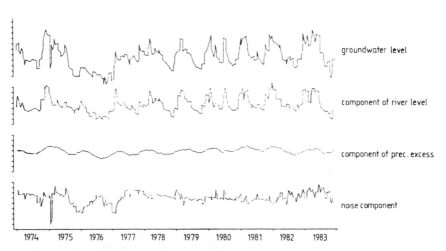

Figure 5 Decomposition of series 33G-83.

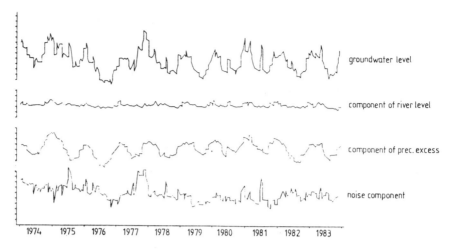

Figure 6   Decomposition of series 33H-27.

Trend detection in Noord-Brabant

The observations in well 51E-2 in the province of Noord-Brabant indicate a moderate downward trend in the groundwater level during the years 1951-1982. A transfer/noise model has been applied to detect whether this trend is caused by changes in regional water management or it is just the result of a natural sequence of dry periods. A very important aspect of this research was the confidence interval of the estimated trend. In other words, whether or not the (artificial) trend is significant.

A transfer/noise model with two inputs has been applied: precipitation excess and the "regional groundwater management".

Measurements of the precipitation and Penman evaporation were available as ten-day totals. The calculated precipitation excess (precipitation - 0.8 * Penman evaporation) has been converted to an input series with measurement frequency of 24 times per year (the groundwater measurement frequency) in a similar way as in the previously described application.

The actual changes in water management are not known, that is: not explicitly. To have an input series for the "regional water management", a linear increasing input series is generated. Of course the actual changes in water management are not known, e.g. the start of a new pumping station generally does not result in a linear increasing input series. But if the inclusion of a linear trend in the transfer/noise shows a significant improvement in comparison with a transfer/noise model without a trend, strong indications for a trend, regardless of the shape, are present. Inference about the shape of the trend demands a more thorough analysis of the water management situation than presented in this paper.

The results of the transfer/noise model are given in Figure 7.

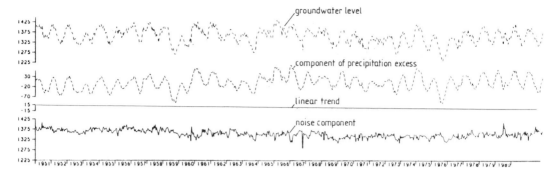

Figure 7   Trend analysis of series 51E-2.

This figure shows a linear trend resulting in a lowering of 27cm over a period of 32 years. The standard error of this estimated lowering is 5 cm.

Conclusions

Applicability

The applications that have been presented in this paper could also be analysed with the technique of linear regression. The model that is used with this method is a special case of the general transfer/noise model. In particular, the time aspect resulting in correlated observations, cannot be dealt with satisfactorily and therefore the more general approach of transfer/noise-analysis is preferable.

For practical problems transfer/noise models are relatively easy to use. The main possibilities of applications are:

(a) Decomposition of groundwater measurement series into natural and artificial components

(b) short time forecasting of groundwater levels

(c) simulation of changes in the groundwater regime

(d) a quick-screening of the data material in preliminary investigations.

Transfer/noise models can be applied if the measurement series of inputs and output are sufficiently long. The period over which measurements should be available depends mainly on the coherence (auto correlation) of the measurement series. For the interpretation of the results a qualitative knowledge of the geohydrologic conditions in situ is required.

Comparison of transfer/noise models and deterministic models

In comparison with the "classical" deterministic models, (e.g. finite elements, finite differences) transfer/noise models require less modelling effort, because no detailed information about the geohydrologic properties need to be collected . Moreover, generally, the dimensions of the computer programs for transfer/noise models are smaller than those for deterministic models. So, transfer/noise modelling can be performed in less time and for lower costs than deterministic models.

Transfer/noise models are restricted to locations, where measurements are available and are not suitable for spatial interpolations.

Remark

In the modelling process of transfer/noise models a number of more or less arbitrary assumptions have been made (as with deterministic geohydrologic models). Therefore basic knowledge of time series analysis as well as of geohydrology is required.

References

Box, G.E.P. and Jenkins, G.M. (1976). Time series Analysis: forecasting and control. Holden-Day, San Francisco.

Defize, P.R. and Rolf, H.L.M. (1985). Statistical analysis of some groundwater measurement series in Easter Gelderland (in dutch). DGV Report No. OS 85-94.

Geer, F.C. van (1986). Lowering of the groundwater level in the Netherlands. Phase 1, Problem verification: Trend analyses of six groundwater measurement series in the period 1951-1984. (In Dutch). DGV Report No. OS 86-25.

The Influence of Climate Change and Climatic Variability on the Hydrologic Regime and Water Resources (Proceedings of the Vancouver Symposium, August 1987). IAHS Publ. no. 168, 1987.

Hydrologic response to an artificial climatic change of rainfall enhancement

Arie Ben-Zvi
Israel Hydrological Service
Jerusalem, Israel
Margarita Langerman
Rainfall Enhancement Unit
Mekorot, Ben-Guryon Airport
Israel

ABSTRACT   The effect of an artificial climatic change on the flow towards a lake is detected and estimated through a comparison of historical records.   The change is generated by the cloud seeding operations in Northern Israel.   The yardstick for the comparison is the depth of the rainfall at the control area for the seeding.   The estimated effect of exogenic man-made changes is included in the comparison.

The change is detected by a comparison of statistical parameters and through a nonparametric test. The magnitude of the change is estimated by a cross application of linear regression models.   The annual flow volume is found to increase by 9%, of which 5% is attributed to the result of climatic fluctuations.   The increase is not significant statistically, but from other research stages it is concluded that a change in the rainfall and in the flow does exist.   Therefore, for water resources management purposes it is recommended to attribute 3% of the total flow to the effect of the rainfall enhancement operations.

Introduction

The evaluation of the hydrologic response to a climatic change depends upon the availability of a quantitative description of the meteorologic change, and of a hydrometeorologic model reliable for both the situations prior to and post to the change.   In an ordinary case of a climatic change the description might be deferred until the occurrence of the change is recognized and distinguished from normal fluctuations.   This process is shorter in the case of an artificial climatic change.   An example of such a change is the project of rainfall enhancement in Israel (Gagin and Neumann 1974, 1981).

Cloud seeding for rainfall enhancement has been carried out in Israel in two modes: operational and experimental.   In the operational mode the clouds are seeded on every day when rain clouds appear, whereas in the experimental mode the clouds are seeded on only randomly pre-selected days.   The evaluation of the meteorologic results of the experiments is based upon comparisons between the rainfalls on seeded and unseeded days, at the target and at the control areas for the enhancement.   The evaluation of the hydrologic results for systems which respond within the experimental time unit can be made by similar comparisons (Ben-Zvi, 1985). Such comparisons are not suitable for systems in which the response time is longer. For these systems the evaluation should rely either on compound models or on comparison of historical records. The compound models require the introduction of internal water balances and their parameters, which, in many cases cannot be estimated accurately enough. On the other hand, the comparison of historical records might involve estimating results of possible exogenic man-made changes and by probable natural fluctuations. The selection between these options depends upon the type of system, the kind of available data, and on personal preference. In the present study we prefer the comparison of historical records.

When comparing historical records, we would like to reduce the effect of exogenic changes and natural fluctuations. In cases where we know the occurrence and magnitude of the exogenic changes we can allow for their influence. The problem of natural fluctuations is more complicated. Gabriel and Petrondas (1983, 1984) have shown through examples that the significance level of the changes in historical records is biased towards the radical conclusion and recommend the use of a correction factor to obtain a less biased conclusion. Howell (1984) recommends reducing the bias by a prior selection of a control which is correlated with the target. These recommendations are considered in the present work.

Rainfall enhancement in Israel

Cloud seeding for rainfall enhancement has been carried out in Israel since 1960. Experimental seeding was conducted in 1960/61 to 1966/67 and in 1969/70 to 1974/75, while operational seeding was conducted in all the other years since 1968/69. Seeding is done from planes flying on fixed tracks and from ground furnaces. The track from 1960 to 1969 followed the Mediterranean coastline and since then follows the foothills of northern Israel, and at several locations in the central region. The locations of the tracks in Northern Israel are shown on Figure 1.

The target areas for the first experiment were the northern and the central regions of Israel. The seeding was carried out on either region, pre-selected randomly, while the other served as a control. The target area for the second experiment was in northern Israel, while the control area was in the coastal plain upwind of the seeding track for this region.

Rainfall and runoff changes

Gagin and Neumann (1974, 1981) have estimated the change in the depth of precipitation due to experimental cloud seeding by comparing four precipitation depths: at the target area on seeded days, at the target area on unseeded days, at the control area on seeded days, and at the control area on unseeded days. The formula for the estimation is a double ratio defined as:

$$D_r = P_{ts} \, P_{cu} \, / \, P_{tu} \, P_{cs} \tag{1}$$

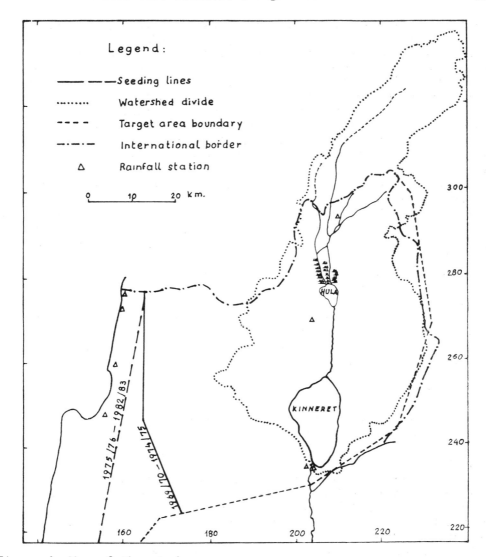

Figure 1   Map of the study area.

in which P is precipitation depth, t denotes target, c denotes control, s denotes seeded, u denotes unseeded, and D_r is the double ratio result.

The estimated increase show geographical variation which reaches a maximum of 15% to 20% over an area located 30 to 50 km downwind from the seeding track. The significance level of the increase was tested by the Mann Whitney Wilcoxson test and by use of numerous random mixings of the observed data.

Gagin and Gabriel (1986) found that the increase in the rainfall depth is accompanied by an increase in the rainfall duration and in the number of rainspell events, but not by an increase in the rainfall intensity.

Sharon (1978) compared the structure of the correlation coefficients between rainfalls at different stations for different historical time series. He found that the coefficients for the seeded periods are generally larger than those for the unseeded periods.

Ben-Zvi (1985) formulated a regression model between daily values of rainfall and direct surface runoff from a small watershed on unseeded days. He found that the model did not describe the relationship for the seeded days. The percentage of the rainfall which reappeared as surface runoff on the seeded days was twice as large as that on the unseeded days.

Ben-Zvi, Massoth and Anderman (1987) compared historical records of the volume of springflow in the target area with respect to the depth of rainfall at the control area and at the target area. With respect to the control, the total volume of the springflow increased by 9%. Yet, a decrease was found in the flow of a number of the studied springs.

Object and methods

The foregoing discussion indicates that cloud seeding in Israel results in a detectable climatic change. The magnitude of the change is small, but the time and space distributions of its occurrence are known. Adjacent to the areas of the enhanced precipitation there are unaffected areas which can serve as a control. Meteorologic and hydrologic data for the target and for the control areas are available both prior to and after the change. Noting these advantages we have selected the hydrology of the target area for rainfall enhancement as the object for the study of the hydrologic results of a climatic change.

We deal here with the flow towards Lake Kinneret which is the largest hydrologic system in the target area, and is a principle component of the Israeli water network. According to Gagin and Neumann (1981), the rainfall over the segment of the Kinneret's catchment, which lies within the target area for the rainfall enhancement in the second experiment, increased by 18%. However, in addition to this change there are other human-made changes in the watershed. The most notable being changes to drainage network and agricultural practices. In order to reduce the effect of these changes we add to the recorded flow into the Lake the net consumption of water in the watershed.

The study of the hydrologic results is composed of two steps: detection of the change and estimation of its magnitude. The detection is carried out by use of statistical comparisons and tests, while the estimation is carried out by use of regression models. The examined hydrologic series is of the net flow volumes towards Lake Kinneret. The yardstick for the examination is the series of rainfall depths at the control area. An intermediate variable, the rainfall over the catchment area, plays a role in the explanation of the changes.

The study area

Lake Kinneret is a fresh water open lake, the average area of its surface is 165 km^2, its maximal depth is about 45m, and its total capacity is about 4.5 km^3. The mean annual volume of the flow into the lake is about 800 x 10^6 m^3, of which 550 x 10^6 m^3 discharge through the Upper Jordan River, and 250 x 10^6 m^3 discharge from other rivers, from peripheral and submarine springs, and as direct rainfall. About 300 x 10^6 m^3 per annum evaporate from the lake, while the mean annual consumption and overflow to the Lower Jordan River is about 500 x 10^6 m^3. The mean net hydrologic input to the lake is about 500 x 10^6 m^3 /annum, and the standard deviation is almost 300 x 10^{6}m^3. The volume of the net inflow is computed from water balances, where it is equal to the sum of the consumptive use from the lake, the overflow and the storage change. These variables are measured with a good accuracy, and therefore the net inflow can also be considered as being accurate.

The catchment area of Lake Kinneret is about 2800 km^2. It extends over portions of Mount Hermon, the Golan Plateau, the Galilee Mountains, and the Hula Valley. The maximal elevation of Mount Hermon is about 2000m, and that of the Galilee Mountains is about 1000m. The typical elevation of the Hula Valley is +70m, and that of the water surface of Lake Kinneret is -210m. The mean annual depth of precipitation ranges from 1200mm at the Hermon, to about 800mm at the higher section of the Galilee, and to about 400mm at the southern Galilee, the southern Golan and on Lake Kinneret. The upper rock layers of the mountains in the catchment are composed of fractured carbonatic facias which allow a high rate of infiltration. According to Goldschmidt (1955) the mean annual volume of the flow through the Upper Jordan River is equal to about 50% of the volume of the precipitation over its catchment.

A large portion of the Hula Valley was covered in the past by a shallow lake and swamps which have been drained in the years 1955 to 1959 and turned into agricultural land. The mean annual volume of evaporation from Lake Hula and the Swamps was estimated by Neumann (1955) as 60 x 10^{6}m^3. Goldschmidt and Kornitz (1954) have estimated the exchange of water between the Jordan River and Lake Hula and the Swamps, and have simulated the daily volumes of the flow of the Jordan River, from 1939/40 to 1952/53, for the situation as if the Hula Lake and Swamps did not exist. Simon and Yatir (1981) have estimated the net consumptive use and the other abstractions from the Hula Valley for the years 1963 to 1978. Their estimation for the mid-seventies is 110 x 10^6 m^3/ annum.

In order to reduce the effect of drainage and other human-made changes we have added the estimated abstractions to the recorded flow to Lake Kinneret. These abstractions are as estimated by Goldschmidt and Kornitz (1954) and by Simon and Yatir (1981). The abstractions for the years 1960 to 1962 and 1979 to 1983 are estimated by us as an extrapolation from those of Simon and Yatir (1981). It should be mentioned here that other increasing abstractions of water in the catchment, are not considered here due to shortage of data for their estimation.

The depth of precipitation over the Kinneret Catchment is represented here by the mean for two stations: Dafna and Ayelet

Hashahar which are located near the Hula Valley, where the maximal effect of the seeding is found for the second experiment (Gagin and Neumann, 1981). The depth of precipitation over the control area is calculated as the mean of four stations: Rosh Hanikra, Naharia, Acco and Haifa Harbour. In addition we consider here a rainfall at the edge of the target area, which is calculated as the mean of two stations: Degania and Kinneret. The locations of the eight rainfall stations are presented on Figure 1. The mean depths of precipitation for the three groups of stations is presented in Table 1. The abstraction corrected volumes of the net flow towards Lake Kinneret are also presented in Table 1. Owing to limitations in the availability of the various data, we have simultaneous records of only 12 years for the unseeded state, and of 13 years for the seeded state.

Table 1   The data

Year	Flow (MCM)	Precipitation (mm)		
		Cont.	Targ.	Edge
40/41	544	556	537	345
41/42	536	315	424	302
42/43	734	832	690	547
43/44	525	569	510	409
44/45	826	749	718	678
45/46	580	646	449	363
46/47	470	501	396	303
49/50	673	618	459	511
50/51	157	364	300	217
51/52	715	688	562	523
52/53	664	650	557	512
59/60	271	435	313	257
69/70	703	529	548	427
70/71	778	798	624	452
71/72	516	566	525	413
72/73	272	386	308	281
73/74	626	633	746	430
74/75	467	616	468	344
75/76	564	583	512	391
76/77	701	739	598	416
77/78	673	669	593	501
78/79	282	381	306	206
79/80	877	751	747	652
80/81	880	661	728	460
81/82	435	427	446	235
82/83	783	740	653	542

Legend: Flow is the volume of flow towards Lake Kinneret, Cont., Targ., and Edge are the mean depths for stations located at the control, target and edge areas of rainfall enhancement.

Detecting the change

The data of the present study have been divided into two groups: prior to and after the change. We ignore here the differences between the operational and the experimental seedings since 1969/70. But, as the seeding track and the geographical location of the effect of the seeding in the years 1960/61 to 1968/69 were different from those of

the later seeding, we do not consider the data for the years 1960/61
to 1968/69.

The changes in the precipitation and hydrologic response due to
seeding is studied from differences in the statistical parameters of
the data, and by use of a significance test. The statistical
parameters of the records are presented on Table 2, while measures of
the comparison are presented on Table 3.

Table 2   Statistical parameters

Variable	Flow		Cont.		Targ.		Edge	
	U	S	U	S	U	S	U	S
Mean	558	611	577	606	493	557	414	411
S.D.	192	198	153	136	130	143	140	119
Median	548	627	570	617	461	548	363	417
Lag 1	-.37	-.17	-.37	-.38	-.15	-.42	-.29	-.52

Table 3   Comparison of variables

Variable	Flow	Cont.	Targ.	Edge
Difference: seeded minus unseeded				
Mean	53	29	64	-3
S.D.	6	-17	13	-21
Median	79	47	87	54
Lag 1	.20	-.01	-.27	-.23
Ratio: seeded over unseeded				
Mean	1.09	1.05	1.13	0.99
S.D.	1.03	0.89	1.10	0.85
Median	1.14	1.08	1.19	1.15
One sided significance level				
MWW	0.26	0.30	0.07	0.54

We find that the increase in the mean runoff of $53 \times 10^6 m^3$ amounts
to 9% of the value for the unseeded state. At the same time the mean
annual depth of the rainfall has increased by 5% for the control
area, and by 13% for the target area, while it has decreased by 1%
for the edge of the catchment. Higher rates of increase are found
for the median values of the series.

The standard deviation of the flow values has increased by 3%, and
that of the rainfall at the target by 10%. The standard deviation of
the control has decreased by 11%, and that of the edge has decreased
by 15%. Thus, the coefficients of variation for the seeded state are
lower than those for the unseeded state.

Concerning the lag-one serial correlation coefficient we observe a
less negative value for the flow, little change for the control, and
more negative values for the target and for the edge after seeding.

The significance of the changes is examined by the one sided Mann

Whitney Wilcoxon test. The significance levels are are presented in Table 3. Following the criteria set by Gabriel and Petrondas (1983, 1984), we conclude that while the change in the rainfall at the target seems statistically significant, the flow change is of undecided significance. The change in the rainfall at the edge seems insignificant, and at the control is of undecided statistical significance.

Estimation of the magnitude

The magnitude of the change is estimated through an application of linear regression models which relate the flow to the rainfall at the control, at the target and at the edge. They relate also the rainfalls at the target and at the edge to the rainfall at the control. The parameters which are obtained for the models are presented in Table 4. The correlations appear a little higher for the seeded state but no general trend is apparent in the regression coefficients.

   The models formulated for the unseeded state have been applied to the mean depths of the rainfall for the seeded state, and vice versa. The results are presented in Table 5 under the column 'Computed'.

Table 4   Regression parameters

Depend	Indep.	Unseeded			Seeded		
		R	A	B	R	A	B
Flow	Cont.	.81	1.01	-24	.86	1.25	-147
Flow	Targ.	.89	1.31	-89	.92	1.27	-95
Flow	Edge	.91	1.26	37	.88	1.46	11
Targ.	Cont.	.85	0.72	75	.83	0.87	32
Edge	Cont.	.86	0.78	-37	.83	0.73	-29
Targ.	Edge	.90	0.84	145	.85	1.02	138

Legend: R is the correlation coefficient, A is slope, and B is the location.

Table 5   Model results

Depend	Indep.	Computed		Increase		Ratio	
		U	S	U	S	U	S
Flow	Cont.	575	587	17	24	1.03	1.04
Flow	Targ.	529	642	-29	-31	0.95	0.95
Flow	Edge	615	555	57	56	1.10	1.10
Targ.	Cont.	532	514	39	43	1.08	1.08
Edge	Cont.	390	437	-24	-26	0.94	0.94
Targ.	Edge	561	490	68	67	1.14	1.14

Legend: U denotes unseeded, S denotes seeded.

The increase due to the rainfall enhancement is estimated by the formulae:

$$I_u = Q_c - Q_m \qquad (2)$$

$$I_s = Q_m - Q_c \qquad (3)$$

in which I is the increase presented in Table 5, Q is the value of the dependent variable (flow, rainfall at the target or rainfall at the edge), u denotes unseeded, s denotes seeded, m denotes measured mean for the denoted state, and c denotes value computed by the model calibrated to the other state and applied to the mean of the indepen- dent variable (control, target or edge) for the denoted state.

The relative increase due to the rainfall enhancement is estimated by the formulae:

$$R_u = Q_c \ / \ Q_m \qquad (4)$$

$$R_s = Q_m \ / \ Q_c \qquad (5)$$

in which R is the increase ratio.

Table 5 shows that, with respect to the control, the net flow towards Lake Kinneret increased by 20 x 10^6 m^3/ annum, a relative increase of 3%. With respect to the rainfall at the target the flow decreased by 5%, while with respect to the edge it has increased by 10%. These differences are explained by the relative increase of the target rainfall by 10%, and the relative decrease of the edge rainfall by 6%.

Discussion

The decrease in the coefficients of variation, and the increase in the medians by values larger than those of the means, seem to result from sampling variations from the unseeded to the seeded historical records.   The negative values of the lag l serial correlations seem to be related to the processes generating the rainfall. Their probable effect on the results of the present work are not evaluated here.

The cross application of the linear regression models, to data different from those used for their calibration, yields pairs of similar results, as is shown in Table 5.   This similarity verifies the predictive power of the simple linear regression model for the present situation, and for the estimation of the changes which are found here.

The rainfall at the target area has increased with respect to the rainfall in the upwind direction (control area) as well as to that in the sidewind direction (edge).   Our data are not sufficient alone to conclude about the cause of this increase, but in view of more detailed studies (Gagin and Neumann, 1974, 1981) we can attribute the increase to the rainfall enhancement operations.

The stations selected to represent the target area are located in the sub-area where the highest increase is found for the rainfall enhancement (Gagin and Neumann, 1981). Therefore, the average increase in the precipitation over the Kinneret Catchment, as it would have been estimated by our methods, should be lower than that presented here for the target. Therefore, in case of a proportional rainfall-runoff relationship the resulted relative increase in the runoff should be lower than that presented here for the rainfall at the target.

The mean change in the flow, which is attributed to the rainfall enhancement program, is only 20×10^6 m^3/ annum. This volume is about 3% of the mean annual volume observed for the unseeded state, and it lies within the accuracy margins of the data. Yet, the higher relative increase of the rainfall at the target, with respect to the control and to the edge, indicates that the increase in the flow is a logical result. Supports for this conclusion are found in the works of Ben-Zvi (1985) and Ben-Zvi Massoth and Anderman (1987) where two components of the flow are studied separately. In the first work the relation of the direct surface runoff to the rainfall on seeded days is found different from that found for unseeded days. In the second work the recharge to groundwater resources, which feed small springs, is found higher, with respect to the control, for the seeded state, than that for the unseeded state.

Conclusions

A change of 9% is observed in the mean annual flow towards Lake Kinneret. Of this change, 3% are attributed to the effect of rainfall enhancement operations, and the rest to natural fluctuations which amount to 5% at the control area. This change is small and not significant statistically, but it follows a relatively higher and more significant change in the rainfall representing the target area for the enhancement. Observed changes in two small components of the flow support the conclusion that cloud seeding does have an effect on the flow. Therefore, we conclude that the rainfall enhancement operations contribute positively to the flow. For management purposes we estimate the mean contribution as 20×10^6 m^3 / annum.

References

Ben-Zvi, A. (1985) A change in the daily rainfall-runoff model of a small basin due to cloud seeding. In Scientific Basis for Water Resources Management (Proc. Jerusalem Symp. Sept. 1985), 31-42, IAHS Publ. no 153.

Ben-Zvi, A., Massoth, S. & Anderman, B. (1987) Changes in springflow following rainfall enhancement. Accepted by Israel J. Earth Sc.

Gabriel, R.K. & Petrondas, D. (1983) On using historical comparison in evaluating cloud seeding operations. J. Clima. Appl. Meteo., 22, 626-631.

Gabriel, R.K. & Petrondas, D. (1984) On using historical comparison in evaluating cloud seeding operations - Reply. J. Clima. Appl. Meteo., 23, 853-854.

Gagin, A. & Gabriel, R.K. (1986) Analysis of recording raingauge data for the Israeli II experiment: I. Effect of cloud seeding on the components of the daily rainfall. Personal communication.

Gagin. A. & Neumann, J. (1974) Rain stimulation and cloud physics in Israel. In Weather and Climate Modification ed. by W.N.Hess, 13,454-494, York.

Gabin, A. & Neumann , J. (1981) The second Israeli randomized seeding experiment: Evaluation of results. J. Appl. Meteo. 20 (11), 1301-1311.

Goldschmidt, M.J. (1955) Precipitation over and runoff from Jordan and Litani Catchments. Hydro. Paper no 1, Israel Hydrological Service, Jerusalem, Israel.

Goldschmidt, M.J. & Kornitz, D. (1954) The flow of the Jordan River as if the Hula Lake does not exist. Unpublished file. Israel Hydrological Service, Jerusalem (in Hebrew).

Howell, W.E. (1984) On using historical comparisons in evaluating cloud seeding operations - Comments. J. Clima. Appl. Meteo. 23, 850-852.

Neumann, J. (1955) On the water balance of Lake Huleh and the Huleh Swamps. Israel Explo. J. 5 (1), 49-58.

Sharon, D. (1978) Rainfall fields in Israel and Jordan and the effect of cloud seeding on them. J. Appl. Meteo. 17, 40-48.

Simon, E. & Yatir, Y. (1981) Synthetic series of streamflow in the Upper Jordan Catchment. Rep. 01/81/102 Tahal, Tel-Aviv, Israel (in Hebrew).

The Influence of Climate Change and Climatic Variability on the Hydrologic Regime and Water Resources (Proceedings of the Vancouver Symposium, August 1987). IAHS Publ. no. 168, 1987.

Will clouds provide a negative feedback in a CO_2-warmed world?

K. McGuffie & A. Henderson-Sellers
Department of Geography
University Of Liverpool
Liverpool, L69 3BX, U.K.

ABSTRACT   Cloud cover records for western Europe, the continental United States and the Indian sub-continent have been analysed in the context of a warming world historical analogue model. It is found that cloud cover has generally increased between the coldest and warmest years this century in each of these three climatologically diverse areas. This result seems to be at variance with the few available climate model predictions which suggest reduced cloud amounts. The degree of cloud amount change is roughly the same although the sign is opposite from that given in model simulations. The real world cloud results may indicate, assuming other cloud properties remain approximately constant, a tendency to reduce initially warmed surface temperatures, i.e. the cloud changes could provide a negative feedback effect for the climate.

Clouds in a warmer world

It seems likely that the increase in the concentration of atmospheric carbon dioxide over pre-industrial levels caused predominantly by the combustion of fossil fuels will lead to increased temperatures (NAS, 1982, WMO, 1986). There are two different and complementary methods which can be employed in trying to understand and predict the response of the climate system to these increased temperatures. One method is the use of large numerical climate models. The other is the study of past climates. Interpretation of palaeoclimatologic data can itself take a wide range of forms (cf. Schneider, 1986). One of these is the construction of historical analogue climate scenarios based on the instrumental record. Analysis of the period of the instrumental record is restricted to the last 100–150 years. A disadvantage of this instrumental scenario method is that this period offers relatively small hemispherical temperature changes compared with those expected to result ultimately (i.e. in an equilibrium change) from future increases in atmospheric carbon dioxide concentration. Thus instrumental scenarios, such as the study described here, should be taken as being indicative of changes and conditions to be anticipated during the early phase of CO_2-induced warming. On the other hand, the gradual, but steady, rise in atmospheric CO_2 levels will cause the climate system to respond relatively slowly. The historical analogue method used here may, fortuitously, incorporate important changes in oceanic and cryospheric boundary conditions.

The atmospheric carbon dioxide concentration is presently about 345 ppmv as compared with various estimates of the pre-industrial concentration ranging from 260 to 290 ppmv (e.g. Brewer, 1978; Chen and Millero, 1979; Wigley, 1983). Predictions of the likely atmospheric concentration of carbon dioxide for the year 2025 range from 440 to 600 ppmv, depending upon the forecast growth rate of energy used (Wigley, 1982). It is for this reason that climate modelling studies generally investigate the effect of doubling the atmospheric concentration of carbon dioxide. Such studies suggest that a doubling of atmospheric CO_2 will produce an increasing global mean temperature of 2-4°C (Gates, 1980; NAS, 1982; WMO, 1986).

Probably the single most difficult climatic feedback process to measure and hence to model is the cloud-radiation feedback (GARP/JOC, 1978). It is well known that clouds affect both the incoming shortwave radiation, generally tending to increase the planetary albedo, and also the emitted terrestrial radiation tending to enhance the "greenhouse effect" (Ohring and Clapp, 1980; Cess et al.,1982). The question of which feature dominates probably varies as a function of cloud type, cloud height and cloud structure (e.g. Schneider, 1972; Shukla and Sud, 1981; Wetherald and Manabe, 1986). The semi-transparent nature of cirrus cloud and the dynamic nature of overlap of layered clouds make even simplified studies of cloud-climate interactions difficult (Stephens and Webster, 1979; Webster and Stephens, 1984).

Many of the climate model investigations of the likely impact of the increased atmospheric CO_2 use specified cloud cover (e.g. Manabe and Wetherald, 1975; Mitchell, 1983). This is clearly an unsatisfactory state of affairs but one which can, perhaps, be viewed sympathetically in the light of the difficulties of parameterizing cloud processes successfully in climate models and particularly the dearth of adequate cloud data. For example, Hansen et al. (1984) note that the physical process contributing the greatest uncertainty to their predicted climate sensitivity on a timescale of 10-100 years appears to be cloud feedback.

The transient nature of the current real-world experiment in which mankind is forcing an increase in the levels of carbon dioxide and other trace gases in the atmosphere (Dickinson and Cicerone, 1986) and the complexity of the cloud radiation feedback combine to make any cloud/climate sensitivity study hazardous. An alternative method of estimating the probable impact of a carbon dioxide induced warming is that of the construction of warm-world analogues from historical meteorologic records (e.g. Flohn, 1977; Wigley et al., 1980; Kellogg and Schware, 1981; Lough et al., 1983). The major advantage of using an historical analogue technique, as employed here, is that the climate (and cloud) change examined are transient. The results presented here are, at worst, a description of an observational data set which climate modellers could usefully exploit. On the other hand, they might indicate that the real-world transient experiment includes a negative feedback effect due to increasing cloud amount.

Analogue model for a warming world: "predicting" cloud change

The construction of analogue models with which to investigate the

probable effects of increasing atmospheric CO_2 was pioneered by Flohn
and Kellogg in 1977 and has been reviewed recently by Pittock and
Salinger (1982). Wigley et al. (1980) composited data from individual
years contrasting a group of warm years with a group of cold years .
More recently Lough et al. (1983) have suggested an improvement on
this basic technique. This analogue method was used successfully by
Henderson-Sellers(1986, 1987) to examine the probable cloud variation
associated with increasing temperatures in Europe and the continental
United States. In this paper, this analogue study has been extended
to include a very different climatological regime, the monsoon by analysis
of data from the Indian sub-continent.

Lough et al. (1983) selected the warmest and coldest twenty-year
periods from the gridded northern hemisphere temperature data produced
by Jones et al. (1982). Only temperature data for the period 1901-
1980 inclusive were considered. In this period the warmest 20 year
period is from 1934-1953 and the coldest from 1901-1920. The northern
hemisphere annual mean surface air temperature differed by 0.4 $^{\circ}$C
between these two periods. Any changes in climatic parameters noted
in going from the cold to the warm period are likely to be associated
with the gradual warming from 1901 to 1983 and may be associated with
the increase in atmospheric CO_2 during the early part of the
twentieth century. Even if the gradual warming examined here is not
the result of increasing atmospheric CO_2, the analogue of slowly changing boundary and temperature conditions is a useful tool with which to
study the transient CO_2-induced changes predicted for the present and
near future (Wigley and Schlesinger, 1985; Hansen et al., 1985)

Cloud observations are not included in the World Weather Record
and thus archive data sources must be sought if the historical record
of cloud amount is to be analyzed. As might be anticipated, quantity
and areal coverage of climatic data increase with time. In this
study, it was essential to obtain similar cloud information for the
cold and warm periods identified: 1901-1920 and 1934-1953. For the
case of Europe, observations of a meteorologic nature had often been
made since the seventeenth century and were commonplace by the early
1900s at many national (often astronomical) observatories. Climatic
data for the early part of this century are somewhat sparse in the
United States; in particular, before the advent of commercial
aircraft in the 1930s, there was little requirement for continuous
weather observations and consequently relatively few were made.
Meteorologic observations are quite extensive for the Indian sub-
continent throughout the first half of this century although there is
some disruption to the record in the 1950s.

Sixty stations were selected from the area of the Indian sub-
continent. Sources of the cloud data were the Indian Weather Review,
The Meteorology of Ceylon and Results of Meteorological Observations
in Ceylon and the Climatological Summary of Burma, all of which are
held by the U.K. National Meteorological Library. Many stations
reported in the Indian Weather Review have a gap in the record from
1949 to 1953 inclusive, with observations beginning again in 1954.
Generally there was only one observation made per day, until about
1933, usually at around 8 a.m. local time although some station
reports include more than one observation in the monthly mean reports.
From 1933 to 1954 there were usually 2 observations per day, typical-

ly at 8 a.m. and 4 p.m. local time although the times differ. The observations seem to have been made in tenths of sky cover throughout the period and monthly means are in tenths.

In this extension to the previous work of Henderson-Sellers (1986,1987) the same convention was used: all the available observations were averaged to produce monthly average cloud amount for the two periods. The one simplifying factor is that all the cloud amount reports contained in the meteorologic records used here are in tenths of sky cover, so no conversions were necessary in this work.

Any use of surface-based observations of cloud amount necessitates assumptions about the accuracy of the observations. It is well established that cloud observations even by trained meteorologists will differ from individual to individual (Meritt, 1966). However there is unlikely to be a systematic bias in the reports used here so that monthly means and hence the seasonal and annual averages analyzed should be fairly representative.

Changing cloud in a warmer world

The temporal trend is of increasing cloud over the study period. Figure 1 compares the northern hemisphere temperature data, in the form of anomalies from 1946-1960 reference period (Figure 1(a)) with the annual total cloud amount for (b) Europe, (c) the continental U.S.A. and (d) The Indian sub-continent. The spatially coherent upward trend in total cloud amount which is observed generally over all three regions strengthens the case for using the historical analogue model. However, caution must be exercised when using the historical analogue method and in particular when applying it to the topic of cloud changes. The fact that hemispheric mean temperature fell by ~0.3 °C in the 30 years (1945-1975) when atmospheric CO_2 concentrations were increasing most rapidly suggests that factors other than CO_2, or combining with CO_2, have at least a comparable effect on temperature. In estimates of the increase in global mean temperature over the last century (Wigley and Schlesinger, 1985) which allow for oceanic thermal inertia, the contribution of increased CO_2 to the observed warming between 1901 and 1953 is only ~0.2 °C, only approximately half the observed warming of 0.4°C. There is no guarantee that, for example long term but, perhaps, eventually random fluctuations in ocean temperature (Folland et al., 1984; Jones et al.,1986) would produce the same changes in cloudiness found with increased CO_2. Despite these caveats this investigation of cloud amounts offers a link between increasing atmospheric CO_2 and cloudiness which is an alternative to numerical modelling results.

Before examining the mean difference in cloud amounts between the cold and warm periods (1901-1920 and 1934-1953) for the annual case and for the 4 seasons it is necessary to establish whether there is any change in the variability between the two periods. The variability in cloud amount within each period about a 1:2:1 binomial filtered trend has been computed so that any changes in variability would be decoupled from possible changes in mean values. This is the same methodology as employed by Lough et al. (1983) and Henderson-Sellers

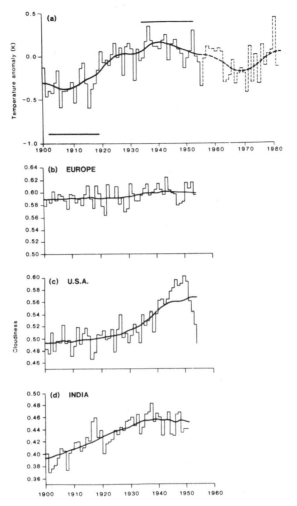

Figure 1 (a) Northern hemisphere mean surface air temperature variations(K) since 1900 shown as anomalies from 1946-1960 reference period. The curve is of 20 year filtered values. The warm and cold periods used here are marked (redrawn from Lough et al., 1983). (b) Europe mean annual cloud amount, 58 station average for the period 1900-1954. The curve is of 20 year filtered values. (c) U.S.A. mean annual cloud amount, 77 station average for the period 1900-1954. The curve is of 20 year filtered values. (d) Indian sub-continent mean annual cloud amount, 60 station average for the period 1900-1954. The curve is of 20 year filtered values.

(1986, 1987). The F test was applied to establish, at a 1% significance level, which stations exhibited a significant change in cloud variability between the cold and warm periods. The significance level for this test was set at 1% in order that a Student's t test applied at the 5% level would be valid on the mean differences as described below. Stations for which the F test failed at 1% level were excluded from Figure 2(a) which shows contours of the Student's t statistic and are shown at the 5% and 0.1% levels for both increasing and decreasing cloudiness. The annual mean differences (warm minus cold) in cloud amount for the three regions are shown in Figure 2(b). The maps show a very considerable degree of consistency. There is generally an increase in total cloud amount of greater than 0.3 tenths of sky cover in the "warming world".

Conclusions - Will clouds cause an important negative feedback on temperature in a warming world?

Cloud amount records for Europe, the U.S.A. and the Indian sub-continent have been analyzed in the context of the "warming world" analogue model described by Lough et al. (1983). Cloud amount increases over practically the entire U.S.A., most of the Indian sub-continent and some areas of Europe in all seasons. These results considerably strengthen the more tentative conclusions of Henderson-Sellers (1986,1987). It is, of course, essential that cloud type (i.e. radiation parameters and cloud height) be predicted successfully (Charlock and Ramanathan, 1985) as well as total cloudiness. In fact it is arguable whether total cloudiness is an issue since the radiative properties depend on the liquid water path not cloud amount (e.g. Somerville and Remer, 1984). The observational data such as those described here offer no information about these other characteristics. Interestingly the results are in contrast to the few numerical model predictions of cloud changes in warming world experiments. A possible, rather tantalizing, conclusion is that current GCM cloud prediction schemes tend to enhance temperature increases through cloud-climate feedback whereas the historical data could suggest a negative feedback. Part, possibly all, of this difference may be the result of the fundamental distinction between the two experimental scenarios: the equilibrium change modelled by

Figure 2   (Next page)
(a) Variation of the Student's t statistic for Europe, U.S.A. and the Indian sub-continent generated from the mean differences(warm minus cold period) for the annual case. A positive difference (warm-cold) suggests an increase in cloudiness in a warmer world. The values at each station have been normalised by the station standard deviation for the 40 years divided by $\sqrt{n}$, where n is the number of years for which valid data were available. While the zones of 5% significance have been shown to be statistically valid an additional test of the variances at a higher significance level would be required to demonstrate the validity of the 0.1% contours shown.
(b)Annual differences in total cloud amount (tenths of sky cover) predicted from the historical analogue study (partly from Henderson-Sellers, 1986, 1987).

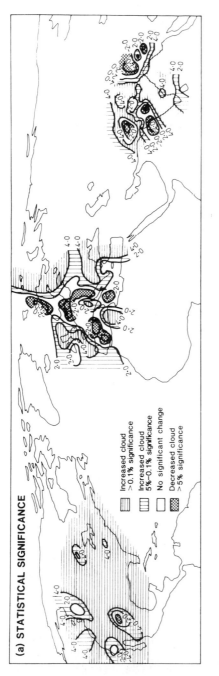

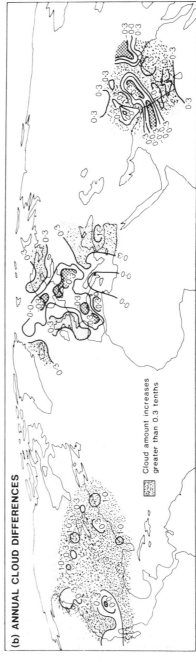

Figure 2 (see p.622 for caption)

GCMs as compared to the smaller transient change represented by the historical analogue. On the other hand the current "real-world" experiment is, itself, a transient change in boundary and atmospheric conditions. At the least, surface-observed cloudiness seems to offer a useful and complementary data source with which to examine one aspect of the performance of numerical climate models.

ACKNOWLEDGEMENTS   This research was sponsored in part by the Air Force Office of Scientific Research, Air Force Systems Command, USAF under grant number AFOSR-86-0118. The U.S. Government authorised to reproduce and distribute reprints for Governmental purposes notwithstanding any copyright thereon.

References

Brewer, P.G., 1978, Direct observations of the oceanic CO_2 increase, Geophys. Res. Lett., 5, 997-1000

Cess, R.D., Briegleb, B.P. & Lian, M.S., 1982, Low-latitude cloudiness and climate feedback: comparative estimates from satellite data, J. Atmos. Sci., 39, 53-59

Charlock, T.P., 1982 cloud optical feedback and climate sensitivity in a radiative-convective model, Tellus, 34, 245-254.

Charlock, T.P. & Ramanathan, V., 1985, The albedo field and cloud radiative forcing produced by a general circulation model with internally generated cloud optics, J. Atmos. Sci., 42, 1408-1429.

Chen, C.T. & Millero, F.J., 1979, Gradual increase of oceanic carbon dioxide, Nature, 277, 205-206

Dickinson, R.E. & Cicerone, R.J., 1986, Future global warming from atmospheric trace gases, Nature, 319, 109-115

Flohn, H., 1977, Climate and energy: a scenario to a 21st century problem, Climatic Change, 1, 5-20.

Folland, C.K., Parker, D.E. & Kates, F.E., 1984, Worldwide Maritime temperature fluctuations 1856-1981, Nature, 310, 670-673.

GARP/JOC, 1978, JOC Study Conference on Parameterization of Extended Cloudiness and Radiation for Climate Models, Oxford, England, GARP Climate Dynamics Sub-programme, WMO, Geneva

Gates, W.L., 1980, A review of modelled surface temperature changes due to increased atmospheric CO_2, Climatic Research Institute Rep. No. 17, Oregon State University, Corvallis, 21pp.

Hansen, J., Lacis, A., Rind, D., Russell, G., Stone, P., Fung, I., Ruedy, R. & Lerner, J., 1984, Climate sensitivity: Analysis of feedback mechanism, in, Climate Processes and Climate Sensitivity, (ed.) J.E. Hansen and K. Takahashi, American Geophysical Union, Washington, D.C., 130-163.

Hansen, J., Russell, G., Lacis, A., Fung, I., Rind, D. & Stone, P., 1985, Climate reponse times: dependence on climate sensitivity and ocean mixing, Science, 229, 857-859.

Henderson-Sellers, A., 1986, Cloud changes in a warmer Europe, Clim. Change, 8, 25-52.

Henderson-Sellers, A., 1987, Increasing cloud in a warmer world, Clim. Change, 8, 267-309.

Jager, J. & Kellogg, W.W., 1983, Anomalies in temperature and rainfall during warm Arctic seasons, Climatic Change, 5,39-60.

Jones, P.D., Wigley, T.M.L. & Kelly, P.M., 1982 Variations in surface air temperature, Part 1: Northern hemisphere, 1881-1980, Mon. Weath. Rev., 110, 59-70.

Jones, P.D., Raper, S.C.B., Bradley, R.S., Diaz, H.F., Kelly, P.M. & Wigley, T.M.L., 1986, Northern Hemisphere surface air temperature variations: 1851-1984, J. Clim. Appl. Meteorol., 25, 161-179

Kellog, W.W., 1977, Effects of human activities on climate, WMO Tech. Note 156, WMO No. 486, World Meteorological Organization, Geneva.

Kellogg, W.W. & Schwarre, R., 1981, Climatic change and Society, Consequences of Increasing Atmospheric Carbon Dioxide, Westview Press, 178pp.

Lough, J.M., Wigley, T.M.L. and Palutikof, J.P., 1983, Climate and climate impact scenarios for Europe in a warmer world, J. Clim. Appl. Meteor., 22, 1673-1684.

Manabe, S. & Wetherald, R.T., 1975, The effects of doubling the CO_2 concentration on the climate of a general circulation model, J. Atmos. Sci., 32, 3-15.

Merritt, E.S., 1966, On the reliability and representativeness of sky cover observations, J. Appl. Meteor., 5,369.

Mitchell, J.F.B., 1983, The seasonal response of a general circulation model to changes in CO_2 and sea temperatures, Quart. J. Roy. Meteor. Soc., 109, 113-152.

National Academy of Science (NAS), 1982, Carbon Dioxide and Climate: A Second Assessment, National Academy Press, Washington, D.C., 72pp.

Ohring, G. & Clapp, P.F., 1980, The Effect of changes in cloud amount on the net radiation at the top of the atmosphere, J. Atmos. Sci., 37, 447-454.

Pittock, A.B. & Salinger, J.M., 1982, Towards regional scenarios for a CO_2-warmed Earth, Climatic Change, 4, 23-40.

Schneider, S.H., 1972, Cloudiness as a global climatic feedback mechanism: The effect on the radiation balance and surface temperature of variations in cloudiness, J. Atmos. Sci., 29, 1413-1422.

Schneider, S.H., 1986, Can modeling of the ancient past verify predictions of future climate?-an editorial, Clim. Change, 8,117-120.

Shukla, J. & Sud, Y., 1981, Effect of cloud radiation feedback on the climate of a general circulation model, J. Atmos. Sci., 38, 2339-2353.

Sommerville, R.C. & Remer, L.A., 1984, Cloud optical thickness feedbacks in the CO_2-climate problem, J. Geophys. Res., 89, 9668-9672.

Stephens, G.L. and Webster, P.J., 1979, Sensitivity of radiative forcing to variable cloud and moisture, J. Atmos. Sci., 36, 1542-1556.

Webster, P.J. & Stephens, G.L., 1984, Cloud-radiation interaction and the climate problem, in The Global Climate (ed) J.T.Houghton, C.U.P., Cambridge, 63-78.

Wetherald, R.T. & Manabe, S., 1986, An investigation of cloud cover change in response to thermal forcing, Climate Change, 8, 5-23.

Wigley, T.M.L., 1982, Energy production and climatic change assessment, Uranium and Nuclear Energy: 1981. Proc. Sixth Int. Symp.,

Uranium Institute, Butterworth Scientific, 289–322.

Wigley, T.M.L., 1983, The pre-industrial carbon dioxide level, Climatic Change, 5, 315–320.

Wigley, T.M.L. & Schlesinger, M.E., 1985, Analytical solution for the effect of increasing CO_2 on global mean temperature, Nature, 315, 649–652.

Wigley, T.M.L., Jones, P.D. & Kelly, P.M. 1980, Scenario for a warm, high-CO_2 world, Nature, 283 17–21.

WMO, 1986, Report of the International Conference on the assessment of the Role of Carbon Dioxide and of other Greenhouse Gases in Climate Variations and Associated Impacts, held in Villach, Austria (October, 1985), WMO No. 661, World Meterological Organization, Geneva, Switzerland, 78pp.

(voir p. 37 S.V.P.)

Les tendances de la précipitation dans
l'Afrique de l'Ouest

Oyediran Ojo
Department of Geography
University of Lagos
Lagos, Nigeria

RESUME   Cet article examine les caractéristiques de   la
variation de la précipitation dans l'Afrique de   l'Ouest
entre 1901 à 1985. Les caractéristiques de la   précipitation, en particulier les tendances, les périodicités, et
les variabilités sont discutées tout en soulignant les
caractéristiques des périodes de dix ans, cinq ans et un
an, et les caractéristiques des saisons. Les conséquences
climato-hydrologiques causées par des changements climatiques, par exemple, les caractéristiques des débits des
rivières, la persistance, sévérité et la nature des
périodes de sécheresse et d'inondation, et les aspects des
variations temporelles et spatiales de certains composants
du bilan hydrique, ont été étudiés. Les conclusions sont
les suivantes:

    a) Ce siècle a commencé par une période de sécheresse
relativement longue (1901-1926). Celle-ci a été suivie par
une période humide relativement longue (1927-60).

    b) Les conditions présentes de sécheresse ont un
caractère persistant depuis 1961, et en particulier depuis
1965.

    c) En dépit des généralités mentionnées ci-dessus, il
n'est pas possible d'observer une régularité dans les
tendances, les périodicités, la persistance des conséquences hydrologiques des variations de la précipitation qui
permette la prédiction des sécheresses et de   leurs
conséquences.

    d) les variations spatiales et temporelles de la
précipitation et les conséquences climato-hydrologiques
rendent nécessaire l'amélioration de la disponibilité et
de la fiabilité des données.

(voir p. 77 S.V.P.)

Les fluctuations climatiques et l'écoulement
des bassins Alpins glacés

David N. Collins
Department of Geography
University of Manchester
Manchester, M13 9PL, UK

RESUME Des mesures hydrométriques concernant des rivières
drainant des bassins dont 35% à 65% de la surface est
glacée dans les Alpes Suisses, ont été analysées pour la
période 1922 à 1983 conjointement avec les données
météorologiques pour permettre la description des
régularités observées dans la variation du climat et pour
déterminer l'influence des facteurs climatiques sur le
ruissellement annuel total. La température moyenne de
l'air de mai à septembre et le débit annuel enregistrent
d'année en annee des variations considérables mais
cependant parallèles. Les étés généralement plus chauds
dans la dernière partie des années 1920 à 1929, le milieu
des années 1930 à 1939 et 1940 à 1949 et dans la dernière
partie des années 1956 à 1959, se traduisent par des
ruissellements plus abondants, avec un maximun en 1947.
Une diminution de la moyenne décennale de la température
de 1° C produit un réduction de quelques 25% dans la
moyenne décennale du ruissellement. Des températures plus
basses et des ruissellements moindres plus bas ont
caractérise les années 1960 à 1969 et 1970 à 1979 avant le
léger retablissement des années 80 à 86. Les moyennes sur
cinq ans des précipitations annuelles ont augmenté entre
1940 à 1979. La surface des glaciers a diminué entre 1920
et 1959 et s'est stabilisée après. On a obtenu des valeurs
élévees de la variance des debits en utilisant une
régression multiple dans laquelle les variables
indépendantes sont la moyenne de la saison d'ablation, la
température de l'air et l'accumulation de la précipitation
du hivernale. Les meilleures corrélations sont obtenues
pour les bassins à large surface glacée.

(voir p. 123 S.V.P.)

Paléocrues et variations climatiques

Victor R. Baker
Department of Geosciences
University of Arizona
Tucson, Arizona 85721 USA

RESUME   Récemment, on a fait des progrès climatiques importants dans la reconstruction et l'interprétation des crues anciennes, en particulier dans l'utilisation des dépôts dans les zones d'eau stagnante et des indicateurs des paléo-niveaux d'équilibre. Pour certains environnements convenables, on peut faire des estimations relativement exactes des débits des paléocrues et les dater sur des périodes de siècles et millénaires. De nouveaux outils statistiques sont disponibles pour extraire un maximum d'information de ces données hydrologiques non-conventionnelles. Les résultats dans le sud-ouest des Etats-Unis indiquent que certains intervalles de temps dans les derniers millénaires ont été caractérisés par l'apparition de larges crues, tandis que d'autres intervalles de temps ont été relativement peu soumis à de tels événements. La cause probable de cette non-stationnarité est le changement climato-hydrologique.

(voir p. 133 S.V.P.)

Variabilité des périodicités présentées par
les données des anneaux des arbres

A. Ramachandra Rao
School of Civil Engineering
Purdue University
W. Lafayette, IN 47907, USA
A. Durgunoglu
Illinois State Water Survey Division
Department of Energy and Natural Resources
2204 griffith Drive
Champaign, IL 61820

RESUME Les anneaux des arbres ont été largement utilisés pour étudier la variabilité climatique. Un résultat intéressant obtenu dans ces études est la "périodicité" de ces séries chronologiques. L'uniformité régionale de ces périodicités est analysée dans la présente étude. Si plusieurs séries d'une région ont le même comportement "périodique", un argument favorisant l'existence d'une variabilité climatique systématique dans la région serait assez fort. Par conséquent, l'étude des "périodicités" dans les données des anneaux des arbres dans une région est importante. Quatre séries de données des anneaux des arbres dans les bassins des rivières Salt et Verde, en Arizona sont analysées dans cette étude. Le test Blackman–Tukey et une variante de l'analyse spectrale par entropie maximale proposée par Marple ont été utilisés. Les données présentent un comportement périodique uniforme, mais la périodicité est faible.

(voir p. 143 S.V.P.)

Utilisation de données paléoclimatiques dans
l'étude des variabilités climato-hydrologiques
Gestion des ressources en eau

William J. Stone
New Mexico Bureau of Mines and Mineral Resources
Campus Station, Socorro, NM 87801 USA

RESUME   Les changements temporels dans la recharge des
nappes souterraines mettent en évidence la variabilité
climato-hydrologique. Les taux de recharge actuels et
anciens peuvent être estimés et datés en utilisant une
méthode de bilan de la masse de chlorure. Dans une région
du sud de l'Australie, il a été démontré qu'il faisait
plus humide il y a 13 500 à 16 000 ans qu'aujourd'hui.
Au centre-ouest du Nouveau-Mexique, les conditions
climatiques étaient plus humides il y a 7 000 à 17 000
ans.   Les données au nord-ouest du Nouveau-Mexique
indiquent que les conditions climatiques étaient plus
humides au cours de la même période générale mais aussi
plus récemment, il y a 1000, 900 et 400 ans.   Des
périodes sèches ont précédé ces intervalles humides (qui
ont eu lieu il y a 1 000 à 2 000 et 10 000 à 14 000
ans).   La plupart de ces périodes correspondent à des
régimes climato-hydrologiques et paléoclimatiques déjà
reconnus dans des études antérieures. Ces différences dans
le taux de recharge sont importantes pour la gestion des
ressources en eau et le contrôle des déchets. Les taux
peuvent être utilisés pour le modelage et pour éviter la
surexploitation des ressources en eau. Les taux de
recharge plus élevés peuvent être considérés comme les
pires valeurs possibles pour l'installation de dépôts
d'ordures.

(voir p. 163 S.V.P.)

Bilan hydrique des glaciers de la Cascade
Nord et implications climatiques

M.S. Pelto
Institute for Quaternary Studies
University of Maine, Orono Maine 04469
Foundation for Glacier & Environmental Research
Seattle, Washington 98109

RESUME   Un inventaire des glaciers de la "Cascade Nord" à
Washington et du bilan hydrique correspondant a été
exécuté pour déterminer leur réponse à des changements
climatiques récents et les changements consécutifs au
ruissellement glacial. La capacité variable des réservoirs
et la bilan hydrique moyen annuel ont été déterminés pour
47 glaciers de la région "Cascade Nord". Pour dix de ces
glaciers, le bilan hydrique annuel a été mesuré en 1984,
1985 et 1986. Le bilan hydrique annuel est fonction de
trois paramètres climatiques: l'activité anticyclonique
pendant la saison d'accumulation, la température pendant
la saison d'ablation et l'activité anticyclonique pendant
l'été. Depuis 1977, une fréquence accrue des conditions
anticycloniques a cáusé une augmentation de 1.1°C de la
température de la saison d'ablation et une diminution de
15% de la précipitation hivernal. Un bilan hydrique de –
0.32m à –0.62m en a été la conséquence pour la période
1977-1986. La diminution du bilan hydrique pendant l'hiver
et l'accumulation de la surface du glacier conduisent
aussi à une diminution de 15 à 24% de la capacité des
réservoirs des glaciers, ce qui correspond à la quantité
d'eau liquide qu'un glacier peut retenir.

(voir p. 173 S.V.P.)

Quantification des tendances à long terme de la
pollution atmosphérique et de l'eutrophication
agricole: Approche par "lake - watershed"

Ian D.L. Foster & John A. Dearing
Geography Department, Coventry Polytechnic
Priory St., Coventry CV1 5FB UK

RESUME Il est difficile d'obtenir des données historiques
qui documentent les tendances de la pollution atmosphérique par les émissions de métaux lourds de l'industrie et
des voitures et l'eutrophication agricole, causée par
l'industrialisation agricole. Cette situation existe,
principalement parce que la surveillance extensive des
divers aspects de la pollution n'a commencé qu'après avoir
identifié les problèmes majeurs. L'évaluation quantitative
directe de telles tendances est importante pour un large
nombre d'environnements non-documentés lorsqu'on examine
les niveaux actuels des nutriments associés avec les
sédiments et les contaminants de l'environnement fluvial
en relation avec les conditions naturelles de la même
région.
    Ce genre d'information peut être l'analyse
d'échantillons isolés obtenus à partir de bassins qui
servent de dépotoirs, mais les données tirées de telles
études sont loin d'être satisfaisantes. Elles indiquent
seulement les taux d'accumulation en un seul point du lac
et ne tiennent pas compte d'un certain nombre de
considérations importantes. Celles-ci incluent: le rapport
lac/bassin versant, les sources des sédiments et les
quantités de contaminants qui s'accumulent au fond du
lac.
    Un excercice comprenant des échantillons multiples,
appliqué pour deux lacs a été utilisé pour obtenir des
informations historiques sur la production de sediments
au centre de l'Angleterre depuis 1965. Ces données ont été
utilisées avec des techniques d'extraction chimique pour
estimer l'apport annuel des contaminants de provenance
atmosphérique et des engrais agricoles dans les sédiments
du lac. La comparaison entre une forêt semi-naturelle et
une région de cultivation intense, indique que des bassins
de lac prudemment sélectionnés peuvent être utilisés pour
obtenir avec suffisamment de fiabilité, des données pour
quantifier la réponse de l'environnement à la pollution de
sources diverses. On affirme, qu'on peut obtenir, par un
analyse précise des sédiments des lacs, non seulement
l'information sur les conditions initiales, mais aussi des
séries de données suffisamment longues pour quantifier la
réponse et le rétablissement d'un environnement assujetti
à des polluants divers. En conclusion, on a trouvé que le
bassin du lac doit être reconnu comme un cadre de travail
hydrologique important pour l'analyse quantitative.
    635

(voir p. 245 S.V.P.)

Estimation consistante des paramètres dans la
famille des modèles de série chronologique à
longue-mémoire

A Ramachandra Rao
School of Civil Engineering
Purdue University
W. Lafayette, IN 47907 USA
Gwo-Hsing Yu
Department of Hydraulic Engineering
Tamkang University
Tamsui, Taipei Hsien, Taiwan R.O.C.

RESUME   Les   modèles   avec des caractéristiques   de
corrélation de courte durée, comme l'AR et l'ARMA,
n'offrent pas beaucoup de fléxibilité dans la modélisation
des caractéristiques des séries chronologiques de courte
et longue durée avec un petit nombre de paramètres. Les
modèles de différence fractionnaire ont été récemment
proposés pour éliminer ce défaut dans les modèles de
corrélation de courte durée et de type Markov. Les
caractéristiques de corrélation d'une série chronologique
de longue ou courte durée peuvent être modélisées de façon
satisfaisante par l'utilisation de ce modèle de différence
fractionnaire.
     Les   caractéristiques   des   modèles de différence
fractionnaire dépendent   du   paramètre de différence
(d) et de la variance de la composante aléatoire. Par
conséquent, il est   important d'estimer ces paramètres
correctement. Deux méthodes ont été déjà proposées pour
ces estimations. Le premier estimateur (Kashyap et Eom,
1984) est fondé sur la méthode des moindres carrés dans le
domaine des   fréquences. Cet estimateur est non-biaisé et
est consistant avec la méthode des moindres carrés. Le
deuxième estimateur (Janacek, 1982) est fondé sur les
caractéristiques   spectrales   du   modèle. Les   deux
estimateurs sont comparés dans cette étude. L'estimateur
de Kashyap et Eom (1984) est recommandé.

(voir p. 269 S.V.P.)

Modélisation spatio-temporelle du processus
précipitation-ruissellement

Kaz Adamowski & Fadil Mohamed
Department of Civil Engineering
Université d'Ottawa
Ottawa, Ontario K1N 6N5
Nicolas R. Delezios
INTERA Technologies Ltd.
785 Carling Avenue
Ottawa, Ontario K1S 5H4

RESUME   Un modèle explicatif appartenant à la classe des
processus de fonction de transfert spatio-temporelle-
bruit (F,T,S,T,B) est présénté. L'article développe une
méthode répétitive en trois stades pour construire ce
modèle   (F,T,S,T,B)   de   processus   précipitation-
ruissellement. Quatre stations pluviométriques et hydrométriques situées dans un bassin versant du sud de
l'Ontario, au Canada, échantillonnées à des intervalles de
15 jours, ont été utilisées dans l'analyse numérique. Le
modèle identifié est du type (F,T,S,T,B) (2,1,0,0,1,0,0).
Les paramètres du modèle ont été estimés par la méthode
polytope, la méthode du pas optimal à pente maximale, une
méthode de gradient conjugué et une méthode pseudo-Newton.
Le modèle spatio-temporel développé s'est avéré -satis-
faisant pour décrire les caractéristiques spatiales et
temporelles des séries chronologiques des précipitations
et des ruissellements.

(voir p. 315 S.V.P.)

Impact des changements climatiques sur la
morphologie des bassins fluviaux

F.H. Verhoog
Division of Water Science
Unesco

RESUME   L'article essaie d'évaluer l'impact des changements climatiques sur la morphologie des bassins fluviaux
au moyen premièrement d'une évaluation de l'impact sur les
précipitations et l'évaporation, deuxièmement sur les éco-
systèmes naturels, troisièmement et enfin sur la morphologie des bassins fluviaux . L'examen essaie d'être
aussi quantitatif que possible mais en raison du manque de
données et de théories vérifiées et expérimentées, les
tableaux figurant dans le texte doivent être considérés
comme seulement hypothétiques.

(voir p. 605 S.V.P.)

Réponse hydrologique à un changement climatique
artificiel correspondant à une amélioration
des précipitations

Arie Ben-Zvi
Israel Hydrological Service
Jerusalem, Israel
Margarita Langerman
Rainfall Enhancement Unit
Mekorot, Ben-Guryon Airport, Israel

RESUME   L'évaluation de la réponse hydrologique à un
changement climatique est dépendante de la disponibilité
d'une description quantitative des changements météorologiques, et d'un modèle hydrométéorologique qui soit
fiable avant et après le changement. Dans le cas d'un
changement climatique naturel, la description peut être
retardée jusqu'au moment où l'apparition des changements
est reconnue et distinguée des fluctuations normales. Ce
processus est plus court dans le cas d'un changement
climatique artificiel. Un exemple d'un tel changement est
le projet d'amélioration de la précipitation en Israël.

(voir p. 617 S.V.P.)

Est-ce que les nuages produisent un effet de
feedback négatif dans un environment
terrestre réchauffé par le CO$_2$

K. McGuffie and A. Henderson-Sellers
Department of Geography
University of Liverpool
Liverpool, L69 3BX, UK

RESUME   Les dossiers sur la couverture nuageuse de
l'Europe de l'ouest, des Etats - Unis continentaux, et du
continent subindien ont été analysés dans le contexte d'un
réchauffement mondial. On a trouvé que la couverture
nuageuse a de façon générale augmenté entre les années les
plus froides et les annees les plus chaudes de ce siècle
dans chacune de ces régions de climatologie différente.
Ce résultat est en désaccord avec les quelques prédictions
climatiques disponibles qui suggèrent une diminution des
masses nuageuses - L'ampli-tude du changement de la
couverture nuageuse est pratiquement la même que celle
obtenue par simulation mais de signe contraire. Les
resultats des observations peuvent indiquer, en supposant
que les autres propriétés des nuages ne changent
pratiquement pas, une tendance à réduire les températures
de la surface terrestre déjà réchauffée. C'est à dire que
les changements de la masse nuageuse produisent une
réaction négative sur le climat.

SEA LEVEL
ICE and
CLIMATIC CHANGE

Edited by Ian Allison

Proceedings of the Canberra Symposium, December 1979
471 + xv pp.; price US$10; published 1981 by the International Association of
Hydrological Sciences as IAHS Publication no. 131

The International Symposium on Sea Level, Ice Sheets, and Climatic Change was held at Canberra, Australia, on 7 and 8 December 1979 as part of the 17th General Assembly of IUGG. The symposium was sponsored by IAHS, IAMAP and IAPSO and was organized by the International Commission on Snow and Ice of IAHS, with support from the Local Organizing Committee for the IUGG General Assembly.

The major objective of the symposium was to review current ideas and recent results on the processes and the effects of interactions between sea level, ice, and climatic change on time scales of 100 to 10 000 000 years. While the cryosphere has been the subject of considerable speculation regarding the climatic past and future, the exact causal relationships between cryosphere phenomena and sea level in the past remain uncertain. Description of those changes in sea level and ice sheets which had causes and effects other than climatic, would hopefully define a residue of features with direct climatic implications, and help to identify interconnections between the three phenomena.

As might be expected with a symposium theme of such general scope, the contributed papers cover a very wide range of topics, and it is hoped that this will highlight the complexity and multidisciplinary nature of the study of relationships between sea level, ice, and climatic change. The papers have been grouped into two major sections, each divided into subsections.

ICE AND SNOW AS ELEMENTS IN THE WEATHER AND CLIMATE SYSTEM AND AS INDICATORS OF CHANGE
The record of climate change in glaciers
The climatic role and environmental effects of snow
Sea ice as a climatic element
Evidence of the past climatic change from large ice sheets

FEATURES AND INTERACTIONS OF SEA LEVEL, ICE AND CLIMATE IN THE QUATERNARY
The global record of the late Quaternary changes of sea level, ice and climate
Processes of interaction between sea level, ice sheet and climate
Sea level, ice, and climatic change: invited summary reviews

The book is available from either the Office of the Treasurer IAHS, 2000 Florida Avenue NW, Washington, DC 20009, USA or the IUGG Publications Office, 39 ter Rue Gay Lussac, 75005 Paris, France. A catalogue of all IAHS publications is available free of charge from either of these addresses or from the IAHS Editorial Office, Institute of Hydrology, Wallingford, Oxon OX10 8BB, UK.